Interdisciplinary Studies on Global Disasters

İsmet Guney
https://orcid.org/0000-0003-4682-4905
Recep Tayyip Erdogan University, Turkey

Mehmet Ali Celik
https://orcid.org/0000-0002-7729-6650
Igdır University, Turkey

IGI Global
Scientific Publishing
Publishing Tomorrow's Research Today

Vice President of Editorial Melissa Wagner
Director of Acquisitions Mikaela Felty
Director of Book Development Jocelynn Hessler
Production Manager Mike Brehm
Cover Design Jose Rosado

Published in the United States of America by
IGI Global Scientific Publishing
701 East Chocolate Avenue
Hershey, PA, 17033, USA
Tel: 717-533-8845 | Fax: 717-533-7115
Website: https://www.igi-global.com E-mail: cust@igi-global.com

Library of Congress Cataloging-in-Publication Data

Names: Guney, İsmet editor | Çelik, Mehmet Ali editor
Title: Interdisciplinary studies on global disasters / edited by İsmet Guney, Mehmet Ali Çelik.
Description: Hershey, PA : IGI Global Scientific Publishing, [2026] | Includes bibliographical references and index. | Summary: "This book looks at different types of natural disasters climatic, geophysical, hydrological, and biological and digs into what drives them. It examines how global forces like climate change, tectonic shifts, and rapid urban growth are raising the stakes"-- Provided by publisher.
Identifiers: LCCN 2025038322 (print) | LCCN 2025038323 (ebook) | ISBN 9798337362748 hardcover | ISBN 9798337362755 paperback | ISBN 9798337362762 ebook
Subjects: LCSH: Disasters | Natural disasters | LCGFT: Cross-cultural studies
Classification: LCC HV553 .I545 2026 (print) | LCC HV553 (ebook)
LC record available at https://lccn.loc.gov/2025038322
LC ebook record available at https://lccn.loc.gov/2025038323

British Cataloguing in Publication Data
A Cataloguing in Publication record for this book is available from the British Library.

Table of Contents

Detailed Table of Contents

Water security has emerged as a critical and multidimensional challenge of the twenty-first century, encompassing biophysical limits, socio-economic dynamics, governance systems, technological resilience, and geopolitical stability. With the global population projected to rise from 7.7 billion to nearly 11.2 billion by 2100, ensuring adequate, safe, and equitable access to freshwater will become increasingly complex. Freshwater resources are fundamental not only for drinking and sanitation but also for sustaining agriculture, ecosystems, energy production, industry, and economic development. Climate change is intensifying blue-water variability and hydrological extremes, while agricultural expansion is altering green-water fluxes and increasing pressure on land–water systems. Strategically important headwater regions, though vital for downstream water supply, remain ecologically fragile and insufficiently protected. Simultaneously, urban and peri-urban areas face growing risks from infrastructure stress, contamination, groundwater depletion, and emerging cyber-physical threats, despite the potential of decentralized solutions such as rainwater harvesting and distributed storage. Water insecurity now reflects both quantitative scarcity and qualitative degradation, with severity varying across regions, particularly in arid and semi-arid zones. This chapter advances an integrated scientific and political understanding of water security, highlighting its multi-scalar, cross-sectoral nature and examining diverse regional experiences to underscore the urgency of resilient and equitable water governance.

This study focuses on the importance of understanding the social dimensions of disasters in the context of natural disaster events and structural social inequalities.

interdisciplinary disaster management issue rather than an engineering problem, and an adaptive management framework is presented for the Mediterranean basin.

The RUSLE model applied in the Ceyhan Basin revealed a heterogeneous erosion pattern driven by topography, land use, and rainfall erosivity. Low R-factor values dominate the northern plains, while Mediterranean influence increases erosivity in the south. Moderate erodibility soils cover most of the basin, yet steep slopes with high LS values form critical hotspots. Protective land cover remains limited, and more than half of the basin lacks conservation practices. As a result, although 60% of the basin experiences low erosion, nearly 13% shows extremely high soil loss, mainly in areas where steep gradients, weak vegetation, and inappropriate land-use practices overlap. These findings highlight the need for targeted conservation strategies and spatially informed land-management planning.

This chapter explores Cambodia's evolving disaster governance as a case of resilient transformation in the Global South. Once seen as highly vulnerable to climate-related disasters, Cambodia has become an active player in disaster risk reduction (DRR), climate adaptation, and regional humanitarian cooperation. The chapter highlights legal reforms, institutional shifts, local preparedness, and cross-border initiatives that reflect a shift from reactive to inclusive, forward-looking planning. It examines how Cambodia integrates DRR with development and climate politics, asserts leadership within ASEAN, and uses disaster governance to build state legitimacy and strengthen diplomacy. Key themes include multi-level governance, gender and youth inclusion, early warning systems, and community-based management. Cambodia's experience shows how resilience-building can drive regional engagement, urging greater recognition of Global South strategies and investment in digital infrastructure, civil society, and cross-border collaboration for lasting climate resilience.

Preface

Global disasters constitute one of the most profound challenges shaping the contemporary world. Intensifying climate change, accelerating urbanization, environmental degradation, technological dependence, and socio-economic inequalities have fundamentally transformed both the frequency and the consequences of disasters. Today's disasters are rarely isolated or purely natural; instead, they emerge from complex interactions between environmental processes, human systems, governance structures, and technological networks. In this context, *Interdisciplinary Studies on Global Disasters* aims to provide a comprehensive interdisciplinary framework for understanding disasters in a rapidly changing global landscape.

This book is grounded in the recognition that disasters must be analyzed as multidimensional phenomena. While hazards such as earthquakes, floods, droughts, landslides, wildfires, and extreme weather events originate from natural processes, their impacts are shaped by exposure, vulnerability, socio-political conditions, economic structures, cultural perceptions, and institutional capacities. Consequently, effective disaster risk reduction and resilience-building require integrated perspectives that transcend disciplinary boundaries.

The volume brings together contributions from earth sciences, geography, environmental studies, engineering, economics, public policy, sociology, gender studies, and data science. It emphasizes both methodological innovation—through GIS, remote sensing, deep learning, and spatial modeling—and critical social perspectives, including governance, inequality, financial literacy, public perception, and community resilience. By combining empirical analyses, theoretical discussions, and policy-oriented insights, the book advances a holistic understanding of global disaster processes.

In the twenty-first century, disasters increasingly operate across scales—from local communities to global systems. Climate change intensifies hydro-meteorological extremes, water scarcity threatens food and energy security, and environmental degradation amplifies disaster risks in coastal, urban, and rural areas alike. At the same time, globalization and technological interdependence mean that disruptions in one region can cascade through economic, ecological, and social systems worldwide.

International frameworks such as the Sendai Framework for Disaster Risk Reduction highlight the necessity of understanding risk in all its dimensions and promoting resilience through science-informed decision-making. Parallel advances in Earth observation technologies, big data, and artificial intelligence have transformed disaster monitoring, early warning, and impact assessment. This book is situated precisely at this intersection—where scientific innovation meets social complexity—offering timely insights into contemporary disaster challenges and future risk trajectories.

The primary audience of this volume includes academic researchers, graduate students, and advanced practitioners working in disaster studies, geography, environmental sciences, climate research, engineering, urban and regional planning, public health, economics, and public policy. The interdisciplinary structure also makes the book highly relevant for policymakers, disaster managers, planners, and professionals in governmental agencies, non-governmental organizations, and international institutions. By bridging theory, data, and practice, the book serves as both a scholarly reference and a practical resource for evidence-based disaster risk reduction.

Chapter 1: From Scarcity to Security: Multidimensional Perspectives on Water in a Changing World

This chapter addresses water scarcity as a global disaster risk, integrating environmental, social, and governance perspectives to frame pathways toward water security.

Chapter 2: Gender and Disasters: The Position of Vulnerable Groups in Disaster Processes

Focusing on social vulnerability, this chapter examines gendered dimensions of disasters and the disproportionate impacts on vulnerable groups. It highlights the importance of equity, inclusion, and gender-sensitive policies in disaster risk reduction.

Chapter 3: What Is the Importance of Wetlands in Reducing Disaster Risks?

Through a review of scientific evidence, this chapter underscores the critical role of wetlands in mitigating floods, storms, and ecological degradation, emphasizing nature-based solutions for disaster resilience.

Chapter 4: Transformation Strategies in Anthropogenic Coastal Erosion

This chapter examines global best practices in managing coastal erosion and presents a roadmap for sustainable coastal transformation, linking local case studies with global lessons.

Chapter 5: Spatial Analysis of Soil Erosion in the Ceyhan Basin (Turkey) Using GIS and RUSLE

This chapter applies geospatial modeling to quantify soil erosion risks, illustrating how spatial analysis supports evidence-based land management and disaster prevention.

Chapter 6: Resilience in Action: Cambodia's Disaster Governance and Regional Cooperation

Focusing on governance, this chapter explores institutional evolution and regional collaboration in disaster management under changing climatic conditions.

Chapter 7: Public Perceptions of Earthquake Risk and Preparedness in Istanbul

Using mixed-methods and digital data sources, this chapter analyzes societal perceptions of earthquake risk, highlighting the role of communication and public awareness.

Chapter 8: Landslides and Mass Movements: Natural and Human-Triggered Processes

This chapter examines landslide dynamics, integrating natural and anthropogenic triggers to improve hazard assessment and mitigation strategies.

Chapter 9: Integrating GIS and Remote Sensing in Early Warning Systems

Focusing on technological innovation, this chapter demonstrates how real-time geospatial data enhance disaster preparedness and early warning capabilities.

Chapter 10: A Special Focus on Natural Disasters in the Field of Earth Sciences: What Do the Reports Tell Us?

This chapter provides a foundational overview of how natural disasters are conceptualized and documented within earth sciences. By synthesizing scientific reports and assessments, it establishes a baseline understanding of hazard dynamics and reporting frameworks that inform subsequent chapters.

Chapter 11: Forest Fires and the Pandemic in Mediterranean Tourism

By linking wildfires, public health crises, and tourism economics, this chapter illustrates the compounded nature of contemporary disasters.

Chapter 12: Extreme Weather Linked to Climate Change: Risk, Vulnerability, and Exposure

This chapter provides a comprehensive analysis of climate-induced extreme events, emphasizing spatial exposure and vulnerability patterns.

Chapter 13: Early Warning Systems and Real-Time Decision Support for Global Preparedness

This chapter advances the discussion on integrated decision-support frameworks, highlighting global applications.

Chapter 14: Climate Change, Disasters, Adaptation, and Trapped Populations in Türkiye

Focusing on immobility and adaptation constraints, this chapter addresses emerging social dimensions of climate-induced disasters.

Chapter 15: Chad's Global Disaster Risk Map

The final chapter presents a national-scale disaster risk mapping effort, illustrating how integrated spatial analysis supports policy and planning.

Interdisciplinary Studies on Global Disasters advances disaster research by integrating scientific, technological, and social perspectives within a single coherent framework. By addressing both physical hazards and human dimensions, the book contributes to a more nuanced understanding of disaster risk and resilience. The editors envision this volume as a catalyst for interdisciplinary dialogue, policy innovation, and future research, ultimately supporting global efforts toward sustainable development and disaster-resilient societies.

Chapter 1

From Scarcity to Security Multidimensional Perspectives on Water in a Changing World:

From Scarcity to Security

Mohsen Saroughi
University of Tehran, Iran

Mohammed Achite
http://orcid.org/0000-0002-7729-6650
Hassiba Benbouali University of Chlef, Algeria

Mehmet Ali Çelik
Iğdır University, Turkey

Kusum Pandey
http://orcid.org/0009-0005-4888-6650
G.B. Pant National Institute of Himalayan Environment, India

ABSTRACT

Water security has emerged as a critical and multidimensional challenge of the twenty-first century, encompassing biophysical limits, socio-economic dynamics, governance systems, technological resilience, and geopolitical stability. With the global population projected to rise from 7.7 billion to nearly 11.2 billion by 2100, ensuring adequate, safe, and equitable access to freshwater will become increasingly complex. Freshwater resources are fundamental not only for drinking and sanitation but also for sustaining agriculture, ecosystems, energy production, industry, and

DOI: 10.4018/979-8-3373-6274-8.ch001

economic development. Climate change is intensifying blue-water variability and hydrological extremes, while agricultural expansion is altering green-water fluxes and increasing pressure on land–water systems. Strategically important headwater regions, though vital for downstream water supply, remain ecologically fragile and insufficiently protected. Simultaneously, urban and peri-urban areas face growing risks from infrastructure stress, contamination, groundwater depletion, and emerging cyber-physical threats, despite the potential of decentralized solutions such as rainwater harvesting and distributed storage. Water insecurity now reflects both quantitative scarcity and qualitative degradation, with severity varying across regions, particularly in arid and semi-arid zones. This chapter advances an integrated scientific and political understanding of water security, highlighting its multi-scalar, cross-sectoral nature and examining diverse regional experiences to underscore the urgency of resilient and equitable water governance.

1. INTRODUCTION

Water is the foundation of life and the structural core of all ecosystems, sustaining agriculture, industry, energy production, public health, and environmental stability (Çelik et al., 2013; Kılıç, 2020). No nation can achieve durable economic growth or political stability without securing reliable and safe water resources. Today, escalating water scarcity and pollution threaten millions of lives, particularly in low-income and climate-vulnerable regions. In this context, aligning water security with the United Nations Sustainable Development Goals (SDGs), especially SDG 6 (Clean Water and Sanitation), is essential. The 2030 Agenda for Sustainable Development (United Nations, 2015) and successive UN-Water World Water Development Reports (2019, 2023) underscore that water security is central to poverty eradication, ecosystem protection, climate adaptation, and global peacebuilding. Positioning water security within this framework strengthens its governance relevance and highlights its cross-sectoral implications for human and environmental security.

The geopolitical dimension of water security is increasingly evident in conflict-affected regions. In Syria, prolonged conflict has severely damaged water infrastructure, including major hydraulic structures such as the Tishrin Dam near Manbij, illustrating how water systems can become both casualties and instruments of war (Gleick, 2019; UNICEF, 2021). Such cases demonstrate the profound water–security nexus, where disruption of supply networks not only exacerbates humanitarian crises but also destabilizes governance and social cohesion. Similar tensions arise in transboundary basins, where upstream–downstream asymmetries create political friction. Disputes among Turkey, Iraq, and Syria over shared rivers exemplify how freshwater resources increasingly influence regional diplomacy and security architecture.

Beyond the Middle East, the intensification of dam construction along the Mekong River provides a compelling comparative case. Large-scale hydropower development has generated environmental degradation, fisheries decline, sediment disruption, and socio-political tensions among riparian states (Molle et al., 2009; Grumbine & Xu, 2011). The Mekong case highlights how infrastructure-driven development, while economically attractive, can amplify ecological vulnerability and geopolitical risk if not governed cooperatively. Water infrastructure, therefore, represents both opportunity and risk—capable of fostering regional integration or exacerbating inequality and instability.

Climate change further compounds these pressures by intensifying droughts, altering rainfall regimes, and threatening vulnerable livelihoods. Water stress, when combined with socio-economic marginalization, disproportionately affects nomadic and pastoral communities, as documented among Kurdish nomads in Mount Agri (Ararat) in Turkey (Sevgi & Yulu, 2025). Globally, drought and scarcity reshape migration patterns, gender roles, and local economies (Hohenthal & Minoia, 2017). Urban areas are equally exposed: the water crisis in Johannesburg and ongoing debates over relocating Iran's capital due to severe shortages in Tehran illustrate the emerging urban dimension of water insecurity. These cases demonstrate that water security is not confined to arid rural landscapes but is a systemic risk affecting megacities, fragile states, and rapidly growing economies alike.

Collectively, these global examples reinforce that water security is multidimensional spanning sustainable development policy, post-conflict reconstruction, transboundary geopolitics, climate vulnerability, and urban resilience. Addressing it requires integrated scientific analysis, cooperative hydro-diplomacy, climate adaptation strategies, and robust institutional frameworks aligned with the SDGs. Without such coordinated action, water insecurity will continue to magnify humanitarian crises, economic disruption, and geopolitical instability in an increasingly water-stressed world.

1.1. Water Security in Dams

Incidents that occur in order to endanger water security in dams are divided into two categories: mankind (including terrorist and military) and natural. In following, these will be described.

1.1.1. Terrorist Incidents in the Dams

A crucial issues in water security is the terrorist and military conflict related to the dam attacks, if it occurs, it would result in heavy losses of financial and human life, and put water security all of the consumer at risk. Several examples are presented below (Table 1):

Table 1. Some examples of terrorist incidents in dams

Section	Dam / Event	Description	Source
New Hampshire Dam	New Hampshire, USA	Gleick (2006) this dam was attacked by local residents in New Hampshire in 1850 and local residents did so because they were dissatisfied with the impact of the dam on the level surface of the water, and also it provided water to downstream factories.	Gleick (2006)
Cahora Bassa Dam	Mozambique	Gleick (2006) this dam was targeted in 1980 by Mozambique's national resistance during the independence of Mozambique, Rhodesia, South Africa, and Zimbabwe which resulted in its destruction.	Gleick (2006)
Inga Dam	Democratic Republic of Congo	Gleick (2006) this dam was attacked in 1998 to falling President of Kabila. In this incident, supply of water and electricity the city of Kinshasa facing with a problem.	Gleick (2006)

1.1.2. Natural Disasters in Dams

Natural disasters in dams can cause severe structural damage such as cracking, overtopping, or total collapse, often triggered by earthquakes, floods, or extreme storms (Table 2). Unlike terrorist or military attacks, these events are not intentional, but their consequences loss of lives, infrastructure destruction, and long-term socio-economic impacts can be equally or even more devastating.

continued on following page

Table 2. Continued

Table 2. Some examples of natural disasters in dams

Section	Dam / Location	Event & Year	Description	Source
Aswan Dam	Egypt (Nile River)	1997–1998, RVF virus outbreak	Pepin et al. (2010) in 1998-1997, the RVF virus outbreak occurred following dam operation. This virus is related to livestock and caused a common disease between humans and livestock. It killed many people and thousands of livestock living on the shores of Lake Dam. The disease had never been seen in this area of Egypt before.	Pepin et al. (2010)
Konya Dam	India (Konya River)	1967, Earthquake magnitude 6.5	Chopra and Chakrabarti (1972) in 1967, the earthquake with 6.5 magnitude damaged the dam and its associated facilities.	Chopra & Chakrabarti (1972)
Zipingpu Dam	China (Min River, Sichuan)	2008, Earthquake magnitude 7.9	Kerr and Stone (2009) this dam collapsed in 2008 as a result of the earthquake with 7.9 magnitude, killing about 80,000 civilians.	Kerr & Stone (2009)
New Orleans Levees	USA (New Orleans)	2005, Hurricane Katrina	Congleton (2006) in 2005, the Katrina storm broke these levees and as a result 80% of the city was submerged. According to estimates, the damages caused by these levees were exceeding the direct damage caused by the storm itself.	Congleton (2006)
Banqiao Dam	China (River Ru)	1976, Severe Storm	Penuel and Statler (2010) in 1976, 145,000 people were killed and almost half a million left homeless as a result of a severe storm.	Penuel & Statler (2010)
Mosul Dam	Iraq (Tigris River, Nineveh Governorate)	Since 1984, ongoing erosion risk	The largest dam in Iraq, built in 1984. Filkins (2016) reports that, according to the US Army Engineering Department's warning, this dam is being destroyed due to early soil erosion beneath its foundations. If it collapses, it will cause a 20m water surge and completely destroy downstream areas including Mosul city (population ~2 million). Investigations show Baghdad may also be exposed to flooding threat.	Filkins (2016)

1.2. Water Security in Water Supply Networks

Water supply networks (WSN), pumping stations, valves, and etc. always provide good conditions for terrorist attacks including availability, insufficient protection, ease of use, and have high losses of human life's and financial. Physical, chemical, biological, and cyber-attacks are the methods that terrorists may carry out against WSN, because if they spread biological agents or contamination to the water supply, treatment or water distribution network, all of the available water in the network will become contaminated, the entire network will be disrupted, and will effect on the lives of thousands people.

1.2.1. Physical Destruction of Supply Networks

Disruptions to critical water infrastructure often result in the immediate collapse of essential services, particularly when key facilities and equipment such as pumping stations, control systems, pipelines, and administrative buildings are damaged or destroyed. Such incidents not only interrupt water supply but also expose urban populations to severe public health and environmental risks. For instance, as documented by Jennifer L. Tierney and Michael Worth (2003), a major water pipeline in Baghdad was deliberately destroyed by insurgents, marking the first direct attack on the city's water system. The explosion caused widespread flooding, submerging nearly ten neighborhoods and severely disrupting potable water distribution. Similarly, according to Susan States (2011), a bomb blast targeted the principal water transmission pipeline in Lusaka, cutting off water supply to nearly three million residents. These cases highlight the vulnerability of centralized water infrastructure to conflict and sabotage, underscoring the urgent need for resilient design, decentralized systems, and enhanced protection of critical water assets to safeguard urban water security.

1.2.2. Intentional Chemical Pollution

As documented by Peter H. Gleick (2006), deliberate contamination of water supply systems has repeatedly demonstrated the grave vulnerability of urban water security networks. In 1972, a plot was uncovered to contaminate the drinking water system of New York City with nerve gas an alarming example of how chemical agents can be weaponized against civilian infrastructure. In 1977, a water storage tank in North Carolina was deliberately contaminated with bacterial agents after safety caps were removed, rendering the water unfit for consumption and forcing authorities to supply potable water through tanker trucks. One of the most severe biological contamination incidents occurred in 1984, when members of the religious group Rajneeshee carried out a large-scale attack in Oregon, targeting public systems and spreading Salmonella. The attack resulted in more than 750 reported cases of illness an extraordinary spike compared to the fewer than five cases typically recorded annually. These incidents underscore the persistent threat of chemical and biological sabotage to municipal water systems and highlight the urgent necessity for strengthened surveillance, infrastructure protection, and emergency response mechanisms to safeguard public health and water security.

1.2.3. Cyber Attacks

According to Peter H. Gleick (2006), the year 2000 witnessed a landmark case of cyber interference targeting urban water infrastructure in Queensland. A former contractor was arrested for attempting to hack into the Maroochy Shire wastewater management system using computers and radio transmitters. By manipulating the supervisory control and data acquisition (SCADA) system, he illegally released untreated wastewater into local parks and rivers, demonstrating how digital vulnerabilities can translate into serious environmental contamination and public health risks. Similarly, Barton Gellman (2002) reported that intelligence assessments in the United States uncovered evidence that al-Qaeda had conducted reconnaissance and surveillance of water control and monitoring facilities. Although no direct attack was executed, the findings highlighted the strategic interest of terrorist groups in critical water infrastructure. Together, these cases underscore the growing convergence of cyber threats and physical water systems, emphasizing the urgent need for enhanced cybersecurity protocols, infrastructure resilience, and proactive monitoring to safeguard water security in the digital age.

1.2.4. Bioterrorism Attacks

As documented by Gleick (2006), the deliberate targeting of drinking water systems has long been recognized as a serious security threat. In 1972, two members of the extremist group Order of the Rising Sun were arrested while attempting to execute a sophisticated plan to contaminate municipal water supplies in major U.S. cities, including Chicago and St. Louis. Authorities recovered multiple biological agents in their possession, including botulism, meningitis pathogens, anthrax, and an estimated 30–40 kilograms of typhoid cultures an alarming indication of the scale of intended harm. In 1973, a German biologist threatened to contaminate water supplies in Germany with biological materials unless he was paid $8.5 million, revealing how water infrastructure can be exploited for extortion. Similarly, in 1991, two cities in Canada—British Columbia and Kelowna—received anonymous threats claiming that their water systems would be infected with biological agents, though no perpetrator was ultimately identified. These incidents collectively underscore the strategic vulnerability of centralized water systems to both ideological extremism and criminal coercion, reinforcing the imperative for robust surveillance, biosecurity safeguards, and rapid response mechanisms.

Beyond direct acts of sabotage, a substantial body of scholarship has examined water security from ecological, socio-economic, technological, and governance perspectives. Vörösmarty et al. (2010) provided a landmark global assessment of threats to human water security and river biodiversity, analyzing 47 of the world's largest river basins. Their framework integrated catchment disturbance, pollution, water resource development, and biological stressors, concluding that nearly five percent of the global population faces high water security risks disproportionately borne by low-income nations lacking infrastructure investment. In Central Asia's Syr Darya Basin, Kai Wegerich et al. (2015) highlighted the fragility of irrigation governance following political transition, emphasizing the need to revitalize water management bureaucracies. Similarly, Shiqiu Jia et al. (2015), in their assessment of the Yellow River Basin, developed an integrated water security index linking hydrological, environmental, and socio-economic subsystems, revealing stark provincial disparities in water security status.

In Saskatchewan River Basin, Howard Wheater and Patricia Gober (2015) emphasized the escalating pressures of climate change and overexploitation, arguing for stronger scientific integration to guide adaptive governance. Urban water security has also received critical attention; Patricia Romero-Lankao and Dan Gnatz (2016) underscored the multidimensional nature of social, economic, technological, ecological, and governance (SETEG) factors in shaping urban resilience. In South Africa's KwaZulu-Natal, Shamiso Sharaunga and Maxwell Mudhara (2016) demonstrated that improved water security does not automatically reduce women's reliance on agricultural income, highlighting the complex socio-economic dynamics at play.

Policy-oriented analyses by Mark Zeitoun and colleagues (2016, 2017) further advanced the discourse by interrogating uncertainty, power asymmetries, and the impacts of armed conflict on urban water systems, including the war-affected city of Basrah. Complementary perspectives range from decentralized rainwater harvesting initiatives in Guatemala (Elgert et al., 2016), to blue–green water footprint assessments in the Savannah River Basin (Veettil and Mishra, 2016), strategic basin protection in South Africa (Nel et al., 2017), legal and sovereignty-based frameworks in Canada (Cook, 2016), and global governance debates within the European Union (Maganda, 2016). Technological innovations, such as nanotechnology for advanced water treatment (Alvarez et al., 2018), further illustrate how scientific advancement can expand non-conventional water sources and strengthen resilience.

Collectively, these studies reveal that water security is not merely a question of supply, but a complex, multi-scalar challenge shaped by environmental change, governance structures, technological capacity, socio-economic inequality, and geopolitical instability. Ensuring sustainable and equitable water security therefore demands integrated strategies that combine infrastructure protection, ecological stewardship, scientific innovation, and inclusive policy reform.

2. DEFINITIONS AND TERMINOLOGY

2.1. Security

Buzan (2008) security is the set of all methods, activities, and actions that ensure that an accident will never occur or reduce the probability of it happening and also, in the event of a dangerous accident, return to normal as quickly as possible and at the lowest possible cost. Depending on what the security reference is, the definition of security will be different. There are two different views on security in this regard, one person referring to the government as security and said "security is static and watch over the government because of its nature" and other person considered the reference security to individual person and said "The only basis for individuals (and not the government) could provide a well-foundation framework for security studies".

2.2. Water Security

Water security is the reliable access to reliable volume and quality water at the right time and on a large scale for water supply systems to the extent that they provides health, finance, and production to the animal and should not be exceed the acceptable risk for water. All of those is a basic precondition for human, food, and environment security, as well as economic growth and sustainable development is impossible without water security. The following are different definitions from the viewpoint of different experts:

- Whiteford and Witter (1999) according to the World Water Partnership Organization, security at every level, from the family to the global dimension, means that every person has a sufficient cost of access to sufficient drinking water for a clean, healthy, and efficient life while maintaining the environment. In this kind of perspective, which is also the most complete, the main goal is to meet basic human needs and to provide them with adequate attention to the environment and the ecosystem. According to proficient and experts, water security without sustainable development is doomed to failure and will pose serious risks to future generations but water security studies have another important dimension besides the scientific dimension that can start the groundwork for local or international conflicts, which is called the political dimension of water security studies.
- Staudinger *et al.* (2006) water engineers have appointed water security as guns, gates and guards to supply drinking water.

- Rijsberman (2006) water security exists when people have access to adequate and affordable water to meet their needs such as drinking, bathing and livelihoods.
- Falkenmark *et al.* (2007) there are exists many associations that only consider water security framework on the need for human beings to water and they state which the security is a situation that only exist sufficient quantity and quality of water and it is necessary at affordable prices for short and long term needs of the nation to be provided to people. It should be noted that this type of approach to water security in many cases can be dangerous, because in this viewpoint the importance of the ecosystem and its role as an inalienable part of water for human neglect, and also stated that the focus on agriculture and water safety of agricultural products to provide enough food for human is very essential and primary goal of water security issues is that.
- Minamyer (2008) America Unites States Environmental Protection Agency define water security as a tools for prevention and protection from pollution and terrorism. For this reason, legislators in the United States is to pay special attention to the security of it's and infrastructure of drinking water.

2.3. Security Sections

Falkenmark *et al.* (2007) security sections focus on quantity and access to water and is often associated with water security assessment tools. Most prominent water security assessment tools are two index of water stress and water scarcity.

2.4. Water Stress Index

Falkenmark and Molden (2008) assessment the relationship use of water and water resources availability such as rivers, aquifers, and etc. As shown as Figure 1. The level of water stress calculated for worldwide as resulted Middle East have maximum water stress in worldwide.

Figure 1. Level of water stress in worldwide

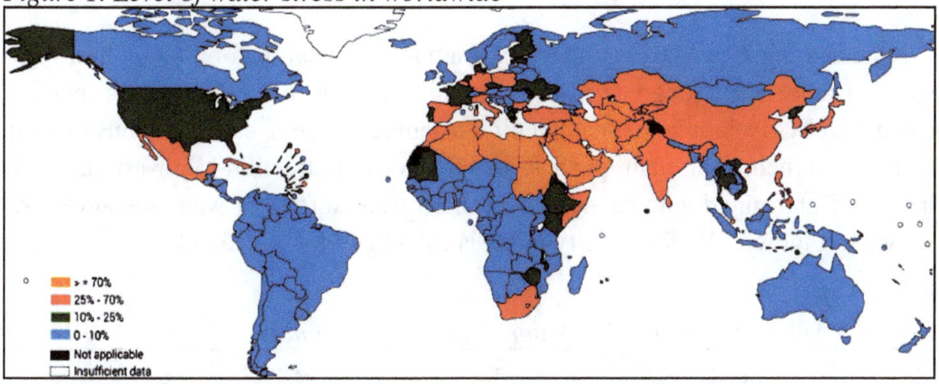

Source: www.unwater.org

2.5. Water Scarcity

Water scarcity is defined by Hinrichsen (2002) as the inability of freshwater resources to meet the demand of a given population or region. Water scarcity does not only mean a physical scarcity of water; it can also arise from the unsuitability of the available water quality for use, the unequal distribution of water spatially or temporally, or socioeconomic and institutional deficiencies that restrict access to water. Therefore, water scarcity is a multidimensional problem that develops as a result of the interaction of both natural and anthropogenic processes.

This concept encompasses two main subcategories: water stress and water crisis. A water crisis is a situation where the amount of potable, unpolluted freshwater in a region decreases to such an extent that it cannot meet the population's basic needs for drinking, sanitation, agriculture, and economic production. It is generally associated with rapid population growth, unregulated urbanization, climate change, pollution, and inadequate infrastructure. In regions experiencing a water crisis, water supply becomes unsustainable, seriously jeopardizing public health, food security, economic activities, and ecosystem integrity.

Water stress, on the other hand, represents a broader, more gradual process. It occurs when access to water resources becomes more difficult due to overexploitation of freshwater resources, falling groundwater levels, water migration out of the basin, reduced surface water flow, or hydroclimatic variability. The fundamental characteristic of water stress is that, even if a full-blown crisis has not yet occurred, access to water becomes increasingly difficult, costly, and unsustainable. This process can develop into a water crisis in the long term, increasing social and environmental pressures, particularly in arid and semi-arid regions.

2.6. Water Poverty Index

Cho and Ogwang (2014) this index is a mathematical data-based tool for measuring water-related degrees in a community, region, or country. This index is calculated by five components including resources, access, capacity, consumption and environment. The main purpose of this index is attention to poverty and this index is Determined who cannot give benefit from sufficient water resources. As shown in Figure 2. Water poverty index is calculated for worldwide.

Figure 2. Sullivan et al. (2000) Water Poverty Index at national scale

2.7. Objective Water Security Index

Shrestha et al. (2018) developed a model that focuses on household water use behavior in urban areas in developing countries and uses physical dimensions to measure water security. Affordability, adaptability, access to drinking water, consumption, alternative water sources, central water system, and social capital are among the seven dimensions included in this index.

2.8. National Water Security Index

Siddiqi (2016) the index for comparing the performance of security in Asia and the Pacific region by "Asian Development Bank" has been developed. This index calculated by five key factors that include Household Water Security, Economic Water Security, Urban Water Security, Environmental Water Security, and Resilience to Water-Related Disasters.

2.9. Water Insecurity Index

Kher *et al.* (2015) this index on a regional scale in India were used and six key factors are used to calculate it that include Limited resources, inadequate access, Limited Capacity, Poor environment, Inadequate use, and Climate stress.

2.10. Shared Water Resources

Surface water or groundwater that existed in the border of two or more countries are called on shared water resources. Rivers, Seas and Lakes are an example of shared waters that will be done cooperating activities and concluding several treaties. Important key issue related to water resources is shared water resources with neighbors. Shahbazbegian *et al.* (2016) groundwater forms a major part of the available freshwater resources and according to statistics collected from around the world, 50% of people life is depend on these resources for drinking water, and also a significant amount of this resources is used for agricultural purposes as well as has made groundwater resources one of the most important shared resources. Governments' competitiveness to own and exploit more and more shared waters has given rise to numerous disputes between states that if the legal and political mechanisms to solve them will not be predicted, it can become a threat to peace and national security.

2.11. Bioterrorism

Bioterrorism means the premeditated release of highly pathogenic microorganisms or their toxins into water resource or non-water sources, causing irreparable damage and extensive death or disease in the human community, plants and animals. Therefore, bioterrorism is a serious threat and its investigation is of great importance because at the same time it threaten both of national security and public policy. Tavakoli *et al.* (2005) recent advances in microbial agents and biological toxins have led to becoming deadly weapons and endangering human life and health. Zare Bidaki and Balali Mood (2015) much evidence shows that biological weapons have a greater potential and more uncontrollable damage capability than chemical weapons and in terms of extend and range of actions, they are not comparable to any weapon. The development of technology and advances in biochemistry and biotechnology have made it easier to produce these weapons, and genetic engineering has played a major role.

2.12. Terrorism

Schmid and Jongman (2017) terrorism is an anxiety-inspiring method of repeated violent acts perpetrated by individual, group or government agents in secret or semi-secretly with politico factional motives or specific beliefs. In terrorist incidents, unlike ordinary crimes, the primary victims are not the main person and they are normal people that chosen at random to convey a message to the enemy by death or injury of them.

Staff (2019) definition of terrorism from the Department of Defense's Viewpoint "Illegal use of violence or threat of violence that is often caused by religious, political, or other incitement ideologies to create fear or force governments or communities to accomplish a specific purpose that is primarily objective that usually they are political issue."

Manning and Goldfrank (2002) defining terrorism from the Federal Emergency Management Agency's viewpoint "the use of force or violence against persons or estate in violation of us criminal law for intimidation, forced or compensated."

2.13. Sustainable Development

Development that satisfies current needs without endangering the capacity of future generations to satisfy their own needs (Shah, 2008).

Sustainable development represents an approach that aims to meet today's economic and social needs without depleting natural resources, destroying ecosystems, or disrupting the environmental balance. This understanding emphasizes that development is not merely about economic growth; it also emphasizes that improving the quality of life, reducing poverty, and meeting basic human needs are at the heart of development. Therefore, sustainability requires a planned development process that ensures the well-being of current generations while not limiting the opportunities of future generations.

At the heart of this approach is the understanding that natural resources must be protected not only for today's society but also for future generations. Because over-consumption or degradation of natural resources such as water, soil, biodiversity, and the climate system directly threatens future social well-being, sustainable development embraces the fundamental principle of not exceeding environmental limits. Within this framework, ensuring resource use aligns with renewable capacity, increasing energy efficiency, and reducing environmental pressure are critical for both ecological and social stability.

Sustainable development also necessitates long-term planning and a balanced management approach. Holistic policies must be developed, taking into account the environmental consequences of decisions made in areas such as energy, agriculture, urbanization, and water management. Such a model aligns economic growth with environmental responsibility while also fostering social equity. Therefore, sustainable development is considered a multidimensional approach that both addresses today's needs and safeguards future living conditions.

2.14. Integrated Water Resources Management (IWRM)

IWRM is a process that facilitates the coordinated development and management of land, water, and related resources with the goal of ensuring the long-term sustainability of the environment while balancing social and economic welfare in an equitable manner.

3. PRINCIPLES AND LOGIC OF THE SUBJECT

After the industrial revolution, human societies have significantly modified the global water system. These changes have occurred while insufficient knowledge is available to fully analyze this complex system and the human intervention feedback on it. Analyzing the issue of water security in the end of 20th and 21st centuries requires attention to management and policy aspects alongside engineering and technology knowledge. The framework of water security first emerged in 1990 and has changed considerably since then. Two decades ago the term water security was specifically related to military security at the time and rarely concerned with environmental security, but in 2000 the Second World Water Forum provided a comprehensive definition of water security that easy access to water, cost-effectiveness of water, human and environmental health needs were also considered. Cook and Bakker (2012) until now, many researchers and politicians have used the term and attributed it to various meanings and definitions. Some of them linked water security to the field of water and the environment and others consider water security as an interdisciplinary approach.

But the most important foundations to form the concept of water security may be found in the status of the world's water resources in the 20th and 21st centuries. The advancement of industry, agriculture, technology, and urbanization made it possible for modern humans to utilize all available environmental capacities that one of the most important of which was available water resources, to achieve maximum efficiency in agricultural and industrial production. For this reason, water such as oil and gold has gradually become a major component of power and progress in a country. On the other hand by taking maximum water reserves in the world and global warming due to climate change in the last two centuries, every day this rare resource has been facing the threat of scarcity and crisis and therefore making water security a place of global security. Lack of attention to water security is the basis of many conflicts and contention that have taken place between countries, which is why world politicians have always placed sufficient attention to water security as the main agenda of their country's development plan.

4. THE IMPORTANCE AND NECESSARY

Kojm (2012) it is predicted that by 2030, the world will have a 40 percent shortage of water demand and supply. This gap will be very serious and means having a small amount or lacking water for drinking, farming, transportation, production, health, habitat and healthy ecosystems for humans, plants.

Water, food, and energy are the resources that are essential for all human activities, in addition these factors have very close and continuous interconnections. Water cannot be delivered to the consumer without consuming a significant amount of energy (for transportation, keep, and etc.) and also, food and energy production and productivity would not be possible without a large volume of water. So, by any definition and evaluation in the lack of water security, development and growth will not be achieved and the guarantee of sustainable development of the world will depend on access to sufficient quality water. Water has played a large role in various environmental aspects of the earth, undoubtedly. For example, the amount of biodiversity reduction in freshwater ecosystems is greater than in oceanic and land ecosystems and in addition, land use degradation and changes are also complex and intertwined with freshwater availability. However, it can be predicted that the trend of water consumption due to population growth will continue as now. So a detailed assessment of the production and consumption process of water seems necessary if global freshwater consumption will exceed the threshold tolerance. Therefore, in this assessment should determine where? For what? And how? Optimize using water for more effectively to make it more profitable to prevent water conflict that will result from water shortages or having a small amount. This subject show the importance of paying attention to water security in today's world.

In general, many statesmen, scholar, and political actors in the Middle East, South Africa, and East Asia are warning of water scarcity as a major and growing problem in their region, which could lead to serious conflicts, thus the importance of water security studies has always been placed in their political, social and economic development plans and also, they try to make their country always better than its neighbors in water and water resources. As show as in Figure 3. There are 655 conflict (Casualty, Trigger or reason to start, and Weapon) on water resource (3000 B.C. - 2018 A.D.).

Figure 3. Conflict of water in world

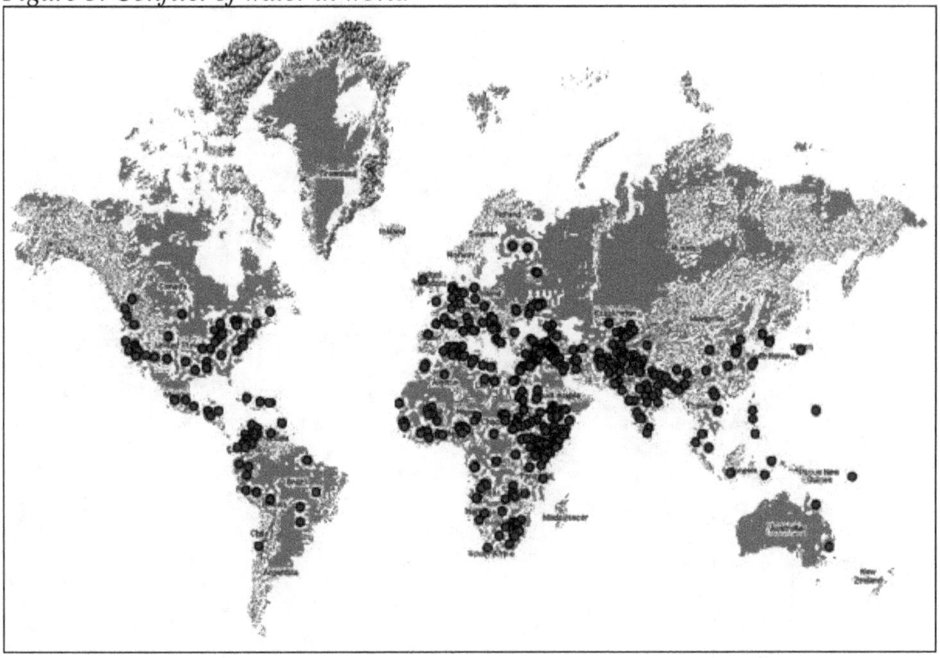

5. METHOD AND FLOWCHART

There are many factors and parameters to consider for water security assessment in each area's that should be taken into account and all of them are in good condition, otherwise there is no water security in the area. If all options are optimized, they will do their best, and water security will be at its optimum. As shown as Figure 4. The water security methodology is examined from two points of view: (1) indices and related issue and (2) attacks on them.

Figure 4. Flowchart of water security

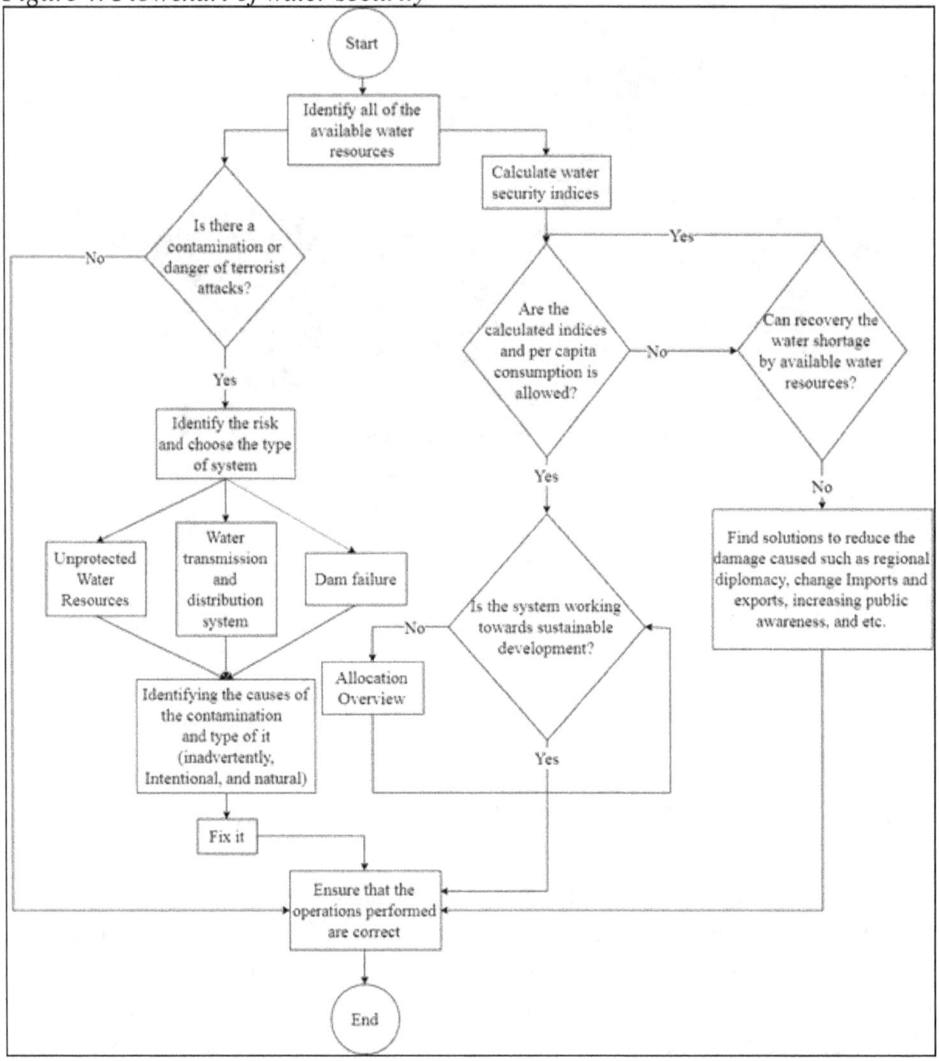

6. DESCRIPTION OF EXAMPLES

6.1. Example One

Amir Ahmadian and Naseri (2013) researched about water security in central Asia and Amu Darya and the Syr Darya basin. Central Asia is a vast landlocked continental climate in the low water belt of the Earth, where water resources are inadequately distributed. Its populous resorts, such as Bukhara, Samarkand and Merv, are among the most critical areas in the world in terms of water issues. The drying up of Aral Lake in the area has become a global issue. The water needs of the five countries - Tajikistan, Kyrgyzstan, Uzbekistan, Turkmenistan and Kazakhstan - are supplied by the two major rivers of the Amu Darya and the Syr Darya in the region. Tajikistan and Kyrgyzstan are in the upper reaches of the rivers and Uzbekistan, Turkmenistan and Kazakhstan are in the lower reaches of these rivers, and they had disagreeing over the division and flow of transboundary rivers after the collapse of the Soviet Union.

The conversion of desert lands to extensive cotton fields during the Soviet Union and the diversion of Amu Darya and the Syr Darya water flows to irrigate these dry lands leading to the establishment of new population centers in this region. The rivers Amu Darya and the Syr Darya originate in Tajikistan and Kyrgyzstan and also supply the water demand of the downstream such as Turkmenistan, Uzbekistan and Kazakhstan.

Before the dissolution of the Soviet Union, the countries of Turkmenistan, Uzbekistan, Kyrgyzstan and Tajikistan were in one economic zone and Kazakhstan was alone in another free economic zone that in coordination with the central government (Moscow), dealing and exchanging with each other. Tajikistan and Kyrgyzstan did not have the energy resources such as oil and gas, and supplied their required hydrocarbon energy from Uzbekistan, Turkmenistan, and Kazakhstan versus supplying the water demanded for the agricultural, industrial and urban consumption of the downstream countries.

Since the dissolution of the Soviet Union, relations between these countries such as regional, political, and economic affairs have been completely changed so that these countries have repeatedly discussed about collaborative sharing and operating of theirs resources. Tajikistan and Kyrgyzstan were among the poorest areas of the Soviet Union and after reaching their independence due to insufficient income were weaker than before but after independence the downstream countries canceled exchange and deal with the upstream countries due to the availability of oil and gas resources and also changed the price of their energy resource exchanges to the level of free market trade and wanted world prices of trade market from upstream countries to supply their energy's.

The upstream countries implemented the plan of sell water to downstream countries due to the lack of sufficient revenue sources to buy the energy demanded, and they attempted to build new dams due to the lack of sufficient reservoirs. In the all of this countries except of Kazakhstan where the industry and services sector is strong, the agricultural sector has a large part of contribution in the employment of the population of these countries because extend agricultural land and shortage or lack of water is facing a crisis to problems such as irrigate agricultural lands, daily livelihoods, water security, food and jobs effect of millions people in the region that This has caused raised tension between the downstream and upstream countries.

China's claim for small and large lakes in the region such as Balkhash Lake in Kazakhstan (downstream) and German's support for projects such as rebuilding small hydroelectric power in Kyrgyzstan (upstream) has increase the water crisis in those countries. The agricultural products in the downstream countries are cotton, wheat and rice and therefore decrease of water resources and the threat to water and food security and economic stability in their country, they are against to the policies and plans being implemented in the upstream countries.

Rittger *et al.* (2017) so far, there have been various attempts at water compromise between the upstream and downstream countries, but all of them have failed. Probability of conflict and threats in the region is increases because high military power and maneuvers in downstream countries have always been one of the main pressure on the upstream countries. For this reason, the water security situation in this region is very critical and severe and it could lead to a bigger crisis than now if no special action are taken to solve this problem.

6.2. Example Two

Kisakye and Van der Bruggen (2018) investigated the impacts of climate change on water conservation and household water security in developing countries. Achieving the goal of "safe, affordable, and accessible water for all" remains a significant challenge, particularly in the face of population growth and the uncertainties introduced by climate change. In many rural areas, centralized public water supply systems have proven inefficient, prompting a shift toward localized alternatives. As a result, rainwater harvesting is increasingly recognized as a viable solution for enhancing household water security.

This study, conducted in Kabarole, Western Uganda—an area with the highest rainfall in the country compared water conservation and water security under historical conditions and six different climate model scenarios (BCC-CSM1.1, ACCESS1.0, CNRM-CM5, HadGEM2-CC, and HadGEM2-ES) for the periods 2025–2055 and 2060–2090.

Water conservation was assessed by dividing the total volume of harvested rainwater by the number of days in each simulated period. Water security was quantified using an index based on the number of days a specific tank size remained empty during the same periods. Although the average per capita water use in the region is around 16 liters per day below the WHO's minimum recommendation of 20 liters the study assumed a usage rate of 50 liters per person per day, with an average household size of five.

Despite variation across the climate models, the findings indicate that climate change is likely to increase water conservation and household water security in the December–February and March–May periods, while reducing both in the June–August and September–November seasons. The March–May season is expected to see the most significant improvement in daily water savings up to 50% by 2070. Overall, the study concludes that households with roofs larger than 50 square meters and rainwater tanks of at least 5 cubic meters, combined with proper water treatment, are likely to experience fewer negative impacts from climate change on water security.

7. DISCUSSION

Our synthesis highlights that water security is inherently multidimensional shaped by biophysical limits, socio-economic demand, governance quality, and geopolitical context and that risks are unevenly distributed across and within countries. Nearly five percent of the global population faces high water-security risk, with wealthier nations often offsetting physical stress through infrastructure investments, while poorer regions remain vulnerable.

A key technical insight is the differentiated behavior of "blue" versus "green" water under climate and land-use change. In the Savannah River Basin, climate change reduces and increases the variability of blue water, whereas agricultural expansion elevates green-water flows illustrating why basin-specific strategies are needed rather than one-size-fits-all interventions. Strategic source areas further complicate this picture: in South Africa, just ~8% of land area supplies ~50% of runoff and underpins a majority of national population and GDP, underscoring the outsized leverage of headwater protection for national water security.

Urban systems remain acutely exposed where institutional fragility intersects with conflict. Long-term degradation of Basrah's drinking-water services driven by skilled-staff flight, chemical shortages, and spare-parts scarcity illustrates how protracted instability can erode service reliability despite existing infrastructure. Decentralized options such as rainwater harvesting can bolster availability, quality, and access in underserved peri-urban and rural settings, but persistence is constrained by cost burden, water-quality consistency, and maintenance demands.

Governance frameworks must therefore move beyond engineering supply to embrace adaptive capacity and equity. Empirical and theoretical work argues that water-security assessment should internalize adaptive capacities and flexible institutions, not only physical metrics. At the same time, the relationship between IWRM and water security is complementary: IWRM provides a technocratic, integrative management scaffold, while "water security" foregrounds societal values, ethics, and power together enabling more holistic decision-making. Yet global norm diffusion is uneven; without alignment by wealthier states, operationalizing rights-based water and hygiene norms risks reproducing existing power imbalances.

Metric choice shapes what we see and fund. Classical tools (water stress/scarcity indices, Water Poverty Index) and newer constructs (Objective Water Security Index, National Water Security Index) capture complementary facets from resource-user ratios and poverty constraints to household affordability, alternative sources, and disaster resilience. Policymakers should triangulate across indices rather than rely on a single score to avoid perverse incentives and misclassification. Additionally, security framings that include protection against pollution and terrorism emphasize the need to harden drinking-water infrastructures and supply chains.

Policy implications follow directly. First, protect and legally designate strategic source areas and manage them as national critical infrastructure, linking headwaters to downstream users to enable cross-sector cooperation. Second, couple demand-side measures (pricing, efficiency, crop shifts) with diversified supplies (reuse, decentralized storage, nature-based solutions) in basins where blue-water deficits are rising and green-water dependence grows with agriculture. Third, strengthen adaptive, equity-oriented governance embedding household-level affordability/access metrics into national indices and aligning IWRM instruments with rights-based commitments. Fourth, in fragile and conflict-affected cities, prioritize institutional capacity (staff retention, procurement of treatment chemicals and parts) alongside physical rehabilitation.

Limitations of the current evidence base include heavy reliance on case studies with heterogeneous methods and time frames (e.g., 1978–2013 for Basrah), which complicates meta-inference and temporal comparability; index construction can also embed normative choices about weights and thresholds. Finally, transboundary governance especially for shared groundwater on which ~50% of people rely for drinking and which is increasingly contested remains under-instrumented, demanding clearer legal and cooperative mechanisms.

Future research should (i) co-produce basin-specific portfolios that jointly optimize blue/green water under climate uncertainty; (ii) fuse cyber-physical risk assessment with equity-focused service metrics for urban systems; (iii) evaluate long-term performance and justice outcomes of decentralized solutions; and (iv) advance transboundary groundwater monitoring and treaty design that integrate ecological-flow safeguards and drought contingency triggers. Together, these steps can align technical, institutional, and social levers to advance durable, just water security at household, city, and basin scales.

8. CONCLUSION

Water is a topic ranging from cells of the human body to complex political and security relationships inside and outside the countries of the world, and access to scarce resources such as water has always been a factor in tensions at home or abroad. However, operating this scarce resource has high effects of economic, security, and geopolitical that governments cannot afford to ignore it. The water resources around the world is in critical condition and the severity of the situation in different regions is different. As shown as Figure 5. Areas such as the Middle East, the coast of sub-Saharan Africa and the region of South Africa are more vulnerable to long-term droughts and water shortages.

Figure 5. Gain et al (2016) aggregated global water security index, calculated using the aggregation of water availability, accessibility, safety and quality, and management indices.

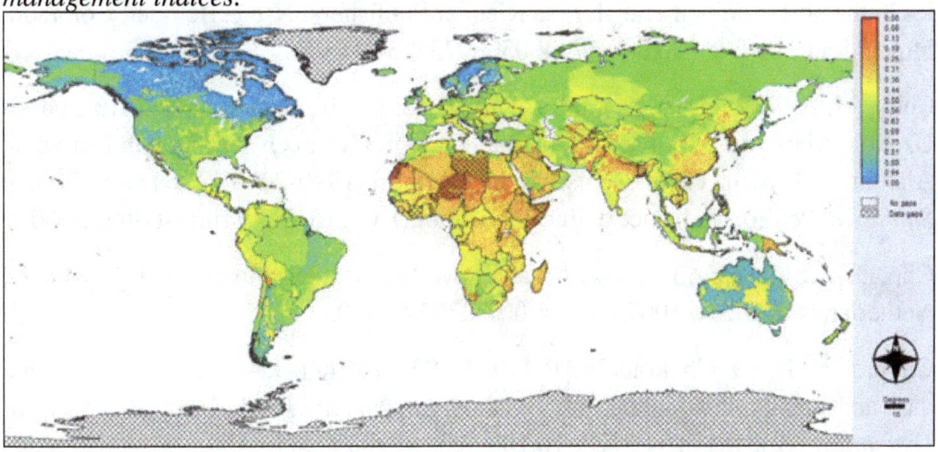

Particularly in developing countries, factors such as rapid population growth, climate change and water mismanagement endanger water security. The nature of water security implies access to adequate and clear water for consumption such as personal, household, energy, industrial, transportation, environmental and biodiversity sustainability, and food security estimates. Today, the relationship between water security and food security is increasingly used in development and consuming more than 70% of water resources to produce food is an important example of this connection. Therefore, it can be concluded that in recent years the concept of water security has gone beyond the consideration of military and war issues and has formed a multifaceted concept. Given the global importance and the interconnected role of water, the issue of water management should be at the forefront of political and

REFERENCES

Alvarez, P. J., Chan, C. K., Elimelech, M., Halas, N. J., & Villagrán, D. (2018). Emerging opportunities for nanotechnology to enhance water security. *Nature Nanotechnology*, *13*(8), 634–641. DOI: 10.1038/s41565-018-0203-2

Amir Ahmadian, B. and Naseri, M. (2013). "Water crisis in central asia with emphasis on regional conflict over water issue." Journal of central eurasia studies, 6(12), 1-20.

Bogardi, J. J., Dudgeon, D., Lawford, R., Flinkerbusch, E., Meyn, A., Pahl-Wostl, C., Vielhauer, K., & Vörösmarty, C. (2012). Water security for a planet under pressure: Interconnected challenges of a changing world call for sustainable solutions. *Current Opinion in Environmental Sustainability*, *4*(1), 35–43. DOI: 10.1016/j.cosust.2011.12.002

Buzan, B. (2008). "People, states & fear: an agenda for international security studies in the post-cold war era." Lynne Rienner Publishers, No. 1, Boulder, Colorado. https://doi.org/DOI: 10.1080/17449050802539344

Çelik, M. A., Kizilelma, Y., Gülersoy, A. E., & Denizdurduran, M. (2013). Farkli Uzaktan Algilama Teknikleri Kullanilarak Aşaği Seyhan Ovasi Güneyindeki Sulak Alanlarda Meydana Gelen Değişimin Incelenmesi (1990-2010). *Electronic Turkish Studies*, *8*(12), 263. Advance online publication. DOI: 10.7827/TurkishStudies.6011

Cho, D. I., & Ogwang, T. (2014). "Water Poverty Index." Springer, No. 1, Dordrecht, Netherlands. DOI: 10.1007/978-94-007-0753-5_3703

Chopra, A. K., & Chakrabarti, P. (1972). The earthquake experience at Koyna dam and stresses in concrete gravity dams. *Earthquake Engineering & Structural Dynamics*, *1*(2), 151–164. DOI: 10.1002/eqe.4290010204

Congleton, R. D. (2006). The story of Katrina: New Orleans and the political economy of catastrophe. *Public Choice*, *127*(1-2), 5–30. DOI: 10.1007/s11127-006-7729-9

Cook, C. (2016). Implementing drinking water security: The limits of source protection. *WIREs. Water*, *3*(1), 5–12. DOI: 10.1002/wat2.1117

Cook, C., & Bakker, K. (2012). Water security: Debating an emerging paradigm. *Global Environmental Change*, *22*(1), 94–102. DOI: 10.1016/j.gloenvcha.2011.10.011

Duda, A. M. (2017). Co-managing land and water for sustainable development. In United Nations Convention to Combat Desertification: Bonn, Germany.

Elgert, L., Austin, P., & Picchione, K. (2016). Improving water security through rainwater harvesting: A case from Guatemala and the potential for expanding coverage. *International Journal of Water Resources Development, 32*(5), 765–780. DOI: 10.1080/07900627.2015.1104499

Falkenmark, M., Berntell, A., Jägerskog, A., Lundqvist, J., Matz, M., & Tropp, H. (2007). "On the verge of a new water scarcity: a call for good governance and human ingenuity." Stockholm International Water Institute (SIWI), No. 1, Sweden.

Falkenmark, M., & Molden, D. (2008). Wake up to realities of river basin closure. *International Journal of Water Resources Development, 24*(2), 201–215. DOI: 10.1080/07900620701723570

Filkins, D. (2016). "A Bigger Problem Than ISIS?" <www.newyorker.com> (July. 27, 2019).

Gain, A., Wada, Y., & Giupponi, C. (2016). Measuring global water security towards sustainable development goals. *Environmental Research Letters, 11*(12), 2–13. DOI: 10.1088/1748-9326/11/12/124015

Gellman, B. (2002). "Cyber-attacks by Al Qaeda feared." <www.washingtonpost.com> (July. 27, 2019).

Gleick, , P. H. (2019). Water as a weapon and casualty of conflict: Freshwater and international humanitarian law. *Water Resources Management, 33*, 1737–1751.

Gleick, P. H. (2006). Water and terrorism. *Water Policy, 8*(6), 481–503. DOI: 10.2166/wp.2006.035

Grumbine, R. E., & Xu, J. (2011). Mekong hydropower development.

Hinrichscn, D. (2002). "Freshwater: lifeblood of the planet." <http://www.peopleandtheplanet.com> (Sep. 23, 2019).

Hohenthal, J., & Minoia, P. (2017). Social aspects of water scarcity and drought. In *Handbook of Drought and Water Scarcity* (pp. 607-625). CRC Press.

Jia, X., Li, C., Cai, Y., Wang, X., & Sun, L. (2015). An improved method for integrated water security assessment in the Yellow River basin, China. *Stochastic Environmental Research and Risk Assessment, 29*(8), 2213–2227. DOI: 10.1007/s00477-014-1012-2

Kerr, R. A., & Stone, R. (2009). A human trigger for the great quake of Sichuan? *Science, 323*(5912), 322–322. DOI: 10.1126/science.323.5912.322

Kher, J., Aggarwal, S., & Punhani, G. (2015). Vulnerability of poor urban women to climate-linked water insecurities at the household level: A case study of slums in delhi. *Indian Journal of Gender Studies*, *22*(1), 15–40. DOI: 10.1177/0971521514556943

Kılıç, Z. (2020). The importance of water and conscious use of water. *International Journal of Hydrology*, *4*(5), 239–241. DOI: 10.15406/ijh.2020.04.00250

Kisakye, V., & Van der Bruggen, B. (2018). Effects of climate change on water savings and water security from rainwater harvesting systems. *Resources, Conservation and Recycling*, *138*, 49–63. DOI: 10.1016/j.resconrec.2018.07.009

Kojm, C. (2012). "Global Water Security." Intelligence Community Assessment, Report No. Washington, DC, USA.

Maganda, C. (2016). Water security debates in 'safe'water security frameworks: Moving beyond the limits of scarcity. *Globalizations*, *13*(6), 683–701. DOI: 10.1080/14747731.2015.1133605

Mahmood, A. N., El-Khazali, R., Çiçek, A., Assi, S. A., Al-Naimi, T. M., Majdi, H. S., & Hussain, A. S. T. (2025). Experimental Analysis of Fresnel Lens-Based Solar Desalination Systems with Copper Receivers for Enhanced Thermal and Electrical Performance. *International Journal of Robotics & Control Systems*, *5*(2), 1191–1210. DOI: 10.31763/ijrcs.v5i2.1852

Manning, F. J., & Goldfrank, L. R. (2002). "Preparing for terrorism: tools for evaluating the metropolitan medical response system program." National Academy Press, No. 1, Washington, DC, USA.

MDH. (2000). "Ministerial declaration of the Hague on water security in the 21st century." Report No. 2, Hague, Netherlands.

Minamyer, S. (2008). Effective crisis communication during water security emergencies. *Journal - American Water Works Association*, *100*(9), 180–184. DOI: 10.1002/j.1551-8833.2008.tb09719.x

Molle, F., Foran, T., & Käkönen, M. (2009). Contested waterscapes in the Mekong Region.

Nel, J. L., Le Maitre, D. C., Roux, D. J., Colvin, C., Smith, J. S., Smith-Adao, L. B., Maherry, A., & Sitas, N. (2017). Strategic water source areas for urban water security: Making the connection between protecting ecosystems and benefiting from their services. *Ecosystem Services*, *28*, 251–259. DOI: 10.1016/j.ecoser.2017.07.013

Penuel, K. B., & Statler, M. (2010). "Encyclopedia of disaster relief." SAGE, No. 2, California, USA.

Pepin, M., Bouloy, M., Bird, B. H., Kemp, A., & Paweska, J. (2010). Rift Valley fever virus (Bunyaviridae: Phlebovirus): an update on pathogenesis, molecular epidemiology, vectors, diagnostics and prevention. *Veterinary Research, 41*(6), 61. DOI: 10.1051/vetres/2010033

Powell, N., Kløcker Larsen, R., de Bruin, A., Powell, S., & Elrick-Barr, C. (2017). Water Security in Times of Climate Change and Intractability: Reconciling Conflict by Transforming Security Concerns into Equity Concerns. *Water (Basel), 9*(12), 934. DOI: 10.3390/w9120934

Rijsberman, F. R. (2006). Water scarcity: Fact or fiction? *Agricultural Water Management, 80*(1-3), 5–22. DOI: 10.1016/j.agwat.2005.07.001

Rittger, K., Bair, E., Racoviteanu, A., Brodzik, M. J., Armstrong, R. L., Jodha Khalsa, S., Raup, B., Painter, T. H., & Dozier, J. (2017). "Quantifying the individual contributions of melt from snow and glaciers in High Mountain Asia river basins: Syr Darya, Amu Darya, Indus, Ganges, and Brahmaputra." *Proceedings of European Geosciences Union General Assembly Conference Abstracts*, Vienna, Austria, April 23–28.

Romero-Lankao, P., & Gnatz, D. M. (2016). Conceptualizing urban water security in an urbanizing world. *Current Opinion in Environmental Sustainability, 21*, 45–51. DOI: 10.1016/j.cosust.2016.11.002

Schmid, A. P., & Jongman, A. J. (2017). "Political terrorism: A new guide to actors, authors, concepts, data bases, theories, and literature." Routledge, No. 1, New York, USA.

Sevgi, , M. AYulu, , A. (2025). The old rains are gone, so are we: Kurdish Nomads of Mount Agri (Ararat). *British Journal of Middle Eastern Studies, 52*(2), 503-529.

Shah, M. M. (2008). "Sustainable Development." Academic Press, No. 1, Oxford, England.

Shahbazbegian, M. R., Turton, A., & Mousavi Shafaee, S. M. (2016). Hydropolitical Self-Organization theory; system dynamics to analyse hydropolitics of Helmand transboundary river. *Water Policy, 18*(5), 1088–1119. DOI: 10.2166/wp.2016.204

Sharaunga, S., & Mudhara, M. (2016). The impact of improved 'water-use security'on women's reliance on agricultural incomes in KwaZulu-Natal Province, South Africa. *Food Security, 8*(6), 1039–1052. DOI: 10.1007/s12571-016-0622-6

Shrestha, S., Aihara, Y., Bhattarai, A. P., Bista, N., Kondo, N., Futaba, K., Nishida, K., & Shindo, J. (2018). Development of an objective water security index and assessment of its association with quality of life in urban areas of developing countries. *SSM - Population Health*, *6*, 276–285. DOI: 10.1016/j.ssmph.2018.10.007

Siddiqi, Y. (2016). "Asian Water Development Outlook 2016: Strengthening Water Security in Asia and the Pacific." Asian Development Bank, Report No. 9292575449, Mandaluyong, Philippine.

Staff, J. (2019). "DoD dictionary of military and associated terms." Joint Publication, No. 1, USA.

States, S. (2011). "Security and Emergency Planning for Water and Wastewater Utilities." American Water Works Association, No. 1, Mumbai, India.

Staudinger, T. J., England, E. C., & Bleckmann, C. (2006). Comparative analysis of water vulnerability assessment methodologies. *Journal of Infrastructure Systems*, *12*(2), 96–106. DOI: 10.1061/(ASCE)1076-0342(2006)12:2(96)

Sullivan, C. A., Meigh, J. R., Simon, S., Lawrence, P., Calow, R., McKenzie, A., Acreman, M. C., & Moore, R. V. (2000). "The Development of a Water Poverty Index." Department for International Development, Report No. 1, Wallingford, UK.

Tavakoli, H.R., Sarafpour, R., and Samadi, M. (2005). "Water, Food and Bioterrorism." Journal of military medicine, 7(1), 75-82.

Ticrncy, J., & Worth, R. F. (2003). "Attacks in iraq may be signals of new tactics." <www.nytimes.com> (Sep. 27, 2019).

Tufail, M., Nasir, M. J., Tariq, A., Ali, Z., & Alarifi, S. S. (2025). Assessing access to safe drinking water in flood-affected areas of District Nowshera, Pakistan: A case study towards achieving sustainable development goal 6.1. *Ecohydrology & Hydrobiology*, *25*(2), 422–437. DOI: 10.1016/j.ecohyd.2024.07.001

UN-Water (2019, 2023). World Water Development Report.

United Nations (2015). Transforming our world: the 2030 Agenda for Sustainable Development.

Veettil, A. V., & Mishra, A. K. (2016). Water security assessment using blue and green water footprint concepts. *Journal of Hydrology (Amsterdam)*, *542*, 589–602. DOI: 10.1016/j.jhydrol.2016.09.032

Vörösmarty, C. J., McIntyre, P. B., Gessner, M. O., Dudgeon, D., Prusevich, A., Green, P., Glidden, S., Bunn, S. E., Sullivan, C. A., Liermann, C. R., & Davies, P. M. (2010). Global threats to human water security and river biodiversity. *Nature*, *467*(7315), 555–561. DOI: 10.1038/nature09440

Wegerich, K., Van Rooijen, D., Soliev, I., & Mukhamedova, N. (2015). Water security in the Syr Darya basin. *Water (Basel)*, *7*(9), 4657–4684. DOI: 10.3390/w7094657

Wheater, H. S., & Gober, P. (2015). Water security and the science agenda. *Water Resources Research*, *51*(7), 5406–5424. DOI: 10.1002/2015WR016892

Whiteford, S., & Witter, S. G. (1999). Water security: The issues and policy challenges. *International Review of Comparative Public Policy*, *11*, 1–25.

Zare Bidaki, m. and Balali Mood, m. (2015). "Bioterrorism and biological warfare, from past to the present: A classic review." Journal Of Birjand University Of Medical Sciences, 22(3), 182-198. https://sid.ir/paper/95436/en

Zeitoun, M., Elaydi, H., Dross, J. P., Talhami, M., de Pinho-Oliveira, E., & Cordoba, J. (2017). Urban Warfare Ecology: A Study of Water Supply in Basrah. *International Journal of Urban and Regional Research*, *41*(6), 904–925. DOI: 10.1111/1468-2427.12546

Zeitoun, M., Lankford, B., Krueger, T., Forsyth, T., Carter, R., Hoekstra, A. Y., Taylor, R., Varis, O., Cleaver, F., Boelens, R., Swatuk, L., Tickner, D., Scott, C. A., Mirumachi, N., & Matthews, N. (2016). Reductionist and integrative research approaches to complex water security policy challenges. *Global Environmental Change*, *39*, 143–154. DOI: 10.1016/j.gloenvcha.2016.04.010

Chapter 2
Gender and Disasters: The Position of Vulnerable Groups in Disaster Processes

Burcu Gököztkut
http://orcid.org/0000-0002-2214-8327
Afyon Kocatepe University, Turkey

ABSTRACT

This study focuses on the importance of understanding the social dimensions of disasters in the context of natural disaster events and structural social inequalities. Disasters not only cause physical destruction but also deepen pre-existing gender-based inequalities in society, leading to multidimensional social crises. The study examines the situation of vulnerable groups in detail through the pre-disaster, disaster and reconstruction phases. This comprehensive analysis is supported by case studies where natural disasters and gender inequality are clearly evident. Consequently, it has been concluded that disasters are a social dynamic that reproduces and deepens existing social inequalities. The study concluded with recommendations for restructuring disaster policies in terms of gender sensitivity.

1. INTRODUCTION

Today, multi-dimensional global problems such as the climate crisis, rapidly increasing disaster events, and existing structural societal inequalities have made disaster management and understanding of the social impacts of disasters a critical issue. Disasters not only cause physical destruction and debilitating effects but also deepen existing gender-based inequalities and vulnerabilities within society, thereby transforming into a social crisis. Groups with structural disadvantages stemming from socio-economic and cultural positions, particularly women, older people, dis-

DOI: 10.4018/979-8-3373-6274-8.ch002

abled people and LGBTQ+ individuals, experience disproportionate vulnerability during disaster processes due to conditions arising from their position in society rather than the direct effects of the disaster itself.

The fact that disaster management and response processes are often shaped by male-dominated and patriarchal structures hinders the inclusion of vulnerable groups' knowledge, experience and needs in decision-making mechanisms, thereby complicating the achievement of social justice in disaster management. Disasters increase the care, reproduction, and household management responsibilities socially attributed to women within vulnerable groups, both quantitatively and qualitatively; this deepens their physical, economic, and psychological vulnerability. The disproportionate impact of disasters on vulnerable groups can be observed across different geographical areas, from Canada to Nepal, revealing that gender inequality and vulnerability are exacerbated during disaster processes. Therefore, the relationship between disasters and gender should be assessed not only in terms of these vulnerable groups being more affected, but also in terms of how disasters reproduce existing social inequalities. This study aims to examine in depth the structural inequalities and multi-layered vulnerabilities faced by women and other vulnerable groups during the pre-disaster, disaster and post-disaster periods, because disaster processes are in no way gender-neutral. In this context, the study assesses the relationship between gender and disasters not only in terms of victimization, but also in terms of how it reveals structural inequalities.

The study is structured systematically to present this comprehensive analysis. The second section presents the fundamental conceptual framework, including the distinction between gender and sex, the construction of social roles, and the relationship with the patriarchal order. The third section examines the situation of vulnerable groups in detail through the three stages of disaster processes: Pre-Disaster Phase (inequality in access to information and male-dominated planning), Impact Moment of Disaster Phase (evacuation difficulties, physical and psychosocial risks) and Post-Disaster and Recovery Process (inequalities in the recovery process, increased burden of responsibility and security gaps). The fourth section addresses the gender inequalities that emerge in the face of natural disasters in different parts of the world. The study concludes with the “Conclusion and Recommendations” section, which presents proposals for restructuring disaster policies in terms of gender sensitivity, based on the findings obtained.

2. GENDER AND DISASTER

Gender, distinct from an individual's biological sex at birth, is a concept used to describe the social and cultural definition of women and men, the ways in which societies distinguish between these two sexes, and the social roles assigned to them. This concept determines the roles, identities, rights and perceptions attributed to individuals by culture and society as a structure that operates at the institutional, interactive and individual levels of social life (Mariki, 2016). Gender reflects the patriarchal order (Mcdowell, 2003) and is based on the unequal social division of labour between the public sphere (production/male) and the private sphere (reproduction/female); this ideology encompasses the belief that men should be breadwinners and women should be mothers and housewives (Tichenor, 1963; Verloo, 2007). These roles are not innate or natural; rather, they vary according to context, situation and time. In this context, gender roles are normative groups of behaviour associated with gender that society expects from individuals. The construction of the concept takes place within the normative functioning of power relations that fix the body to specific gender and identity categories (Arpacı, 2013). However, gender should not be understood merely as cultural meanings added to a biological foundation. According to Butler (2005), however, the function of gender goes even further and views sex, which is accepted as "natural" and "biological," as a discursive (linguistic and narrative) and cultural tool used to produce even itself. This tool allows biological sex to be perceived as if it were outside of culture (pre-cultural), neutral, and a reality given by nature. However, Butler argues that this perception of naturalness and neutrality is also continuously constructed through social discourses.

Disasters exacerbate existing gender-based inequalities and vulnerabilities within society, leading to gender-specific impacts. This situation demonstrates that gender distinctions become more pronounced during disaster processes. Research indicates that women are in a more disadvantaged position compared to men in terms of access to education, health, employment, and development services before a disaster, and are therefore structurally considered among the vulnerable groups (Gündüz & Atalay, 2025; Sarı Karademir & Şahin, 2025). These existing inequalities become even more pronounced during disaster periods; women's socially attributed responsibilities for care and reproduction increase, which in turn deepens their physical, economic, and psychological vulnerability. As a result of their social roles and responsibilities, women experience disasters in different ways than men; they bear a heavier physical, economic, and psychological burden of disasters, particularly due to their obligations in care, reproduction, and household management. (Reyes & Lu, 2016). Within this framework, disasters cause women to experience a multi-layered process of vulnerability and victimization due to their low socio-economic status, limited representation in the public sphere, and the intensity of care responsibilities within the household.

Since disaster management and response processes are frequently shaped by patriarchal social relations and male-dominated institutional structures, the knowledge, experience, and contributions of women and other vulnerable groups do not become sufficiently visible in decision-making mechanisms (Danielsson & Eriksson, 2022). This situation complicates the achievement of social justice in disaster management and leads to the needs of vulnerable groups being overlooked. Therefore, the relationship between gender and disasters cannot be explained solely by the disproportionate impact on women; it must also be assessed in terms of how disasters reproduce social inequalities. The high vulnerability of women and other vulnerable groups during disaster processes is closely linked to the continuity of structural inequalities, limited access to resources, discriminatory practices, and patriarchal cultural norms (Baig & Sharif, 2013; Fatema et al., 2023; Ural et al., 2024).

This increased burden during disaster processes leads to a higher prevalence of mental health issues such as post-traumatic stress disorder and depression among women, while also paving the way for gender-based forms of violence to emerge in different ways (Demirci & Avcu, 2021; Slick & Hertz, 2024; Thapa & Pathranarakul, 2019). However, the concept of vulnerable groups is not limited solely to women and children; it also includes older people, disabled individuals, and LGBTQ+ individuals. These groups are affected in different ways across all phases of disasters due to multi-layered disadvantages such as structural inequalities, social exclusion, and limited access to resources. As a common denominator, these groups are subjected to exclusion and discrimination across social, economic, and cultural spheres even in the pre-disaster period. Therefore, disasters are not only considered a process of physical destruction but also a social dynamic through which existing social inequalities are reproduced, and in some cases, further deepened (Cutter, 1995; Çıtak & Toker, 2025; Dominey-Howes et al., 2014).

3. THE ROLE AND POSITION OF VULNERABLE GROUPS WITHIN THE DISASTER PROCESS

Gender differences and vulnerabilities become prominently visible during disaster processes. Several factors influence the resilience of vulnerable groups against disasters throughout these processes. One of these factors is income inequality. Income inequality leads to women being predominantly employed in low-wage, hourly jobs, often lacking social protection and benefits, which limits their economic independence and fails to ensure economic stability, largely because the income earned is often spent on childcare (Gündüz & Atalay, 2025). Another factor affecting vulnerable groups during disaster processes is gender-based violence. Gender-based violence is observed globally, regardless of development levels, with women and girls bearing the heaviest burden of this violence in conflict and disaster zones. This situation directly affects women's capacity for adaptation and recovery from disasters. Another factor affecting vulnerable groups is large-scale population movements. Especially during the journey, women and young girls face risks of exploitation, sexual abuse, and human trafficking; meanwhile, returning or repatriated refugees are disproportionately and negatively affected during the reconstruction process (Cutter, 2017).

Women are in a position of increased vulnerability during disaster processes due to taking on more practical responsibilities compared to men. For instance, while it has been found that women in Bangladesh have made significant progress in areas such as health, social participation, education, and economy within the scope of disaster resilience indicators, they still lag behind men in terms of structural variables like social status, overall economic standing, and access to information, representing persistent inequalities in achieving complete disaster resilience parity (Kadir, 2021). In their study examining the long-term effects of flood damage on rural employment in India, Chowdhury et al. (2022) reveal that women's employment in the agricultural sector was more negatively affected compared to men's employment. The study identified the intense dependence of rural women workers on climate-sensitive agriculture and their limited flexibility to transition to non-agricultural sectors as the primary reasons for this situation, thereby empirically proving the structural vulnerability of these groups. Similarly, an empirical study in Metro Manila, the Philippines, revealed the distinction between the de facto responsibilities of women during a disaster and the socially recognized roles of men. The study's findings indicate that women undertook essential care and survival tasks such as food provision, psychosocial support (including maintaining household morale), and care for the sick/injured, while men were symbolically recognized more often in the disaster context with roles such as household head and leader positions. This situation demonstrates how gender roles perpetuate the inequality between visible leadership and invisible labor, even in times of crisis (Reyes & Lu, 2016).

In this context, the situation of vulnerable groups during disaster processes is examined in this section across three stages: the pre-disaster phase, the impact moment of disaster phase, and the post-disaster and recovery phase. Accordingly, the analysis aims to reveal how gender-based vulnerabilities differentiate across these phases, which is subsequently illustrated with global examples.

3.1. Pre-Disaster Phase

When the pre-disaster phase is examined from a gender perspective, it becomes clearer how disaster risks are produced and from which structural conditions the vulnerability of specific groups originates. Three core areas of inequality stand out in this phase: inequality in access to information, the exclusionary nature of participation mechanisms, and the predominantly masculine and heteronormative character of planning processes. These are critical factors that directly determine the fairness and effectiveness of disaster management. Disparities in access to information create significant gaps in disaster awareness and preparedness levels by limiting the access of women and other vulnerable groups to early warning systems, risk reduction programs, and capacity building trainings. Furthermore, the exclusionary nature of decision-making and planning mechanisms prevents the inclusion of the unique experiences and needs of women, LGBTQ+ individuals, disabled people, and other vulnerable groups in preparedness processes, thereby generating serious representation problems in disaster planning. The fact that planning processes are generally shaped through traditional masculine roles and a normative citizen profile leads to the formulation of disaster response and preparedness strategies within a homogeneous and non-inclusive framework that is far from being gender-sensitive.

The relationship between disasters and vulnerability should be assessed within the framework of structural and socially constructed inequalities that cause specific social groups to become susceptible to harm, long before the disaster event itself occurs. Vulnerability is concretized during the preparedness and mitigation phases of the pre-disaster period through inequalities in access to resources and limited coping capacities. In this process, gender intersects with other axes of social differentiation such as class, age, and status, leading vulnerability to acquire an intersectional nature (Ajibade et al, 2013; Alexander, 2023; Arora, 2022; K.C. & Hilhorst, 2022; Reyes & Lu, 2016; True, 2013; Yumarni et al., 2014). The aforementioned intersectional structure of vulnerability clearly demonstrates the inadequacy of disaster risk management policies that focus solely on the physical dimension of the hazard, revealing the necessity for preparedness and mitigation efforts to be redesigned to target these structural inequalities.

According to research, the primary reason for the existence of gender inequality in the pre-disaster period is patriarchal structures and the resulting unequal access to resources (Ajibade et al., 2013; Alexander, 2023; Arora, 2022; True, 2013; Danila, 2023; Dominey-Howes et al., 2014; Goldsmith, Raditz & Mendez, 2022; Gorman-Murray et al., 2014; Gürkan, 2024; K.C. & Hilhorst, 2022; Moreno & Shaw, 2018; Yumarni et al, 2014). Accordingly, structural disadvantages across social, economic, political, legal, and cultural statuses lead to women and other vulnerable groups being at a higher risk against disasters. Mechanisms such as the gendered division of labor, the unequal distribution between paid and unpaid work, and limited access to property and financial resources both reduce preparedness capacity and weaken coping mechanisms in the pre-disaster period. The core dynamic behind this vulnerability is the patriarchal culture and ideologies that reproduce male domination, which concentrate control over power and resources in the hands of specific groups, thereby preventing the full participation of women and gender-disadvantaged individuals in economic and political life. Exclusion from decision-making processes therefore constitutes a critical layer of inequality in the context of disaster management; the limited participation of women, LGBTQ+ individuals, and other minority groups in disaster governance and political processes leads both to the disregard of their knowledge and experience and to the failure of their needs to be reflected in planning processes. For disabled people, this exclusion is even more pronounced; inaccessible physical environments, inadequate service infrastructure, and their invisibility in disaster planning severely reduce their pre-disaster preparedness levels. For example, research by Aslan and Şahinöz (2023) on the İzmir earthquake in Türkiye indicates that disabled individuals faced significant obstacles during evacuation processes and were excluded from decision-making mechanisms due to their low preparedness levels. In addition, the deprivation of women's property rights, citizenship documents, or other legal statuses not only undermines their economic independence in the pre-disaster phase but also severely restricts their access to post-disaster support and resources. Consequently, these multi-layered inequalities produced by patriarchal structures emerge as a fundamental determinant in the deepening of gender-based vulnerabilities during the pre-disaster phase.

The second reason for the existence of pre-disaster gender inequality is socio-economic vulnerability and the resulting unequal distribution of livelihoods (Danila, 2023; Enarson, 2007; Morena & Shaw, 2018; Reyes & Lu, 2016; Yumarni et al., 2014). Economic status and income level constitute structural pressures that directly influence disaster preparedness and risk mitigation capacity of vulnerable groups. This situation is further deepened by the fact that women and girls constitute a large portion of the low-income population in developing countries. The predominant employment of women in low-wage and precarious informal sectors limits both their ability to secure a regular income and their capacity to diversify livelihoods. The fact that approximately 90% of women working in non-agricultural sectors in India are engaged in informal employment serves as a striking example of this structural inequality (Ajibade et al., 2013). A critical dimension of socioeconomic vulnerability is housing conditions: the residence of impoverished groups in structurally fragile housing located in high-risk disaster zones not only threatens physical safety but also signifies the loss of livelihoods for women who generate income through home-based production or small-scale enterprises. Furthermore, women's limited access to education and capacity development opportunities weakens their ability to enhance their resilience against climate and disaster-induced shocks. The study conducted by Ajibade et al. (2013) in Bangladesh indicates that, in addition to poverty, the lack of vital skills such as swimming increases women's vulnerability to climatic disasters. Furthermore, economic vulnerabilities, such as the pre-disaster debt burden, create an additional disadvantage factor that complicates the recovery process when a disaster occurs. All these factors collectively demonstrate that socio-economic vulnerabilities intersect with gender inequalities, significantly escalating the risk and exposure levels of women and other disadvantaged groups during the pre-disaster period.

The third reason for the existence of pre-disaster gender inequality is the physical vulnerability resulting from established gender roles (Ajibade et al., 2013; Arora, 2022; Danila, 2023; Enarson, 2007; Reyes & Lu, 2016; Tan et al., 2023; Yumarni et al, 2014). Traditional gender roles determine women's positions and responsibilities in daily life, thereby creating insecure conditions that render them more susceptible to physical dangers during a disaster. The gendered division of labor structurally confines women predominantly to the home; this situation leads to women being more severely affected by destruction, particularly when sudden-onset hazards like earthquakes occur during specific hours of the day. Their higher probability of being in domestic spaces during house chores, meal preparation, or care activities is a structural mechanism that increases women's physical risk at the moment of a disaster. In addition to this, the assignment of responsibility for the care of children, the elderly, and the sick to women exacerbates physical and emotional exhaustion even before a disaster occurs, thereby limiting their mobility and capacity for rapid decision-making in an emergency. The economic invisibilization of unpaid care work contributes both to the disempowerment of women and to the weakening of their pre-disaster resilience. Moreover, groups such as pregnant women, elderly women, and disabled women face a more pronounced physical vulnerability during the pre-disaster period due to factors such as limited mobility, inaccessible buildings, and restricted access to health services. Pre-disaster physical vulnerability is a combination of multi-layered disadvantages produced by patriarchal social structures, socio-economic inequalities, and cultural norms regarding care work, rather than solely the hazard itself. Therefore, a comprehensive gender-based analysis of the pre-disaster phase is of vital importance not only for enhancing social resilience against disasters but also for developing a more equitable, inclusive, and transformative disaster management approach by making structural inequalities visible.

3.2. The Impact Phase of Disaster

The moment of disaster is a critical period where socially constructed vulnerabilities are most clearly revealed, and inequalities become concrete. In this process, where physical hazards intensify, psychosocial risks also become prominent. Vital elements, such as access to resources, security, access to information, and evacuation processes, are much more limited and problematic for individuals who are disadvantaged due to their social standing the chaos experienced in collective shelter and distribution areas increases the possibility of vulnerable groups being exposed to physical and psychological violence. Incidents such as pushing or assault during food distribution are clear indicators of this situation (Ural et al., 2024). The increase in cases of gender-based violence during this period reveals that crisis environments reinforce existing gender-based inequalities. Irregularities in evacuation processes, disruptions in access to basic services, and the weakening of social support mechanisms particularly increase the security risks for vulnerable individuals and cause the reproduction of the fragility cycle. Therefore, the moment of disaster is not only a period of physical destruction but also a social crisis area where social inequalities are deepened in a layered manner.

Women and girls are among the groups most exposed to the acute effects of gender-based inequalities during disaster processes. Gender-based upbringing processes restrict women's access to life-saving skills such as swimming, climbing, or moving quickly; and limited physical mobility also complicates avoidance and evacuation processes during a disaster (Gündüz, 2022). Due to the structure of the social division of labor that confines women to the domestic sphere, women are more exposed to intra-household risks, especially when sudden disasters occur at certain times of the day (Enarson, 2007). Pregnant women, particularly in the later stages of pregnancy due to significantly reduced mobility, may experience delays in avoiding danger and moving towards safe areas during rapidly developing disasters like earthquakes and floods; this leads to their increased exposure to the adverse effects of the disaster. This vulnerability is fueled not only by the restriction of physical mobility but also by the additional pressures created by stress, panic, and crowd movement during the disaster. Factors such as reduced capacity for balance, speed, and changing direction in the final stages of pregnancy; difficulty accessing suitable evacuation means, and the inability to meet privacy and health needs in collective shelter spaces make pregnant women more vulnerable, both physiologically and psychosocially.

People with disabilities and the elderly are among the groups with the highest rates of death and injury in disasters. Findings indicating that people with disabilities face a two to four times higher risk of death compared to the general population in disasters (Aslan & Şahinöz, 2023; Arı Kovancı, 2024) reveal the disproportionate vulnerability faced by this group. Approximately 80% of people with disabilities report experiencing serious difficulties during the evacuation process in sudden-onset disasters, with some stating that they were stranded for long periods until help arrived (Bhuiya, Shao & Jones, 2023). During evacuation processes, people with disabilities face serious difficulties and experience delays due to physical mobility restrictions, limited access to evacuation and early warning information, and non-inclusive infrastructure such as unramped or stair-filled evacuation routes. This directly affects safe evacuation and their survival probabilities during a disaster, while the necessity of relying on others complicates timely intervention and increases the risk of death. Furthermore, the difficulties experienced by individuals with sensory or cognitive disabilities in perceiving emergency warning messages, alarms, or directional signs (Karma et al., 2024) make access to safe areas even more problematic. Similarly, elderly individuals face serious disadvantages in disasters due to low mobility, chronic health problems, cognitive decline, and social isolation (Aslan & Şahinöz, 2023; Arı Kovancı, 2024). Both limited physical mobility and difficulties in accessing healthcare services exponentially increase the risk of death and illness for the elderly in disasters; seriously restricting their capacity to protect themselves or reach safe areas during a moment of danger.

LGBTQ+ individuals who do not conform to gender identity norms are another vulnerable group frequently overlooked and rendered invisible in disaster settings. Heteronormative social structures and emergency organizations often fail to recognize the needs of these individuals during times of crisis; this creates significant barriers to accessing safe shelter, healthcare services, psychosocial support, and protection mechanisms (Goldsmith, Raditz & Mendez, 2022). The sharpening of social exclusion during a disaster leads to LGBTQ+ individuals remaining in physically and psychologically insecure conditions. All these findings demonstrate that disaster processes are in no way gender-neutral; instead, they possess a character that further deepens already existing structural inequalities. Therefore, it is clear that a safe and inclusive approach, which centers the unique and intersectional needs of vulnerable groups, is of vital importance in disaster response.

3.3. Post-Disaster and Recovery Phase

The post-disaster period is not merely about physical damage repair; it is also a complex recovery process in which vulnerable social structures, existing inequalities, and power relations become visible again. The post-disaster recovery (rehabilitation) and reconstruction periods are complex processes in which vulnerable groups are disproportionately and profoundly affected by disasters, and their existing social and economic vulnerabilities are further exacerbated. These processes are far from being equitable. While those who had greater access to resources before the disaster can access resources more easily in the post-disaster period and recover quickly (Sert et al., 2024; Thapa & Pathranarakul, 2019), socially disadvantaged groups become trapped in resource loss cycles, and their recovery processes are prolonged (Lightfoot et. al., 2020; Salari et. al., 2025).

Vulnerable groups face multi-dimensional risks and structural barriers in post-disaster processes. For vulnerable segments such as women, the elderly, and people with disabilities, this period involves intense difficulties in accessing safe shelter, utilizing healthcare services, re-establishing livelihood strategies, and ensuring personal safety. The damages and losses of vulnerable groups reach quite high dimensions due to their low housing quality; furthermore, post-disaster temporary shelter conditions and shortcomings in aid policies expose these groups even more to threats of violence, discrimination, and insecurity. Alarming increases in sexual and gender-based violence (rape, abuse, harassment) have been reported in chaotic and insecure temporary camp settings (Arı Kovancı, 2024; Demirci & Avcu, 2021; Gündüz, 2022; Ural et al., 2024). The lack of privacy in these overcrowded shelter areas makes women and girls vulnerable to sexual abuse and violence (Gorman-Murray, et al., 2014), while the lack of food security and inequality in aid distribution can lead to some women being exposed to sexual exploitation in exchange for food. Additionally, detachment from social networks creates extra risks, particularly for women migrating from rural to urban areas, increasing their vulnerability to sexual abuse, prostitution, and human trafficking. A study conducted in Nepal after a disaster (Bradley et al., 2023), in which 57% of female participants described violence against women as common or very common, strikingly reveals the extent of post-disaster vulnerability.

Gender-blindness in aid policies and deep-seated socioeconomic vulnerabilities are among the key factors that make inequalities more visible and structural in the post-disaster period. The systematic classification of individuals by official aid mechanisms solely under the categories of "woman, man, girl, or boy" systematically prevents LGBTQ+ individuals from accessing humanitarian aid. Particularly marginalized groups such as transgender women are more exposed to the risk of harassment and ill-treatment due to the lack of privacy in temporary shelter areas (Dominey-Howes et al., 2014; Gorman-Murray et al., 2014).

The post-disaster reconstruction phase indicates that the burden of responsibility on vulnerable groups further increases. The risks and uncertainties characterizing urban poverty during the reconstruction process are also distinctly gendered; many factors, ranging from personal safety to working conditions, and from health risks to restrictions on daily mobility, are experienced differently by women, men, and children. Women, who are primarily held responsible for meeting the family's needs due to the gender-based division of labor, experience these deprivations both directly, such as collecting water and firewood or searching for safe areas for sanitation needs, and indirectly, through health risks, increased care burden, and restricted livelihood opportunities. Women are forced to continue their traditional roles related to the care of household and family members (including the elderly and children) in the post-disaster process, a situation which is studied under the theme of "burden of responsibility" and disproportionately increases their psychological and workload (Arı Kovancı, 2024; Çıtak & Toker, 2025; Demirci & Avcu, 2021; Gürkan, 2024; Ural et al., 2024). In this context, disasters further reveal existing inequalities and increase the burden on vulnerable groups. For instance, in the post-2015 Nepal earthquake reconstruction process, women had to undertake the primary responsibility for the reconstruction of damaged houses, in addition to high-risk tasks such as salvaging usable items from the rubble and setting up temporary shelters. This multi-faceted and heavy burden led approximately 59% of the women affected by the earthquake to state that the inability to perform domestic work to adequate standards triggered male violence, concretely demonstrating how gender-based violence is reproduced in the post-disaster period. (Bradley et al., 2023).

Post-disaster reconstruction frequently neglects the essential needs and active participation of vulnerable groups, including women and people with disabilities. Gender-insensitive reconstruction policies exacerbate existing inequalities and hinder sustainable development. While accessibility problems persist for people with disabilities, the interruption of treatment for chronic patients and difficulties in accessing necessary medical devices and care services increase physical vulnerability. Furthermore, in developing and least developed countries, post-disaster malnutrition has severe consequences, particularly for pregnant and lactating women and children, such as increased risk of infection, growth retardation, and learning difficulties. In addition to this, the fact that water scarcity directly prevents the education of girls, who shoulder the responsibility of water collection (Cutter, 2017), and the fact that discrimination against women with disabilities increases their risk of exposure to violence, further highlights the multi-layered structure of post-disaster vulnerabilities.

4. CASE STUDIES

2010 Haiti Earthquake

On January 12, 2010, Haiti was shaken by a devastating 7.0 magnitude earthquake that occurred approximately 40 kilometers southwest of the capital, Port-au-Prince. The earthquake led to a large-scale humanitarian crisis with estimated casualties ranging from 50,000 to 230,000, and more than 300,000 people injured. Approximately 97,000 homes were completely destroyed, and more than 1.5 million people were displaced, severely impacting mass sheltering and basic service needs (Amnesty International, 2011; Behrman & Weitzman, 2016; Llorente-Marrón et al., 2021). Research indicates that the impacts of disasters are not solely caused by natural factors but are shaped by human decisions and deep socio-economic structures (Horton, 2012; Llorente-Marrón et al., 2020). In the case of Haiti, earthquake-related losses were further exacerbated due to long-standing structural problems such as its dependent position within the global economy, its colonial past, a tradition of authoritarian governance, and deep class, race, and gender inequalities.

Even before the 2010 earthquake, Haiti was the country with the highest maternal mortality rate in the Western Hemisphere, and the infrastructure for women's reproductive health had long suffered from severe inadequacies. With the earthquake, women's reproductive health was even more negatively affected, particularly in the areas where the tremors were most intensely felt. The massive collapse of the healthcare system severely limited access to essential services such as antenatal care, family planning services, and medical support following gender-based violence. As a result, there were significant increases in unwanted pregnancies, miscarriages, and various epidemics. Furthermore, a marked decrease was observed in the use of injectable contraception, the most common modern contraceptive method in Haiti; this significantly increased both the probability of existing pregnancy and the risk of unwanted pregnancies. (Behrman & Weitzman, 2016; Gündüz, 2022).

A significant increase in the rates of sexual violence was reported in the post-earthquake period (Amnesty International, 2011; Campbell et al., 2016; Tøraasen, 2020). Within the displaced population concentrated in post-disaster camps, women remained extremely vulnerable to sexual violence and abuse due to insecure living spaces, insufficient lighting, and weak security mechanisms. Specifically, the transfer of aid distribution mechanisms to the control of male-dominated committees within the camps turned access to basic necessities into a bargaining arena where sexual favors could be demanded, thus institutionalizing power imbalances against women in the disaster setting. In addition to this, official data and field studies revealed that a large portion of the violence was perpetrated by current or former partners rather than strangers, thus demonstrating a marked increase in post-disaster domestic and partner-based violence (Campbell et al., 2016; Horton, 2012). This vulnerability also translated into an increased risk of violence, abuse, and exclusion for LGBT+ individuals already struggling with stigma and marginalization.

Following the earthquake, the major redirection of donor resources towards international non-governmental organizations significantly weakened the capacity of the Haitian state and local women's organizations. During this process, the specific needs of women were often overlooked, and their visibility and voices in decision-making mechanisms were largely excluded. Furthermore, the loss of three prominent leaders of the Haitian women's movement (Myriam Merlet, Magalie Marcelin, and Anne Marie Coriolan) in the earthquake created a serious gap in women's rights advocacy in the country. Despite this, women leaders and activists mobilized against this exclusion; in response to the shortcomings in the Interim Haiti Recovery Commission, they published the "Gender Shadow Report" and conducted strong advocacy at the local level to make women's needs visible by establishing solidarity networks (Campbell et al., 2016; Horton, 2012; Tøraasen, 2020). These initiatives clearly revealed the vital roles of women during crisis periods in organizing, establishing solidarity networks, and sustaining care work. However, the Haiti earthquake still demonstrated that disasters not only reflect existing inequalities but also multiply the risks faced by women, particularly in the areas of economic and social security.

Flood Disasters in Nigeria

Flooding is one of the most common and devastating environmental disasters in Nigeria, emerging under the pressure of climate change; the resulting overflows, which exceed the capacity of water channels, deeply shake social and economic life (Echendu, 2021). In this sense, in coastal regions like Lagos, floods have recurred almost annually since 1968 (excluding the drought year of 1973), intensifying particularly during the rainy season between July and October each year (Ndimele et al., 2024; Nkwunonwo et al., 2015). Major disasters that stand out in the country's recent history include the 2012 floods, which affected approximately 7.7 million people, and the 2022 floods, which claimed 603 lives across 33 states and displaced 1.4 million people (Ajaero, 2017; Ejem & Ben-Enukora, 2025; Lawanson et al., 2020).

Women's heightened vulnerability to flood disasters is primarily driven by the intersection of traditional social roles and structural economic inequalities, as women often hold lower socio-economic statuses and disproportionately bear domestic care responsibilities. In Nigeria, where patriarchal norms are strong, this situation is even more pronounced: women are disproportionately exposed to the negative impacts of disasters like floods, both due to gender-based inequalities and their reliance on fragile economic activities that are dependent on the natural environment for their livelihoods. Ejem and Ben-Enukora's (2025) research on the 2022 floods clearly reveals that women's domestic and societal roles made them more vulnerable to flood hazards. The study shows that approximately 75% of women undertook domestic responsibilities, such as caregiving and housework; consequently, they were often at home during the flood, placing them at greater physical risk. Furthermore, it is noteworthy that approximately 50% of women were employed in informal and precarious jobs, such as small-scale farming, firewood collection, or street vending, while this rate was only 11.1% for men. The fact that 37.4% of men were employed in formal, more secure jobs further highlights why flood disasters disproportionately destroy women's livelihoods.

Similarly, the study by Ajibade et al. (2013) on the 2011 Lagos flood illustrates how gender and class-based inequalities deepen the disaster experience. According to this study, women, particularly in the low-income Badia neighborhood, were severely affected both economically and emotionally, forced to single-handedly shoulder the responsibilities of domestic care and providing household income; the damage to food and commercial goods caused by floodwaters inundating their homes further exacerbated income losses. Deficiencies in urban infrastructure, sexist health policies, and difficulties in accessing social support mechanisms made women's lives even more precarious, sharpening post-disaster inequalities.

Following the flood disasters in Nigeria, gender inequality has also been observed to persist. The unsanitary and cramped nature of the temporary accommodation areas following the disaster created an environment conducive to the rapid spread of waterborne, insect-borne, and airborne diseases, severely threatening women's health. All these conditions paved the way for the emergence of various secondary health problems, including sexually transmitted infections such as HIV, chlamydia, and gonorrhea. Furthermore, an increase in cases of gender-based violence, such as sexual assault, physical abuse, and human trafficking, has been reported in these temporary shelter environments (Ndimele et al., 2024).

In disasters, women are often disadvantaged in accessing post-disaster economic and social opportunities due to societal roles and limitations in educational opportunities. This situation was clearly observed following the 2012 flood in Anambra; while men were able to diversify their economic activities by turning to non-agricultural jobs like carpentry, most women remained stuck in gender roles that deemed such jobs "unsuitable for women," due to societal expectations and lack of education (Ajaero, 2017). Furthermore, they were forced to struggle with long-term psychological and mental health issues such as depression, anxiety, and sleep disorders due to the trauma caused by the flood.

2015 and 2023 Nepal Earthquakes

Major disasters in Nepal, such as the 2015 Gorkha and 2023 Jajarkot earthquakes, have exposed the devastating consequences of existing gender inequalities. While disasters affect individuals differently based on factors like age, disability, and geography, women and girls disproportionately bear the brunt during and after crises due to pre-existing socio-economic and patriarchal conditions.

Pre-disaster structural inequalities existing in Nepalese society limit the capacity for crisis preparedness and response against women. A high level of male labor migration is observed in Nepal; for instance, the labor migration gender ratio in the West Rukum region has been recorded as 30 men to 1 woman (Chaudhary et al., 2024). The reduction of the male population due to migration abroad leads to the burden of domestic and community-based responsibilities falling largely on the shoulders of the women and elderly individuals who remain behind. This situation significantly increases the burden on women, especially during post-disaster reconstruction and recovery processes, creating a more complex struggle both physically and emotionally (Gurung, 2019). Traditional social norms in Nepal legitimize the internal allocation of limited household resources (water, food, cash, fuel) primarily toward male family members during disasters; this practice systematically limits women's access to basic necessities. Furthermore, closed and hierarchical information sharing processes inherent in patriarchal social structures largely restrict women's participation in pre-disaster preparedness training and awareness campaigns. This multi-layered exclusion weakens women's capacity to be actively involved in search-and-rescue and evacuation processes during a disaster, positioning them outside the response mechanisms.

During the disaster, the tasks stemming from women's gender roles increased their likelihood of exposure to life-threatening dangers. In the 2015 Gorkha earthquake, women and girls accounted for 55% to 56.1% of the total fatalities (Gurung, 2019). Similarly, in the 2023 Jajarkot earthquake, the number of women and children's fatalities in West Rukum was higher than that of men (Chaudhary et al., 2024). The fact that the 2015 earthquake occurred during the midday hours, when intense domestic activities were underway, coincided with the time periods when women were busy with domestic responsibilities such as traditional meal preparation, cleaning, and childcare. This synchronicity limited women's capacity to quickly and safely evacuate their homes in the face of danger; consequently, it significantly elevated their likelihood of injury or fatality. In the post-disaster phase, with the increase in psychosocial and economic stress caused by the crisis, a significant and statistically meaningful increase in intimate partner violence was observed, particularly in urban areas (Khanna & Fujii, 2024).

The reconstruction and recovery processes have generated multi-layered spatial and socio-economic vulnerabilities for women. The temporary shelter solutions rapidly established after the earthquake not only offered limited physical protection but also severely weakened women's access to services such as security, privacy, nutrition, mental health, and reproductive health. The fact that more than 3,147 women and 8,854 children were left homeless in West Rukum following the 2023 Jajarkot Earthquake is a striking indicator of this spatial vulnerability (Chaudhary et al., 2024). Furthermore, shared temporary toilets negatively impacted basic health conditions by compromising women's privacy, particularly concerning bathing, laundry, and menstrual hygiene needs.

Women faced significant obstacles in reconstruction efforts due to gender disparities in access to resources, decision-making authority, and legal property rights. For example, they were excluded from aid mechanisms and compensation was often controlled by men, due to lacking official documents or property registrations in their own names. Furthermore, the increased care burden (water provision, domestic responsibilities) and the unemployment/migration of partners following the earthquake created great economic and psychological pressure on women. The slow progress of recovery efforts due to political conflicts and governance shortcomings further deepened the challenges. Most importantly, women were underrepresented and excluded from post-disaster reconstruction and recovery decisions; this led to the disregard of gender sensitivity due to the traditional patriarchal culture and a lack of awareness. Consequently, the specific needs of women and their active roles in the crisis remained invisible (Thapa & Pathranarakul, 2019).

Field studies demonstrate that natural disasters in Nepal reflect the impacts of social, economic, and political structural inequalities. The vulnerability faced by women is exacerbated by the intersection of factors such as poverty, lack of property ownership, and ethnic background, producing an exclusion that is explained by the concept of “disaster justice” (Titz, 2021).

2023 Kahramanmaraş (Türkiye) Earthquakes

The twin earthquakes centered in Kahramanmaraş that occurred on February 6, 2023, in the southeastern and Mediterranean regions of Türkiye, have gone down in history as the most devastating natural disaster faced by the Republic of Türkiye in its last century (Gürkan, 2024; UN Women, 2023a; 2023b). The first earthquake, a magnitude 7.7 (Mw) centered in Pazarcık, was followed by a second major earthquake, a magnitude 7.6 (Mw) centered in Elbistan, on the same day, and the region continued to be shaken by thousands of aftershocks (Altıparmak & Birel, 2024; Avrupa Konseyi, 2023; Sakarya et al, 2024; Sert et al., 2024). These earthquakes, impacting primarily Kahramanmaraş and Hatay, affected 11 provinces with a combined population of roughly 14 million people (16.4% of the total population of the country). Over 50,000 people lost their lives, and over 107,000 have been injured A substantial proportion of the built environment has incurred extensive destruction, encompassing educational facilities, health care institutions, and a wide array of critical public infrastructure, including transportation networks (roads, airports, and ports), energy transmission systems, oil terminals, and water and wastewater services. Consequently, an estimated 2.7 million individuals have been displaced from the affected region, relocating to other areas of the country (UN Women, 2023a). This large-scale disaster disproportionately affected the physical, psychological, and social well-being of women, thus deepening gender inequalities (Çıtak & Toker, 2025).

In the 11 provinces affected by the Kahramanmaraş earthquakes, the structural disadvantages women face in the economic, social, and political spheres are clearly visible. The fact that women's labor force participation rates (21–33%) in these provinces remain significantly below the national average limits women's access to social security and deepens their economic dependency. Furthermore, the fact that approximately 72% of women in the region remain outside of employment leads to the concentration of unpaid domestic care labor and household responsibilities largely on women, in line with traditional gender norms (Başkan, 2024; Sarı Karademir & Şahin, 2025).

Women's active participation in disaster risk reduction and preparedness processes is severely limited, particularly due to regional socio-economic and cultural constraints. Structural inequalities, such as women's low average educational attainment and the prevalence of early marriage, weaken their capacity to build resilience against disasters and participate in preparedness programs. The political barriers reinforcing this situation—namely, the low political representation of women in affected metropolitan areas and male-dominated local governance structures—lead to the exclusion of women from disaster-related decision-making processes and spatial planning activities. This educational, political, and economic dependency triangle reduces women to ineffective "objects" in measures taken at both macro and micro levels and limits the applicability of gender-focused policies in disaster management (Başkan, 2024; Sarı Karademir & Şahin, 2025; TSKB, 2024).

The fact that the vast majority of women (72%) in the disaster-affected region were outside of employment significantly increased their risk of being trapped under the rubble during the moment of destruction, as they spent a considerable portion of their daily time within the home environment. However, the most critical vulnerability difference stemmed from the care responsibilities attributed by gender roles. During the earthquake, women prioritized the evacuation of children, the elderly, and disabled individuals (both those with pre-existing disabilities and those newly disabled due to injury), placing their own personal safety as secondary; this societal duty severely limited their capacity for quick escape and evacuation (Altıok et al., 2025; Başkan, 2024). Another significant area of challenge can be explained by the impact of traditional clothing codes and cultural norms on evacuation processes. The tendency of some women to adhere to socially imposed practices of covering up rather than prioritizing rapid evacuation during the disaster demonstrates how emergency behavior is shaped by cultural norms. Such cultural expectations can restrict women's mobility and lead to loss of time, particularly in situations requiring sudden evacuation. Furthermore, a distinctly male-dominated structure was observed in critical response processes such as search-and-rescue, debris removal, and aid distribution carried out during and after the disaster. Obstacles encountered regarding women's demands for active participation in these operations indicate that the response mechanisms functioned within a sexist framework. According to the statements of women volunteers, “men were in charge of everything”; this is a situation indicating that women were not allowed to intervene (Ural et al., 2024). These findings reveal that the moment of disaster pushed women into a more passive and at-risk position due to gender roles, cultural norms, and spatial constraints. Thus, it is understood that gender-based barriers in emergency management limited not only women's capacity to be involved but also the overall effectiveness of the disaster response.

The post-earthquake recovery and temporary shelter phases produced a multilayered crisis environment for women by rendering the existing inequalities they already faced more visible and intense. Studies conducted in the earthquake zone demonstrated that women continued to fulfill their domestic social, childcare, and family care roles under more challenging and constrained conditions in tent cities or container settlements (Altıok et al., 2025; Altıparmak & Birel, 2024; Ural et al., 2024). The care burden significantly increased due to the rise in family members who were injured or became disabled in the earthquake. 92.4% of women entrepreneurs reported experiencing great difficulties in balancing work and family life during this period (UNDP, 2025; UN Women, 2023b).

The temporary shelter areas established after the disaster contained serious structural deficiencies in terms of access to necessities and security, generating distinct risks, especially for women and girls. Lack of security, privacy, and hygiene were among the fundamental issues in temporary shelter areas. Women and girls reported feeling more vulnerable to Gender-Based Violence, citing an increased risk of sexual harassment and rape (Council of Europe, 2023). Furthermore, the serious inadequacy of the hygiene and sanitation infrastructure restricted women's access to essential personal hygiene products, while the lack of safe and appropriate toilet and bathing areas created significant difficulties in meeting daily needs. The profound effects of these negative conditions on physical and psychosocial health are manifested in the fact that some women reported intentionally eating and drinking less to reduce their need to use the toilet (Sarı Karademir & Şahin, 2025). This situation highlights the devastating consequences of the lack of hygiene and privacy on bio-psychosocial well-being.

Significant gender-based inequalities also emerged in financial and legal processes. The payment of housing damage compensation predominantly to men, who are generally considered the "head of the household," even when the property was registered in the woman's name, severely limited the ability of women, particularly those undergoing divorce, to benefit from financial support. This practice demonstrates that post-disaster economic recovery mechanisms reproduce patriarchal norms. Similarly, women who were victims of violence also experienced structural disadvantages because aid distribution was organized based on the "family unit"; this approach, in some cases, led to women being forced to stay in the same shelter area as their abusive partners (Avrupa Konseyi, 2023). These conditions clearly reveal how post-disaster legal and administrative arrangements create additional risks and vulnerabilities when they are not gender-sensitive.

The 2023 Kahramanmaraş earthquakes constitute a striking example of how existing social inequalities and patriarchal cultural structures are sharply intensified within the context of a disaster. Women faced multi-layered vulnerabilities, including an increased care burden stemming from gender-based roles, security gaps (risk of violence and lack of privacy), the disregard of specific hygiene requirements, and exclusion from decision-making processes.

5. CONCLUSION AND RECOMMENDATIONS

Disaster processes are in no way gender-neutral; on the contrary, they render existing inequalities more visible and heavier for vulnerable groups such as women, the elderly, individuals with disabilities, and LGBTQ+ people. As demonstrated throughout this study, multi-layered issues deepen the vulnerability of these groups:

limited access to resources, socio-economic fragility, and patriarchal norms in the pre-disaster phase; security risks, evacuation difficulties, and gender-based violence during the disaster; and loss of livelihoods, increased care burden, and exclusion from decision-making mechanisms in the post-disaster phase. Therefore, disasters are not merely a physical threat but also a social process that reproduces existing societal inequalities.

A holistic analysis of the social, economic, and spatial conditions that determine disaster risk emerges as a critical necessity for making visible how and on which social groups vulnerability is concentrated. Considering that susceptibility to disasters is not random but a result of specific living conditions, inequalities in access to services, and settlement patterns, planning processes must be carried out with an approach that systematically questions these inequalities. Such an analysis will enable disaster policies to be developed within a more just and inclusive framework, focusing not only on managing the hazard but also on transforming the structural mechanisms that convert the hazard into vulnerability.

In the long term, post-disaster recovery and reconstruction processes must be carried out through inclusive policies that focus not only on physical infrastructure but also on reducing social inequalities. The implementation of programs supporting the economic empowerment of vulnerable groups, the expansion of technical training and employment opportunities, and the establishment of social support mechanisms that acknowledge care labor are essential for a fairer and more sustainable reconstruction. This holistic approach will both increase communities' resilience to disasters and carry the potential to break the cycles of social inequality produced by disasters.

REFERENCES

Ajaero, C. K. (2017). A gender perspective on the impact of flood on the food security of households in rural communities of Anambra state, Nigeria. *Food Security, 9*(4), 685–695. DOI: 10.1007/s12571-017-0695-x

Ajibade, I., McBean, G., & Bezner-Kerr, R. (2013). Urban flooding in Lagos, Nigeria: Patterns of vulnerability and resilience among women. *Global Environmental Change, 23*(6), 1714–1725. DOI: 10.1016/j.gloenvcha.2013.08.009

Alexander, R. (2023). A Critical Introduction to gender and disaster: Learning from women survivors in northeast Japan. (19-40). Editor: Ronni Alexander & Siriporn Wajjwalku, *Making Disaster: SaferA Gender and Vulnerability Approach,* Singapore: Springer Pte Ltd.

Altıok, B., Arıcı, A., & Özüçelik, D. N. (2025). 6 Şubat 2023 Kahramanmaraş depremlerinin kadınlar ve çocuklar üzerindeki etkileri ve sosyal hizmet müdahaleleri. *Socrates Journal of Interdisciplinary Social Researches, 11*(52). Advance online publication. DOI: 10.5281/zenodo.15244885

Altıparmak, İ. B., & Birel, E. (2024). 6 Şubat 2023 depremi sonrasında sosyal yaşamın toplumsal cinsiyet bağlamında değerlendirilmesi: Nurdağı örneği, *Hacettepe Üniversitesi. Edebiyat Fakültesi Dergisi, 41*(1), 113–134. DOI: 10.32600/huefd.1304991

Amnesty International. (2011). *Aftershocks women speak out against sexual violence in Haiti's camps*. Amnesty International Ltd.

Arı Kovancı, Y. (2024). Afet ve kırılgan gruplar. *Siirt Üniversitesi Sosyal Bilimler Enstitüsü Dergisi, 12*(1), 13–30. DOI: 10.53586/susbid.1474653

Arora, S. (2022). Intersectional vulnerability in postdisaster contexts: Lived experiences of Dalit women after the Nepal earthquake, 2015. *Disasters, 46*(2), 329–347. DOI: 10.1111/disa.12471 PMID: 33314308

Arpacı, M. (2013). Modernitenin eşiğinde toplumsal cinsiyet rejimi: Pastoral iktidar, beden politikaları ve evlilik, *Doğu-Batı Düşünce Dergisi*, Yıl:16. *Sayı, 63*, 131–146.

Aslan, R., & Şahinöz, S. (2023). The experiences of people with disabilities in the 2020 Izmir earthquake: A phenomenological research. *International Journal of Disaster Risk Reduction, 95*, 103868. DOI: 10.1016/j.ijdrr.2023.103868

Avrupa Konseyi (2023). Deprem bölgelerinde toplumsal cinsiyet analizi: Kadınların adalete ve adli yardıma erişimi, (Haz: Prof. Dr. Gülriz Uygur), Aralık, 2023

Baig, M. N., & Sharif, R. (2013). Gender perspective considerations in disasters like earthquakes and floods of Pakistan. *International Journal of Business, Human and Social Sciences, Vol*, *7*(6), 1487–1490.

Başkan, B. (2024). Deprem, toplumsal cinsiyet ve depremzede kadınlar. İ. Erdoğan Tarakçı (Ed.), *Disiplinlerarası Yaklaşımlarla Kadın Çalışmaları 3* (ss. 51-70) içinde. İstanbul: Efe Akademi Yayınları.

Behrman, J. A., & Weitzman, A. (2016). Effects of the 2010 Haiti earthquake on women's reproductive health. *Studies in Family Planning*, *47*(1), 3–17. https://www.jstor.org/stable/24720394. DOI: 10.1111/j.1728-4465.2016.00045.x PMID: 27027990

Bhuiya, M. M. R., Shao, W., & Jones, S. (2023). Measuring accessibility of movement challenged persons during earthquake evacuation of Dhaka City, Bangladesh: A participatory GIS-Based approach. *Applied Geography (Sevenoaks, England)*, *156*, 1–16.

Bradley, T., Martin, Z., Upreti, B. R., Subedu, B., & Shrestha, S. (2023). Gender and disaster: The impact of natural disasters on violence against women in Nepal. *Journal of Asian and African Studies*, *58*(3), 354–371. DOI: 10.1177/00219096211062474

Butler, J. (2005). *Cinsiyet belası: Feminizm ve kimliğin altüst edilmesi*. Metis Yayınları.

Campbell, D. W., Campbell, J. C., Yarandi, H. N., O'Connor, A. L., Dollar, E., Killion, C., Sloand, E., Callwood, G. B., Cesar, N. M., Hassan, M., & Gary, F. (2016). Violence and abuse of internally displaced women survivors of the 2010 Haiti earthquake. *International Journal of Environmental Research and Public Health*, *61*(8), 981–992. DOI: 10.1007/s00038-016-0895-8 PMID: 27624625

Chaudhary, R., Sharma, B. D., Shrestha, S., Gautam, S., & Dhital, R. (2024). Impact of improper shelters on women on the aftermath of Jajarkot earthquake in West Rukum, Nepal. *Journal of Environmental Sciences (China)*, *10*(1), 154–162. DOI: 10.3126/jes.v10i1.67033

Chowdhury, J. R., Parida, Y., & Agarwal, P. (2022). How flood affects rural employment in India: A gender analysis. *International Journal of Disaster Risk Reduction*, *73*, 102881. Advance online publication. DOI: 10.1016/j.ijdrr.2022.102881

Çıtak, G., & Toker, S. (2025). Exploring gender-based challenges after the earthquake: A phenomenological study of women in Turkey. *BMC Public Health*, *25*(1), 3555. DOI: 10.1186/s12889-025-23907-6 PMID: 41121101

Cutter, S. L. (1995). The forgotten casualties: Women, children, and environmental change. *Global Environmental Change*, *5*(3), 181–194. DOI: 10.1016/0959-3780(95)00046-Q PMID: 12320249

Cutter, S. L. (2017). The forgotten casualties redux: Women, children, and disaster risk. *Global Environmental Change*, *42*, 117–121. DOI: 10.1016/j.gloenvcha.2016.12.010

Danielsson, E., & Eriksson, K. (2022). Women's invisible work in disaster contexts: Gender norms in speech on women's work after a forest fire in Sweden. *Disasters*, *46*(1), 141–161. DOI: 10.1111/disa.12464 PMID: 32895981

Danila, M. (2023). Mainstreaming gender into disaster recovery policy and practice: The Case Study of post-earthquake in Yogyakarta special province, Indonesia. (41-60). Editor: Ronni Alexander & Siriporn Wajjwalku, *Making Disaster: SaferA Gender and Vulnerability Approach,* Singapore: Springer Pte Ltd.

Demirci, K., & Avcu, T. (2021). Afet süreçlerinde kadın bireylerin yaşadığı sorunlar ve çözüm önerileri: İzmir ili örneği. *Batman University Journal of Life Sciences*, *11*(1), 86–105.

Dominey-Howes, D., Gorman-Murray, A., & McKinnon, S. (2014). Queering disasters: On the need to account for LGBTI experiences in natural disaster contexts. *Gender, Place and Culture*, *21*(7), 905–918. DOI: 10.1080/0966369X.2013.802673

Echendu, A. J. (2021). Women, development, and flooding disaster research in Nigeria: A scoping review. *Environmental and Earth Sciences Research Journal*, *8*(4), 147–152. DOI: 10.18280/eesrj.080401

Ejem, A. A., & Ben-Enukora, C. A. (2025). Gendered impacts of 2022 floods on livelihoods and health vulnerability of rural communities in select Southern states in Nigeria. *Discover Social Science and Health*, *5*(67), 67. Advance online publication. DOI: 10.1007/s44155-025-00211-7

Enarson, E. (2007). Toward a feminist perspective on women and disaster. *Women's Rights Law Reporter*, *28*(1), 3–12.

Fatema, S. R., East, L., Islam, S., & Usher, K. (2023). Gender-based vulnerabilities for women during natural disasters in Bangladesh, *Sec. Frontiers in Communication*, *8*, 1180406. Advance online publication. DOI: 10.3389/fcomm.2023.1180406

Goldsmith, L., Raditz, V., & Mendez, M. (2022). Queer and present danger: Understanding the disparate impacts of disasters on LGBTQ+ communities. *Disasters*, *46*(4), 946–973. DOI: 10.1111/disa.12509 PMID: 34498778

Gorman-Murray, A., McKinnon, S., & Dominey-Howes, D. (2014). Queer domicide. *Home Cultures, 11*(2), 237–261. DOI: 10.2752/175174214X13891916944751

Gündüz, F. (2022). Afetlerde kadın ve toplumsal cinsiyet perspektifi ile çıkarılması gereken dersler (Haiti ve Japonya depremi örneği). *IBAD Sosyal Bilimler Dergisi*, (12), 440–461. DOI: 10.21733/ibad.1039215

Gündüz, F., & Atalay, E. (2025). Determining the effect of women's earthquake fear on earthquake risk perception. *Fırat Üniversitesi Sosyal Bilimler Dergisi, 35*(2), 801–813. DOI: 10.18069/firatsbed.1599856

Gürkan, S. (2024). The feminine and masculine dimensions of earthquakes: A gender-focused sociological research on the victims of the 2023 Kahramanmaraş / Türkiye earthquakes. *Sted, 33*(5), 351–362. DOI: 10.17942/sted.1456836

Gurung, D. (2019). Through women's eyes: A study of vulnerability in 2015 earthquake. *Dhaulagiri, 13*, 76–83. DOI: 10.3126/dsaj.v13i0.23361

Horton, J. (2012). After the earthquake: Gender inequality and transformation in post-disaster Haiti. *Gender and Development, 20*(2), 295–308. DOI: 10.1080/13552074.2012.693284

K. C., L. & Hilhorst D. (2022). Gendered experience of disaster: Women's account of evacuation, relief and recovery in Nepal. *International Journal of Disaster Risk Reduction, 72*, 102840. DOI: 10.1016/j.ijdrr.2022.102840

Kadir, S. B. (2021). Viewing disaster resilience through gender sensitive lens: A composite indicator based assessment. *International Journal of Disaster Risk Reduction, 62*, 102398. Advance online publication. DOI: 10.1016/j.ijdrr.2021.102398

Karma, . (2024). *People vulnerability before, during and after a disaster: A dynamic taxonomic approach* (Di Pietro, A., Ed.). Climate Change and Risk Management - Strategies, Analysis, and Adaptation., DOI: 10.5772/intechopen.1005205

Khanna, A., & Fujii, T. (2024). Do natural disasters cause domestic violence?: A study of the 2015 Nepal earthquake. Available at *SSRN*: https://ssrn.com/abstract=5084508 or http://dx.doi.org/DOI: 10.2139/ssrn.5084508

Lawanson, O. I., Proverbs, D., & Ibrahim, R. L. (2020). The impact of flooding on poor communities in Lagos State, Nigeria: The case of the Makoko urban settlement. *Journal of Flood Risk Management*. Advance online publication. DOI: 10.1111/jfr3.12838

Lightfoot, E. S., Lesen, A. E., & Ferreira, R. J. (2020). Gender and resilience in Gulf Coast communities: Risk and protective factors following a technological disaster. *International Journal of Disaster Risk Reduction, 50*, 101716. Advance online publication. DOI: 10.1016/j.ijdrr.2020.101716

Llorente-Marrón, M., Díaz-Fernández, M., Méndez-Rodríguez, P., & Arias, R. G. (2020). Social vulnerability, gender and disasters: The case of Haiti in 2010. *Sustainability (Basel), 12*(9), 3574. DOI: 10.3390/su12093574

Llorente-Marrón, M., Fontanil-Gómez, Y., Díaz-Fernández, M., & García, P. S. (2021). Disasters, gender, and HIV infection: The impact of the 2010 Haiti earthquake. *International Journal of Environmental Research and Public Health, 2021*(18), 7198. DOI: 10.3390/ijerph18137198 PMID: 34281135

Mariki, S. B. (2016). Social impacts of protected areas on gender in West Kilimanjaro, Tanzania. *Open Journal of Social Sciences, 4*(3), 220–235. DOI: 10.4236/jss.2016.43028

McDowell, L. (2003). *Gender, identity and place: Understanding feminist geographies* (2nd ed.). University of Minnesota Press.

Moreno, J., & Shaw, D. (2018). Women's empowerment following disaster: A longitudinal study of social change. *Natural Hazards, 2018*(92), 205–224. DOI: 10.1007/s11069-018-3204-4

Ndimele, P. E., Ojewole, A. E., Mekuleyi, G. O., Badmos, L. A., Agosu, C. M., Olatunbosun, E. S., Lawal, O. O., Shittu, J. A., Joseph, O. O., Ositimehin, K. M., Ndimele, F. C., Ojewole, C. O., Abdulganiy, I. O., & Ayodele, O. T. (2024). Vulnerability, resilience and adaptation of Lagos coastal communities to flooding. *Earth Science Systems and Society, 4*(1), 10087. DOI: 10.3389/esss.2024.10087

Nkwunonwo, . (2015). Relevance of social vulnerability assessment to flood risk reduction in the Lagos Metropolis of Nigeria. *British Journal of Applied Science and Technology, 8*(4), 366–382. DOI: 10.9734/BJAST/2015/17518

Reyes, D. D., & Lu, J. L. (2016). Gender dimension in disaster situations: A case study of flood prone women in Malabon City, Metro Manila. *International Journal of Disaster Risk Reduction, 15*, 162–168. DOI: 10.1016/j.ijdrr.2015.11.001

Sakarya, A., Akın, O., & Kılıç, A. (2024). Investigating the damage from the February 6, 2023 earthquakes in terms of building characteristics in the Antakya and Defne districts of Hatay province. *International Journal of Disaster Risk Reduction, 112*, 104783. DOI: 10.1016/j.ijdrr.2024.104783

Salari, M., Abbasiyazdi, A., & Shabanikiya, H. (2025). Geospatial analysis for identifying socially vulnerable areas to natural disasters. *Disaster Medicine and Public Health Preparedness*, *19*(e208), 1–11. DOI: 10.1017/dmp.2025.123 PMID: 40709522

Sarı Karademir, B., & Şahin, S. B. (2025). Gender and disaster management in Turkey: A strategic-relational analysis of the state response to the February 2023 earthquakes. *Southeast European and Black Sea Studies*, 1–19. Advance online publication. DOI: 10.1080/14683857.2025.2473169

Sert, et. al. (2024). The disaster of the century, Kahramanmaras earthquake: A qualitative study on the experiences of the earthquake victims and search and rescue teams, *International Journal of Disaster Risk Reduction*, 107(2024) 104462, https://doi.org/DOI: 10.1016/j.ijdrr.2024.104462

Slick, J., & Hertz, G. (2024). Gender and gender-based violence in disaster contexts in Canada: A systematic review of the literature. *International Journal of Disaster Risk Reduction*, *108*, 104529. Advance online publication. DOI: 10.1016/j.ijdrr.2024.104529

Tan, K. W., Huang, Y. F., & Loh, P. N. (2023). Toward climate-resilient development and management: Developing indicators of climate hazards in Malaysia. (163-180). Editor: Ronni Alexander & Siriporn Wajjwalku, *Making Disaster: Safer A Gender and Vulnerability Approach*, Singapore: Springer Pte Ltd.

Thapa, V. & Pathranarakul, P. (2019). Gender inclusiveness in disaster risk governance for sustainable recovery of 2015 Gorkha Earthquake, Nepal, International Journal of Disaster Risk Reduction, 34 (2019), 209–219. https://doi.org/DOI: 10.1016/j.ijdrr.2018.11.019

Tichenor, V. J. (1963). Earning more and getting less. New Brunswick, New Jersey, and London: Rutgers University Press

Titz, A. (2021). Geographies of doing nothing–internal displacement and practices of post-disaster recovery in urban areas of the Kathmandu Valley, Nepal. *Social Sciences (Basel, Switzerland)*, *10*(3), 110. DOI: 10.3390/socsci10030110

Tøraasen, M. (2020). Women's status in Haiti ten years after the earthquake, CMI BRIEF 2020:07.

True, J. (2013). Gendered violence in natural disasters: Learning from New Orleans, Haiti and Christchurch. *Aotearoa New Zealand Social Work*, *25*(2), 78–89. DOI: 10.11157/anzswj-vol25iss2id83

TSKB (2024). The fault line deep down: Gender inequality, March-2024.

UNDP. (2025). Kahramanmaraş'ta kadın mikro işletmelerinin iyileşme süreci ve afet dayanıklılığı: afet sonrası ihtiyaç analizi. Ocak 2025, İstanbul.

Ural, , Yalçın, E., İlkin Aydın, E., & Höbek Akarsu, R. (2024). Women's experiences in the aftermath of the Kahramanmaraş earthquakes in Türkiye: A phenomenological study. *International Journal of Disaster Risk Reduction, 110*, 104617. DOI: 10.1016/j.ijdrr.2024.104617

Verloo, M. (2007). *Multiple meanings of gender equality: A critical frame analysis of gender policies in Europe*. Central European University Press. DOI: 10.1515/9786155211393

Women, U. N. (2023a). Un Women brief on earthquake in Türkiye: impacts and priorities for women and girls, April 2023, Türkiye

Women, U. N. (2023b). Depremin ardından kadınlar ve kız çocukları: Türkiye'deki depremlerin etkisi: kadın ve kız çocuklarının hakları için çalışan sivil toplum kuruluşlarının deneyim ve bakış açıları, Mayıs 2023, Türkiye

Yumarni, T., Amaratunga, D., & Haigh, R. (2014). Assessing gender vulnerability within post-earthquake reconstruction: Case study from Indonesia. *Procedia Economics and Finance, 18*, 763–771. DOI: 10.1016/S2212-5671(14)01000-4

Chapter 3

What Is the Importance of Wetlands in Reducing Disaster Risks?

What Do Scientific Reports Tell Us?

Emrah Çelik
http://orcid.org/0000-0003-1274-4122
Igdir University, Turkey

Erkan Azizoğlu
Hakkari University, Turkey

ABSTRACT

This chapter provides a bibliometric analysis of 960 scientific publications from 1978 to 2024, focusing on the connection between wetlands and disaster risk reduction. The study highlights the sharp growth of research in recent decades, underlining wetlands' essential roles in flood mitigation, climate regulation, and the provision of ecosystem services. It identifies major trends such as the use of remote sensing technologies, the integration of climate adaptation strategies, and the need for interdisciplinary collaboration. The findings emphasize the importance of including wetlands in policy frameworks, advancing restoration practices, and promoting stakeholder participation. Overall, the chapter concludes that wetlands are vital natural assets that enhance resilience and sustainability, offering valuable insights for shaping future research, policy, and practical action.

DOI: 10.4018/979-8-3373-6274-8.ch003

1. INTRODUCTION

Disasters are often destructive events that test the resilience of human societies and natural ecosystems (Smith and Petley, 2009; Walz et al., 2021). These events include recurring and extreme natural disasters related to earthquakes, floods (Ferreira et al., 2023), droughts (De Feudis et al., 2024), heat waves, tropical cyclones, volcanic eruptions, and earthquakes. Beyond floods and coastal hazards, wetlands are also relevant in multi-hazard and compound-risk settings. For example, drought–heat interactions can intensify ecological stress and water scarcity in wetland-dependent landscapes, while wildfire risk is particularly salient for peatlands, where drying and burning can create cascading impacts through carbon emissions, smoke, infrastructure disruption, and livelihood losses. These compound and cascading dynamics underscore that wetland–DRR linkages should be assessed across alternating and interacting hazards (e.g., drought followed by flooding), not only single-event flood protection. Every year, we witness striking similarities in the images of losses and destruction caused by disasters. Disasters consistently undermine local and national development efforts, disrupting efforts to support livelihoods, promote economic growth, and ensure overall human well-being (Cutter et al., 2003). Additionally, ecosystem losses during these disasters are generally not reported (Walz et al., 2021). This lack of reporting can lead to an underestimation of the long-term ecological impacts, which may, in turn, reduce the effectiveness of post-disaster recovery strategies and policies (Kehoe et al., 2017; Turetsky et al., 2015; Zscheischler et al., 2018).

Understanding and coping with the effects of disasters requires focusing on the harmonious functioning of ecosystems with nature (Graveline and Germain, 2022). The relationship between disaster management and wetlands is a complex and vital issue that requires in-depth examination from both ecological and social perspectives (Moomaw et al., 2018). In this context, natural ecosystems like wetlands play a critical role in both mitigating the effects of disasters and increasing community resilience (Sutton-Grier and Sandifer 2019). Wetlands, by enhancing ecosystem services such as water purification, groundwater recharge, and climate regulation, provide not only direct protection but also co-benefits that contribute to overall environmental health and disaster risk reduction (Gardner and Finlayson, 2018; Davidson et al., 2019).

Wetlands act as a natural buffer with functions such as water diversion, flood control, and carbon storage (Costanza et al., 1997). Additionally, these ecosystems contribute to environmental stability by providing a significant habitat for biodiversity (Mitsch and Gosselink, 2000). With these characteristics, wetlands are considered an integral part of disaster management strategies. Understanding the benefits these services provide to human communities in disaster situations is of enormous importance for both ecological and social sustainability. It is known that wetlands act as a protective barrier against natural disasters; for example, mangrove forests and swamps protect coastal areas by absorbing wave energy during storms (Barbier et al., 2011). Furthermore, evidence shows that the restoration and conservation of wetlands can significantly reduce disaster recovery costs and enhance the adaptive capacity of both human and natural systems in the face of increasing climate-related hazards (Das and Vincent, 2009; Spalding et al., 2014). Although much of the wetland–disaster literature has traditionally emphasized floods and coastal storm-surge impacts, wetlands are also directly relevant to a wider set of hazard profiles that are increasingly prominent under climate change. In drought- and heat-stressed periods, wetlands can support risk reduction through water storage, groundwater recharge, and hydrological buffering that sustains ecosystems and livelihoods (Gardner and Finlayson, 2018; Cunha et al., 2024). In addition, wetland landscapes are often embedded in multi-hazard settings where risks may compound or cascade (e.g., alternating flood–drought regimes, or indirect impacts transmitted through infrastructure, food systems, and health). Framing wetlands within these broader hazard contexts strengthens the rationale for evidence synthesis and helps align wetland management with climate adaptation and integrated disaster risk governance (Moomaw et al., 2018; Sutton-Grier and Sandifer, 2019).

The direct and indirect effects of disasters play a decisive role in the structure and functioning of wetlands. Sudden changes in hydrological cycles can affect both the biotic and abiotic components of these ecosystems (Dudgeon et al., 2006). Hydrological disasters such as floods, storms, and tsunamis can immediately impact wetland ecosystems, leading to consequences such as water quality degradation, vegetation destruction, and animal population decline (Bennett et al., 2009). For example, the tsunami that occurred in the Indian Ocean in 2004 severely impacted many coastal wetlands in Southeast Asia, temporarily weakening their ecological functions (Srinivas and Nakagawa, 2008; Gilbreath, 2011). Such disasters increase pressure on the biodiversity of wetlands while also presenting opportunities for their rehabilitation and reconstruction (Middleton, 1999). However, these rehabilitation processes are often complex and require long-term planning.

Ecological restoration projects are designed to restore wetlands to their former functions after disasters, and these processes often require extensive ecological and hydrological assessments (Zedler and Kercher, 2005). Recent studies emphasize that successful ecological restoration must integrate adaptive management approaches, considering future climate projections and the cumulative impacts of multiple stressors on wetland resilience (Erwin, 2009; Suding, 2011). The impact of disasters on wetlands creates significant opportunities and challenges in both the conservation and post-disaster recovery processes of these areas. Moreover, there is growing recognition of the need to involve local communities and stakeholders in wetland restoration, as their traditional knowledge and participation can enhance the effectiveness and sustainability of recovery efforts (Reed, 2008). While wetlands face direct and indirect pressures from disasters as sensitive ecosystems that harbor biodiversity (Mitsch and Gosselink, 2000) and maintain environmental balance, they also stand out for the services they provide in terms of disaster management (Alongi, 2008; Barbier et al., 2011). This two-way relationship necessitates not only the conservation of wetlands but also the active assessment of their capacity to mitigate the effects of natural disasters. Evaluating these capacities is crucial for the integration of wetlands into national and regional disaster risk reduction strategies (Gardner and Finlayson, 2018). These ecosystem services hold both environmental and economic value. Wetlands contribute to significant cost savings during the post-disaster recovery process. For instance, the water storage and flood control functions of wetlands in a particular area can greatly lower the expenses associated with human-made infrastructure, such as dams and levees (Costanza et al., 2008). Furthermore, the economic valuation of wetland ecosystem services highlights their role as natural capital, justifying investments in their protection and restoration as cost-effective measures for disaster risk management (TEEB, 2010; Davidson et al., 2019).

To strengthen synthesis, the literature on wetlands and disaster risk reduction can be interpreted through three complementary strands.

Conceptual contributions. Foundational work frames wetlands as multi-functional socio-ecological systems that generate ecosystem services (e.g., flood attenuation, water purification, groundwater recharge, and climate regulation) and thereby support resilience and climate adaptation. This conceptual framing positions wetlands as "natural infrastructure" within disaster risk reduction strategies and emphasizes co-benefits alongside risk mitigation (Costanza et al., 1997; Gardner and Finlayson, 2018; Davidson et al., 2019; Sutton-Grier and Sandifer, 2019).

Empirical case evidence. A substantial body of hazard- and place-based studies documents protective functions of wetlands, particularly in coastal and floodplain contexts (e.g., mangroves attenuating storm surge and wave energy), as well as post-event recovery and cost-reduction benefits associated with conservation and restoration. These empirical findings have been reinforced by evidence from major disaster events and subsequent recovery efforts (Barbier et al., 2011; Das and Vincent, 2009; Spalding et al., 2014; Srinivas and Nakagawa, 2008; Gilbreath, 2011).

Methodological advances. Recent scholarship increasingly operationalizes wetland–DRR linkages through advances in ecological and hydrological assessment, adaptive restoration planning, and monitoring technologies. In parallel, evidence-synthesis approaches (including systematic reviews and bibliometric science mapping) support more transparent and scalable consolidation of a rapidly expanding, interdisciplinary literature (Zedler and Kercher, 2005; Erwin, 2009; Suding, 2011; Pullin and Knight, 2009; Haddaway et al., 2015).

The conservation and restoration of wetlands is an ecological imperative and a crucial strategy for economic and social sustainability. Addressing complex issues related to wetland conservation and restoration requires a systematic analysis of existing scientific knowledge and the use of interdisciplinary approaches. Such approaches are vital for developing integrated policies that balance ecological priorities with human needs, especially in the context of climate change and increasing anthropogenic pressures (Moomaw et al., 2018; Gardner and Finlayson, 2018 Çelik et al., 2021; Duran and Çelik, 2022; Yulu et al., 2023).

At this stage, gaining a thorough understanding of the scope and trends in scientific research can provide a strategic edge to decision-makers and practitioners. The necessity for a detailed analysis of the literature becomes increasingly evident when examining the complex nature of topics like wetlands and disaster management. It is crucial to recognize how such analyses can be employed not only to assess the generation of scientific knowledge but also to pinpoint gaps in research areas. A systematic evaluation of research outputs facilitates evidence-based policymaking and enhances the effectiveness of conservation interventions (Pullin and Knight, 2009; Haddaway et al., 2015). Accordingly, the following sections consolidate this literature to provide a clear rationale for the bibliometric mapping undertaken in this chapter.

Bibliometric analyses serve as powerful tools for examining scientific productivity and development trends within specific research areas (Chiu and Ho, 2007; Çelik, 2020; Kulak et al., 2019; Çelik et al., 2021; Çelik and Sarıboğa, 2023; Donthu et al., 2021; Katakidis et al., 2023; Bulut and Yıldız, 2024; Çelik and Çelik, 2024; İli and Bakar, 2024). These analyses assist in identifying key themes in the literature, the most frequently cited studies, significant research centers, and leading scientists (Pritchard, 1969). Additionally, bibliometric studies provide guidance for researchers and policymakers by highlighting knowledge gaps and potential future research opportunities in a given field (Ellegaard, 2018). Such insights are increasingly important for fostering international scientific cooperation and mobilizing resources towards underexplored areas, especially in rapidly evolving fields like wetland disaster resilience (Aria and Cuccurullo, 2017; Donthu et al., 2021).

To consolidate the breadth of scholarship more clearly, we synthesize prior work on wetlands and disaster risk reduction into three complementary strands. First, conceptual contributions frame wetlands as multi-functional "natural infrastructure" within resilience, ecosystem services, and climate adaptation debates, emphasizing how ecological integrity and governance shape risk outcomes. Second, empirical case evidence demonstrates hazard-specific and place-based performance of wetlands (e.g., mangroves and coastal wetlands reducing storm and flood impacts; inland wetlands supporting water storage, attenuation, and recovery), while highlighting the importance of context, exposure, and capacity. Third, methodological advances increasingly rely on remote sensing, GIS, and modeling to quantify wetland change, exposure, and post-event impacts, alongside review and bibliometric approaches that enable systematic mapping of rapidly expanding literatures. (Barbier et al., 2011; Sutton-Grier and Sandifer, 2019).

Positioning of the bibliometric contribution. Existing narrative and systematic reviews provide essential synthesis of mechanisms and case evidence, yet they can be limited in revealing the intellectual structure and evolution of a large, interdisciplinary field. Our bibliometric mapping complements these reviews by (i) identifying dominant and emerging thematic clusters, (ii) tracing keyword and thematic evolution over time, and (iii) highlighting collaboration structures and geographic concentration in knowledge production, thereby supporting more targeted research agendas and applied learning. (Aria and Cuccurullo, 2017; Donthu et al., 2021).

In this study, the existing literature on the relationship between disaster management and wetlands was reviewed. A search of the Scopus database using the keywords "disaster," "disasters," and "wetland" resulted in 960 studies, which were evaluated through bibliometric analysis. The research examined various bibliometric indicators, including the number of publications over time, the most cited articles, keyword networks, and geographical distributions. The results demonstrate a growing scientific interest in the intersection of wetlands and disaster management, with notable increases in publications following major global natural disasters and international policy developments (Endter-Wada et al., 2018; Sutton-Grier and Sandifer, 2019). These analyses underscore the significance of scientific collaborations and multidisciplinary approaches while also identifying key trends and focus areas within the research field. Ultimately, bibliometric mapping not only tracks the evolution of research landscapes, but also supports the development of targeted strategies for future research and policy formulation (Aria and Cuccurullo, 2017).

Although narrative and systematic reviews have provided valuable qualitative syntheses on wetland ecosystem services and hazard-mitigation functions, the evidence base remains dispersed across disciplinary silos and is often organized around specific wetland types, hazards, or regions. To complement these review-based syntheses, this chapter applies a systematic, reproducible bibliometric mapping of the Scopus-indexed literature at the intersection of wetlands and disasters (n = 960). Unlike narrative-only approaches, bibliometric science mapping helps (i) reveal the field's intellectual structure and major knowledge clusters, (ii) trace thematic evolution and research frontiers over time, and (iii) identify collaboration networks and geographic concentration patterns that shape what is studied and where (Pullin and Knight, 2009; Haddaway et al., 2015; Aria and Cuccurullo, 2017; Ellegaard, 2018; Donthu et al., 2021).

Recent advances in disaster risk reduction have led to an increased acknowledgement of wetlands as a type of "natural infrastructure". Nevertheless, the translation of a mounting body of research into actionable decisions regarding the allocation of investments continues to present a significant challenge. The evidence collated in this chapter suggests that the protection and restoration of wetlands has the potential to enhance engineered measures by facilitating flood mitigation and storm protection. Concurrently, this can result in the provision of ancillary benefits such as carbon sequestration and biodiversity preservation (Barbier et al., 2011; Spalding et al., 2014). Concurrently, the pervasive utilisation of remote sensing and associated analytical methodologies within the extant literature signifies a pragmatic approach to the practical implementation of these concepts. The capacity to monitor the condition, fragmentation and inundation dynamics of wetlands in real time is rapidly advancing, thereby enabling the utilisation of the resulting information to directly inform risk screening and post-disaster impact assessments (Wu et al., 2022; Shih et al., 2023). The body of evidence pertaining to the economic and societal advantages under discussion lends support to the policy argument that investing in wetlands could be economically viable for mitigating disaster-related losses while yielding enduring benefits through ecosystem services (Das and Vincent, 2009; Davidson et al., 2019).

Bibliometric network analysis was conducted to address the following questions:

- How have the number of documents related to "disaster management" and "wetlands" changed over the years?
- Which countries have published the most documents on these topics?
- What are the primary research themes in the fields of "disaster management" and "wetlands"?
- What keywords are most frequently used in studies concerning these subjects?
- How do the thematic maps generated through bibliometric analyses represent the main themes and research trends?

2. METHODOLOGY

2.1. Dimension Reduction and Bibliometric Studies

Bibliometric analyses are effective in utilizing dimensionality reduction methods to interpret large and complex datasets (Kulak et al., 2019; Celik et al., 2021). Dimensionality reduction enhances clarity by presenting underlying structures and relationships more effectively while reducing dataset size. These methods facilitate a simpler and more comprehensible analysis of the intricate, high-dimensional relationships among variables in the dataset. Specifically, bibliometric studies employ dimensionality reduction to identify key themes in research areas, author collaborations, and patterns of scientific productivity. By streamlining the academic literature, we can analyze research trends and knowledge gaps more efficiently.

We employed bibliometric mapping to summarize a rapidly expanding, interdisciplinary literature and to visualize themes, collaboration structures, and temporal evolution. All visualizations were generated using Bibliometrix (R) and VOSviewer (v1.6.17). Scopus records were exported in CSV and BibTeX formats to support descriptive indicators and network-based analyses. These studies serve as powerful tools for understanding the dynamics of scientific productivity and academic collaborations. Compared to traditional compilations, bibliometric analyses have the advantage of systematically and objectively examining large datasets. While traditional reviews typically use a qualitative approach, bibliometric studies utilize quantitative methods, enabling more comprehensive and verifiable results. Furthermore, bibliometric analyses are recognized as a quicker and more effective means of identifying major trends and knowledge gaps within the literature. All visuals and graphs derived from the dataset employed in the study were generated using Bibliometrix (https://www.bibliometrix.org/home/), a free online software based on R-Studio, as well as VOSviewer version 1.6.17. For content analysis, BibTeX and CSV file formats of Scopus documents were acquired. (Aria and Cuccurullo, 2017; Van Eck and Waltman, 2010).

2.1. Research Sources and Data Processing

In the contemporary era, a plethora of databases are at one's disposal for the purpose of accessing information and conducting bibliographic or bibliometric searches (Kulak et al., 2019; Celik et al., 2021). Among these databases, Web of Science (WoS), Scopus, Google Scholar, PubMed, and MEDLINE are considered to be of the utmost importance (Gavel and Iselid, 2008; Chen, 2017). The acquisition of the most comprehensive collection of documents for researchers constitutes one of the study's primary objectives; consequently, the capacity to make comparisons across various databases is of critical importance. A subsequent investigation into publications containing the same keywords revealed that Scopus contained a greater number of documents related to the topic than other databases. A search was conducted in the Scopus database using the 'TITLE-ABS-KEY' filter for the keywords 'disaster' OR 'disasters' AND 'wetland' (accessed: November 28, 2024). Upon completion of the search, a total of 960 documents were retrieved (see Table 1). The documents were saved in CSV format for subsequent analysis. Various visual aids, including tables and graphs, were created using the Bibliometrix and VOSviewer programs for visualization and analysis purposes.

Table 1. Types and numbers of documents obtained

Documents Types	Numbers
Article	578
Conference Paper	167
Book Chapter	95
Review	51
Conference Review	21
Note	17
Book	11
Short Research	9
Editorial /Editorial Letter	13
Data	1

3. RESULTS AND DISCUSSION

A comprehensive search of the Scopus database yielded 960 documents examining the relationship between "disaster" and "wetland." The first scientific documents on this topic were published in 1978 as conference proceedings (n=2). The findings of both of these studies provide valuable insights into the role and importance of wetlands in environmental risks (Cowell, 1978; Nelson-Smith, 1978). A comprehensive analysis of data from 535 diverse sources was conducted in conjunction with the aforementioned studies, which were published between 1978 and 2024 on the subjects of disaster management and wetlands. The mean annual growth rate was calculated to be 8.07% (see Figure 1).

Figure 1. Main information about the documents

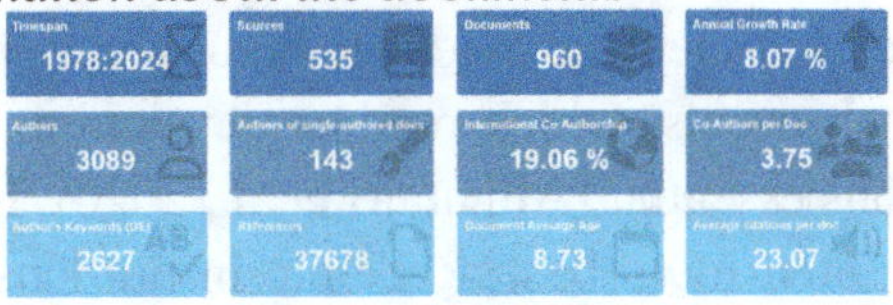

Figure 2 illustrates annual publication trends in the wetlands–disaster risk reduction literature. While early contributions were sporadic, the marked increase in output after 2010 suggests a maturation of the field toward more interdisciplinary and policy-relevant work, supported by the growing recognition of wetlands as nature-based solutions for risk reduction and by enabling advances in remote sensing and

GIS. This surge also motivates closer examination of where expertise and evidence are concentrated (e.g., leading countries and authors) and where thematic gaps remain, which we address in the subsequent analyses. (Sutton-Grier and Sandifer, 2019; Ruan et al., 2022).

Figure 2. Annual scientific production related to the topic between 1978 and 2024 (Source: Derived from the annual document production analysis of documents obtained from Scopus using the R-based Bibliometrix software)

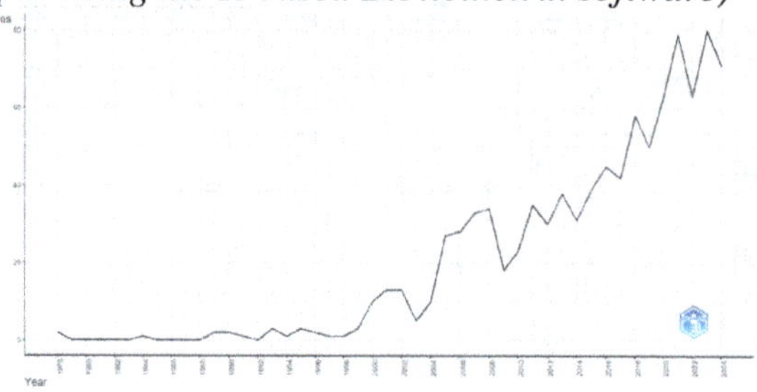

The distribution of scientific knowledge is not only an indicator of research intensity but also a reflection of countries' scientific infrastructure, regional needs, and the effectiveness of their environmental policies. In this context, the distribution of scientific publications among countries exhibits significant disparities. The United States America (USA) is the country that has published the most documents on wetlands and disaster management (n=236). The subsequent ranking comprises China (n=218), India (n=61), Indonesia (n=59), and the United Kingdom (n=50) (Figures 3 and 4). The underlying factors contributing to these variations include the strength of each nation's scientific infrastructure, regional specific needs, and the efficacy of environmental policies implemented. It has been observed that countries, particularly those along the ocean coast, are increasingly prioritizing research in disaster management due to their heightened vulnerability to natural disasters. This trend is supported by scientific evidence highlighting that coastal areas face greater risks from such events, which renders developing disaster management policies in these regions essential (Almutairi et al., 2020). In contrast, the United States' high research intensity, robust academic institutions, and comprehensive environmental policies also contribute to this focus. Combining these factors leads to a more pronounced disparity in the number of research documents produced by different countries.

Figure 3. Authorship collaboration between countries (Dark blue = Countries with the most document publications and the most collaborations)

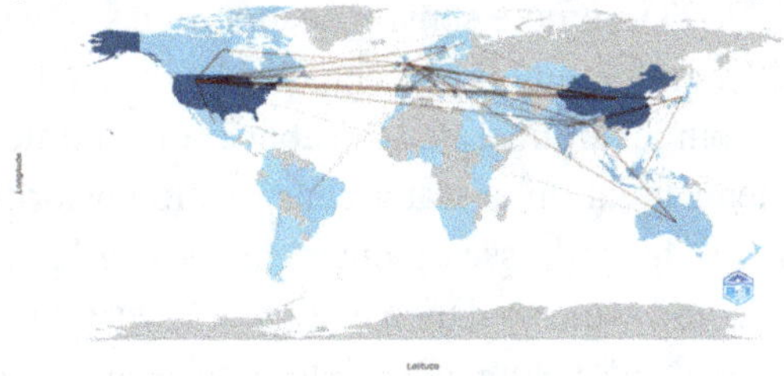

Figure 4. Top 10 countries publishing the most documents (1978-2024) (This graph is taken from the SCOPUS database. Accessed: 28.11.2024)

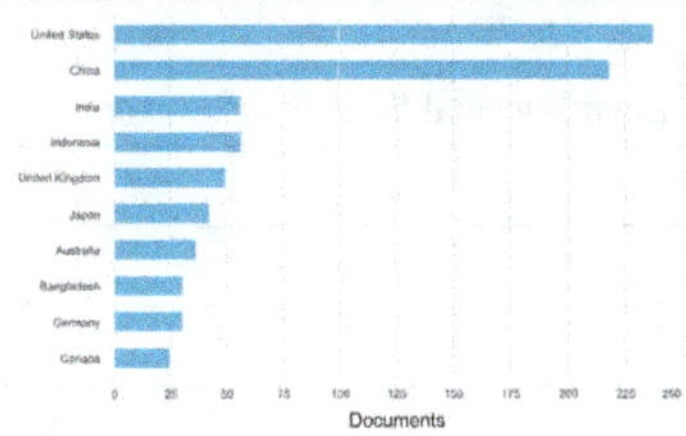

Scientific publications concerning the relationship between disaster-wetland systems and the natural and social sciences place significant emphasis on the value of interdisciplinary research in the fields of environmental science and disaster management. In this context, research findings published in various academic journals provide important insights for the development of sustainable management strategies.

In this context, scientific publications promote interdisciplinary approaches by exploring various aspects of the disaster-wetland relationship. The journal IOP Conference Series: Earth and Environmental Science, which has the highest number of publications on this topic, features studies covering a range of subjects, including water conservation and watershed management (Anggraini et al., 2024), climate change and migration (Ghani & Hidayati, 2024), flood hazard and hydraulic analysis (Pratiwi et al., 2024), and water balance analysis (Fibriantika et al., 2021) related to the disaster-wetland relationship (see Table 2, Figure 5).

The journal “Science of The Total Environment” is a multidisciplinary and international academic journal published in the field of environmental sciences. This journal covers the relationships between the environmental sciences and the atmosphere, hydrosphere, biosphere, lithosphere, and anthroposphere. The studies published in the journal cover topics such as environmental impacts on wetlands

(Hernández et al., 1999; Queiroz et al., 2018), studies evaluating soil organic carbon levels (Mendes et al., 2024), research developing digital elevation models for wetlands (Shih et al., 2023), remote sensing technologies – wetlands (Ruan et al., 2022; Wu et al., 2022), and simulation techniques – wetlands (Song et al., 2021) (Table 2). The journal sources enumerated in Table 2 concentrate on critical research subjects such as the determination of water conservation priorities, the combatting of climate change, flood hazard assessments, changes in land use, the design of resilient infrastructure, and water balance analyses. These publications make significant contributions to the development of sustainable management strategies by conducting in-depth analyses of the connections between environmental science and disaster management.

Table 2. Main Sources

Sources	Number of Documents
IOP Conference Series: Earth and Environmental Science	35
Science of the Total Environment	20
Shengtai Xuebao	17
Science	16
Environmental Management	13
Nature	13
International Journal of Disaster Risk Reduction	12
Marine Pollution Bulletin	11
Proceedings of Spie - The International Society for Optical Engineering	11
International Geoscience and Remote Sensing Symposium (IGARSS)	10

Figure 5. Network map of the journals publishing the most documents related to the topic (disaster-wetland).

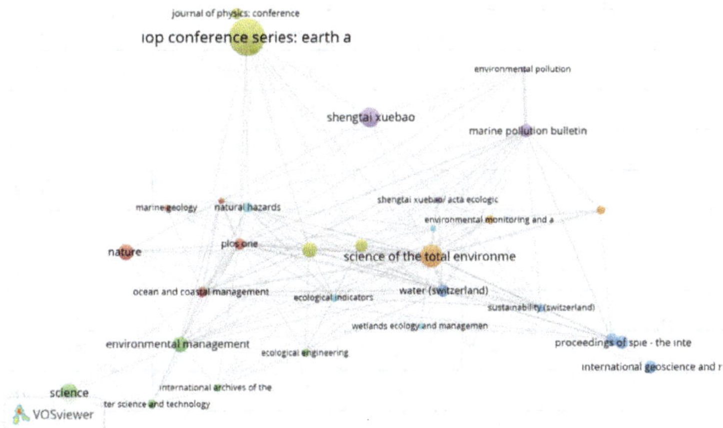

Lotka's law outlines the productivity patterns of authors within a specific field of study (Lotka, 1926). This law suggests that an analysis of author productivity concerning studies on the disaster-wetland relationship reveals a discernible order in scientific output. As indicated by the data in Table 3, the relationship between the number of publications and the number of authors adheres to a power law distribution. The findings illustrate that the majority of authors have published only one article (89.6%); the proportion of authors with two publications is 7.3%, those with three publications is 2%, and those with four publications is 0.5%. The number of authors who have published five or more works remains relatively low.

The regression analysis yielded the coefficients b = 2.871637 and c = 0.830315, confirming that the majority of authors produced only a small number of publications, while a limited number of authors contributed extensively. The Kolmogorov-Smirnov goodness-of-fit test further indicated that there was no significant difference between the observed and expected number of authors ($p > 0.05$). These findings demonstrate that scientific productivity on the relationship between disasters and wetlands conforms to Lotka's law, revealing that the number of authors decreases inversely with the number of publications (Figure 6)

Table 3. Author Productivity of Studies on the Disaster-Wetland Relationship

Number of documents	Number of authors	Ratio of authors
1	2767	0,896
2	226	0,073
3	62	0,02
4	14	0,005
5	7	0,002
6	6	0,002
7	3	0,001
8	1	0
9	1	0
10	1	0
17	1	0

Figure 6. Author productivity with Lotka's law

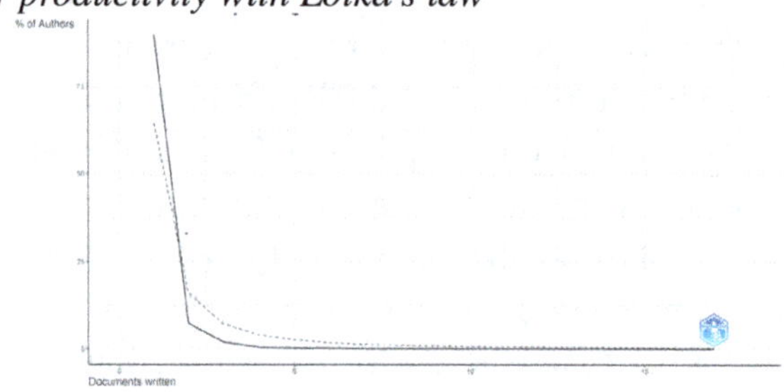

Figure 7 shows the citation network among the countries most cited in studies on the disaster–wetland relationship. The most cited country is the United States of America (USA) with 10,077 citations, highlighting its dominant role and well-established research infrastructure in this field. This is followed by Australia with 4,079 citations, reflecting its strong focus on coastal and wetland conservation, particularly in the context of disaster resilience. China ranks third with 3,572 citations, indicating its rapidly growing research capacity and increasing global influence in environmental science.

Other countries with notable citation counts include the United Kingdom, Canada, Germany, Japan, India, and Bangladesh, each contributing significantly to the scientific discourse on disaster impacts and wetland ecosystems. The presence of these countries in the citation network underscores the international scope and collaborative nature of research in this domain.

Figure 7. Citation network map between countries

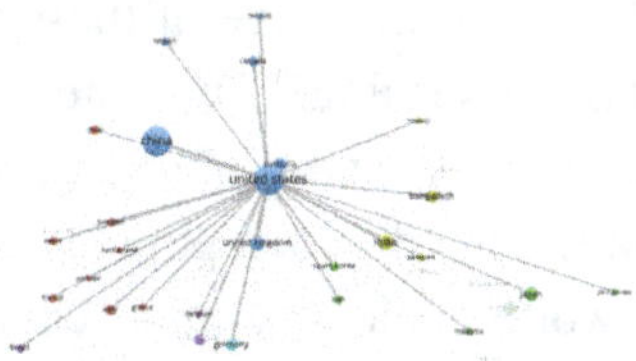

Table 4. Total number of documents and citations related to the topic (disaster-wetland) in countries

Country	Documents	Citations	Total link strength
USA	236	10077	75
China	218	3572	34
India	61	757	20
Indonesia	59	379	12
United Kingdom	50	2427	62
Japan	44	624	24
Australia	37	4079	27
Germany	31	1178	48
Bangladesh	30	722	29
Canada	24	1280	20

The concentration of high-output and high-impact research in a limited set of countries (e.g., USA, China, Australia, United Kingdom) has direct implications for disaster risk reduction policy and wetland management. First, it indicates an uneven global distribution of analytical capacity, data infrastructure, and investment in wetland–risk research, even though many highly exposed regions have substantial wetland-dependent livelihoods and rapidly changing hazard profiles. Second, it cautions against one-size-fits-all transfer of interventions, because the effectiveness of wetland-based measures (e.g., mangrove restoration, floodplain reconnection, peatland management) is highly context-dependent in terms of geomorphology, governance, and social vulnerability. Therefore, the collaboration links highlighted by the citation network (Figure 7) can be operationalised as a basis for targeted knowledge transfer, capacity-building, and comparable monitoring frameworks—especially for underrepresented regions—so that wetland-based DRR decisions are supported by locally relevant evidence and implementable indicators. (Seddon et al., 2020; Sebesvari et al., 2023).

Figure 8 shows the keywords, the most frequently used terms in studies on the disaster-wetland relationship, and the usage of these terms over time. Among the keywords, “wetlands,” “remote sensing,” “climate change,” “flooding,” “natural disaster,” and “ecosystem services” stand out. Keyword analysis helps us understand how certain topics have gained or lost popularity over time. For instance, the increased utilisation of keywords such as “climate change” and “remote sensing” in recent years signifies an escalating level of importance attributed to these subjects by researchers (Ruan et al., 2022; Wu et al., 2022; Bu et al., 2024; Cunha et al., 2024; Phan et al., 2024; Wang et al., 2024). In a similar vein, the pervasive utilisation of terms such as “flooding” (Ghosh et al., 2024; Sun et al., 2024) and “natural disaster” (Deasy et al., 2023; Grant and Cooker, 2023) signifies that research endeavours concerning the disaster-wetland relationship are centred on these concerns.

Figure 8. **A:** *Co-occurrence of keywords,* **B:** *Trend of keywords*

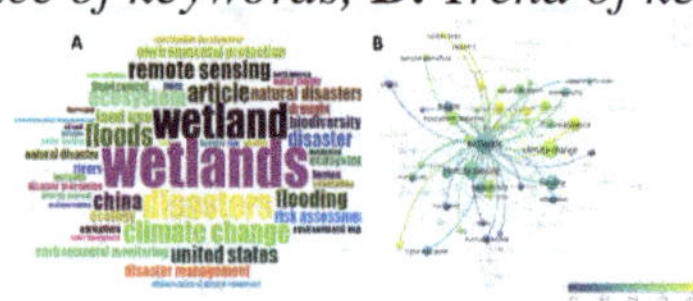

A review of the evolution of studies addressing the relationship between disasters and wetlands over time reveals that, in the early stages, the focus was on basic ecological concepts (e.g., “population dynamics,” “hydrology”). However, as time progressed, themes such as climate change, risk management and sustainable development gained prominence, with a broader interdisciplinary approach being adopted. While the importance of wetlands in terms of ecosystem services was emphasised with concepts such as “environmental monitoring,” “ecosystem,” and “biodiversity” during the period of 2007-2013, technological and solution-oriented approaches such as “remote sensing,” “spatiotemporal analysis,” and “disaster prevention” have dominated the literature since 2014. The increase in keywords indicates that the strategic importance of wetlands in the context of disaster prevention and management is increasingly recognised, and a global and policy-oriented framework is being adopted in this area. These studies, with a particular focus on concepts such as climate change and risk assessment, underscore the imperative for enhanced utilisation of interdisciplinary approaches and technological tools. From an operational standpoint, the rise of remote sensing and climate-risk terminology suggests that wetland-based DRR is moving toward deployable monitoring and decision-support systems (e.g., risk screening, early warning support, post-event impact assessment). In parallel, the increasing prominence of climate change indicates that wetland conservation and restoration should be framed within adaptive planning horizons, where design and prioritisation are informed by scenario-based risk governance rather than static historical baselines. (Ruan et al., 2022; Wu et al., 2022).

Figure 9 shows the thematic distribution of studies on the relationship between "disaster" and "wetland." Niche Themes, Motor Themes, Emerging or Declining Themes, and Basic Themes. Niche Themes include topics with a high degree of development but low importance, covering specific areas such as "sustainable development" and "machine learning."

Motor Themes, on the other hand, have both high development and high importance, and include topics such as "climate change adaptation," "governance," "risk," and "water management," which form the main focus areas of the research. Emerging or Declining Themes, which have low development and low importance, include topics such as "environment," "ecosystem services," and "land use." Basic Themes include topics with a low level of development but high importance, such as "hydrology," "wetlands," "remote sensing," and "disaster," which are located in this region. This thematic map plays a critical role in identifying research trends and determining which themes require further investigation (Alkhammash, 2023). The analyses conducted provide important clues for determining the direction of the research, revealing which topics researchers should focus on more and which themes require further investigation.

Practical implications for DRR policy and wetland management. Beyond describing thematic structure, the bibliometric results offer actionable signals for implementation. First, the prominence of "governance," "risk," and "water management" as Motor Themes (Figure 9) indicates that policy and institutional design is a central pathway through which wetland benefits are translated into disaster risk reduction outcomes. In practical terms, this supports (i) risk-informed land-use planning and zoning instruments that protect wetland buffers and regulate floodplain development (e.g., wetland protection overlays, set-back lines, and floodplain zoning), (ii) prioritization of nature-based solutions portfolios in which wetland protection/restoration is evaluated alongside engineered options using ecosystem-service and cost-effectiveness evidence (Costanza et al., 2008; TEEB, 2010; Davidson et al., 2019), and (iii) risk-informed restoration targeting that directs investments toward locations where wetlands most effectively reduce exposure and support climate adaptation (Spalding et al., 2014; Gardner and Finlayson, 2018).

Second, the strong emergence of "remote sensing" and "climate change" among frequent and rising keywords (Figure 8) highlights an operational route for monitoring, early warning support, and post-event assessment in wetland landscapes (Ruan et al., 2022; Wu et al., 2022; Bu et al., 2024). Remote sensing and GIS-based monitoring can be used to track wetland extent and condition over time, assess compliance in protected areas, and support adaptive management by linking observed ecosystem changes to hazard dynamics and policy interventions. Together, these applied mechanisms demonstrate how the chapter's bibliometric mapping can inform implementable DRR and wetland-governance instruments, rather than remaining only descriptive.

Figure 9. Thematic map based on density (y-axis) and centrality (x-axis).

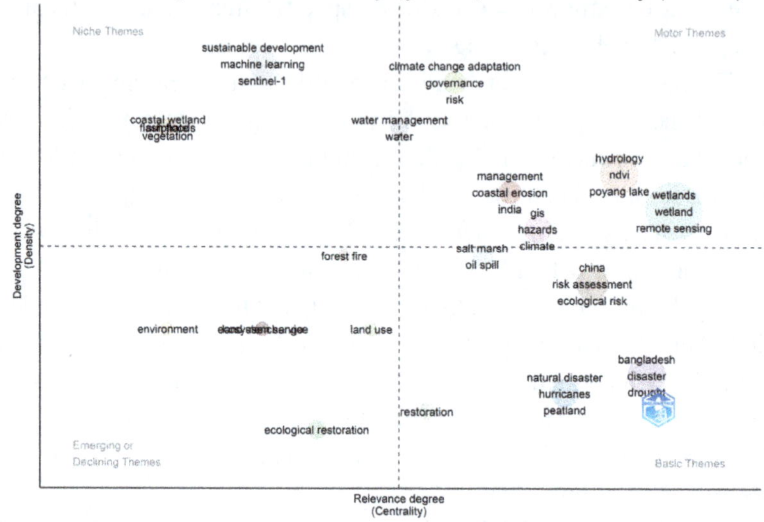

Equity, social vulnerability, and knowledge pluralism. While wetlands can reduce hazard impacts, disaster risk reduction outcomes are also shaped by who is exposed, who benefits from ecosystem services, and whose priorities are reflected in planning and restoration decisions. Social vulnerability and equity considerations therefore function as cross-cutting determinants of effectiveness, particularly where wetland-dependent livelihoods, informal settlements, or uneven access to resources influence adaptive capacity (Cutter et al., 2003; Almutairi et al., 2020). Notably, terms linked to equity, vulnerability, and indigenous/local knowledge do not appear among the most frequent keywords highlighted in Figure 8, suggesting that these dimensions remain less visible in dominant research trends despite their practical relevance. Strengthening future research and policy linkages will benefit from more explicit integration of local and indigenous knowledge systems, participatory decision processes, and distributional impact assessment in wetland-based DRR strategies (Reed, 2008; Sutton-Grier and Sandifer, 2019).

The thematic structure elucidated by bibliometric mapping has practical effects for the formulation of wetland-based disaster risk reduction strategies. Moreover, the positioning of "climate change adaptation" among the motor themes reinforces that wetland-based DRR measures should be embedded in long-term adaptation governance, including adaptive management, periodic reassessment of risk baselines, and cross-sector planning. The integration of the themes of "governance," "risk," and "water management" as predominant themes signifies that successful implementation increasingly relies on institutional coordination, cross-sector planning, and risk-informed decision-making processes, rather than solely on ecological factors (Almutairi et al., 2020; Endter-Wada et al., 2018). Consequently, the conservation and restoration of wetlands should be incorporated into broader strategies for enhancing resilience, such as watershed management, land-use regulation, and emergency planning. This strategy would ensure that ecological functions, including storage, attenuation, and buffering, are explicitly incorporated as elements of risk reduction portfolios (Spalding et al., 2014; Ghosh et al., 2024).

Concurrently, the emergence of "remote sensing" and associated technological terminology indicates a discernible trajectory from research to implementation. Conventional earth-observation products can facilitate spatial prioritisation, early warning systems, and post-event impact assessment by monitoring wetland flooding patterns, shoreline changes, and fragmentation pressures (Wu et al., 2022; Shih et al., 2023). Empirical studies consistently demonstrate the significance of mangroves and coastal wetlands in mitigating storm impacts and reducing human fatalities in coastal regions (Das and Vincent, 2009; Barbier et al., 2011). This underscores the notion that restoration should be regarded as a risk-sensitive investment strategy, as opposed to a mere conservation initiative. The extant evidence suggests that community-based preparedness and locally suitable management strategies are beneficial in the context of inland floodplains and wetland-centric rural communities. These actions contribute to the maintenance of stable wetland hydrology and connectivity, thereby enhancing the wetlands' adaptability (Deasy et al., 2023; Ferreira et al., 2023).

4. CONCLUSION

1. In the study, a total of 960 documents were examined by searching the Scopus database using the keywords "disaster" and "wetland."
2. Among the documents examined, the most common publication type was articles (n=578), followed by conference papers (n=167) and book chapters (n=95).
3. Overall, the citation network reveals a globally distributed research effort on disaster–wetland interactions, led by the United States with the highest citation impact. Australia and China follow as key contributors, while countries such as the United Kingdom, Canada, Germany, Japan, India, and Bangladesh also play significant roles. This international engagement highlights the collaborative nature and growing scientific attention toward wetland resilience in the face of environmental disasters.

4. Among the frequently used keywords in the studies are "wetlands", "remote sensing", "climate change", "flooding", "natural disaster", and "ecosystem services". In recent years, key words focused on technological and environmental management, such as "climate change" and "remote sensing," have particularly come to the forefront.
5. According to the results of the "Thematic Map" analysis of the 960 studies examined, 4 themes were identified. The topics addressed in these themes include: core themes: research topics of high importance and sophistication, such as "climate change adaptation," "risk," "governance," and "water management" (see Figure 9); basic themes: topics of critical importance in disaster-wetland relationships but addressed less in-depth, such as "hydrology," "wetlands," and "remote sensing"; niche themes: topics with high sophistication but less widespread interest, focusing on specific sub-areas such as "sustainable development" and "machine learning"; and evolving themes: while basic ecological concepts (e.g., "population dynamics," "hydrology") were emphasized in the early years, technology and solution-oriented approaches (e.g., "remote sensing," "spatiotemporal analysis") came to the forefront after 2014.
6. Most of the studies are multi-authored and based on international collaboration. The high disaster risk, especially in countries along the ocean coast, has increased cooperation in these areas. The United States and China are in a leading position in both document production and citation networks.

5. FUTURE OUTLOOK AND THE WAY FORWARD

Looking ahead, the strategic integration of wetlands into disaster risk reduction frameworks is paramount for both ecological resilience and sustainable development. As scientific evidence increasingly highlights the critical functions of wetlands—including flood regulation, water storage, and biodiversity support—there is an urgent need for policymakers to embed ecosystem-based approaches into national and international disaster management strategies, in line with global frameworks such as the Sendai Framework and the Sustainable Development Goals (Gardner & Finlayson, 2018). To support this integration, advancing research through the adoption of innovative technologies like remote sensing and GIS will be essential for monitoring wetland health and evaluating their effectiveness in mitigating disaster risks. In parallel, fostering participatory governance that empowers local communities and leverages traditional knowledge will enhance the legitimacy and sustainability of wetland restoration and management efforts (Reed, 2008).

In addition, equity-oriented planning is essential to ensure that wetland-based interventions do not unintentionally shift risk or exclude vulnerable groups. Incorporating social vulnerability assessment and participatory, knowledge-plural governance can improve legitimacy and increase the long-term effectiveness of DRR outcomes (Cutter et al., 2003; Reed, 2008).

Scaling up restoration initiatives—backed by robust financial mechanisms and multi-sectoral partnerships—can further strengthen the adaptive capacity of both ecosystems and societies, delivering long-term economic and social benefits (Spalding et al., 2014; TEEB, 2010). At the same time, interdisciplinary and international collaboration should be prioritized, given that wetland-disaster interactions are complex and often transboundary in nature, requiring coordinated research, policy, and management responses (Moomaw et al., 2018; Aria & Cuccurullo, 2017). Finally, establishing comprehensive monitoring and evaluation systems will enable adaptive management and evidence-based learning, while targeted capacity-building programs will ensure that best practices are implemented at all levels. Collectively, these actions will help position wetlands as vital natural infrastructure at the core of disaster risk reduction and sustainable development agendas worldwide.

REFERENCES

Alkhammash, R. (2023). Bibliometric, network, and thematic mapping analyses of metaphor and discourse in COVID-19 publications from 2020 to 2022. *Frontiers in Psychology, 13*, 1062943. DOI: 10.3389/fpsyg.2022.1062943

Almutairi, A., Mourshed, M., & Ameen, R. F. M. (2020). Coastal community resilience frameworks for disaster risk management. *Natural Hazards, 101*(2), 595–630. DOI: 10.1007/s11069-020-03875-3

Alongi, D. M. (2008). Mangrove forests: Resilience, protection from tsunamis, and responses to global climate change. *Estuarine, Coastal and Shelf Science, 76*(1), 1–13. DOI: 10.1016/j.ecss.2007.08.024

Anggraini, E., Yuliastuti, N., & Budihardjo, M. A. (2024, April). Analysis of Water Conservation Priority Locations in Upper Sub Watershed. [). IOP Publishing.]. *IOP Conference Series. Earth and Environmental Science, 1323*(1), 012001. DOI: 10.1088/1755-1315/1323/1/012001

Aria, M., & Cuccurullo, C. (2017). bibliometrix: An R-tool for comprehensive science mapping analysis. *Journal of Informetrics, 11*(4), 959–975. DOI: 10.1016/j.joi.2017.08.007

Barbier, E. B., Hacker, S. D., Kennedy, C., Koch, E. W., Stier, A. C., & Silliman, B. R. (2011). The value of estuarine and coastal ecosystem services. *Ecological Monographs, 81*(2), 169–193. DOI: 10.1890/10-1510.1

Bennett, E. M., Peterson, G. D., & Gordon, L. J. (2009). Understanding relationships among multiple ecosystem services. *Ecology Letters, 12*(12), 1394–1404. DOI: 10.1111/j.1461-0248.2009.01387.x

Bu, J., Wang, Q., Wang, Z., Fan, S., Liu, X., & Zuo, X. (2024). Land Remote Sensing Applications Using Spaceborne GNSS Reflectometry: A Comprehensive Overview. *IEEE Journal of Selected Topics in Applied Earth Observations and Remote Sensing, 17*, 12811–12841. DOI: 10.1109/JSTARS.2024.3415754

Bulut, C., & Yıldız, F. (2024). Bibliometric analysis of laser powder bed additive manufacturing of high-entropy alloys: Where are we and how far have we come? *Progress in Additive Manufacturing*. Advance online publication. DOI: 10.1007/s40964-024-00840-5

Celik, E., Durmus, A., Adizel, O., & Nergiz Uyar, H. (2021). A bibliometric analysis: What do we know about metals (loids) accumulation in wild birds? *Environmental Science and Pollution Research International*, *28*(8), 10302–10334. DOI: 10.1007/s11356-021-12344-8

Çelik, E., & Sarıboğa, L. (2023). Assessment of the role of birds in seed distribution: a bibliometric analysis of studies on this subject. Recent advances in natural and engineering Sciences. Livre de lyon, Lyon, 29-53.

Çelik, M. A. (2020). Kuraklık Araştırmalarında Yeni Eğilimler, Kullanılan Teknikler ve Kavramlar Üzerine Bibliyometrik Ağ Analizi. International Journal of Geography and Geography Education, (42), 602-630.

Çelik, M. A., & Çelik, E. (2024). Are Urbanisation and Biodiversity Antithetical? A Bibliometric Analysis. *Coğrafya Dergisi*, (48), 121–135.

Çelik, M. A., Kopar, İ., & Çelik, E. (2021). Doğubayazıt sazlığının (Ağrı-Türkiye) arazi örtüsü deseninde meydana gelen değişimlerin ekolojik sonuçları üzerine bir analiz. *Doğu Coğrafya Dergisi*, *26*(46), 193–210.

Chen, C. (2017). Science mapping: A systematic review of the literature. *Journal of Data and Information Science*, *2*(2), 1–40. DOI: 10.1515/jdis-2017-0006

Chiu, W. T., & Ho, Y. S. (2007). Bibliometric analysis of tsunami research. *Scientometrics*, *73*(1), 3–17. DOI: 10.1007/s11192-005-1523-1

Costanza, R., d'Arge, R., de Groot, R., Farber, S., Grasso, M., Hannon, B., Limburg, K., Naeem, S., O'Neill, R. V., Paruelo, J., Raskin, R. G., Sutton, P., & van den Belt, M. (1997). The value of the world's ecosystem services and natural capital. *Nature*, *387*(6630), 253–260. DOI: 10.1038/387253a0

Costanza, R., Perez-Maqueo, O., Martinez, M. L., Sutton, P., Anderson, S. J., & Mulder, K. (2008). The value of coastal wetlands for hurricane protection. *Ambio*, *37*(4), 241–248. DOI: 10.1579/0044-7447(2008)37[241:TVOCWF]2.0.CO;2

Cowell, E. B. (1978). Ecological effects of dispersants in the United Kingdom. In *Chemical Dispersants for the Control of Oil Spills*. ASTM International. DOI: 10.1520/STP35727S

Cunha, A. P. M., Buermann, W., & Marengo, J. A. (2024). Changes in compound drought-heat events over Brazil's Pantanal wetland: An assessment using remote sensing data and multiple drought indicators. *Climate Dynamics*, *62*(1), 739–757. DOI: 10.1007/s00382-023-06937-x

Cutter, S. L., Boruff, B. J., & Shirley, W. L. (2003). Social vulnerability to environmental hazards. *Social Science Quarterly*, *84*(2), 242–261. DOI: 10.1111/1540-6237.8402002

Das, S., & Vincent, J. R. (2009). Mangroves protected villages and reduced death toll during Indian super cyclone. *Proceedings of the National Academy of Sciences of the United States of America*, *106*(18), 7357–7360. DOI: 10.1073/pnas.0810440106

Davidson, N. C., Van Dam, A. A., Finlayson, C. M., & McInnes, R. J. (2019). Worth of wetlands: Revised global monetary values of coastal and inland wetland ecosystem services. *Marine and Freshwater Research*, *70*(8), 1189–1194. DOI: 10.1071/MF18391

De Feudis, M., Trenti, W., & Manfredi, P.. (2024). Negative impact of alluvial sediments on physical properties of agricultural soils affected by flooding in May 2023 in Emilia Romagna Region (Northern Italy). *Journal of Soils and Sediments*. Advance online publication. DOI: 10.1007/s11368-024-03931-6

Deasy, A., Puji, H. K., Herry, P., & Nugroho, P. (2023). Disaster-resilient villages: Strengthening community capacity in flood disasters managing in wetland areas. *Disaster Advances*, *16*(4), 1–7.

Donthu, N., Kumar, S., Mukherjee, D., Pandey, N., & Lim, W. M. (2021). How to conduct a bibliometric analysis: An overview and guidelines. *Journal of Business Research*, *133*, 285–296. DOI: 10.1016/j.jbusres.2021.04.070

Dudgeon, D., Arthington, A. H., Gessner, M. O., Kawabata, Z. I., Knowler, D. J., Lévêque, C., Naiman, R. J., Prieur-Richard, A.-H., Soto, D., Stiassny, M. L. J., & Sullivan, C. A. (2006). Freshwater biodiversity: Importance, threats, status and conservation challenges. *Biological Reviews of the Cambridge Philosophical Society*, *81*(2), 163–182. DOI: 10.1017/S1464793105006950

Duran, V., & Çelik, M. A. (2022). Çeşitli Yöntemler Kullanılarak İklim Değişikliği ve Çevre Sorunlarına Yönelik Tutum Analizi. *International Journal of Geography and Geography Education*, (46), 79–103. DOI: 10.32003/igge.1069665

Ellegaard, O. (2018). The application of bibliometric analysis: Disciplinary and user aspects. *Scientometrics*, *116*(1), 181–202. DOI: 10.1007/s11192-018-2765-z

Endter-Wada, J., Kettenring, K. M., & Sutton-Grier, A. E. (2018). Sustaining wetlands to mitigate disasters and protect people. *Frontiers in Ecology and the Environment*, *16*(8), 431. DOI: 10.1002/fee.1959

Erwin, K. L. (2009). Wetlands and global climate change: The role of wetland restoration in a changing world. *Wetlands Ecology and Management, 17*(1), 71–84. DOI: 10.1007/s11273-008-9119-1

Ferreira, C. S., Kašanin-Grubin, M., Solomun, M. K., Sushkova, S., Minkina, T., Zhao, W., & Kalantari, Z. (2023). Wetlands as nature-based solutions for water management in different environments. *Current Opinion in Environmental Science & Health, 33*, 100476. DOI: 10.1016/j.coesh.2023.100476

Fibriantika, E., Mujahidin, A., Maryamah, S., & Rasyid, I. A. (2021, July). Analysis of water balance conditions as the impact of climate and land-use changes in Kapuas Watershed, Indonesia. [). IOP Publishing.]. *IOP Conference Series. Earth and Environmental Science, 824*(1), 012018. DOI: 10.1088/1755-1315/824/1/012018

Gardner, R. C., & Finlayson, C. (2018, October). Global wetland outlook: state of the world's wetlands and their services to people. In Ramsar convention secretariat (pp. 2020-5). DOI: 10.69556/GWO-2018-eng

Gavel, Y., & Iselid, L. (2008). Web of Science and Scopus: A journal title overlap study. *Online Information Review, 32*(1), 8–21. DOI: 10.1108/14684520810865958

Ghani, M. W., & Hidayati, I. (2024, March). Adapting to sea level rise: Migration decisions and disaster mitigation strategies in small-scale fishing communities. [). IOP Publishing.]. *IOP Conference Series. Earth and Environmental Science, 1314*(1), 012013. DOI: 10.1088/1755-1315/1314/1/012013

Ghosh, P., Sudarsan, J. S., & Nithiyanantham, S. (2024). Nature-based disaster risk reduction of floods in Urban Areas. *Water Resources Management, 38*(6), 1847–1866. DOI: 10.1007/s11269-024-03757-4

Gilbreath, D. (2011). *The Indian Ocean tsunami: The global response to a natural disaster*. University Press of Kentucky.

Grant, A., & Cooker, M. J. (2023). Reductions in water level over coastal wetlands during storm surges and tsunamis: An analytical result and a critical review of the literature. *Coastal Engineering, 183*, 104328. DOI: 10.1016/j.coastaleng.2023.104328

Graveline, M. H., & Germain, D. (2022). Disaster Risk Resilience: Conceptual Evolution, Key Issues, and Opportunities. *International Journal of Disaster Risk Science, 13*(3), 330–341. DOI: 10.1007/s13753-022-00419-0

Haddaway, N. R., Woodcock, P., Macura, B., & Collins, A. (2015). Making literature reviews more reliable through application of lessons from systematic reviews. *Conservation Biology, 29*(6), 1596–1605. DOI: 10.1111/cobi.12541

Hernández, L. M., Gómara, B., Fernández, M., Jiménez, B., González, M. J., Baos, R., Hiraldo, F., Ferrer, M., Benito, V., Suñer, M. A., Devesa, V., Muñoz, O., & Montoro, R. (1999). Accumulation of heavy metals and As in wetland birds in the area around Donana National Park affected by the Aznalcollar toxic spill. *The Science of the Total Environment*, *242*(1-3), 293–308. DOI: 10.1016/S0048-9697(99)00397-6

İli, B., & Bakar, H. (2024). Veri Gazeteciliğine Yönelik Araştırmaların Bibliyometrik Analizi. *Erciyes İletişim Dergisi*, *11*(2), 607–626. DOI: 10.17680/erciyesiletisim.1430017

Katakidis, A., Kodonas, K., Fardi, A., & Gogos, C. (2023). A scientometric, bibliometric, and thematic map analysis of hydraulic calcium silicate root canal sealers. *Restorative Dentistry & Endodontics*, *48*(4), e41. DOI: 10.5395/rde.2023.48.e41

Kehoe, L., Romero-Muñoz, A., Polaina, E., Estes, L., Kreft, H., & Kuemmerle, T. (2017). Biodiversity at risk under future cropland expansion and intensification. *Nature Ecology & Evolution*, *1*(8), 1129–1135. DOI: 10.1038/s41559-017-0234-3

Kulak, M., Ozkan, A., & Bindak, R. (2019). A bibliometric analysis of the essential oil-bearing plants exposed to the water stress: How long way we have come and how much further? *Scientia Horticulturae*, *246*, 418–436. DOI: 10.1016/j.scienta.2018.11.031

Lotka, A. J. (1926). The frequency distribution of scientific productivity. *Journal of the Washington Academy of Sciences*, *16*(12), 317–323.

Mendes, R. G., do Valle, R. F.Junior, Feitosa, T. H. S., de Melo, M. M. A. P., Fernandes, L. F. S., Pacheco, F. A. L., & Valera, C. A. (2024). Carbon footprints of tailings dams' disasters: A study in the Brumadinho region (Brazil). *The Science of the Total Environment*, *949*, 175026. DOI: 10.1016/j.scitotenv.2024.175026

Middleton, B. A. (1999). *Wetland Restoration, Flood Pulsing, and Disturbance Dynamics*. John Wiley & Sons.

Mitsch, W. J., & Gosselink, J. G. (2000). *Wetlands*. John Wiley & Sons.

Moomaw, W. R., Chmura, G. L., Davies, G. T., Finlayson, C. M., Middleton, B. A., Natali, S. M., Perry, J. E., Roulet, N., & Sutton-Grier, A. E. (2018). Wetlands in a changing climate: Science, policy and management. *Wetlands*, *38*(2), 183–205. DOI: 10.1007/s13157-018-1023-8

Nelson-Smith, A. (1978). Effects of dispersant use on shore life. In *Chemical dispersants for the control of oil spills*. ASTM International. DOI: 10.1520/STP35725S

Phan, L. T., Jou, S. C., Chiou, C. R., & Yang, C. J. (2024). Is it because of climate change? Social-ecological system analysis of wetland protected area in Vietnamese Mekong Delta. *Wetlands Ecology and Management*, *32*(1), 33–57. DOI: 10.1007/s11273-023-09960-1

Pratiwi, A. P., Rohmat, F. I. W., Nugroho, E. O., Farid, M., & Kusuma, M. S. B. (2024, March). Flood hazard comparison based on geomorphic flood index and hydraulic HEC-RAS (Case study in Ciliwung Watershed, Jakarta). [). IOP Publishing.]. *IOP Conference Series. Earth and Environmental Science*, *1314*(1), 012084. DOI: 10.1088/1755-1315/1314/1/012084

Pullin, A. S., & Knight, T. M. (2009). Doing more good than harm–Building an evidence-base for conservation and environmental management. *Biological Conservation*, *142*(5), 931–934. DOI: 10.1016/j.biocon.2009.01.010

Queiroz, H. M., Nóbrega, G. N., Ferreira, T. O., Almeida, L. S., Romero, T. B., Santaella, S. T., Bernardino, A. F., & Otero, X. L. (2018). The Samarco mine tailing disaster: A possible time-bomb for heavy metals contamination? *The Science of the Total Environment*, *637*, 498–506. DOI: 10.1016/j.scitotenv.2018.04.370

Reed, M. S. (2008). Stakeholder participation for environmental management: A literature review. *Biological Conservation*, *141*(10), 2417–2431. DOI: 10.1016/j.biocon.2008.07.014

Ruan, L., Yan, M., Zhang, L., Fan, X., & Yang, H. (2022). Spatial-temporal NDVI pattern of global mangroves: A growing trend during 2000–2018. *The Science of the Total Environment*, *844*, 157075. DOI: 10.1016/j.scitotenv.2022.157075

Sebesvari, Z., Sudmeier-Rieux, K., & Walz, Y. (2023). Towards a typology of nature-based solutions for disaster risk reduction. *Nature-Based Solutions*, *3*, 100057. DOI: 10.1016/j.nbsj.2023.100057

Seddon, N., Chausson, A., Berry, P., Girardin, C. A. J., Smith, A., & Turner, B. (2020). Understanding the value and limits of nature-based solutions to climate change and other global challenges. *Philosophical Transactions of the Royal Society of London. Series B, Biological Sciences*, *375*(1794), 20190120. DOI: 10.1098/rstb.2019.0120

Shih, S. S., Hsu, W. C., & Hsu, Y. W. (2023). Waterline digital elevation model development to quantify inundation duration and coastal protection of tidal wetlands. *The Science of the Total Environment*, *874*, 162519. DOI: 10.1016/j.scitotenv.2023.162519

Smith, K., & Petley, D. N. (2009). *Environmental Hazards: Assessing Risk and Reducing Disaster*. Routledge. DOI: 10.4324/9780203884805

Song, K., Choi, Y. E., Han, H. J., & Chon, J. (2021). Adaptation and transformation planning for resilient social-ecological system in coastal wetland using spatial-temporal simulation. *The Science of the Total Environment*, *789*, 148007. DOI: 10.1016/j.scitotenv.2021.148007

Spalding, M. D., McIvor, A. L., Beck, M. W., Koch, E. W., Möller, I., Reed, D. J., Rubinoff, P., Spencer, T., Tolhurst, T. J., Wamsley, T. V., van Wesenbeeck, B. K., Wolanski, E., & Woodroffe, C. D. (2014). Coastal ecosystems: A critical element of risk reduction. *Conservation Letters*, *7*(3), 293–301. DOI: 10.1111/conl.12074

Srinivas, H., & Nakagawa, Y. (2008). Environmental implications for disaster preparedness: Lessons learnt from the Indian Ocean Tsunami. *Journal of Environmental Management*, *89*(1), 4–13. DOI: 10.1016/j.jenvman.2007.01.054

Suding, K. N. (2011). Toward an era of restoration in ecology: Successes, failures, and opportunities ahead. *Annual Review of Ecology, Evolution, and Systematics*, *42*(1), 465–487. DOI: 10.1146/annurev-ecolsys-102710-145115

Sun, D., Yang, T., Li, S., Goldberg, M., Kalluri, S., Helfrich, S., & Miralles, F. R. (2024). Hazard or non-hazard flood: Post analysis for paddy rice, wetland, and other potential non-hazard flood extraction from the VIIRS flood products. *ISPRS Journal of Photogrammetry and Remote Sensing*, *209*, 415–431. DOI: 10.1016/j.isprsjprs.2024.02.013

Sutton-Grier, A. E., & Sandifer, P. A. (2019). Conservation of wetlands and other coastal ecosystems: A commentary on their value to protect biodiversity, reduce disaster impacts, and promote human health and well-being. *Wetlands*, *39*(6), 1295–1302. DOI: 10.1007/s13157-018-1039-0

TEEB. (2010). *The Economics of Ecosystems and Biodiversity: Mainstreaming the Economics of Nature: A Synthesis of the Approach, Conclusions and Recommendations of TEEB*. Progress Press.

Turetsky, M. R., Benscoter, B., Page, S., Rein, G., van der Werf, G. R., & Watts, A. (2015). Global vulnerability of peatlands to fire and carbon loss. *Nature Geoscience*, *8*(1), 11–14. DOI: 10.1038/ngeo2325

Van Eck, N. J., & Waltman, L. (2010). Software survey: VOSviewer, a computer program for bibliometric mapping. *Scientometrics*, *84*(2), 523–538. DOI: 10.1007/s11192-009-0146-3

Walz, Y., Janzen, S., Narvaez, L., Ortiz-Vargas, A., Woelki, J., Doswald, N., & Sebesvari, Z. (2021). Disaster-related losses of ecosystems and their services. Why and how do losses matter for disaster risk reduction? *International Journal of Disaster Risk Reduction*, *63*, 102425. DOI: 10.1016/j.ijdrr.2021.102425

Wang, L., Cheng, J., Jiang, Y., Liu, N., & Wang, K. (2024). Prediction of Adaptability of Typical Vegetation Species in Flood Storage Areas under Future Climate Change: A Case in Hongze Lake FDZ, China. *Sustainability (Basel)*, *16*(15), 6331. DOI: 10.3390/su16156331

Wu, W., Zhi, C., Gao, Y., Chen, C., Chen, Z., Su, H., Lu, W., & Tian, B. (2022). Increasing fragmentation and squeezing of coastal wetlands: Status, drivers, and sustainable protection from the perspective of remote sensing. *The Science of the Total Environment*, *811*, 152339. DOI: 10.1016/j.scitotenv.2021.152339

Yulu, A., Albayrak, F., Işık, Z., Aras, S., & Taşçı, H. E. (2023). Türkiye'nin Çatısında Çevresel Kirlilik: Balıkgöl Örneği (Ağrı). *Doğu Coğrafya Dergisi*, *28*(49), 94–104. DOI: 10.5152/EGJ.2023.22046

Zedler, J. B., & Kercher, S. (2005). Wetland resources: Status, trends, ecosystem services, and restorability. *Annual Review of Environment and Resources*, *30*(1), 39–74. DOI: 10.1146/annurev.energy.30.050504.144248

Zscheischler, J., Westra, S., van den Hurk, B. J. J. M., Seneviratne, S. I., Ward, P. J., Pitman, A., AghaKouchak, A., Bresch, D. N., Leonard, M., Wahl, T., & Zhang, X. (2018). Future climate risk from compound events. *Nature Climate Change*, *8*(6), 469–477. DOI: 10.1038/s41558-018-0156-3

Chapter 4
Transformation Strategies in Anthropogenic Coastal Erosion:
Global Successes and a Roadmap for Bogaçay

Muhammed Ernur Akiner
http://orcid.org/0000-0002-5192-2473
Akdeniz University, Turkey

ABSTRACT

Coastal erosion, which is increasing globally, is a multidimensional disaster triggered by human interventions. This study compares the erosion crisis at the Antalya Boğaçay Project with successful intervention examples worldwide and proposes sustainable solutions. Environmentally friendly approaches such as the Dutch "Sand Engine" project, sediment bypass systems in the US, and Australia's geotube applications have achieved 70-90% success rates in coastal stabilization. Dynamic coastal management, artificial reefs, and targeted sediment recharge techniques are critical solutions for Boğaçay. The study analyzes the applicability of these strategies to the Turkish coast through hydrodynamic modeling, ecosystem-based design, and disaster risk management perspectives. Consequently, it emphasizes that coastal erosion is an interdisciplinary disaster management issue rather than an engineering problem, and an adaptive management framework is presented for the Mediterranean basin.

DOI: 10.4018/979-8-3373-6274-8.ch004

INTRODUCTION

Konyaaltı Beach is one of the Mediterranean's most significant coastal ecosystems with considerable recreational and ecological value (Akiner et al., 2025a; Durap & Balas, 2024). The beach remained in natural balance for hundreds of years due to sediments brought in by the riverside, Boğaçay. However, the Boğaçay Project, which was initiated in 2017, disrupted sediment transport by lowering the riverbed below sea level, and thus, coastal dynamics have been altered to a great extent. Hence, coastal erosion has occurred, and the beach has become narrow.

Any coastal engineering works affecting natural processes may produce permanent changes, especially when the sediment balance is disturbed. Understanding how the Boğaçay Project affected coastal evolution is a key consideration for planning similar projects.

This dynamic nature of coastal systems is the product of natural processes interacting with human interruptions. Usually, these systems attain a state of more or less equilibrium through factors such as sediment transport, wave energy dissipation, and any variation in sea level. However, the increasing number of human projects on the coast erodes this natural balance, with many hitting the coastline in recent decades. The Boğaçay Project is a critical instance for the Konyaaltı coast and the management of similar ecosystems. Any restriction in the natural sediment flow induces a chain of effects leading to physical regression and ecological losses (Bruun, 1962; Dionísio António et al., 2023; McCall et al., 2010).

Coastline morphology is directly related to basin-scale processes. Infrastructure interventions in such sensitive areas can have an effect locally and all along the coastline. So, while gaining insight into the morphological consequences of the Boğaçay Project, basin-coastal integrity must also be evaluated (Akiner 2025, Arabi et al. 2008, and Akiner and Lačnjevac 2022).

Since the physical changes are only one aspect of coastal management, ecosystem services, biodiversity, and socioeconomic impacts have an equal role to play. Konyaaltı Beach is a recreational area and a tourism destination that offers economic opportunities to the locals. Thus, the great shrinking occasioned by coastal erosion poses serious threats to business, residential, and agricultural lands alongside the coast (Das & Swain, 2024). This entire threat structure denotes that coastal engineering projects should never be judged on a purely physical basis (Koç et al., 2015; Ferreira et al., 2022).

The slow death of coastal ecosystems under global climate change and anthropogenic pressure runs parallel to environmental degradation and monetizes these scenarios into disasters for all socioeconomic systems. A prime example is the Boğaçay Project, carried out in 2017 on the Konyaaltı coast of Antalya. By dredging the river bed of Boğaçay to 1.5 m below sea level, the project disrupted the sediment transport and set in motion a multidimensional disaster spiral in the region (Akiner et al., 2025a).

While anthropogenic disturbances remain the first-order pressure applied to coastal systems, climate change-induced sea level rise importantly acts as a multiplier of existing risk. Vousdoukas et al. (2018) predict that risk pertaining to climate change-induced flooding of European coasts shall rise drastically in the coming decades, with socioeconomic costs running into hundreds of billions of euros. Recent trends in the literature bring up the complex relationship between climate changes and coastal coastline erosion processes along with suggestive modeling approaches (Pang et al., 2023). In this respect, the already destabilized Boğaçay coastline, deformed by human interference on the local scale, has turned highly susceptible to the global threat of sea level rise. Present erosion exists not merely because of sediment disruption but due to the deadly combination of factors brought about by global climate change and local disturbances.

Human interference and climate conditions are both responsible for current coastal erosion, but the long-term coastal management strategies will have to include climate change impacts that cannot be avoided. The IPCC scenarios predict that there will be a rise in sea levels in the Mediterranean basin, an increase in the number and intensity of extreme weather events, and alterations in the paths of storms. These changes will have multiple impacts on the Boğaçay coast, namely: (1) according to the Bruun rule, the rise in sea level will result in higher rates of coastal retreat, (2) the more intense and frequent storms will lead to greater erosion during the winter months and thereby change the seasonal balance, and (3) the altered precipitation patterns will discharge less sediment from the basin and at a later time, thus diminishing the coast's resupply ability. For this reason, the management strategies suggested (e.g., sediment recharge volumes, structural designs) should be rooted in the dynamic scenarios that can be periodically reviewed and adapted to climate projections, rather than the static ones.

The main components of disasters in the region can be categorized as follows:

- Physical Disaster: Morphological collapse manifested by a coastal retreat of 21.7 meters (Akiner, 2025)
- Ecological Disaster: Collapse of Posidonia oceanica populations and loss of benthic habitat
- Socioeconomic Disaster: Decline in tourism revenues and salinization of agricultural land

In the past few years, a clear trend has been noticed moving away from single-model methods to the incorporation of multiple and interdisciplinary modeling frameworks which are integrating models like Pelnard-Considère, SWAT, and XBeach

(Ferreira et al., 2022; Goldstein et al., 2019). The integration of these three models provides a particularly strong method, since the Pelnard-Considère model gives an analytical foundation to the long-term shoreline changes, SWAT (Soil and Water Assessment Tool) working at the large basin scale helps to understand the sediment and nutrient load transport, and XBeach simulates short-term wave hydrodynamics and morphodynamics at fine scales (Akiner, 2025). The ongoing studies in the domain of coast management and adaptation are bringing out the possibilities of machine learning methods, together with such integrated modeling, in the areas of reducing the uncertainties and capturing complex patterns of the traditional physics-based models (Goldstein et al., 2019). This paper is in line with these contemporary trends by making practical the abovementioned model integration and putting the need for hybrid (physics-ML) early warning systems in future research into the spotlight.

This research is based on the integrated modeling framework (SWAT-XBeach-Pelnard-Considère) which was developed and validated in Akiner's (2025) research and quantitatively demonstrated the morphodynamic impacts of the Boğaçay Project, and its principal conclusions. The findings of that research such as the 21.7-meter coastal retreat, the sediment deficit, and the changes in water quality are considered in this chapter not as a new analysis but as an already available and approved database. The section's major contribution and innovation is to add to these technical findings by placing them in a multidisciplinary disaster risk management perspective; comparing them with global and regional case studies from the Netherlands, the USA, Italy, Australia, and the Mediterranean; analyzing socio-economic impacts, community participation, and governance difficulties in detail; and finally, issuing practical, incremental, and flexible policy recommendations and action strategies for decision-makers. In doing so, this study intends to go beyond the current model results, providing a user-oriented roadmap and conceptual expansion instead of an academic synthesis.

Our study aims to provide a framework for how coastal engineering projects can be integrated with disaster management principles by addressing this multilayered disaster structure from an interdisciplinary perspective.

This paper investigates the issue of coastal erosion through the lens of a multi-layered disaster management framework using the Boğaçay case study. Initially, the problem's physical, ecological, and socioeconomic aspects will be delineated and subjected to a quantitative analysis through the integration of various models (SWAT, XBeach, Pelnard-Considère). The model results demonstrate the changing coastline and the degree of the erosion problem through Figures 3 and 4. Afterwards, there will be a discussion of regional cases from the Mediterranean and global success stories from the Netherlands, USA, Italy, and Australia to create a link between the models' outputs and practical solutions. The last parts will offer technical findings, global lessons, and short, medium, and long-term action recommendations as well as governance reforms for policymakers and implementers. These reforms call for substantial changes in policy, which the next section will address in detail. The conclusion will highlight the most important findings through the lens of the necessity for interdisciplinary and adaptive governance. The prompt here is to grasp the issue holistically and thus propose an interdisciplinary solution.

This research mainly concentrates on the multi-layered disaster structure and also opens up the floor for more discussions on coastal management under the backlight of current trends and integrated methods. The chapter will deeply investigate the long-term effects, integrated technologies, economic nature-based versus engineering solution comparisons, social vulnerability, community participation, adaptive management monitoring systems, and finally the necessity of international cooperation in the Mediterranean area, using the Boğaçay case as an example and linking it to policy recommendations. The new framework thus proposed is meant to give a comprehensive view of the issue and assist decision-making.

Coastal erosion is a multifaceted problem that must be addressed from a disaster management perspective rather than an engineering perspective. The 21.7-meter coastal retreat triggered by the Boğaçay Project is not merely a morphological change; it is also a chain of ecological and socioeconomic disasters, including a decline in tourism revenues, salinization of agricultural lands, and a decline in the Posidonia oceanica population. This study aims to synthesize integrated modeling results with global success examples to offer disaster risk reduction strategies for Antalya.

Similar coastal disasters have occurred in the Netherlands (Dutch), the Chesapeake Bay in the US, and the Gold Coast in Australia, but these were controlled by 70-90% with adaptive management strategies. In the Mediterranean basin, interventions in the Sulina Channel (Romania) and the Yeşilırmak Delta (Türkiye) have produced outcomes similar to those of Boğaçay. This study develops a unique roadmap for Antalya based on these global experiences.

The global examples examined demonstrate that the success of combating coastal erosion lies in engineering solutions, a strong legal framework, and inter-institutional coordination. For example, the Dutch "Sand Engine" project was implemented within a national "Delta Program" framework with long-term funding. However, many projects in Mediterranean countries, such as Boğaçay, are limited to short-term actions by local governments and lack basin-wide planning. This results in temporary interventions and postponement of disasters. The fundamental problem here is not an engineering failure but rather a weakness in coastal governance.

Coastal erosion is more than just a physical environmental problem; it threatens local people's livelihoods, cultural heritage, and social fabric. The decline in fishing in Boğaçay, the loss of income for beach operators, and the salinization of agricultural lands highlight the humanitarian dimension of the disaster. Therefore, a successful response strategy should aim not only to restore sediment transport but also to enhance the resilience of this socio-ecological system. A transparent and participatory planning approach that engages the community will significantly increase the effectiveness and sustainability of technical solutions.

Adaptive coastal management inevitably requires a high initial cost. However, global experience has shown how this cost can be offset through innovative financing mechanisms such as the build-operate-transfer model, environmental bonds, and the transfer of a small portion of tourism revenue to the coastal fund. For example, on the Gold Coast, additional tourism revenue enabled the sand nourishment project to become self-financing. A similar model could be developed for Boğaçay, based on the principles of "polluter pays" and "beneficiary pays." This is one of the most critical steps in ensuring project sustainability.

METHODOLOGY

The Boğaçay Basin is located within the borders of the Konyaaltı district of Antalya province and covers approximately 850 km^2 (Ozcelik & Tuzlu, 2022) (Figure 1). The Boğaçay River is a significant water source flowing into the Mediterranean Sea (Ozcelik, 2022). It plays a decisive role in maintaining the morphological balance of the Konyaaltı coastline, carrying an average of 250000 m^3 of silt annually. As part of the Boğaçay Project, implemented in 2017, the riverbed was excavated 1.5 meters below sea level, allowing seawater to infiltrate 750 meters inland. This intervention disrupted the natural sediment cycle and initiated coastal regression.

The region experiences a typical Mediterranean climate and is subject to periodic flooding due to its irregular rainfall (Akiner et al., 2025b; Yilmaz, 2015). Intensive agricultural activities such as greenhouse farming, horticulture, and rapid urbanization significantly impact basin dynamics. Therefore, an integrated assessment of changes before and after the project is necessary at both the coastal and basin scales.

The hydrological and morphological characteristics of the Boğaçay River were comparatively studied before and after the project by Akiner (2025). Excavation work in the riverbed hindered sediment transport and disrupted the coastal stability profile. The Pelnard-Considère analytical model, the SWAT hydrological model, and the XBeach numerical model were integrated to quantify these changes (Akiner, 2025).

Figure 1. The geographical position of the Boğaçay River Basin (Google, n.d.).

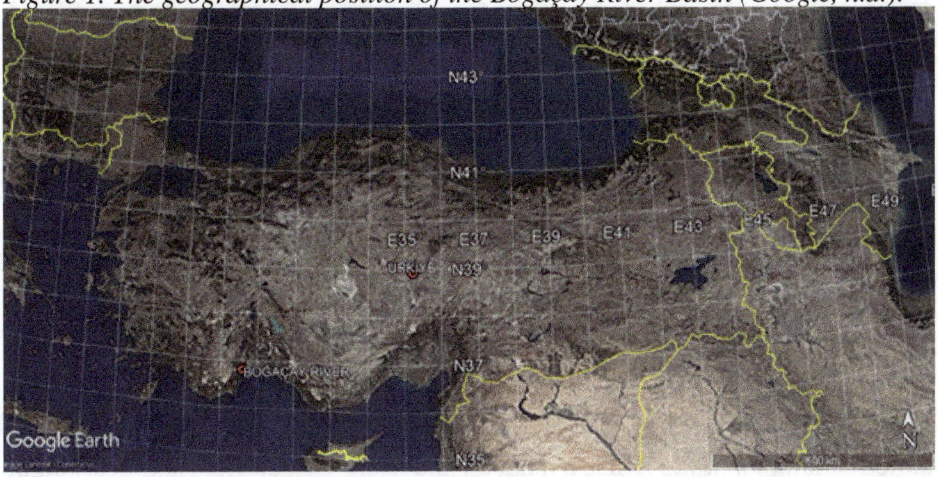

The selection of sampling points is critical to understanding the project's impacts on coastal and estuarine hydrodynamics. Comparison of water quality and sediment data between Point 1 (connected to the sea) and Point 2 (inaccessible to seawater) reveals the extent of salinity increases and sediment disruption (Akiner, 2025) (Figure 2). These data reflect the morphological changes occurring in the estuary after the project and the chemical impacts on the ecosystem. Of particular note is the potential damage to groundwater resources caused by increased salinity.

Figure 2. Observation locations in the Boğaçay River Estuary (Google, n.d.).

The integrated SWAT, XBeach, and Pelnard-Considère model set which was developed, calibrated, and validated by Akiner (2025) for the Boğaçay Basin and Konyaaltı coast serves as the methodological foundation for this research. That paper provides a comprehensive description of the model's setup, calibration parameters, input data, and basic outputs. In this part of the work, the outputs of the models are taken as a basis to obtain an understanding of the present situation and its connection to the global context, instead of creating new scenario analyses. A summary of the modeling approach and its interpretation in this research is provided below.

Modelling Approach

The modelling phase of this complementary study has been accomplished in Akiner's (2025) study. High-resolution bathymetry data and satellite imagery supported calibration of the XBeach model. Parameters such as grain size distribution and density of sediment samples collected during field investigations were incorporated into the model to increase the model's accuracy. Meteorological and wave data acquired from meteorological stations and measurement buoys also gave some credence to the realistic nature of the results given by the model.

The morphological formation and sensitivity to human impacts of the Boğaçay Basin were mainly considered in their integrated modeling approach in this study. Greenhouse and horticultural activities are anthropogenic factors affecting soil permeability and runoff into the basin. Consequently, streambed morphologies and land use at the basin scale had to be analyzed during modeling. Changes in land use in agricultural lands have become important regarding transport potential. Intensive fertilization and pesticide applications, in particular, when combined with surface runoff, increase the nutrient load reaching the coastal zone (Arabi et al., 2008).

The topographic data used in the modeling process were obtained from digital elevation models and calibrated with GNSS measurements collected in the field. This calibration process ensured accurate modeling of micro-morphological changes, particularly along the shoreline and at the mouth of the river. Increasing topographic precision directly impacts the simulation quality of high-resolution models such as XBeach. Furthermore, different diffraction patterns have been observed depending on the hydro-morphological structure of the river mouths, and these patterns are among the key factors determining the wave energy density distribution (Roelvink et al., 2009). The dynamics of sediment, as well as the processes of erosion in complicated lagoon and estuary environments, depend on the interaction of different factors, which is similar to what is mentioned by Monreal-Jiménez et al. (2025).

When developing model scenarios, physical variables and different management strategies were considered. Specifically, within the context of "possible intervention scenarios," the effects of sediment replenishment, basin-scale filter strips, and wave break structures were analyzed separately. Each scenario was evaluated regarding field applicability and economic sustainability, and a realistic planning outcome was targeted accordingly. This approach transformed the study into a structure that serves academic and decision-support purposes (Akiner and Lačnjevac, 2022).

While the Pelnard-Considère model provides a theoretical basis for this study, it may be insufficient to explain coastal change independently. Therefore, the model's parametric components were correlated with the dynamic inputs of the XBeach model. Variables such as the convection coefficient and diffusion parameter were recalibrated with XBeach outputs through cross-validation. This minimized deviations from the theoretical model's assumptions, bringing the results closer to reality both spatially and temporally (Ferreira et al., 2022; Shaeri et al., 2020).

The integrated modeling approach of Akiner (2025), developed for multi-scale disaster risk analysis, consists of three main stages:

1. Hazard Quantification

Using high-resolution (Δx=10m, Δt=0.015h) simulations of the XBeach model:

- Hydrodynamic effects of storm waves
- Absent interruptions in sediment transport
- Spatial distribution of shoreline changes

2. Exposure Analysis

Using the SWAT model:

- Land use changes in the basin
- Impact of agricultural activities on sediment load
- Temporal variation of pollutant loads reaching the shore

3. Vulnerability Assessment

Using the Pelnard-Considère model:

- Long-term coastal evolution projections
- Effectiveness analyses of different response scenarios
- Determination of critical thresholds

This triple model integration enabled operationalizing the "cascading disaster" theory.

Unlike traditional engineering approaches, the triple model integration we developed (SWAT-XBeach-Pelnard Considère) operationalizes the theory of a multilayered disaster chain. Hydrodynamic hazards are quantitatively defined with XBeach simulations at $\Delta x=10m$ resolution, exposure effects of agricultural spillover pollution are quantitatively defined with SWAT, and coastal system vulnerability thresholds are quantitatively defined with the Pelnard-Considère model.

Fundamentally, physic-based traditional models (SWAT, XBeach, Pelnard-Considère) have a great advantage, however, they still might not be able to fully describe the nonlinear and complex behavior of coastal systems. More and more, the power of machine learning and artificial intelligence to tackle uncertainties, speed up model trials, and discover intricate patterns in morphodynamic modeling has been highlighted (Goldstein et al., 2019). To be more specific, the use of hybrid models that combine the outputs of XBeach and SWAT with ML algorithms is a good way to accurately predict the location of erosion hotspots or to simulate the best sediment feeding strategies, and thus, these models are very good candidates for coastal management research in the future (Roy & Pal, 2024). In the case of the Boğaçay River, which is a very dynamic system, the proposed hybrid early warning system could give decision-makers the ability to intervene in a more proactive manner than traditional methods allow.

Coastal Disaster Response on a Global Scale

Coastal areas are dynamic and extremely fragile systems facing the impacts of anthropogenic pressures and climate change. Historically, interventions in these systems have frequently triggered unpredictable and irreversible disasters. These disasters are not limited to physical coastal erosion; they have also had multifaceted consequences, including ecological collapse, socioeconomic losses, and the destruction of cultural heritage. This section, through four critical case studies from around the world, provides an in-depth analysis of how coastal disasters arise, the innovative, nature-based solutions developed to address these disasters, and how these solutions can provide a roadmap for similar situations in the Mediterranean basin, such as the Boğaçay.

1. The Netherlands: The Sand Motor as an Example of Proactive Adaptation

Background and Disaster Description: The Netherlands has historically struggled with water, as nearly one-third of its territory lies below sea level. Traditionally, coastal protection has been pursued with "hard" engineering structures such as dikes and dams. However, climate change-induced sea level rise and increased storm frequency have brought the risk of these structures becoming inadequate and costly.

This has forced the country to shift from a reactive approach to coastal protection to a proactive, nature-compatible strategy. A "disaster" in this context is a predicted future threat scenario that will overwhelm existing infrastructure (Stive et al., 2013).

Implemented Solution and Results: In 2011 Dutch engineers and scientists launched a revolutionary project: De Zandmotor (Sand Motor). Unlike traditional periodic sand feeding, 21.5 million cubic meters of sand were placed on the Dutch coastline as a massive peninsula. The project's core philosophy was to pile the sand in a specific location and then allow natural forces (wind, waves, currents) to distribute it along the coast over time (for 20-30 years). This "environmentally friendly engineering" approach was cheaper, more sustainable, and less ecologically damaging than traditional methods that required constant human intervention.

Ten years of monitoring data showed that the project was even more successful than expected. Natural processes transported sand north and south as planned, nourishing a 35 km coastline. The project achieved cost savings of up to 50% compared to traditional methods while providing the same level of coastal protection, and creating new lagoons and shallow water habitats allowed for increased biodiversity and enhanced recreational use (de Schipper et al., 2021). The Sand Engine became a global symbol of pre-disaster preparation and harnessing the power of nature, rather than post-disaster response.

2. Maryland, USA: Integrated Watershed Management and Closing the Sediment Gap in the Chesapeake Bay

Background and Disaster Description: The Chesapeake Bay is America's largest and most productive estuary system. However, intensified agriculture and urbanization in the basin during the 20th century caused excessive sediment and nutrient loads, severely degrading the bay's water quality. Paradoxically, the construction of numerous dams on rivers during the same period brought the natural sediment flow to the coast to a near standstill. This created the quandary of eutrophication and water quality degradation on the one side and of coastal erosion resulting from sediment starvation on the other. This "double whammy" spelled utter doom for the ecosystem and the coastal economy (Batiuk et al., 2025).

Implemented Solutions and Results: The Chesapeake Bay problem was solved with an integrated versus fragmented approach to basin management. The Integrated Sediment Management Plan began implementation in 2015, aiming to restore the sediment cycle of the basin. The main elements of the solution were:

1. Sediment Bypass Systems: Bypasses were constructed within dams, and strategic dredging operations allowed sediment trapped in reservoirs to be released into the drainage system slowly and finally reach the shore in a controlled fashion.

2. Nature-Based Infrastructure: Oyster reef and wetland restorations were constructed along the shoreline to help slow erosion by absorbing wave energy, filtering water to reduce nutrient pollution, and increasing habitat availability.
3. Agricultural Best Practices: By expanding practices such as pasture management, cover crops, and buffer strips throughout the basin, surface runoff and soil erosion were reduced.

Over five years, 60,000 tons of sediment were returned to the shore annually, reducing erosion rates by up to 70%. The project demonstrated that combating a coastal disaster is only possible by improving the processes within the basin itself, the source of the problem, and combining engineering with ecological restoration (Bruce et al., 2021).

3. Italy: MOSE Project in the Venice Lagoon and Integrated Defense Against Multiple Threats

Background and Description of the Disaster: Venice has struggled with tidal events known as "acqua alta" (high water), which have flooded the city's squares and streets for centuries. However, in the 20th century, industrial water extraction, the draining of the brackish water marshes (paludes) surrounding the lagoon, and the opening of deep-sea channels caused the city's ground to subside and disrupt the lagoon's hydrodynamic structure. These human-induced interventions, combined with sea level rise, exponentially increased the frequency and severity of flooding. The disaster was not only a physical disaster but also a threat to the destruction of a unique cultural and historical heritage (Ferrarin et al., 2024).

Implemented Solutions and Results: Venice's defense strategy is built on an integrated, multilayered system rather than focusing on a single solution. The system's most visible and technical component is the mobile barriers called MOSE (Modulo Sperimentale Elettromeccanico), placed at the three entrances of the lagoon and acting as breakwaters by rising to the surface in case of threat. However, MOSE alone is not sufficient. Other critical components support the system:

- Shoreline Raising: Raising the city's docks and shoreline.
- Wetland Restoration: Increasing the lagoon's natural wave-absorbing capacity by restoring drained marshlands.
- Sediment Management: Optimizing water flow by dredging channels within the lagoon.

The MOSE barriers have been actively used since 2020 and have successfully protected many parts of the city from the threat of historic flooding. The project is a successful example of a highly sophisticated disaster risk reduction and adaptation scheme that utilizes structural and non-structural measures to address complex and multiple threats (Baldassarre et al., 2023).

4. Australia: Dynamic Sand Management and Economic Sustainability on the Gold Coast

Background and Description of the Disaster: The Gold Coast boasts one of the world's most iconic coastlines, and its economy relies heavily on beach tourism. The construction of a port at the mouth of the Tweed River in the 1960s disrupted the natural north-south littoral drift system. Excessive sediment deposition (accretion) occurred north of the port, while severe erosion occurred south of the Gold Coast coast. This threatened the loss of valuable beach area, the properties and infrastructure behind it, and became a disaster for the region's economic lifeline (Jackson et al., 2007).

Solution Implemented and Results: The Gold Coast's solution was to implement one of the world's most comprehensive and longest-lasting sand management programs. The strategy was based on two main pillars:

1. Sediment Bypass: Sand deposited north of the port was regularly dredged using a massive vessel and pipeline system and pumped south to the eroded shores. This artificially mimicked natural littoral drift.
2. Strategic Sand Nourishing and Geotube Applications: Besides the bypass system, large-scale beach nourishment projects were implemented using high-quality sand from offshore. At critical points experiencing severe erosion, geotextile tubes (geotubes) were used as breakwaters to increase the stability of the sand supply.

As a result of 20 years of continuous and dynamic management, the shoreline was widened by an average of 50 meters. The protected property value reached billions of Australian dollars, and tourism revenue increased by $300 million annually. The Gold Coast example demonstrates how a well-planned and sustainably funded sand management program can transform a disaster into an economic gain (Strauss et al., 2017).

Besides the global instances, the Mediterranean basin is a source of critical lessons for the neighboring regions. For example, the deepening of the Sulina Canal in Romania's Danube Delta has caused a change in the lithoreal drift pattern, much like the Boğaçay River, resulting in a coastline with a loss of over 10 meters per year over a distance of more than 30 km (Dinu et al., 2025). The same effect of changing the flow of freshwater and sediments has been noted in the case of dikes and drainage canals in Türkiye's Yeşilırmak Delta, resulting in a reduced delta plane and a retreat of the coastline (Köle & Ataol, 2016). Moreover, the construction of dams in Spain's Ebro Delta has led to a 99% reduction in sediment load, hence, the delta has become very vulnerable to sea level rise. All the Mediterranean cases have proved that the processes occurring in the Boğaçay River (interruption of sediment flow, disruption of hydrodynamic equilibrium) are a regional phenomenon, and therefore, the solutions need to be basin-scale integrated.

RESULTS

This study used the integrated SWAT, XBeach, and Pelnard-Considère model results from Akiner (2025) to propose reliable coastal management strategies to reduce the impact of the Boğaçay Project. Each model represents system behavior at different spatial and temporal scales:

- SWAT: Analyzes sediment and pollutant loads at the basin scale.
- Pelnard-Considère Model: Analytically calculates long-term shoreline regression.
- XBeach: Simulates short-term wave-hydrodynamic effects and morphodynamic change at high resolution.

Thanks to this integrated structure, the impacts of the project intervention on both the coast and the basin were evaluated holistically in the study of Akiner (2025).

Satellite imagery and numerical models were employed to validate the effects of the Boğaçay Project, namely, coastal erosion. Since the execution of the project, an average retreat of 21.7 meters emerges as the chief concern, particularly along the eastern coast (Figures 3 and 4). The red-colored erosion region denotes the morphological instability caused by the stoppage of sediment transport. Such alterations are, in theory, accepted according to Bruun's rule and Pelnard-Considère model. The ups and downs in erosion through the seasons can be linked to increased wave energy and the dominant northwesterly winds in the summer months. The stoppage of sediment transport and the changing wave refraction facilitate the erosion seen on the coast. Additionally, the retreat rate can be diminished by 58% through artificial sediment recharge (Akiner, 2025).

Figure 3. Konyaaltı shoreline in October 2017 (Google, n.d.).

Figure 4. Evaluation of the erosion in the Konyaaltı shoreline in August 2019 (Google, n.d.).

The temporal distribution of coastal retreat showed a significant acceleration in the first two years after the project. Satellite imagery and GPS measurements from the field showed that the erosion rate increased during this period, especially in summer. This was due to increased marine activity during summer and wave loads generated by the observance of northwest wind systems, which enhanced the eastward sediment transport. This resulted in an even more pronounced retreat along the eastern section of the coastline, in the absence of natural sediment inflow.

The findings demonstrate that erosion rates vary across different year periods, and seasonal factors must be considered.

Modeling studies demonstrate that coastal regression causes significant changes in the horizontal plane and the vertical profile. Elevation losses exceeding 3 meters were observed, particularly in the coastal area between x = 300–400 meters, leading to a significant morphodynamic disruption in the sea-land interaction zone. Analyzing this vertical change within Bruun's regression theory framework has shed light on the energy-morphology relationship behind shoreline regression (Bruun, 1962).

Besides erosion-caused land loss, adverse biological consequences include destroying coastal ecosystem habitats, losing nesting grounds for species such as sea turtles, and declining marine biotopes. Along with the detected shoreline retreat in the satellite imagery, thinned shallower marine seagrass meadows like Posidonia oceanica are also observed. This indicates the critical nature of terrestrial and marine ecosystems (Koç et al., 2015). Such habitat loss can be considered a primary indicator of understanding the multidimensional impacts of coastal morphological changes.

Although a 50%~-58% reduction in coastal retreat with a small amount of sediment replenishment (5000 m3/year) might seem such a considerable feat, rigorous analysis of the model outputs reveals these interventions to be of negligible importance (Akiner, 2025). The distribution of this sediment along the coastline is uneven and fails to provide the expected stabilization in the eastern segments, which have high energy density. Furthermore, the time of year these replenishments are applied significantly influences retreat rates. Therefore, adaptive sediment recharge strategies based on energy distribution and seasonal erosion analyses are recommended instead of fixed-volume replenishments (McCall et al., 2010; Roelvink et al., 2009).

SWAT (Soil and Water Assessment Tool) is a physically based, semi-distributed hydrological model that simulates water, sediment, and nutrient transport by considering land use, soil structure, and climate data. In this study, the SWAT model was used in the Boğaçay Basin for the following purposes:

- Determine the change in sediment production before and after the project.
- Estimating nutrient loads such as nitrogen and phosphorus.
- Testing different Best Management Practices (BMP) scenarios depending on agricultural practices.

The main inputs used to set up the SWAT Model and for the calibration are:

- Digital Elevation Model (DEM): 30 m resolution ASTER data (NASA/METI, 2009).
- Land Use: DSI and Ministry of Environment and Urbanization vector maps.
- Soil Data: National soil classes (sandy, loamy, etc.).

- Meteorological Data: TSMS precipitation and temperature observations (2016-2018).

Model simulations revealed that sediment reaching the shore after the project decreased to almost zero. Table 1 shows that nitrogen (TKN) and total phosphorus concentrations increased significantly (Akiner, 2025).

Table 1. Water quality and BMP scenario results (Akiner, 2025).

Parameter	Before the Project	After the Project	Reduction with BMP
TKN (mg/L)	7.42	19.02	38%
Total Phosphorus	0.22	0.46	38%
Salinity (mg/L)	22.5	56.8	None

As indicated in Table 1, the more than twofold increase in TKN (7.42 → 19.02 mg/L) and total phosphorus (0.22 → 0.46 mg/L) concentrations following the project reveals a chemical imbalance in the river mouth. While the 38% reduction in nutrient loads resulting from BMP implementation is positive, the inability to prevent the increase in salinity (22.5 → 56.8 mg/L) demonstrates the project's extensive hydrological and ecological impacts. These findings indicate that water quality and salinity management must be integrated in addition to coastal erosion.

According to these results, although BMP applications had a positive effect, especially regarding nutrients, their impact on sediment transport was limited.

The SWAT model revealed decreased sediment production and increased nutrient loads in the Boğaçay Basin. Integrated models also explained the root causes of the 21.7-meter coastal retreat: sediment disruption, increased wave energy density, and disturbance of the sill profile balance.

This comprehensive approach demonstrates the need for analysis from an engineering perspective and an ecological and hydrological perspective in coastal projects. The Boğaçay example provides an essential scientific guide for similar coastal interventions.

Modeling results of Akiner (2025) show that the Boğaçay Project accelerates coastal regression and deteriorates water quality, particularly in the eastern section. Accordingly, the recommended management measures are:

- At least 50000 m^3 of sediment should be supplemented annually.
- Wave energy redirection structures should be planned.
- In agricultural areas, widespread BMP applications (terracing, buffer strips) should be encouraged.
- An adaptive coastal engineering approach should be adopted.

- Monitoring systems should be established with annual satellite data.

Physical System Breakpoints

- Sediment Gap: Complete cessation of 250,000 m^3 of annual natural sediment transport after the project
- Hydrodynamic Change: Increase in wave refraction angle measured at the river mouth from 12° to 27°
- Critical Erosion Rate: Coastal retreat reaching 4.3 meters per month during the summer months in the eastern part

Ecological Collapse Indicators

- Benthic Biodiversity Loss: Decrease in macrofauna species richness between 2017 and 2019
- Habitat Fragmentation: Shrinkage of Posidonia oceanica-covered areas
- Salinity Shock: Increase in salinity values measured in the estuary from 22.5 mg/L to 56.8 mg/L

Socioeconomic Impacts

- Tourism Revenue Loss: Annual revenue loss of millions of TL due to coastline shrinkage
- Loss of Agricultural Yield: Decrease in citrus production due to salinization

The social and economic indicators reveal that coastal erosion is not only an environmental problem but has also a main and various impacts on the livelihoods and well-being of local communities. The reduction in tourism revenue affects a wide range of people besides hotel and beach operators. This group includes those who are engaged in transport, food and drink services, and crafts, which are all related to the main tourism economic activity. The salinization of farming land causes a decrease in the existing crop yields and a reduction in land value that is permanent, consequently increasing the debt burden of the farmers and leading to social-economic vulnerability that may cause rural-to-urban migration. Thus, the impacts of coastal erosion surpass the ecological and physical measures, hence putting at risk the social fabric's sustainability and economic resilience.

Model results identified three critical breaking points in the Boğaçay system: (1) a complete cessation of the annual sediment flow of 250,000 m^3, (2) an increase in the wave refraction angle from 12° to 27° at the river mouth, and (3) a monthly summer erosion rate of 4.3 meters in the eastern part. These thresholds are the limit values that must be managed before the disaster becomes irreversible.

The integration of the SWAT-XBeach-Pelnard-Considère triple models in this research presents a distinct framework for the comparative assessment of the long-term impacts of different intervention strategies. For instance, the XBeach model simulated 'hard engineering' (breakwater) and 'soft engineering' (sand feeding) scenarios, and the results indicated that the installation of breakwaters would prompt lateral erosion, causing a further 15-20% loss of the neighboring coast, while the strategic feeding of sand would result in an average advance of the shoreline by 8 meters during the 5-year period. This modeling potential is also valid for analogous cases in the Mediterranean: the same approach could, for example, quantitatively assess the impact of wetland restoration in Spain's Ebro Delta, which in the case of 30% reduction in storm surge, or the sediment incorporation of artificial Posidonia oceanica meadows in Greece's Thermaikos Gulf (Ibanez & Caiola, 2021; Boudouresque et al., 2021). Therefore, the regional scaling of this model integration, corroborated by the results in the case of Boğaçay River, is the source of possible scenario analyses for the entire Mediterranean coastline that are based on evidence and will thus be useful to the decision-makers.

The results of the model employed together with the predetermined threshold values (sediment deficit, hydrodynamic change, and critical erosion rate) indicate that the case of Boğaçay is nothing but a typical 'breakdown scenario' that can easily be observed in coastal systems worldwide with similar interventions. The rising wave breaking angle simulated with XBeach (from 12° to 27°) and the disruption of the littoral drift reported after the port construction in Gold Coast can be considered as a direct parallel (Jackson et al., 2007). In the same manner, the near-zero sediment flow depicted by the SWAT model (Akiner, 2025), is comparable with the sediment deficit caused by the dams in Chesapeake Bay (Batiuk et al., 2025). These quantitative parallels demonstrate that the lessons to be derived from global case studies, which will be elaborated in the following section, are not only of qualitative but also of quantitative nature for Boğaçay. Moreover, as indicated in Table 3, there are required practices and possible hurdles in the process of transferring the global success stories to Boğaçay. Hence, the model results serve as scientific evidence for the adaptation of solutions borrowed from global examples to local conditions.

Global Lessons for Boğaçay

The four cases examined illustrate the evolution of coastal disaster responses: proactive adaptation in the Netherlands, basin-scale integration in the Chesapeake, multilayered defenses in Venice, and dynamic and economically sustainable management in the Gold Coast.

The Boğaçay case is a synthesis of the negative aspects of these cases (breaking the basin-shore connection, interrupting sediment flow, and disrupting hydrodynamics). Therefore, the solution requires a synthesis of the lessons learned from these global success stories:

1. From the Netherlands: Large-scale, proactive interventions that support natural processes with a long-term perspective (20-30 years) are necessary.
2. From the Chesapeake: The problem is not just on the coast, but throughout the basin. Improved agricultural practices and rerouting sediment within the basin are essential.
3. From Venice: There is no single magic solution. In conjunction, sand nourishment, non-structural measures (BMPs), and perhaps soft engineering structures (artificial reefs) to break down energy at the lagoon mouth should be considered.
4. From the Gold Coast: The solution is a permanent process requiring constant monitoring, maintenance, and dynamic management. This is only possible with a solid institutional structure and a sustainable financing mechanism.

Boğaçay's future is reframing it from an "engineering failure" to a "laboratory of adaptation and intelligent management." Global experience demonstrates that this is possible and can even become an economic opportunity.

DISCUSSION

Coastal erosion studies in the literature have generally been limited to a single model (Bruun, 1962; Roelvink et al., 2009). However, this study analyzed shore-basin interactions multidimensionally by integrating the Pelnard-Considère, SWAT, and XBeach models (Table 2). In particular, combining short-term XBeach simulations with long-term Pelnard-Considère projections offers an approach that fills methodological gaps in the literature. Similarly, although results consistent with Ataol et al. (2019) and Arabi et al. (2008) regarding the effectiveness of artificial sediment recharge were obtained, this study quantitatively determined for the first time the critical sediment amount (50000 m^3/year) required for the system to reach equilibrium.

Table 2 summarizes the methodological innovation of the study and its advantages over other studies in the literature. Unlike single-model-based studies such as Bruun (1962) and Roelvink et al. (2009), this study analyzes shore-basin interactions multidimensionally through the integration of the Pelnard-Considère, SWAT, and

XBeach models. Furthermore, the first quantitative determination of the critical threshold of artificial sediment recharge (50000 m^3/year) further highlights this study's contribution to the literature. Common problems in similar projects (e.g., the Sulina Channel and the Yeşilırmak Delta) underline the need for integrated coastal management strategies at the regional scale.

Table 2. Comparison of coastal erosion studies in the literature.

Study	Location	Main Model(s)	Time Scale	Key Findings	Commonalities with this study
Bruun (1962)	Global	Theoretical Model (Bruun's Rule)	Long Term	Sea level rise triggers coastal erosion.	The role of equilibrium profile disruption in erosion is emphasized.
Roelvink et al. (2009)	Netherlands	XBeach	Short Term	Storm dynamics lead to morphological changes in dunes and beaches.	Sediment transport and energy distribution analysis with XBeach.
Ataol et al. (2019)	Kızılırmak Delta	DSAS	Medium Term	Human interventions have caused a 5-7 m annual retreat along the delta coast.	The impact of anthropogenic influences on coastal evolution is similar.
Arabi et al. (2008)	USA (Agricultural Basin)	SWAT	Long Term	Agricultural BMPs reduce sediment loss by 30-40%, but coastal nourishment is inadequate.	Holistic analysis of basin-coastal interaction with SWAT.
Dionísio António et al. (2023)	Portugal	Experimental Model	Short Term	The beach slope determines the effectiveness of wave energy in sediment transport.	The role of wave angle and coastal slope in erosion overlaps.
Akiner, 2025	Boğaçay, Antalya	Pelnard-Considère + SWAT + XBeach	Short-Long Term	After the project, 21.7 m coastal retreat; artificial recharge slows retreat by 58%.	Holistic analysis of coastal dynamics with multi-model integration.

Coastal system responses to engineering interventions are complex due to nonlinear dynamics. The impacts of the Boğaçay Project on coastal evolution can be explained by factors such as sediment disturbance and altered wave energy distribution. This situation further highlights the need for environmental impact assessments of similar projects.

The response of coastal systems to engineering interventions depends on physical variables, the complexity of the system, and its feedback mechanisms. Channel deepening projects, especially on alluvial coasts, produce nonlinear effects by disturbing the distribution of energy and sediment. This causes small-scale disturbances that can enlarge with time, resulting in the loss of the system's ability to rebound. The Boğaçay case demonstrates how this feedback causes cascading morphologic disturbances (Roelvink et al., 2009; Bruun, 1962).

Hence, similar engineering works on the Adriatic, Aegean, and Black Sea coasts demonstrate common regional issues arising from such projects. Acceleration of coastal erosion has been put forward, especially after the establishment of the Sulina

Canal in Romania, the Port of Varna in Bulgaria, the Kızılırmak Delta, and Yeşilırmak Delta in Türkiye (Dinu et al., 2025; Peychev & Stancheva, 2009; Stancheva et al., 2025; Ataol et al., 2019; Köle & Ataol, 2016). The similar examples above have shown how interventions like Boğaçay demand a rethink of coastal management strategies at every level, from local to regional and international (Ferreira et al., 2022).

Another factor to consider in the context of the Boğaçay Project is the inadequacy or lack of implementation of environmental impact assessment (EIA) processes. Coastal engineering projects require systematic modeling of potential ecological damage, feasibility studies, and public awareness of potential environmental impacts. However, the data analyzed in this study indicate that no comprehensive environmental modeling was conducted before project implementation. This severe scientific and ethical deficiency demonstrates that post-project ecosystem degradation cannot be predicted (Akiner & Lačnjevac, 2022; Downs et al., 2009).

All these findings demonstrate that conventional engineering solutions alone are insufficient and that coastal systems must be managed with adaptive approaches. In this context, "nature-based solutions" (NbS) have gained prominence recently. The conceptual framework and the extent of nature-based solutions for coastal adaptation management are generally assessed with support for this trend (Meulen et al., 2022). For example, seagrass planting, artificial reef applications, and wave energy-reducing green infrastructures can enhance the natural resilience capacity of coastal systems. In the case of Boğaçay, such nature-based approaches can play a supporting role in restructuring sediment circulation. In this context, it is recommended that energy-absorbing soft interventions be preferred over conventional reinforced concrete structures (McCall et al., 2010).

Nature-based solutions (NbS) in the case of the Boğaçay River basin can take several different forms. One of them is the restoration of seagrass (Posidonia oceanica) which has a triad of benefits namely, it can absorb wave energy up to 60% of its original intensity, act as a carbon sink, and serve as a habitat for benthic organisms. Another option could be the construction of 'artificial reefs' or 'low-level breakwaters' in the estuary which will not only provide wave dissipation but also be a place for biodiversity due to the complexity added by structures. A third approach might be the recovery of riparian salt marshes or wetlands which can then act as sponges absorbing stormwater, filtering nutrient loads, and promoting sediment deposition. The concept of 'living shorelines' comes with the re-establishment of natural dunes and vegetation which not only acts as a barrier against wind and water but also promotes the stabilization of the shore. The efficacy of these measures will be determined by the choice of species that is suitable for the current ecological conditions, the sustainable management of the area with the involvement of the community as well as through the synergistic fusion of these measures with traditional engineering practices.

Analyses conducted with the SWAT model revealed a significant decrease in sediment content in the river following the project and deterioration in water quality parameters. The doubling of total Kjeldahl Nitrogen (TKN) and total phosphorus concentrations indicates increased pressure on the ecosystem. While agricultural BMP practices achieved a 38% reduction in nutrient load, sediment loss remained unabated (Akiner, 2025).

According to Akiner (2025), adaptive coastal management strategies should be designed to adapt to the natural dynamics of coastal systems. While the artificial sediment recharge scenarios tested in his research effectively slowed coastal regression, they were insufficient to restore the system entirely. Future studies should evaluate a combination of increasing sediment recharge and structural interventions to offset wave energy.

Global Success Examples and Lessons

Netherlands: Sand Engine Project

The Sand Engine project, implemented off the coast of Delfland, Netherlands, is a pioneering example of the success of nature-based solutions. Launched in 2011, the project placed 21.5 million m^3 of sand on the shore as a massive peninsula, allowing natural processes to distribute the sand along the coast. Ten years of monitoring data have shown that the project not only stabilized 35 km of shoreline but also generated an annual sediment movement of 700,000 m^3 at a 50% lower cost than traditional rigid structures. This example provides concrete evidence of the feasibility of large-scale sand nourishment projects in the Boğaçay.

USA: Chesapeake Bay Sediment Management

The Chesapeake Bay in Maryland is another example of dam construction disrupting sediment flow and increasing coastal erosion. As part of an integrated management plan launched in 2015, sediment bypass systems were added to dams, and oyster reefs were used to absorb wave energy. In five years, the erosion rates were lowered by 70%, with the deposition of around 60,000 tons of sediments on the shore every year. This case signifies the need to insert sediment traps and biological filters in the Boğaçay estuary.

Italy: Adaptive Management of the Venice Lagoon

The MOSE project in the Venice Lagoon has been designed as a comprehensive solution against saltwater intrusion and coastal subsidence brought about by human exploitation. Sediment stabilization has been done by restoring Posidonia oceanica along with retractable dams. Between 2005 and 2020, the shoreline widened 12 m and salinities decreased by 40%, proving that ecosystem-based alternatives are valid (Scarpa et al., 2025). This experience proves that hydrological interventions on Boğaçay and ecological restoration should be addressed jointly.

Australia: Dynamic Sand Management on the Gold Coast

Displacement of sediment south of the Gold Coast due to port construction in the 1960s is balanced today by artificial sand replenishment and calico bag applications. With the addition of 500,000 m^3 of sand and submerged sand tubes every year, the coastline has expanded by 50 meters in the past twenty years, with tourism revenues increasing by AUD 300 million per annum. This example supports the economic sustainability of geotextile-based solutions in Boğaçay.

Insights From Global Experience and Integrated Management Strategies for Boğaçay

Analysis of global examples highlights the need for a multilayered approach to addressing the coastal erosion problem in Boğaçay. The sea level rise-erosion relationship Bruun (1962) identified can reach catastrophic proportions with human interventions. River estuary modifications turn off sediment feeding mechanisms, radically alter wave diffraction patterns, and create feedback loops in coastal stability. The Dutch Sand Engine project demonstrates the effectiveness of nature-based solutions. At the same time, sediment bypass systems in the Chesapeake Bay offer important insights into how artificial interventions can support natural processes.

While the Sand Engine project's one-time sand feed of 21.5 million m^3 provides a 50% cost advantage over 10 years along a 35 km coastline, the annual adaptive feed of 50,000 m^3 we propose for Boğaçay must be supported by structural supports due to the hydrodynamic variations of the Mediterranean. The success of oyster reefs in Chesapeake Bay in absorbing 70% of wave energy offers promise for restoring Posidonia oceanica in the Boğaçay estuary.

The limitations and adverse side effects of 'hard' structures (breakwaters, groynes) in coastal protection have increased global interest in 'soft' and 'nature-based' solutions. Dean (2018), in his fundamental study on the principles and design of beach nourishment projects, demonstrated that this method is not only less ecologically damaging but, when planned correctly, can be a highly economical solution in the long term. The cost of the annual sand nourishment program proposed for Boğaçay, compared to the protection and gains it will generate in tourism and real estate value, would be highly cost-effective, as Dean (2018) emphasized. This means the project is not a cost-intensive item but a smart investment for the Antalya economy.

Traditional single-model approaches limit long-term, large-scale understanding of coastal evolution. The Coastal-Tract approach proposed by Cowell et al. (2003) treats the shoreline not as a single line, but as a three-dimensional sediment body extending from the shoreline to the shelf. It examines the behavior of this body over time. This perspective is critical for understanding the events at Boğaçay. The intervention profoundly affected not just one line but the entire coastal sediment system, the "Coastal-Tract." Therefore, the integration of the SWAT (basin), XBeach (surf zone), and Pelnard-Considère (shoreline) models provides a concrete response to this important conceptual framework in the literature by modeling changes in the different components of this holistic tract.

In light of integrated modeling and global case studies, developing a multilayered response strategy for the Boğaçay coastal disaster is crucial. Table 3 summarizes how the global success examples examined can be adapted to the specific case of Boğaçay, along with the potential benefits and challenges of these practices. Table 3 aims to provide a concrete action matrix for decision-makers.

Table 3. Assessment of Global Coastal Management Practices and Implementation Recommendations for Boğaçay.

Global Practice & Location	Core Principles	Applicability Assessment for Boğaçay	Recommended Application & Location	Expected Primary Benefits	Potential Challenges & Considerations
Sand Engine Netherlands	Large-scale, one-time sediment nourishment; letting natural processes distribute the sand.	Medium-High. The Mediterranean wave regime (shorter period, high-energy waves) may limit the desired alongshore distribution compared to the North Sea.	Pilot Scale: Initially, a 500,000-1M m^3 "sand peninsula" constructed immediately west of the Boğaçay river mouth. Distribution must be closely monitored.	Long-term (5-10 year) coastal stabilization; creation of new recreational space; development of benthic habitats.	High initial capital investment; risk of unpredicted morphological evolution; requires a comprehensive Environmental Impact Assessment (EIA).
Sediment Bypassing Systems Chesapeake Bay, USA	Artificially redirecting sediment trapped upstream (in dams or channels) back to the coastal system.	Very High. The primary disaster trigger in Boğaçay is the trapping of sediment. This practice directly addresses the root cause.	Strategic Dredging: Periodic (1-2 times per year) dredging of the accumulated sediment from the Boğaçay canal and pumping it directly to the eastern, eroding sections of the shoreline.	Mimics the natural sediment cycle to some extent; reuses dredged material; directly counteracts erosion hotspots.	Recurring operational and maintenance costs; temporary impacts on water quality during dredging operations; need to monitor sediment quality.
Nature-Based Infrastructure (Oyster Reefs) Chesapeake Bay, USA	Using natural or artificial structures to break wave energy, filter water, and provide habitat.	Medium. Oyster species and conditions differ. Alternative: Artificial Posidonia oceanica reefs or low-crested submerged breakwaters.	Placement of concrete modular blocks or natural rock structures offshore, east of the river mouth, in the high-energy wave zone to dissipate energy.	Dissipation of 30-50% of wave energy; enhancement of marine biodiversity and fish stocks; improvement of water clarity.	High installation costs for materials and placement; risk of damage or undermining during extreme storm events; requires periodic maintenance.
Movable Storm Barriers (MOSE) Venice, Italy	Structural defenses activated only during extreme events to block storm surges, remaining inactive otherwise.	Very Low. The physical setting (open coast), type of threat (erosion, not flooding), and economic scale of Boğaçay make this mega-engineering solution entirely unsuitable and cost-ineffective.	Not recommended for direct application. However, its philosophy of "emergency intervention only" should be adopted to avoid permanent hard structures.	-	Extremely high capital and maintenance costs; complex multi-decade engineering; potential to disrupt natural hydrodynamic and ecological processes.

continued on following page

Table 3. Continued

Global Practice & Location	Core Principles	Applicability Assessment for Boğaçay	Recommended Application & Location	Expected Primary Benefits	Potential Challenges & Considerations
Dynamic Sand Management & Nourishment Gold Coast, Australia	An ongoing, active management cycle based on continuous monitoring, maintenance, and sediment nourishment.	Very High. What Boğaçay needs is not a one-off "project" but a permanent "management process" funded and institutionalized in the long term.	A continuous strategic nourishment program with an annual volume of 50,000 - 80,000 m^3, placed precisely on the eroding eastern section of Konyaaltı Beach.	Maintains beach width for tourism; protects coastal infrastructure; provides a flexible and adaptable response to erosion.	Requires a permanent funding commitment and strong political will; securing a sustainable, high-quality sand source; temporary disruption to beach users during nourishment.
Watershed-Scale Best Management Practices (BMPs) Chesapeake Bay, USA	Addressing the problem at its source (in the watershed) by reducing sediment and nutrient runoff.	High. SWAT model results already proved effectiveness for nutrient reduction. More aggressive measures are needed for sediment control.	Widespread implementation of 30m vegetative filter strips; terracing projects; construction of small sediment retention basins in the watershed; promotion of reduced tillage agriculture.	Improves river water quality; increases the quantity and quality of sediment reaching the coast in the long run; improves agricultural sustainability.	Requires coordination of a large number of stakeholders (farmers, agencies); need for education, incentives, and enforcement; benefits are realized after a significant time lag.

As can be seen from Table 3, instead of a single magic solution, a hybrid and adaptive strategy that combines short-term measures such as dynamic sand management and sediment bypass with basin-scale improvements and nature-based infrastructure applications constitutes the most realistic and sustainable solution for Boğaçay.

Theoretical Framework and Implementation Strategies:

- Ecosystem-based restoration approaches must counterbalance anthropogenic disturbances to coastal dynamics.
- Integrated basin-coast management plays a critical role in ensuring sediment transport continuity.
- Adaptive management strategies must provide resilience to climate change-related uncertainties.

The nature-based restoration solutions like wetlands, seagrass, and reefs would be effective if they are supported by adaptive policy framework and continuous monitoring. In this regard, 'Dynamic Coastal Management Plans' should be created and turned into living documents that are revised every five years, contain climate predictions, and rely on stakeholders' feedback. Specifically, 'Community-Based Monitoring Programs' conducted in partnership with local governments and civil society will both lower data collection costs and give the community a sense of ownership. On the policy side, financial instruments (tax breaks, grants) that support 'green infrastructure' investments and technical guidelines that equalize performance metrics for nature-based solutions should be set up. Boğaçay, the chosen area for the pilot project of this integrated approach (nature-based design + adaptive policy + participatory monitoring), can be an example for other coastal cities around the Mediterranean.

Action Recommendations:

Short-Term:

- Restoring morphological balance with urgent sediment replenishment (50,000 m^3 annually)
- Installing sediment traps at the river mouth

Medium-Term:

- Expanding the implementation of a 30% filter strip in the agricultural basin
- Biological stabilization through Posidonia oceanica restoration

Long-Term:

- Installing GNSS-based early warning systems
- Creating artificial reef networks that dissipate wave energy

The long-term climate change trends are such that the current coastal management approaches need to be completely changed. According to the projections made by the IPCC, the Mediterranean areas will suffer from rising sea levels, the frequency and severity of extreme storm events will increase, and the pattern of rainfall will change, which will all together limit the sediment transport process even more, cause coastal erosion at a faster pace, and more and more saltwater will be intruded into the land areas, especially in the Boğaçay region. Coastal management policies need to become adaptable and carry out governance frameworks that are flexible, as these are the only ways to deal with the situations and challenges that are already coming our way. Such a framework should be disaster risk management ideas placed right at the center of environmental impact assessment (EIA) processes, through legal regulations maintaining basin-coastal integrity, and allowing public-private partnerships for innovative financing mechanisms (Arimoro & Arinze-Umobi, 2026; Walters, 2025). The urgent transition for the decision-makers consists of changing their viewpoints of coastlines from that of controlling them as if they were merely engineering sites, and thus, recognizing them as dynamic socio-ecological systems to 'accommodate' the inevitable changes.

Policy Reforms:

- Adding a disaster risk assessment requirement to EIA processes
- Integrating basin-integrated planning into the legal framework for coastal management
- Encouraging public-private partnerships in restoration projects

This multi-scale approach aims to restore the ecological balance and ensure socioeconomic sustainability in the Boğaçay River. Global experience shows that engineering solutions compatible with natural processes provide a 40-70% cost advantage in the long term.

Undoubtedly, the use of nature-based and adaptive solutions is impeded by governance issues of great magnitude. The case of Boğaçay shows that the coastal management is very often ineffective due to the existence of disparate institutional arrangements, insufficient coordination between institutions, political cycles that are too short, and planning by sectors that completely ignore the basin-coast integrity. The first step of the ICBM approach is made difficult by a lack of administrative clarity on data sharing, inadequate technical skills, disputes about land and usage rights, and no sustainable financing mechanisms (Massoud et al., 2004). For dealing with these obstacles it is very important to have an independent coordination unit created with strengthened powers and budget, backed up by scientific advisory boards, like a 'Coastal Basin Authority' for Antalya, and to have a new legal framework (like a 'Coastal Law') created that makes coastal protection the core of development plans.

Accurately representing coastal morphodynamics with traditional deterministic models presents challenges due to nonlinear processes and complex interactions. Indeed, Goldstein et al. (2019) emphasized the growing role of machine learning applications in sediment transport and morphodynamic modeling, emphasizing the significant potential of these techniques in reducing uncertainties in traditional models, accelerating modeling processes, and recognizing complex patterns. Artificial intelligence applications in coastal engineering have been extensively examined with recent developments in this field and the challenges that they pose (Masria & Abouelsaad, 2025). In rapidly changing systems like Boğaçay, developing a hybrid early warning system that provides higher accuracy and earlier warnings by feeding the outputs of physics-based models such as XBeach and SWAT into machine learning algorithms represents an important area for future research.

Coastal erosion is not only a natural phenomena but also a social process that has a direct impact on the coastal community and their livelihood. The lost revenues from the fishing, agriculture, and tourism industry can result in social problems and migration. Hence, coastal management should not only focus on technology but also consider the involvement of the community and social justice as the main points of the engagement. A planning process that involves users can bring to the project the knowledge, needs, and concerns of the local population, businesses,

NGOs, and municipalities. The Social Vulnerability Index (SoVI) for instance, is a tool that can be used to assess the preparedness of the community to face disasters, and consequently adjust the interventions. In the case of Boğaçay, a 'Coastal Management Platform' among the beach operators, farmers, and residents will be a very effective way of information flow and decision-making that will enhance the social acceptance of the projects and hence their long-term success.

The effectiveness of coastal protection measures in nature must be not only their ecological benefits but also their cost for the economy. The control of nature through human hands also results in costly construction and maintenance, and it kills the aesthetics of the place, making it less attractive to people. Going the other way, NBS and soft engineering, albeit the initial investment and conducting, through much less maintenance costing, increased ecosystem services (fisheries, biodiversity) and consequently sustained tourism revenue and the like, can in the end turn out to be considerably cheap. For instance, the sand management program in Gold Coast has got itself financed by the additional tourism revenue generated through it. For the Boğaçay River, an economic analysis should take into account the different alternatives in terms of life cycle costs (e.g., 50,000 m^3 of sand replenishment annually, filter strips, artificial reefs), the property and infrastructure losses that are now avoided, and the increase in tourism revenue that is now likely to happen. Eco-friendly financing mechanisms drawing on the 'polluter pays' and 'beneficiary pays' principles (e.g., environmental bonds, coastal funds) can be very crucial in making such projects economically sustainable.

CONCLUSION

The study employs three main approaches to decipher the intricate coastal dynamics: (i) the SWAT model that estimates the quantity of soil and pollutants entering rivers via agricultural and urban runoff; (ii) the XBeach model that reproduces the physical transformations (erosion, sedimentation) caused by waves along the shore; and (iii) the Pelnard-Considère model that foresees the coastline's progression over several decades. The primary cause of the 21.7 meters coastal retreat at Boğaçay is the stoppage of the natural sand/gravel transport and the alteration of wave energy distribution resulting from the riverbed lowering. Such technical results receive backing from showcasing of eco-friendly methods like 'Sand Engine' in the Netherlands, innovative methods like integrated watershed management in the US Chesapeake Bay, and financially viable methods like dynamic sand management at Australia's Gold Coast. The watershed message is: The management of coastal erosion must not only be through engineering but also through a multidisciplinary, flexible strategy incorporating ecology, economics, and society.

This study has demonstrated that the findings presented in the case of Boğaçay serve as a general warning for similar coastal cities in the Mediterranean basin. These developments demonstrate that traditional "hard engineering" solutions have become too fragile to protect our coasts. Climate change and excessive human pressure have permanently altered coastal dynamics. Therefore, what is needed is a flexible and dynamic management paradigm that "adapts to change," not one that seeks to "control the coast." This paradigm shift necessitates an interdisciplinary approach that combines engineering, ecology, economics, and sociology.

The situation in Boğaçay is a representation of the wider and long-term trends occurring in the Mediterranean. Climate-related problems, for instance, rising sea levels, escalating storms, and altered water flows are bringing about cumulative effects in coasts where similar measures were taken, like the Sulina Canal (Romania), the Yeşilırmak Delta (Türkiye), or the Port of Varna (Bulgaria). The local effects of single projects are interplaying with each other on a regional scale, thus, gradually, withdrawing the sediments, disturbing the ecological passages, and weakening the whole basin's resistance to outside forces. As a result, short-term and localized interventions will only be effective if they are part of a 'climate-smart coastal management' strategy implemented regionally. The solutions proposed for Boğaçay should be looked at alongside the influences in other montages around the Mediterranean and backed up with policy tools from the likes of the European Green Deal or the Barcelona Agreement.

This study evaluated the morphological and ecological impacts of the Boğaçay Project on Konyaaltı Beach using an integrated modeling approach. The findings revealed the critical importance of sediment balance for coastal stability. Protecting natural processes during the project design phase is crucial to prevent irreversible damage.

Based on available data, the Boğaçay Project's impacts on coastal erosion indicate a permanent trend. Due to the disruption of sediment transport, the coastline is unlikely to ever return to its natural equilibrium. The deepened river mouth's hydrodynamic changes have permanently altered wave refraction, accelerating erosion, particularly in the eastern reaches. While measures such as artificial sediment recharge offer temporary solutions, radical engineering interventions are required for the system to reach its original transport capacity.

The findings clarify that coastal engineering projects should not ignore natural processes. Coastal system responses to engineering projects are not limited to short-term morphological changes; they also have long-term consequences on ecosystem structures, groundwater levels, and impacts on coastal agriculture. The Boğaçay Project should be addressed as a coastal erosion problem and an integrated environmental crisis. The physical decline of coastal zones also increases social and economic vulnerability, putting pressure on the livelihoods of local people. Therefore, coastal engineering projects should consider technical parameters and socio-ecological systems.

The Boğaçay example can be considered a guide to the stages of planning that require caution in future similar projects. A detailed sediment budget analysis, particularly during the pre-project phase, is essential, along with a quantitative determination of natural carrying capacity and multivariate modeling of secondary factors affecting this carrying capacity (wind, waves, basin use, etc.). Advanced models in these processes provide decision-makers with scenario-based forecasts, enabling more informed infrastructure investments. This creates a strategic advantage not only in terms of environmental protection but also in terms of investment sustainability.

The findings of this study demonstrate the necessity of a multilayered planning approach for similar coastal landscaping projects. Sediment budget analyses must be conducted before the project, hydrodynamic models must be implemented that take seasonal variability into account, and the effects of structural interventions on ecological balance must be simulated. Alternative designs for protecting sediment transport pathways should be evaluated, particularly in estuarine interventions.

This study also highlights the need for integrated, multilayered approaches to modeling morphodynamic processes. Analyses based solely on a single model or parameter are insufficient to understand the complexity of coastal systems. In particular, the combined use of the SWAT, XBeach, and Pelnard-Considère models provided a holistic perspective on coastal systems across both spatial and temporal scales. Such model integrations should serve as a scientific reference point for future coastal engineering applications and should be actively considered by policymakers.

Finally, coastal zone protection should no longer solely focus on "intervention" but on "adaptation and monitoring." For highly sensitive coastal systems like Boğaçay, AI-powered morphodynamic monitoring systems, updated at least annually with satellite imagery, should be developed. Such systems will identify changes in the coastline early, enabling decision-makers to intervene proactively. This will protect coastal areas and allow them to be managed harmoniously with natural dynamics. This approach serves as a model for the Turkish coast and all Mediterranean coastal countries under the pressure of climate change.

Shoreline retreat has triggered an ecological transformation that goes beyond physical changes. The loss of habitat in shallow waters has reduced the habitats of benthic organisms and disrupted coastal food chains. The changes observed in dune ecosystems and beach shrinkage have secondary effects that threaten the region's biodiversity. Such ecosystem responses are critical for understanding the environmental costs of coastal engineering projects.

Adopting dynamic, data-driven approaches to coastal management is essential to ensure sustainability. This study demonstrates that the results obtained from the Boğaçay example can serve as a universal guide for similar coastal projects. In future studies, machine learning-based morphodynamic modeling techniques could help better understand the complex behavior of coastal systems.

Modeling studies have demonstrated that coastal systems respond to human disturbances nonlinearly. It has become clear that small-scale changes can lead to major morphological transformations over time. This highlights the need to develop early warning systems and continuous monitoring mechanisms in coastal management. Experience gained during model validation can guide the improvement of data collection strategies in similar studies.

The Boğaçay example demonstrates that coastal engineering projects must move beyond the traditional "engineering solutions" perspective and instead be approached within a "disaster risk management" paradigm. A three-stage framework is recommended for similar projects planned in the Mediterranean basin, specifically:

1. Pre-Project: Sediment budget analyses and disaster scenario modeling
2. During implementation: Real-time morphodynamic monitoring systems
3. Post-Project: Adaptive management strategies

This approach offers a holistic disaster management model considering physical risks and ecological and socioeconomic vulnerabilities.

This chapter enlarges the scope of technical and quantitative model results disclosed by Akiner (2025), and positions them within an interdisciplinary and policy-oriented context. We mainly contribute by repositioning coastal erosion as a 'multilayered disaster' problem rather than just a morphodynamic one. Within the proposed framework, we suggest an adaptable and sustainable management roadmap inspired by global experiences.

Coastal erosion is an issue which countries need to face together and it is especially true in a semi-enclosed basin like the Mediterranean. The Sulina Canal (Romania) and the Yeşilırmak Delta (Türkiye) are examples whereby similar interventions have led to learning a regional approach that is necessary. Each country involved in international collaboration benefits the most from it by the exchange of established practices, scientific information, modeling techniques, and even new funding methods. The current avenues of the European Union's 'Coastal Management Strategy' or 'Mediterranean Action Plan (MAP)' can help in sharing the experiences gained from the Boğaçay River and in implementing the integrated basin-coastal management principles all over the region (De Benedetti & Marzano, 2024; Rodríguez-Rojo et al., 2025). The establishment of monitoring programs and the conduct of research projects will jointly give rise to the capacity of the Mediterranean coastline to withstand climate change impacts.

The methodological expansion we recommend for future studies should include the following elements that will strengthen the disaster management perspective:

Social Vulnerability Analysis: Development of a survey-based social vulnerability index (SoVI) to measure the impacts of coastal erosion on local communities. This methodology will quantify the economic impacts of erosion on tourism revenues and agricultural production.

Multi-Disaster Interaction Model: Using a system dynamics approach, modeling the cascade effects between coastal erosion and other disasters such as salinization, flooding, and sea level rise. Projections compatible with IPCC scenarios are particularly critical for long-term risk management.

Participatory Early Warning System: GNSS-based monitoring data is processed with artificial intelligence algorithms to produce understandable risk notifications for local governments and the public. This system could increase pre-disaster forecasting capacity by at least 40% (as a methodological extension of McCall et al., 2010). Accordingly, the role of participatory approaches in evaluating climate change adaptation measures comes into focus (Coelho et al., 2023).

Comprehensive and continuous post-project monitoring program is the primary requirement due to the dynamic nature of coastal systems. Coastal Management Adaptive is a process that goes in cycles and it is to enable the flexible change of plans based on the real-time and monitoring data. The planned monitoring system for Boğaçay will have to track shoreline, and bathymetric changes using satellite imagery and LiDAR scans every year, then displacement with GNSS stations, and salinity and nutrient load using automated water quality measurement stations, along with tracking of economic and social indicators by means of social surveys. The data that is collected should be evaluated regularly (for instance, annually) and converted into a 'Management Scorecard' that will guide decisions, which will include changing sediment feed volumes, positioning structures, or introducing new scenarios of intervention. In this way, management will switch from being reactive to being proactive and driven by data.

This methodological framework can potentially set a new standard in disaster risk management for Boğaçay and similar coastal systems in the Mediterranean basin.

Policy Recommendations:

Ultimately, the path to disaster risk reduction in Boğaçay and similar coastal systems lies in adopting a holistic planning approach known as "Integrated Coastal Basin Management" (ICBM). Along this continuum, extending from the mountain-top to the depths of the sea, every intervention made at one point directly impacts the other.

Using this integrated method necessitates a significant change in current policy frameworks. To start, the EIA processes must be extended to include not only the project-based approaches but also the valley-wide strategic environmental assessments (SEA) and climate scenarios. Secondly, the institutions that are responsible for agriculture, urbanization, water management, and coastal planning are to share data and make joint decisions legally. Thirdly, the existing fragmented governance structure should be replaced by creating an independent coordination body with good technical support and budget, for example, a 'Coastal Basin Authority'. Finally, innovative financing instruments that use the 'Polluter Pays' and 'Beneficiary Pays' principles (such as, environmental bonds, coastal funds) should be created in order to ensure the long-term upkeep of nature-based solutions. Boğaçay might be used as a testbed for such a revolutionary policy reform.

Therefore, agricultural policies, urbanization strategies, water management, and coastal planning cannot be considered independently. This study invites decision-makers to develop a new governance model (e.g., a "Coastal Basin Authority") that will enable this integration and to view our coasts not as "engineering sites" but as "living systems." The future lies in solutions that work in harmony with nature and collaborate with it.

Policymakers and implementers can certainly derive actionable measures from the results of this study. In the short run (1-2 years): the top-up of emergency sediment will consist of 50,000 m^3/year while the river mouth will have temporary sediment traps to slow down the erosion rates. In the medium term (3-5 years): the current filter strips and terracing will be expanded by 30% in the basin; the pilot areas for seagrass (Posidonia oceanica) planting will be established; and the coastal land use plans will be reviewed. In the long run (5+ years): a fully integrated early warning and monitoring system using both satellites and ground stations will be set up; wave energy will be mitigated by the deployment of artificial reef networks; and a 'Coastal Restoration Fund' based on the 'Polluter Pays' principle will be created. The most important factor for these steps to succeed is that they are under the management of a 'Coastal Basin Authority' with full jurisdiction and a considerable budget, which is a collaboration of the Ministries of Agriculture and Forestry, Environment, Urbanization and Climate Change, and Culture and Tourism.

This study mandates three fundamental reforms in coastal engineering projects: (1) Requiring disaster risk assessment in EIA processes, (2) Integrating basin-coastal management into the legal framework, and (3) Incentive mechanisms for public-private partnerships in restoration projects. The Boğaçay experience serves as an early warning for coastal disasters in the Mediterranean, which will increase with climate change.

Author Contributions: M.E.A. solely contributed to the study's conception and design, performed material preparation, data collection, and analysis, wrote the first draft of the manuscript, commented on previous versions, and read and approved the final manuscript.

Funding: This research received no specific grant from any funding agency.

Data Availability: No datasets were generated or analyzed during the current study.

Acknowledgments: Special thanks to the Republic of Türkiye Governorship of Antalya, the State Hydraulic Works (DSİ), and the Ministry of Environment, Urbanization, and Climate Change (MEUCC) for providing the database used in this study. Thanks to Grammarly Premium for helping correct grammatical and spelling errors. The free version of the DeepL translator was used to translate the article into English.

Conflicts of Interest: The authors declare no conflict of interest.

REFERENCES

Akiner, M. E. (2025). Integrated Modeling Approach to Predict the Morphodynamic Impacts of the Boğaçay Project on Konyaaltı Beach. International Journal of Environment. *Engineering and Education*, *7*(2), 97–108. DOI: 10.55151/ijeedu.v7i2.215

Akiner, M. E., Akiner, N., & Akiner, İ. (2025a). From Historical Management to Modern Sustainability: A Deep Dive Into Antalya's Environmental Evolution and Climate Resilience. In Community Climate Justice and Sustainable Development (pp. 181-230). IGI Global Scientific Publishing. DOI: 10.4018/979-8-3373-0619-3.ch009

Akıner, M. E., Akıner, N., & Akıner, İ. (2025b). Interplay Between Climate Change and The Migration-Induced Population Growth: A Case Study in Konyaaltı, Antalya, Turkey. *Sürdürülebilir Çevre Dergisi*, *5*(1), 1–18. DOI: 10.62816/cevder.1581432

Akiner, M. E., & Lačnjevac, Č. (2022). Managing Saltwater Intrusion and Agricultural Practices along the Boğaçay River, Turkey: Effects from Excavation and Land Source Pollution. *Journal of Coastal Research*, *38*(3), 567–577. DOI: 10.2112/JCOASTRES-D-21-00101.1

Arabi, M., Frankenberger, J. R., Engel, B. A., & Arnold, J. G. (2008). Representation of agricultural conservation practices with SWAT. *Hydrological Processes*, *22*(16), 3042–3055. DOI: 10.1002/hyp.6890

Arimoro, A. E., & Arinze-Umobi, C. (2026). Sustainable public-private partnerships in Sub-Saharan Africa: A conceptual framework for low carbon development and domestic financing. *Public Works Management & Policy*, *31*(1), 40–76. DOI: 10.1177/1087724X251356462

Ataol, M., Kale, M. M., & Tekkanat, İ. S. (2019). Assessment of the changes in shoreline using digital shoreline analysis system: A case study of Kızılırmak Delta in northern Turkey from 1951 to 2017. *Environmental Earth Sciences*, *78*(19), 579. DOI: 10.1007/s12665-019-8591-7

Baldassarre, L., Natali, V., De Pascale, F., Vezzi, A., Banchi, E., Bazzaro, M., Relitti, F., Tagliapietra, D., & Cibic, T. (2023). The impact of MOSE (Experimental electromechanical module) flood barriers on microphytobenthic community of the Venice Lagoon. *Microorganisms*, *11*(4), 936. DOI: 10.3390/microorganisms11040936 PMID: 37110359

Batiuk, R. A., Blankenship, K., Boesch, D. F., Dennison, W. C., Harrison, V., Hershner, C., Hoagland, R. A., Horton, T. W., Matuszeski, W., Miller, T. J., Swanson, A. P., Todd, A. H., & Wardrop, D. H. (2025). Restoration of the Chesapeake Bay: Lessons for other ecosystems. *Frontiers in Marine Science, 12*, 1581261. DOI: 10.3389/fmars.2025.1581261

Boudouresque, C. F., Blanfuné, A., Pergent, G., & Thibaut, T. (2021). Restoration of seagrass meadows in the Mediterranean Sea: A critical review of effectiveness and ethical issues. *Water (Basel), 13*(8), 1034. DOI: 10.3390/w13081034

Bruce, D. G., Cornwell, J. C., Harris, L., Ihde, T. F., Kellogg, M. L., Knoche, S., . . . Vogt, B. (2021). A synopsis of research on the ecosystem services provided by large-scale oyster restoration in the Chesapeake Bay. NOAA Tech. Memo. NMFS-OHC, 8, 52. https://spo.nmfs.noaa.gov/tech-memos/ (Accessed 19 August 2025)

Bruun, P. (1962). Sea-level rise as a cause of shore erosion. *Journal of the Waterways and Harbors Division, 88*(1), 117–132. DOI: 10.1061/JWHEAU.0000252

Coelho, C., Lima, M., Alves, F., Roebeling, P., Pais-Barbosa, J., & Marto, M. (2023). Assessing Coastal Erosion and Climate Change Adaptation Measures: A Novel Participatory Approach. *Environments (Basel, Switzerland), 10*(7), 110. DOI: 10.3390/environments10070110

Cowell, P. J., Stive, M. J., Niedoroda, A. W., Swift, D. J., de Vriend, H. J., Buijsman, M. C., & de Boer, P. L. (2003). The coastal-tract (part 2): Applications of aggregated modeling of lower-order coastal change. *Journal of Coastal Research*, •••, 828–848. Retrieved August 19, 2025, from https://www.jstor.org/stable/4299223

Das, A., & Swain, P. K. (2024). Navigating the sea level rise: Exploring the interplay of climate change, sea level rise, and coastal communities in India. *Environmental Monitoring and Assessment, 196*(11), 1010. DOI: 10.1007/s10661-024-13191-z PMID: 39361047

De Benedetti, L., & Marzano, M. (2024). Coastal Protection in the Mediterranean Sea: An Overview of International and European Coastal Defense Regulations. In Maccarrone, V., & Fadzil Akhir, M. (Eds.), *Coastal Sustainability. Coastal Research Library* (Vol. 39, pp. 273–296). Springer., DOI: 10.1007/978-3-031-75749-5_13

De Schipper, M. A., Ludka, B. C., Raubenheimer, B., Luijendijk, A. P., & Schlacher, T. A. (2021). Beach nourishment has complex implications for the future of sandy shores. *Nature Reviews. Earth & Environment, 2*(1), 70–84. DOI: 10.1038/s43017-020-00109-9

Dean, R. G. (2018). Principles of beach nourishment. Handbook of coastal processes and erosion, 217-232.

Dinu, I., Gràcia, V., García-León, M., Lin-Ye, J., Stănică, A., & Sánchez-Arcilla, A. (2025). Feasibility of artificial reefs as coastal protection measures at the Danube Delta coast, Romania. *Frontiers in Environmental Science*, *13*, 1568364. DOI: 10.3389/fenvs.2025.1568364

Dionísio António, S., van der Werf, J., Horstman, E., Cáceres, I., Alsina, J., van der Zanden, J., & Hulscher, S. (2023). Influence of beach slope on morphological changes and sediment transport under irregular waves. *Journal of Marine Science and Engineering*, *11*(12), 2244. DOI: 10.3390/jmse11122244

Downs, P. W., Cui, Y., Wooster, J. K., Dusterhoff, S. R., Booth, D. B., Dietrich, W. E., & Sklar, L. S. (2009). Managing reservoir sediment release in dam removal projects: An approach informed by physical and numerical modelling of non-cohesive sediment. *International Journal of River Basin Management*, *7*(4), 433–452. DOI: 10.1080/15715124.2009.9635401

Durap, A., & Balas, C. E. (2024). Towards sustainable coastal management: A hybrid model for vulnerability and risk assessment. *Journal of Coastal Conservation*, *28*(4), 66. DOI: 10.1007/s11852-024-01065-y

Ferrarin, C., Bonaldo, D., Bergamasco, A., & Ghezzo, M. (2024). Sea level and temperature extremes in a regulated Lagoon of Venice. *Frontiers in Climate*, *5*, 1330388. DOI: 10.3389/fclim.2023.1330388

Ferreira, M., Coelho, C., & Silva, P. A. (2022). Littoral drift gradients on the Portuguese coastal sector Esmoriz-Nazare: Past and future trends. *Coastal Engineering Proceedings*, (37), 22–22. DOI: 10.9753/icce.v37.papers.22

Goldstein, E. B., Coco, G., & Plant, N. G. (2019). A review of machine learning applications to coastal sediment transport and morphodynamics. *Earth-Science Reviews*, *194*, 97–108. DOI: 10.1016/j.earscirev.2019.04.022

Google. (n.d.). Boğaçay River and Konyaaltı Shoreline digital imaginary, satellite map. https://earth.google.com/web/@36.85211757,30.62297089,1512.2599538a,0d,35y,-0.6805h,2.0574t,-0r/data=CgRCAggBQgIIAEoNCP__________wEQAA?utm_source=earth7&utm_campaign=vine&hl=en (accessed 19 August 2025).

Ibanez, C., & Caiola, N. (2021). Sea-level rise, marine storms and the resilience of Mediterranean coastal wetlands: Lessons learned from the Ebro Delta. *Marine and Freshwater Research*, *73*(10), 1246–1254. DOI: 10.1071/MF21140

Jackson, A., Corbett, B., & Tomlinson, R. (2007). Palm Beach Protection Strategy Update on recommendations as part of the Gold Coast Shoreline Management Plan. http://hdl.handle.net/10072/93914 (Accessed 19 August 2025).

Koç, K., Koşun, E., & Karaman, M. E. (2015). Kıyı Sedimentlerinde Tane Boyunu Etkileyen Faktörler: Antalya Körfezi'nde Konyaaltı ve Lara Plajlarının (Antalya) Karşılaştırılması. *Türkiye Jeoloji Bülteni*, *58*(2), 85–102. DOI: 10.25288/tjb.298589

Köle, M. M., & Ataol, M. (2016). Coastal Erosion of Yeşilırmak Delta Between 2000–2016/Yeşilırmak Deltası'nda 2000-2016 Yılları Arasında Gözlenen Kıyı Erozyonu. In 4th International Geography Symposium-GEOMED (pp. 460-468). DOI: 10.17482/uumfd.1248184

Masria, A., & Abouelsaad, O. (2025). Artificial intelligence applications in coastal engineering and its challenges – A review. *Continental Shelf Research*, *286*, 105425. DOI: 10.1016/j.csr.2025.105425

Massoud, M. A., Scrimshaw, M. D., & Lester, J. N. (2004). Integrated coastal zone and river basin management: A review of the literature, concepts and trends for decision makers. *Water Policy*, *6*(6), 519–548. DOI: 10.2166/wp.2004.0034

McCall, R. T., van Thiel de Vries, J. S. M., Plant, N. G., van Dongeren, A. R., Roelvink, J. A., Thompson, D. M., & Reniers, A. J. H. M. (2010). Two-dimensional time dependent hurricane overwash and erosion modeling at Santa Rosa Island. *Coastal Engineering*, *57*(7), 668–683. DOI: 10.1016/j.coastaleng.2010.02.006

Meulen, F., IJff, S., & van Zetten, R. (2022). Nature-based solutions for coastal adaptation management, concepts and scope, an overview. *Nordic Journal of Botany*, *2023*(1), e03290. DOI: 10.1111/njb.03290

Monreal-Jiménez, R., Carbajal, N., Contreras-Tereza, V. K., & Salas-Monreal, D. (2025). Sediment Dynamics and Erosion in a Complex Coastal Lagoon System in the Southern Gulf of Mexico. Water, 17(16), 2408. https://doi.org/10.3390/w17162408

NASA/METI. (2009). ASTER Global Digital Elevation Model V2 [30m resolution DEM]. https://doi.org/DOI: 10.5067/ASTER/ASTGTM.002

Ozcelik, M. (2022). Comparison of the environmental impact and production cost rates of aggregates produced from stream deposits and crushed rock quarries (Boğaçay Basin/Antalya/Turkey). *Geoheritage*, *14*(1), 18. DOI: 10.1007/s12371-022-00659-y

Ozcelik, M., & Tuzlu, F. (2022). Environmental impacts of abandoned marina construction: An example Boğaçay River and Marina Project (Antalya, Turkey). *Arabian Journal of Geosciences*, *15*(23), 1721. DOI: 10.1007/s12517-022-11026-0

Pang, T., Wang, X., Nawaz, R. A., Keefe, G., & Adekanmbi, T. (2023). Coastal erosion and climate change: A review on coastal-change process and modeling. *Ambio, 52*(12), 2034–2052. DOI: 10.1007/s13280-023-01901-9 PMID: 37405570

Peychev, V., & Stancheva, M. (2009). Changes of sediment balance at the Bulgarian Black Sea coastal zone influenced by anthropogenic impacts. *Compt. Rend. Acad. Bulg. Sci, 62*(2), 277–285.

Rodríguez-Rojo, C. N., Pérez-Cayeiro, M. L., & Chica-Ruiz, J. A. (2025). Adaptation to Climate Change in Coastal Countries of the European Union—An Evaluation of Plans and Strategies. *Applied Sciences (Basel, Switzerland), 15*(11), 6281. DOI: 10.3390/app15116281

Roelvink, D., Reniers, A., van Dongeren, A., van Thiel de Vries, J., McCall, R., & Lescinski, J. (2009). Modeling storm impacts on beaches, dunes and barrier islands. *Coastal Engineering, 56*(11–12), 1133–1152. DOI: 10.1016/j.coastaleng.2009.08.006

Roy, P., & Pal, S. C. (2024). Modeling soil erosion susceptibility considering morphometric analysis and SWAT application: Policy recommendation to achieve SDGs. *Modeling Earth Systems and Environment, 10*(4), 5735–5752. DOI: 10.1007/s40808-024-02089-x

Scarpa, G. M., Davison, S., Manfè, G., Lorenzetti, G., Zaggia, L., & Braga, F. (2025). Suspended sediment dynamics at the inlets of Venice Lagoon: Unraveling the effects of storm surges and mobile barrier operations. *Journal of Hydrology (Amsterdam), 651*, 132588. DOI: 10.1016/j.jhydrol.2024.132588

Shaeri, S., Etemad-Shahidi, A., & Tomlinson, R. (2020). Revisiting longshore sediment transport formulas. *Journal of Waterway, Port, Coastal, and Ocean Engineering, 146*(4), 04020009. DOI: 10.1061/(ASCE)WW.1943-5460.0000557

Stancheva, M., Ramieri, E., Stanchev, H., Bocci, M., Markovic, M., & Roberts, T. (2025). Exploring integration of land-sea interactions in maritime spatial planning in the Bulgarian context, Black Sea. *Marine Policy, 171*, 106416. DOI: 10.1016/j.marpol.2024.106416

Stive, M. J. F., De Schipper, M. A., Luijendijk, A. P., Ranasinghe, R. W. M. R. J. B., Van Thiel de Vries, J., Aarninkhof, S., . . . Marx, S. (2013, June). The sand engine: a solution for vulnerable deltas in the 21st century. In Proceedings, Coastal Dynamics 2013: 7th International conference on coastal dynamics.

Strauss, D., Murray, T., Harry, M., & Todd, D. (2017). Coastal data collection and profile surveys on the Gold Coast: 50 years on. In *Australasian Coasts & Ports 2017: Working with Nature: Working with Nature* (pp. 1030–1036). Engineers Australia, PIANC Australia and Institute of Professional Engineers New Zealand.

Vousdoukas, M. I., Mentaschi, L., Voukouvalas, E., Bianchi, A., Dottori, F., & Feyen, L. (2018). Climatic and socioeconomic controls of future coastal flood risk in Europe. *Nature Climate Change*, *8*(9), 776–780. DOI: 10.1038/s41558-018-0260-4

Walters, S. (2025). Recommendations for Managing Loans and Grants: A Policymaker Perspective. In Aid Dynamics in Jamaican Development: Empowering Communities through Loans and Grants (pp. 135-162). Palgrave Macmillan, Cham: Springer Nature Switzerland. DOI: 10.1007/978-3-031-94885-5_5

Yilmaz, A. G. (2015). The effects of climate change on historical and future extreme rainfall in Antalya, Turkey. *Hydrological Sciences Journal*, *60*(12), 2148–2162. DOI: 10.1080/02626667.2014.945455

ADDITIONAL READINGS

Abanades, J., Greaves, D., & Iglesias, G. (2014). Wave farm impact on the beach profile: A case study. *Coastal Engineering*, *86*, 36–44. DOI: 10.1016/j.coastaleng.2014.01.008

Bacopoulos, P., & Clark, R. R. (2021). Coastal erosion and structural damage due to four consecutive-year major hurricanes: Beach projects afford resilience and coastal protection. *Ocean and Coastal Management*, *209*, 105643. DOI: 10.1016/j.ocecoaman.2021.105643

Del Río, L., Gracia, F. J., & Benavente, J. (2013). Shoreline change patterns in sandy coasts. A case study in SW Spain. *Geomorphology*, *196*, 252–266. DOI: 10.1016/j.geomorph.2012.07.027

Hsu, T. W., Lin, T. Y., & Tseng, I. F. (2007). Human impact on coastal erosion in Taiwan. *Journal of Coastal Research*, *23*(4), 961–973. DOI: 10.2112/04-0353R.1

Kirk, R. M. (1991). River-beach interaction on mixed sand and gravel coasts: A geomorphic model for water resource planning. *Applied Geography (Sevenoaks, England)*, *11*(4), 267–287. DOI: 10.1016/0143-6228(91)90018-5

Lim, C., Kim, T. K., Lee, S., Yeon, Y. J., & Lee, J. L. (2021). Assessment of potential beach erosion risk and impact of coastal zone development: A case study on Bongpo–Cheonjin Beach. *Natural Hazards and Earth System Sciences*, *21*(12), 3827–3842. DOI: 10.5194/nhess-21-3827-2021

Tsoukala, V. K., Katsardi, V., Hadjibiros, K., & Moutzouris, C. I. (2015). Beach erosion and consequential impacts due to the presence of harbours in sandy beaches in Greece and Cyprus. *Environmental Processes*, *2*(S1, Suppl 1), 55–71. DOI: 10.1007/s40710-015-0096-0

KEY TERMS AND DEFINITIONS

Boğaçay Project: The comprehensive infrastructure and environmental development project undertaken in Antalya significantly impacts morphodynamic processes and sediment transport in the area where the Boğaçay River flows into the sea.

Coastal Erosion: The process of shoreline regression over time due to wave action, currents, and human activities. It is characterized by sediment loss and changes in coastal morphology.

Coastal Management Practices: It is the name given to the whole of structural, non-structural, nature-based, and institutional intervention strategies implemented to ensure the sustainable use of coastal areas, the protection of natural resources, and the reduction of disaster risks.

Coastal Morphodynamics: The scientific discipline that studies the relationship between topographic changes occurring in coastal areas and physical processes (e.g., wave and sediment interaction).

Human Intervention: Structural or non-structural interventions made to the coastal system through human-induced engineering practices such as harbors, dikes, channel openings, or reclamation.

Sediment Transport: The process of sedimentary material moving from one location to another via waves, currents, or rivers. It directly affects the location of the shoreline and beach morphology.

SWAT (Soil and Water Assessment Tool): The physically based software that models hydrological processes such as surface runoff, erosion, and sediment transport at the basin scale.

The Pelnard-Considère Model: The analytical model used to calculate long-term shoreline changes. It predicts sediment accumulation and loss based on linear wave theory.

XBeach Model: An open-source numerical model that simulates coastal morphodynamics and sediment transport under storm conditions. It integrates processes such as wave refraction, entrainment, and flood effects.

Chapter 5
Spatial Analysis of Soil Erosion in the Ceyhan Basin (Turkey) With GIS and the RUSLE Model

Ersin Ateş
http://orcid.org/0000-0002-1956-2017
Ankara University, Turkey

ABSTRACT

The RUSLE model applied in the Ceyhan Basin revealed a heterogeneous erosion pattern driven by topography, land use, and rainfall erosivity. Low R-factor values dominate the northern plains, while Mediterranean influence increases erosivity in the south. Moderate erodibility soils cover most of the basin, yet steep slopes with high LS values form critical hotspots. Protective land cover remains limited, and more than half of the basin lacks conservation practices. As a result, although 60% of the basin experiences low erosion, nearly 13% shows extremely high soil loss, mainly in areas where steep gradients, weak vegetation, and inappropriate land-use practices overlap. These findings highlight the need for targeted conservation strategies and spatially informed land-management planning.

INTRODUCTION

Healthy soil structure is essential for the functioning of ecosystems and for meeting fundamental human needs (Borrelli et al., 2017). Although soil is vital for the continuation of human life, soil erosion induced by human activities threatens food security, land productivity, and socioeconomic systems worldwide (Borrelli et al.,

DOI: 10.4018/979-8-3373-6274-8.ch005

2017). Soil provides approximately 95% of global food supply, making it one of the most critical components of human well-being (FAO, 2016). The increase in global population further intensifies pressure on soil resources by expanding demands on natural resources and agricultural lands (Foley et al., 2011).

A global assessment showed that soil erosion reached 35 billion tons per year in 2001, and increased by 2.5% to 35.9 billion tons per year by 2012 (Borrelli et al., 2017). In 2001, the highest soil erosion rates were observed in South America (3.53 tons/ha/year) and Asia (3.47 tons/ha/year), while by 2012 the most significant increase occurred in Africa, where erosion rates rose to 3.88 tons/ha/year (Borrelli et al., 2017, 2020). A United Nations study conducted in 2016 reported that approximately 33% of the world's soils are moderately to highly degraded. The same report indicated that agricultural lands lose an estimated 75 billion tons of soil annually on a global scale, and that erosion alone results in an annual loss of 7.6 million tons in cereal production (FAO, 2016).

Turkey has a predominantly rugged and mountainous topography, with an average elevation of 1,250 m, and 62% of its land surface exhibiting slopes greater than 15% (Imamoglu and Dengiz, 2017; TOKB, 1987). Each year, approximately 642 million tons of soil are displaced in Turkey—equivalent to about 2.5 times the total area of the country's agricultural lands (24 million ha)—which corresponds to 20 tons of soil becoming unusable every second (Erpul et al., 2018). Nationwide assessments indicate that 60.28% of Turkey's surface area experiences very slight water erosion, 19.13% slight, 7.93% moderate, 5.97% severe, and 6.7% very severe erosion (Erpul et al., 2018, 2020). These values reveal that 41% (99,400 km^2) of cultivated agricultural lands are subject to severe or very severe water erosion, posing a significant threat to agricultural sustainability. In terms of land-use categories, water erosion in agricultural areas predominantly occurs at very slight (44.3%) and slight (28.78%) levels, while moderate (11.91%), severe (8.2%), and very severe (6.37%) erosion are relatively limited. In contrast, rangeland areas exhibit notably higher proportions of moderate (13.59%), severe (11.57%), and especially very severe (17.2%) erosion, indicating that rangelands are more vulnerable to erosion processes (Erpul et al., 2018).

According to the GASEMT database, which compiles 24 different erosion models (Borrelli et al., 2021), the RUSLE model is the most widely used soil erosion estimation method globally, with a utilization rate of 17.1% (Bezak et al., 2021). Empirical models such as RUSLE have been employed extensively for assessing erosion risk in large river basins (Burrough, 1986). Advances in geographic information systems (GIS) and remote sensing (RS) have enabled the development of more cost-effective and scalable approaches for environmental analyses, facilitating applications over larger areas with improved accuracy (Imamoglu and Dengiz, 2017; Millward and Mersey, 1999; Wischmeier and Smith, 1978). The RUSLE approach is preferred because it quantifies multiple interacting erosion processes using GIS-based raster datasets and allows the estimation of average annual soil loss using input factors that are accessible from publicly available data sources (Laflen et al., 1991; Lu et al., 2004; Martin et al., 2003). Comprising six numerical factors, each representing a distinct component of the erosion process, the RUSLE model enables the generation of erosion risk maps even in data-limited environments and is widely adopted due to its broad applicability and methodological robustness (Panagos, Borrelli, Poesen et al., 2015; Renard et al., 1997).

The topographic complexity, intensive agricultural activities, and pronounced climatic variability of the Ceyhan Basin indicate a high sensitivity to soil erosion. In this context, the primary objective of this study is to apply the Revised Universal Soil Loss Equation (RUSLE) model within a Geographic Information Systems (GIS) environment to the Ceyhan Basin. The study aims to spatially estimate the average annual potential soil loss (A, ton/ha/year) at a watershed scale with 100-meter spatial resolution and to analyze the relative influence of the controlling factors (R, K, LS, C, and P). Previous research on the Ceyhan Basin has often been constrained by limited spatial resolution or incomplete factor representation; therefore, this study provides a high-resolution assessment of erosion potential and establishes a scientific basis for developing sustainable land-use and management strategies in the region. The RUSLE model applied in the Ceyhan Basin also contributes to identifying erosion-prone areas and supports the development of localized strategies for mitigating soil loss. By incorporating essential factors such as slope, land use, rainfall, and soil properties, the RUSLE model offers a spatially explicit framework for assessing the sustainability of agricultural lands. Furthermore, the results provide critical information for local authorities in the planning of agricultural activities and in the development of decision-support systems. In regions like the Ceyhan Basin, where agricultural land use is highly intensive, the use of such models plays a crucial role in preventing environmental degradation.

STUDY AREA

The Ceyhan Basin is located in the southern part of Turkey and is bounded by distinct natural features: the Tahtalı Mountains to the west, the Uzunyayla Plateau to the north, the Malatya and Nur Mountains to the east, and the Mediterranean Sea to the south. Administratively, the basin includes parts of the provinces of Adana, Osmaniye, Kahramanmaraş, Kayseri, Malatya, Adıyaman, Gaziantep, and Hatay. Covering an area of 21,573 km^2, the Ceyhan Basin accounts for approximately 2.76%

of Turkey's total surface area. Together with its tributaries, the Ceyhan River has a total channel length of 724.41 km within the basin (Figure 1).

The southern portion of the basin encompasses Çukurova and the Upper Plain, while the Dibek and Binboğa Mountains lie to the west, the Elbistan Plain to the north, and the Nurhak Mountains and Sağlık Plain to the east. There are five major dams along the Ceyhan River: Menzelet, Kılavuzlu, Sır, Berke, and Aslantaş. Based on long-term average temperatures for the period 1979–2013, the warmest areas of the basin are located in the southern sections, particularly within Çukurova and the Upper Plain, whereas the coldest areas are found in the higher elevations of the Binboğa Mountains. An examination of average annual precipitation for the same period indicates that the wettest areas occur around Osmaniye and the Dibek Mountains, while the driest region is the Elbistan Plain. Elevations within the basin range from sea level along the Mediterranean coast to approximately 3,300 m. Slopes are generally low in Çukurova, the Upper Plain, the Elbistan Plain, and the Sağlık Plain, but increase sharply in the Nurhak Mountains as well as in the Dibek and Binboğa Mountain ranges (Figure 2).

Figure 1. Location map of the Ceyhan Basin

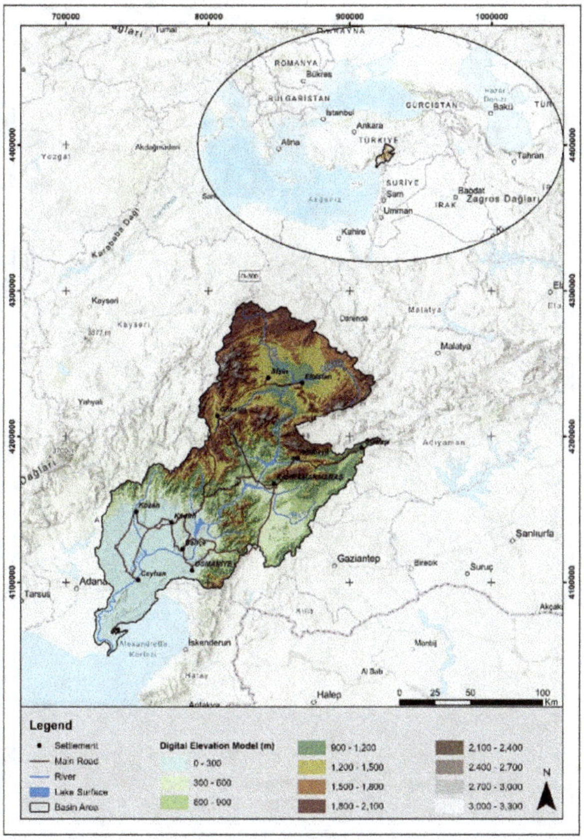

Figure 2. DEM of the Ceyhan Basin

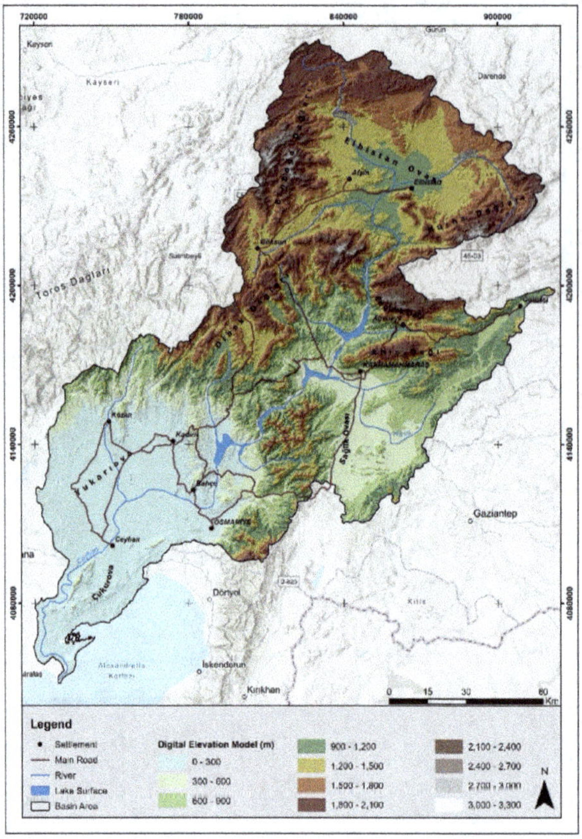

Within the Ceyhan Basin, there are 2 national parks, 1 protected area designated under the RAMSAR Convention, 5 nature parks, 2 nature conservation areas, and 1 wildlife development area. Protected areas collectively account for 3.22% of the total basin area. Among these, wetlands constitute the largest share of protected land, representing 68.73% (477.89 km^2) of the total protected area. National parks account for 22.05% (153.34 km^2), followed by wildlife development areas with 4.46% (31.01 km^2), nature parks with 4.06% (28.25 km^2), and nature conservation areas with 0.69% (4.82 km^2) (Table 1; Figure 3).

Table 1. Protected Areas within the Ceyhan Basin

Name of Protected Area	Protection Status	Protected Area (km²)
Karatepe Aslantaş	National Park	41.45
Yumurtalık Lagünü	National Park	111.88
Yumurtalık Lagünleri RAMSAR Alanı	RAMSAR Wetland Boundary	105.242
Yumurtalık Lagünü Sulak Alanı RAMSAR Tampon Bölge Sınırı	RAMSAR Wetland Buffer Zone	197.08
Yumurtalık Lagünü Sulak Alanı RAMSAR Tampon Bölge Sınırı	RAMSAR Wetland Buffer Zone	197.08
Gölbaşı Gölleri Tabiat Parkı	Nature Park	20.87
Gölbaşı Gölleri Sulak Alanı Tampon Bölgesi	Designated Wetland Conservation Area	175.56
Dağlıcak	Nature Park	0.23
Huzurlu	Nature Park	1.45
Kapıçam	Nature Park	1.79
Yavşan Yaylası	Nature Park	3.41
Çiftmazı	Nature Park	0.50
Körçoban	Nature Conservation Area	4.81
Tekkoz-Kengerlidüz	Nature Conservation Area	0.01
Osmaniye Zorkun Yaylası	Wildlife Development Area	31.02

According to CORINE 2018 data, forests and semi-natural areas constitute the largest land-cover class within the basin, accounting for 49.83% of the total area. Agricultural lands represent 46.17%, while artificial surfaces and wetlands cover 2.10% and 1.17% of the basin, respectively (Table 2) (EEA, 2018). Within the portion of the Ceyhan Basin that lies in Adana Province, 81% of the land consists of arable fields, 11% of orchards, 6% of vegetable gardens, and 2% of fallow land. In Kahramanmaraş Province, which also falls partly within the basin, 76% of the area is composed of cultivated agricultural land, 14% of orchards, 2% of vegetable gardens, and 8% of fallow land. In Osmaniye Province, 89% of the land within the basin consists of cultivated agricultural areas, 8% of orchards, 3% of vegetable gardens, and less than 1% of fallow land (SYGM, 2018) (Figure 4).

Table 2. Land Use Types and Their Percentages within the Ceyhan Basin

CORINE Land Use	Area within the Basin (km²)	Percentage within the Basin (%)
Agricultural Land	9961.80	46.17
Forest and Semi-Natural Areas	10762.13	49.89
Artificial Surfaces	453.84	2.10
Wetlands	253.42	1.17
Water Bodies	0.27	0.27

Figure 3. Protected Areas within the Ceyhan Basin

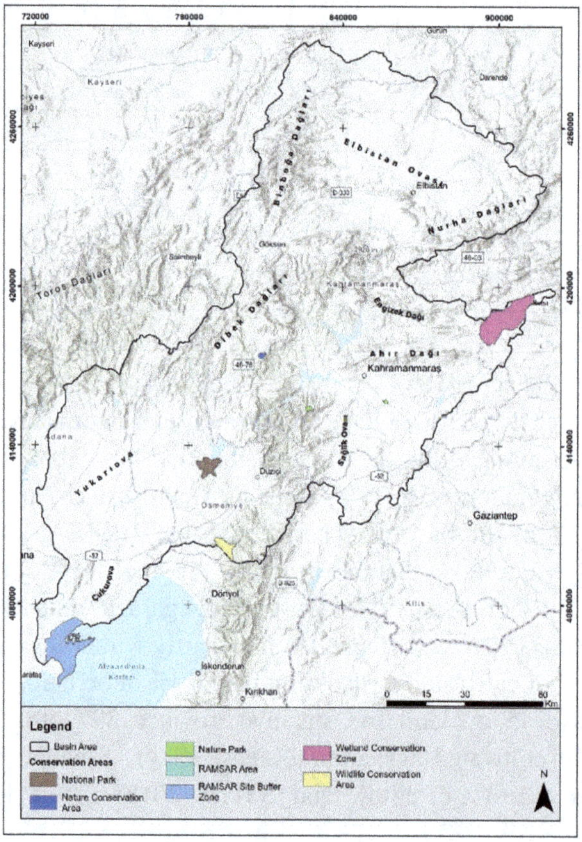

Figure 4. CORINE land use map of the Ceyhan Basin

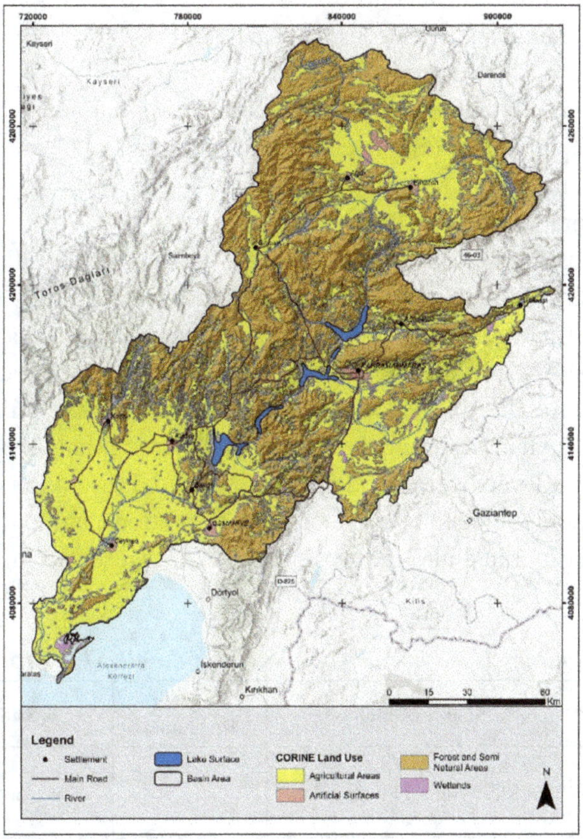

MATERIALS AND METHODS

This study was conducted in the Ceyhan Basin, located in southern Turkey, which is characterized by substantial topographic variation (0–3300 m), diverse land-use patterns, and highly variable precipitation and temperature regimes. To model erosion processes driven by surface runoff influenced by rainfall characteristics, land use, and soil properties, the Revised Universal Soil Loss Equation (RUSLE) was employed. The RUSLE model is an enhanced, digitally adapted version of the Universal Soil Loss Equation (USLE), and estimates the average annual soil loss (A, tons/ha/year) caused by sheet and rill erosion (Renard et al., 1997; Wischmeier and Smith, 1978). The model is expressed by the following equation:

$$A = R \times K \times LS \times C \times P$$

A: Average annual soil loss (tons/ha/year)
R: Rainfall erosivity factor (MJ·mm·ha^{-1}·h^{-1}·year^{-1})
K: Soil erodibility factor (ton·ha·h/ha·MJ·mm)
LS: Topographic factor (slope length and steepness)
C: Cover-management factor (C)
P: Support practice factor (P)

One of the major advantages of the RUSLE erosion model is that its input variables are available through open-access data sources and are fully compatible with geographic information system (GIS) data structures (Ganasri and Ramesh, 2015; Renard et al., 1997). For the RUSLE-based erosion assessment conducted in the Ceyhan Basin, ArcGIS 10.6.1 and ArcGIS Pro software were utilized. Due to the large size of the Ceyhan Basin and the varying spatial resolutions of publicly available datasets, all raster layers were standardized to a spatial resolution of 100 meters.

The data types, source databases, original resolutions, and purposes of use for each dataset incorporated into the erosion model are presented in Table 3.

Table 3. Data Sources Used in the Application of the RUSLE Erosion Model

Data Type	Source / Database	Original Resolution	Kullanım Amacı
Precipitation (P)	Karger et al. (1979–2013 Mean Annual Precipitation Data)	-	Calculation of the R Factor
Soil Properties (Clay, Sand, Silt, SOC)	SoilGrids (15–30 cm depth)	250m	Calculation of the K Factor
Digital Elevation Model (DEM)	ALOS DEM (JAXA)	30m	Calculation of the LS Factor (Slope and Flow Accumulation)
Land Use / Land Cover (LULC)	CORINE Land Cover 2018 (EEA)	100 m (Standardized)	Assignment of C Factor Coefficients
Conservation Practices (LULC-based)	CORINE Land Cover 2018 (EEA)	100 m (Standardized)	Assignment of P Factor Coefficients

R Factor: The R factor quantifies the impact of rainfall frequency and intensity on soil erosion (Renard et al., 1997). Although the R factor is typically calculated using the EI30 method (rainfall energy and maximum 30-minute intensity), this approach requires high-resolution, time-series precipitation data. In regions where such detailed datasets are unavailable, empirical equations based on mean annual precipitation (mm) are commonly used (Cruz et al., 2025; Panagos, Borrelli, Poesen et al., 2015; Shaheemath Suhara et al., 2023). In this study, mean annual precipitation

(P) data for the years 1979–2013 were utilized (Karger et al., 2017, 2020). Due to data limitations, the empirical equation R = k · P was applied for the computation of the R factor. The coefficient k = 0.5 was selected considering the structure of the available dataset and the climatic characteristics of the region, and is consistent with recommendations for Mediterranean climatic conditions. The resulting R-factor values ranged between 171.43 and 595.46 $MJ \cdot mm \cdot ha^{-1} \cdot h^{-1} \cdot year^{-1}$ across the basin. At the European scale, studies report R-factor values for Mediterranean climatic zones typically within the range of approximately 300–700 $MJ \cdot mm \cdot ha^{-1} \cdot h^{-1} \cdot year^{-1}$ (Panagos, Borrelli, Poesen et al., 2015). Thus, the empirical equation used in this study produces results that are consistent with regional climatic conditions.

$$R = k \times P$$

k: Selected coefficient (0.5)

P: Mean annual total precipitation (mm)

K Factor: The soil erodibility (K) factor was calculated using organic matter (OM) content together with the clay, silt, and sand fractions. The empirical equation used for the computation of the K factor has been revised for European and Mediterranean basins and is widely recognized as a reliable approach. This equation quantifies the resistance of soil to erosion based on its intrinsic physical properties (Panagos et al., 2014; Renard et al., 1997; Torri et al., 1997). The K factor represents the amount of soil loss ($ton \cdot ha^{-1} \cdot MJ^{-1} \cdot mm^{-1}$) under a standardized rainfall event for a given slope length and steepness (Renard et al., 1997).

The soil properties required for the Ceyhan Basin were obtained from the SoilGrids database. SoilGrids provides global soil information at 250 m resolution, generated through machine-learning models trained on more than 240,000 soil profile observations and over 400 environmental covariates, including vegetation, climate, landform morphology, geology, and hydrology (Poggio et al., 2021). For the Ceyhan Basin, soil properties corresponding to the 15–30 cm depth interval were extracted from the SoilGrids database. The soil parameters used for the Ceyhan Basin are as follows:

Clay (%): Clay fraction

Sand (%): Sand fraction

Silt (%): Silt fraction

Soil Organic Carbon (SOC): Soil organic carbon content (g/kg)

The 250-m resolution soil data obtained from the SoilGrids database were resampled to 100-m spatial resolution to match the dataset used in this study, and subsequently converted into percentage values before being incorporated into the analysis. Soil properties obtained from the SoilGrids database have a spatial resolution of 250 m. These data have been resampled to a resolution of 100 m to ensure compatibility with other spatial layers (DEM, CORINE) used in the study. This process aims only to achieve cellular compatibility between raster datasets, not to generate new spatial information about soil properties. During the resampling process, the nearest neighbour method was preferred, taking into account the categorical and discontinuous nature of soil texture components (clay, silt and sand). This approach prevents the creation of artificial intermediate values and physically meaningless soil fractions that linear or cubic interpolation methods may cause, and is a common practice recommended in the literature for soil properties (Panagos et al., 2014; Poggio et al., 2021). To determine the organic matter content, the following equation was applied (Panagos et al., 2014; Torri et al., 1997).

$$OM = SOC \times 0.1 \times 1.724$$

Soil erodibility was determined using the following empirical equation, in which organic matter (OM) is calculated based on the clay, silt, and sand fractions (Panagos et al., 2014; Torri et al., 1997).

$$K = 50 \times \left(0.0021 \times (12 - OM) \times \left(\frac{Silt \times (100 - Clay)}{10000}\right)^{1.14}\right)$$

After applying the above equation, the resulting soil erodibility (K-factor) values at 100-m spatial resolution ranged between 0.10 and 0.45, which is consistent with the value ranges reported in previous studies (Panagos et al., 2014; Panagos, Borrelli, Poesen et al., 2015).

LS Factor: For the calculation of the LS factor, the ALOS DEM (30 m) dataset produced by the Japan Aerospace Exploration Agency was used (Japan Aerospace Exploration Agency, 2021). Based on this digital elevation model (DEM), slope (in degrees) and flow accumulation rasters were derived. Flow accumulation represents the number of upstream cells that contribute flow into a given cell. The formula used to evaluate the influence of slope steepness and slope length on erosion requires converting the slope raster from degrees to radians (Desmet and Govers, 1996; Moore and Burch, 1986).

$$\left(\theta_{radyan} = \theta_{derece} \times \frac{\pi}{180}\right)$$

This conversion and the subsequent LS calculation were performed using the Raster Calculator tool in ArcGIS Pro and ArcGIS Desktop 10.6.1. By analyzing the digital elevation model together with the slope (in degrees) and flow accumulation rasters, the LS factor was derived in accordance with the formulation proposed by Moore and Burch (1986)..

$$LS = \left(\frac{Flow\ Acc \times pixel\ size}{22.13}\right)^{0.4} \cdot \left(\frac{sin(\theta)}{0.0896}\right)^{1.3}$$

Flow Acc: Represents the flow accumulation raster.

θ: Represents the slope raster (converted from degrees to radians).

C Factor: In soil erosion modeling, the cover-management factor (C factor) is one of the most sensitive and influential parameters affecting model performance (Guo et al., 2023; Renard et al., 1997). The C factor incorporates land cover and land management practices into the erosion equation, thereby representing the influence of vegetation type and surface conditions on erosion processes (Guo et al., 2023). As one of the most variable factors in the RUSLE model, the C factor strongly influences the rate of soil detachment and transport, since land cover directly affects rainfall splash, surface runoff, and sediment transport by water and wind. Consequently, misclassification or inaccurate assignment of C-factor values may substantially distort model outputs (Almagro et al., 2019; Guo et al., 2023).

Distinguishing between forests, grasslands, rangelands, and agricultural areas—as well as accounting for temporal changes in land management (e.g., fallow practices, crop rotation, shifts in vegetation cover)—significantly improves the accuracy of erosion analyses. Remote sensing datasets and geographic databases such as CORINE are particularly valuable for detecting these spatial and temporal changes (Benavidez et al., 2018; Panagos, Borrelli, Meusburger et al., 2015).

In this study, to incorporate land cover information into the model, the CORINE Land Cover 2018 dataset—widely recognized and used across European countries—was employed (EEA, 2018). The C-factor values were assigned based on coefficients reported in the literature for each CORINE land-cover class (Table 4) (Benavidez et al., 2018; Guo et al., 2023; Panagos, Borrelli, Poesen et al., 2015; Renard et al., 1991; Wischmeier and Smith, 1978). Coefficients approaching 1 indicate low soil protection, while coefficients approaching 0 reflect high soil protection.

Table 4. CORINE Land Use Classes and Assigned C-Factor Coefficients Used for the Ceyhan Basin

Land Cover Type (CORINE 2018)	C Factor Value	Land Cover Type (CORINE 2018)	C Factor Value
Non-irrigated arable land	0.3	Beaches, dunes, sands	0.01
Permanently irrigated land	0.2	Airports	0.003
Dump sites	0.2	Sport and leisure facilities	0.003
Complex cultivation patterns	0.18	Industrial or commercial units	0.003

continued on following page

Table 4. Continued

Land Cover Type (CORINE 2018)	C Factor Value	Land Cover Type (CORINE 2018)	C Factor Value
Vineyards	0.15	Green urban areas	0.003
Olive groves	0.15	Discontinuous urban fabric	0.003
Land principally occupied by agriculture, with significant areas of natural vegetation	0.14	Continuous urban fabric	0.003
Agro-forestry areas	0.12	Road and rail networks and associated land	0.003
Construction sites	0.12	Salt marshes	0.002
Fruit trees and berry plantations	0.12	Broad-leaved forest	0.001
Transitional woodland-shrub	0.05	Mixed forest	0.001
Sclerophyllous vegetation	0.05	Coniferous forest	0.001
Natural grasslands	0.05	Bare rocks	0.001
Inland marshes	0.05	Coastal lagoons	0
Sparsely vegetated areas	0.02	Sea and ocean	0
Pastures	0.01	Water bodies	0
Mineral extraction sites	0.01	Water courses	0

P Factor: The P factor, one of the key components of the Revised Universal Soil Loss Equation (RUSLE), is a dimensionless coefficient that quantifies the effectiveness of soil conservation practices in reducing erosion (Renard et al., 1997). This coefficient represents the degree to which the applied conservation measures enhance the soil's resistance to erosion. The P value ranges between 0 and 1, where a value of 1.0 indicates the absence of any conservation practice, while lower values signify the presence of effective erosion-control measures (Wischmeier and Smith, 1978).

The P factor quantifies the effectiveness of soil conservation practices—such as terracing, strip cropping, and contour plowing—in reducing soil erosion. Due to the lack of high-resolution spatial inventory data on conservation practices across the basin, the P factor was classified based on CORINE Land Cover 2018 data and assigned according to conservation effectiveness values widely accepted in the literature. In this study, the P factor was determined using a land use/land cover (LULC)-based approach. The CORINE Land Cover 2018 dataset was used at 100-m spatial resolution, and each land-use class was assigned a P-factor coefficient consistent with commonly referenced values in the literature (Table 5) (Feng et al., 2025; Kimberlin and Moldenhauer, 1977; Renard et al., 1997; Wischmeier and Smith, 1978).

Table 5. P-Factor Coefficients Assigned to Land Use/Land Cover Classes within the Ceyhan Basin

Land Use / Land Cover	P-factor coefficient
Forest and Natural Vegetation	0.3
Contour Plowing in Agricultural Areas	0.5
Downslope (Slope-Parallel) Tillage in Agricultural Areas	0.6
Steppe, Bare Soil, Urban/Settled Areas	1.0

RESULTS

The application of the RUSLE erosion model in the Ceyhan Basin clearly revealed the influence of topographic structure, precipitation regime, land cover, and land-use dynamics on soil transport processes. Through the spatial analysis of the R, K, LS, C, and P factors, erosion potential maps were generated, enabling the identification of areas with high susceptibility to erosion as well as regions predominantly shaped by human activities.

R Factor: The rainfall erosivity index (R factor) shows considerable spatial variability across the basin. Very low erosivity values (171.4–250) occur predominantly in the Elbistan Plain, covering 20.6% of the basin. Low erosivity zones (250–350), largely situated in the central and eastern parts of the basin, account for 34.1% of the total area. Moderate erosivity values (350–450), extending from the Mediterranean coastline toward the basin's central regions, constitute 40.1% of the basin. High erosivity zones (450–500), located mainly south of Kahramanmaraş and east of Osmaniye, represent 3.4%, while very high erosivity areas (>500) cover 1.3% of the basin (Figure 5).

K Factor: The soil erodibility factor (K) reveals that the lowest soil resistance to erosion occurs in the southern parts of Kahramanmaraş and areas near the Mediterranean coast. Very low erodibility values (0–0.1) cover 0.02% of the basin. Low erodibility classes (0.1–0.2), located in similar southern regions, account for 8.35% of the basin. Moderate erodibility values (0.2–0.3), concentrated mainly in the central portions of the basin and representing the most widespread class, cover 55.2% of the total area. High erodibility values (0.3–0.4), predominantly in the western sections of the basin, constitute 36.07% of the area. Finally, very high erodibility zones (0.4–0.5), occurring in small fragmented patches in the central basin, account for 0.3% of the basin (Figure 6).

Figure 5. R-Factor Distribution in the Ceyhan Basin

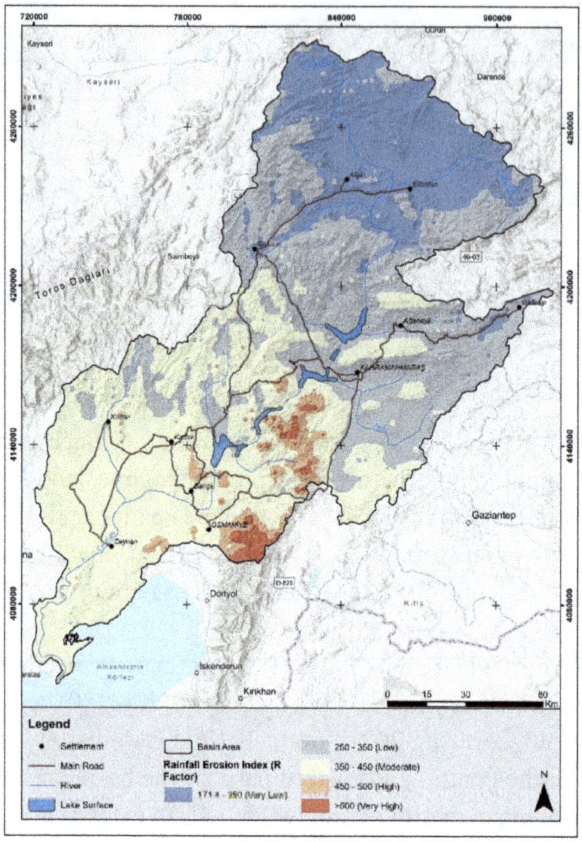

Figure 6. K-Factor Distribution in the Ceyhan Basin

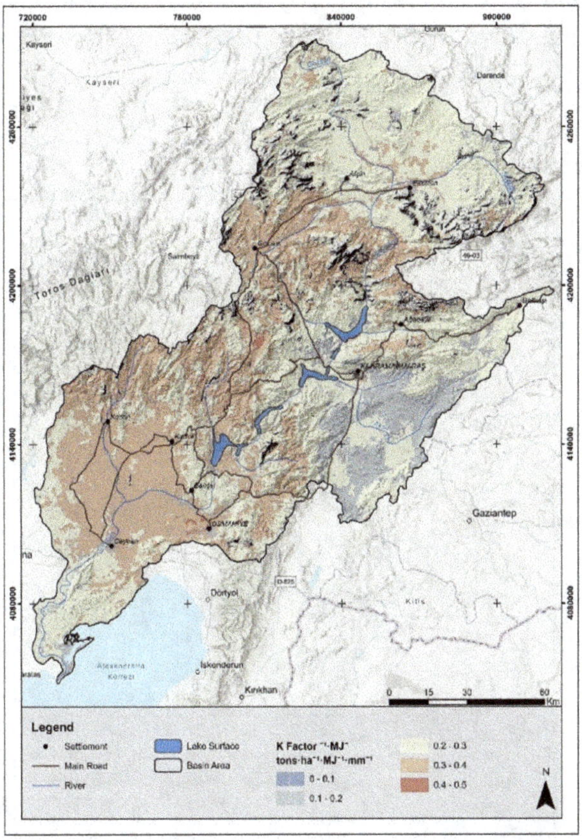

LS Factor: The influence of slope steepness and slope length on soil erosion was evaluated using the LS factor. The lowest LS values occur in flat or nearly flat areas such as the Elbistan Plain and the Sağlık Plain. LS values between 0–2 cover 55.07% of the basin. Areas with LS values of 2–5 constitute 10.34%, while those with values between 5–10 cover 12.01% of the basin. LS values of 10–20 account for 13.22%, and areas with LS values between 20–50 represent 8.27%. The highest LS values (≥50) cover 1.04% of the basin (Figure 7).C faktörü: toprak üzerindeki kullanım şeklinin RUSLE erozyon modeline dahil edilmiş şeklidir. 0.001-0.01 arasındaki çok yüksek korunan bölgeler havzanın %15.62'sini, 0.01-0.1 arasındaki yüksek korunan bölgeler havzanın %16.22'sini, 0.1-0.25 arasindaki orta düzey korunan bölgeler havzanın %56.56'sını, 0.25-0.45 arasındaki düşük koruma bölgeleri havzanın %11.59'unu oluşturmaktadır (Şekil 8).

C Factor: The C factor represents the influence of land use and land cover on soil susceptibility within the RUSLE model. Highly protected areas with C values between 0.001–0.01 cover 15.62% of the basin, while areas with high protection (0.01–0.1) account for 16.22%. Moderately protected zones with C values of 0.1–0.25 constitute the largest portion of the basin at 56.56%. Low-protection areas with C values between 0.25–0.45 cover 11.59% of the basin (Figure 8).

Figure 7. LS Factor Distribution in the Ceyhan Basin

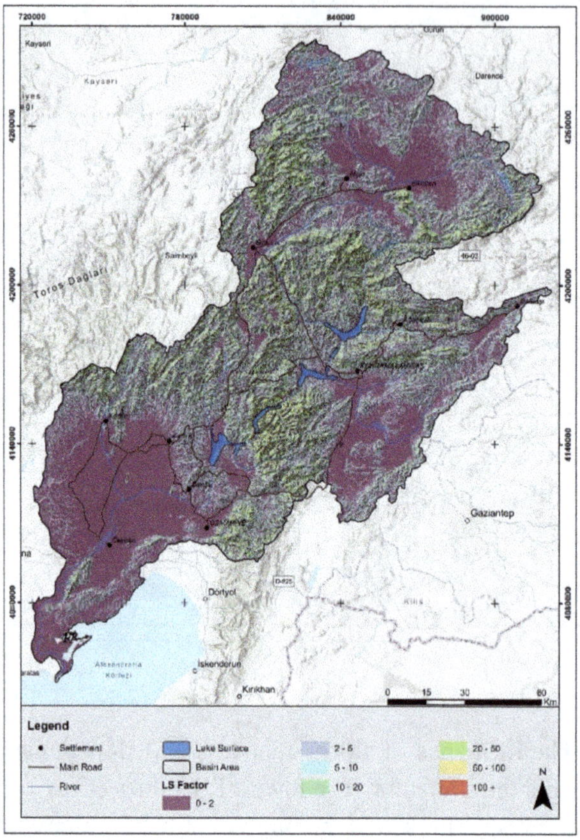

Figure 8. C-Factor Distribution in the Ceyhan Basin

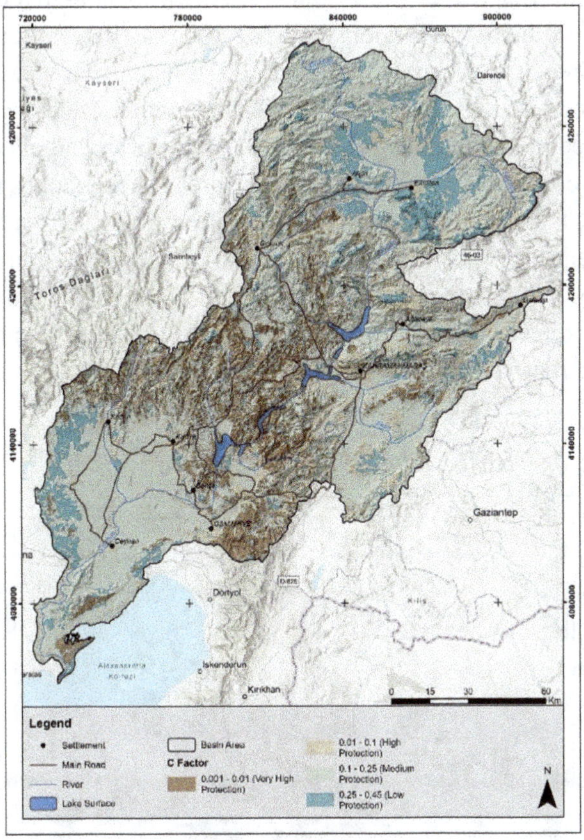

P Factor: Using the P factor, the degree to which human activities (e.g., terracing) protect the soil surface was identified based on land-use data and classified into four main categories. Areas with the highest level of protection (P = 0.3) cover 14.49% of the basin. Effectively protected regions (P = 0.5) account for 7.74%, while moderately protected areas (P = 0.6) represent 23.34% of the basin. Areas with no protection against erosion (P = 1.0) constitute the largest share, covering 54.20% of the basin (Figure 9).A (RUSLE): 5 faktörün bir araya gelmesi ile yapılan analiz sonucunda Ceyhan havzasının ton/ha/yıl cinsinden değerleri bulunmuştur. Buna yılda 0-5 ton/ha/yıl kaybeden bölgeler havzanın %60.76'sını, 5-25 ton/ha/yıl havzanın %11.25'ini, 25-50 ton/ha/yıl havzanın %7.48'ini, 50-100 ton/ha/yıl havzanın 7.64'ünü ve 100 ton/ha/yıl'dan fazla toprak kaybı yaşayan alanlar havzanın %12.85'ini oluşturmaktadır (Şekil 10).

A (RUSLE Soil Loss): The integration of all five RUSLE factors resulted in the estimation of annual soil loss values (tons/ha/year) for the Ceyhan Basin. Areas experiencing 0–5 tons/ha/year of soil loss account for 60.76% of the basin. Soil loss values of 5–25 tons/ha/year correspond to 11.25%, while 25–50 tons/ha/year correspond to 7.48% of the basin. Regions with 50–100 tons/ha/year represent 7.64% of the basin, and areas experiencing more than 100 tons/ha/year of soil loss constitute 12.85% of the total area (Figure 10). Areas where annual soil loss exceeds 100 tons/ha/year are concentrated in fields where high LS values are largely accompanied by low vegetation cover (high C factor). In these areas, steep slopes increase surface runoff velocity, facilitating the transport of soil particles; agricultural use, sparse vegetation cover, or bare surfaces cause rainfall to impact the soil surface directly. Additionally, the fact that the P factor in these areas is mostly in the range of 0.6–1.0 indicates that protective measures are inadequate and that human-induced land management increases the risk of erosion.

Figure 9. P-Factor Distribution in the Ceyhan Basin

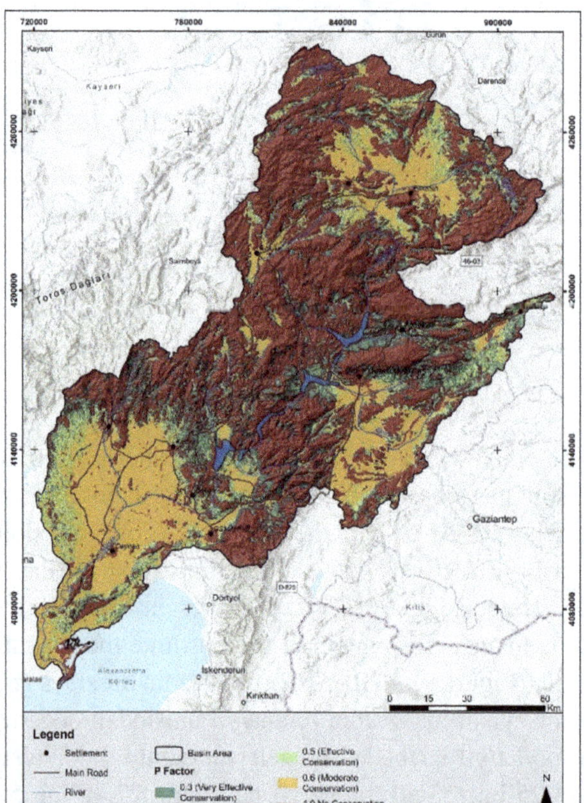

Figure 10. RUSLE Soil Loss (tons/ha/year) in the Ceyhan Basin

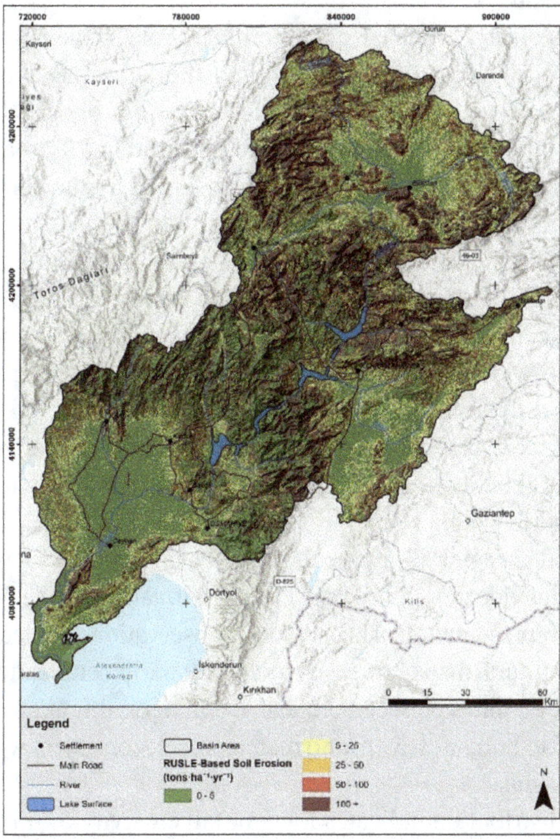

DISCUSSION

This study provides a significant contribution to the literature by comprehensively applying the RUSLE model in one of Turkey's most topographically complex and land-use-intensive basins, the Ceyhan Basin. The integration of high-resolution GIS infrastructure, SoilGrids-based soil properties, CORINE 2018 land-use classes, and LS components derived from ALOS DEM enabled a detailed factor-based spatial analysis of erosion dynamics at the watershed scale. Methodologically, the study is consistent with large-scale erosion assessments (Panagos et al., 2015; Borrelli et al., 2017; Imamoğlu and Dengiz, 2017) while filling an important quantitative gap for the Ceyhan Basin.

Model outputs clearly emphasize the dominant role of the LS factor in shaping erosion susceptibility. High LS values around the Dibek, Binboğa, and Nurhak Mountains demonstrate the controlling influence of topography on high-risk zones, consistent with the findings of Moore and Burch (1986) and Renard et al. (1997). Erosion susceptibility in the basin is governed by the combined effects of topography and climate. Extensive flat areas with low slopes in the northern and northwestern sections limit erosion, whereas rapidly increasing slopes over short distances in mountainous areas intensify surface runoff and promote high erosion potential. Climatically, lower rainfall erosivity prevails in the north under continental conditions, while short-duration, high-intensity Mediterranean rainfall in the south significantly accelerates erosion processes, particularly on steep terrain. Regarding the C factor, forests and semi-natural areas provide effective protection, whereas agricultural lands—especially rainfed areas and those with complex cropping patterns—exhibit higher erosion susceptibility. This confirms the reliability of CORINE-based C-factor assignments at the basin scale, in line with Panagos et al. (2015) and Benavidez et al. (2018). The classification of the P factor into four categories allows representation of broad-scale conservation practices; however, local variability in agricultural management introduces some generalization. Nevertheless, the dominance of P values between 0.6 and 1.0 in agricultural areas highlights the basin's sensitivity to human-driven land management, consistent with findings from similar Turkish basins.

The R-factor distribution indicates low rainfall erosivity over the Elbistan Plain and substantially higher values in Mediterranean-influenced southern sections, particularly around Çukurova and Osmaniye, consistent with Panagos et al. (2015). The empirical approach ($R = 0.5 \cdot P$) is widely applied in data-scarce regions and produces values within ranges reported in European and Turkish studies. K-factor results show that soils with moderate erodibility dominate the basin. Although geological structure was not included as an explicit model factor, geological influences are indirectly reflected through the K factor via SoilGrids inputs that incorporate geomorphological and geological variables (Poggio et al., 2021). While resampling SoilGrids data from 250 m to 100 m introduces some generalization, the resulting K-value ranges are consistent with Panagos et al. (2014) and Torri et al. (1997).

The integrated RUSLE output (A factor) demonstrates that erosion patterns result from the combined influence of natural conditions and land-use practices. Areas with soil losses exceeding 100 tons/ha/year are concentrated on steep slopes with sparse vegetation, whereas more than 60% of the basin experiences low erosion rates due to the protective effect of forest and semi-natural covers.

Overall, the findings provide robust scientific and managerial insights into erosion-risk dynamics in the Ceyhan Basin. The results offer a strong spatial decision-support framework for improving conservation practices, refining land-use planning on steep slopes, and developing sustainable, data-driven watershed and agricultural management strategies.

CONCLUSION

The RUSLE erosion model applied in the Ceyhan Basin revealed the combined influence of the basin's topographic, climatic, and land-use characteristics on soil loss. The spatial modeling of the R, K, LS, C, and P factors demonstrated that both natural and anthropogenic drivers play critical roles in shaping the distribution of erosion risk within the basin. The markedly lower R-factor values in the north and higher values in the south confirm that the Mediterranean climate is the dominant determinant of rainfall erosivity in the region. K-factor values indicate that soils with moderate erodibility are widespread throughout the basin, highlighting erosion as a significant threat particularly in areas where agricultural activities are concentrated.

The LS factor, identified as one of the most influential components in the basin, clearly demonstrates that steep slopes act as the primary drivers of erosion processes. The C-factor results show that forests and natural vegetation substantially reduce erosion, whereas rainfed agricultural areas and fragmented cultivation patterns are associated with higher erosion risk. The P factor further indicates that conservation practices are largely insufficient across much of the basin; the finding that more than 50% of the basin falls within the category lacking any conservation measures represents a critical concern for sustainable land management.

The A values derived from the RUSLE model indicate that erosion levels are low across approximately 60% of the basin; however, very high erosion occurs in 12–13% of the area. These zones correspond primarily to steep slopes where sparse vegetation cover and inappropriate land-use decisions coincide. Overall, the findings show that the Ceyhan Basin exhibits a heterogeneous erosion pattern, with priority intervention required in areas where steep topographic gradients and intensive agricultural pressure overlap. The results provide a robust decision-support basis for watershed management, agricultural planning, and the development of environmental protection policies.

REFERENCES

Almagro, A., Thomé, T. C., Colman, C. B., Pereira, R. B., Junior, J. M., Rodrigues, D. B. B., & Oliveira, P. T. S. (2019). Improving Cover and Management Factor (C-factor) Estimation Using remote Sensing Approaches for Tropical Regions. *International Soil and Water Conservation Research*, *7*(4), 325–334. Advance online publication. DOI: 10.1016/j.iswcr.2019.08.005

Benavidez, R., Jackson, B., Maxwell, D., & Norton, K. (2018). A Review of the (Revised) Universal Soil Loss Equation (R/USLE) in the Context of Land Use and Management Changes. *Hydrology and Earth System Sciences*, *22*(11), 4001–4025. https://doi.org/https://doi.org/10.5194/hess-22-6059-2018

Bezak, N., Mikoš, M., Borrelli, P., Alewell, C., Alvarez, P., Anache, J. A. A., Baartman, J., Ballabio, C., Biddoccu, M., Cerdà, A., Chalise, D., Chen, S., Chen, W., De Girolamo, A. M., Gessesse, G. D., Deumlich, D., Diodato, N., Efthimiou, N., Erpul, G., & Panagos, P. (2021). Soil Erosion Modelling: A Bibliometric Analysis. *Environmental Research*, *197*(March), 111087. Advance online publication. DOI: 10.1016/j.envres.2021.111087 PMID: 33798514

Borrelli, P., Alewell, C., Alvarez, P., Anache, J. A. A., Baartman, J., Ballabio, C., Bezak, N., Biddoccu, M., Cerdà, A., Chalise, D., Chen, S., Chen, W., De Girolamo, A. M., Gessesse, G. D., Deumlich, D., Diodato, N., Efthimiou, N., Erpul, G., Fiener, P., & Panagos, P. (2021). Soil Erosion Modelling: A Global Review and Statistical Analysis. *The Science of the Total Environment*, *780*, 146494. Advance online publication. DOI: 10.1016/j.scitotenv.2021.146494 PMID: 33773346

Borrelli, P., Robinson, D. A., Fleischer, L. R., Lugato, E., Ballabio, C., Alewell, C., Meusburger, K., Modugno, S., Schütt, B., Ferro, V., Bagarello, V., Van Oost, K., Montanarella, L., & Panagos, P. (2017). An assessment of the global impact of 21st century land use change on soil erosion. *Nature Communications*, *8*(1), 2013. Advance online publication. DOI: 10.1038/s41467-017-02142-7 PMID: 29222506

Borrelli, P., Robinson, D. A., Panagos, P., Lugato, E., Yang, J. E., Alewell, C., Wuepper, D., Montanarella, L., & Ballabio, C. (2020). Land use and climate change impacts on global soil erosion by water (2015-2070). *Proceedings of the National Academy of Sciences of the United States of America*, *117*(36), 21994–22001. DOI: 10.1073/pnas.2001403117 PMID: 32839306

Burrough, P. A. (1986). Principles of Geographical. In Information systems for land resource assessment.... https://www.academia.edu/download/2438559/9fjg8q78n4wux4l.pdf

Cruz, A. M. D., Maniquiz-Redillas, M. C., Tanhueco, R. M., & De Leon, M. P. (2025). Estimation of the Rainfall Erosivity Factor (R-Factor) for Application in Soil Loss Models. *Water (Basel), 17*(6), 837. https://doi.org/https://doi.org/10.3390/w17060837. DOI: 10.3390/w17060837

Desmet, P. J., & Govers, G. (1996). A GIS Procedure for Automatically Calculating the USLE LS Factor on Topographically Complex Landscape Units. *Journal of Soil and Water Conservation, 51*(5), 427–433. DOI: 10.1080/00224561.1996.12457102

EEA. (2018). Copernicus Land Monitoring Service. European Environmental Agency. https://land.copernicus.eu/en/products/corine-land-cover/clc2018#download

Erpul, G., İnce, K., Demirhan, A., Küçümen, A., Akdağ, M. A., Demirtaş, İ., Sarıhan, B., Çetin, E., & Şahin, S. (2020). *Su Erozyonu İl İstatistikleri Toprak Erozyonu Kontrol Stratejileri*. Çevre Şehircilik ve İklim Bakanlığı, Çölleşme ve Erozyonla Mücadele Müdürlüğü.

Erpul, G., Şahi, S., İnce, K., Küçümen, A., Akdağ, M. A., Demirtaş, İ., & Çetin, E. (2018). *Türkiye Su Erozyonu Atlası*. Çölleşme ve Erozyonla Mücadele Genel Müdürlüğü.

FAO. (2016). Report of the Forty-fourth Session of the Codex Committee on Food Labelling. FAO.

Feng, S., Zhao, W., Yan, J., Xia, F., Wang, H., Zhou, A., Pereira, P. (2025). Vegetation Restoration Contributes to a Reduction in Wind and Water Erosion in China's Drylands. Applied Geography, 176(May 2024). DOI: 10.1016/j.apgeog.2025.103517

Foley, J. A., Ramankutty, N., Brauman, K. A., Cassidy, E. S., Gerber, J. S., Johnston, M., Mueller, N. D., O'Connell, C., Ray, D. K., West, P. C., Balzer, C., Bennett, E. M., Carpenter, S. R., Hill, J., Monfreda, C., Polasky, S., Rockström, J., Sheehan, J., Siebert, S., & Zaks, D. P. M. (2011). Solutions for a cultivated planet. *Nature, 478*(7369), 337–342. DOI: 10.1038/nature10452 PMID: 21993620

Ganasri, B. P., & Ramesh, H. (2015). Geoscience Frontiers Assessment of Soil Erosion by RUSLE Model Using Remote Sensing and GIS - A Case Study of Nethravathi Basin. *Geoscience Frontiers, 7*(6), 1–9. DOI: 10.1016/j.gsf.2015.10.007

Guo, G., Pan, Y., Kuai, J., Dai, Q., Zhu, Y., Xu, Y., Jia, X., Zeng, G., & Lin, J. (2023). A More Accurate Approach to Estimate the C-factor of RUSLE by Coupling Stratified Vegetation Cover Index in Southern China. *Forest Ecology and Management, 541*(120979), 120979. https://doi.org/https://doi.org/10.1016/j.foreco.2023.120979. DOI: 10.1016/j.foreco.2023.120979

Imamoglu, A., & Dengiz, O. (2017). Determination of Soil Rrosion Risk Using RUSLE Model and Soil Organic Carbon Loss in Alaca Catchment (Central Black Sea Region, Turkey). *Rendiconti Lincei. Scienze Fisiche e Naturali*, *28*(1), 11–23. DOI: 10.1007/s12210-016-0556-0

Japan Aerospace Exploration Agency. (2021). ALOS Global Digital Surface Model. Distributed by OpenTopography. https://doi.org/Japan Aerospace Exploration Agency (2021). ALOS World 3D 30 meter DEM. V3.2, Jan 2021. Distributed by OpenTopography. DOI: 10.5069/G94M92HB

Karger, D. N., Conrad, O., Böhner, J., Kawohl, T., Kreft, H., Soria-Auza, R. W., Zimmermann, N. E., Linder, H. P., & Kessler, M. (2017). CHELSA V2.1: Technical specification. *Scientific Data*, •••, 4.

Karger, D. N., Schmatz, D. R., Dettling, G., & Zimmermann, N. E. (2020). High-resolution monthly precipitation and temperature time series from 2006 to 2100. *Scientific Data*, *7*(1), 1–10. DOI: 10.1038/s41597-020-00587-y PMID: 32703947

Kimberlin, L. W., & Moldenhauer, W. C. (1977). Predicting Soil Erosion. In ASAE Publ (Issues 4–77, pp. 31–42).

Laflen, J. M., Lane, L. J., & Foster, G. R. (1991). WEPP: A New Generation of Erosion Prediction Technology. *Journal of Soil and Water Conservation*, *46*(1), 34–38. https://doi.org/https://doi.org/10.1080/00224561.1991.12456572. DOI: 10.1080/00224561.1991.12456572

Lu, D., Li, G., Valladares, G. S., & Batistella, M. (2004). Mapping Soil Erosion Risk in Rondônia, Brazilian Amazonia: Using RUSLE, Remote Sensing and GIS. *Land Degradation & Development*, *15*(5), 499–512. DOI: 10.1002/ldr.634

Martin, A., Gunter, J. T., & Regens, J. L. (2003). Estimating Erosion in a Riverine Watershed: Bayou Liberty-Tchefuncta River in Louisiana. *Environmental Science and Pollution Research International*, *10*(4), 245–250. DOI: 10.1065/espr2003.05.153 PMID: 12943008

Millward, A. A., & Mersey, J. E. (1999). Adapting the RUSLE to Model Soil Erosion Potential in a Mountainous Tropical Watershed. *Catena*, *38*(2), 109–129. DOI: 10.1016/S0341-8162(99)00067-3

Moore, I. D., & Burch, G. J. (1986). Physical Basis of the Length-Slope Factor in the Universal Soil Loss Equation. *Soil Science Society of America Journal*, *50*(5), 1294–1298. https://doi.org/https://doi.org/10.2136/sssaj1986.03615995005000050042x. DOI: 10.2136/sssaj1986.03615995005000050042x

Panagos, P., Borrelli, P., Meusburger, K., Alewell, C., Lugato, E., & Montanarella, L. (2015). Estimating the Soil Erosion Cover-Management Factor at the European Scale. *Land Use Policy*, *48*, 38–50. DOI: 10.1016/j.landusepol.2015.05.021

Panagos, P., Borrelli, P., Poesen, J., Ballabio, C., Lugato, E., Meusburger, K., Montanarella, L., & Alewell, C. (2015). The New Assessment of Soil Loss by Water Erosion in Europe. *Environmental Science & Policy*, *54*, 438–447. DOI: 10.1016/j.envsci.2015.08.012

Panagos, P., Meusburger, K., Ballabio, C., Borrelli, P., & Alewell, C. (2014). Science of the Total Environment Soil Erodibility in Europe : A high-Resolution Dataset Based on LUCAS. *The Science of the Total Environment*, *479*(480), 189–200. DOI: 10.1016/j.scitotenv.2014.02.010 PMID: 24561925

Poggio, L., Sousa, L. M. D., Batjes, N. H., Heuvelink, G. B. M., Kempen, B., & Ribeiro, E. (2021). SoilGrids 2. 0 : Producing Soil Information for the Globe with Quantified SpatialUncertainty. SOIL, 7(1995), 217–240. https://doi.org/https://doi.org/10.5194/soil-7-217-2021

Renard, K. G., Foster, G. R., Weesies, G. A., McCool, D. K., & Yoder, D. C. (1997). *Predicting Soil Erosion by Water: a Guide to Conservation Planning With the Revised Universal Soil Loss Equation (RUSLE).* US Department of Agriculture, Agricultural Research Service.

Renard, K. G., Foster, R., Weesies, G., & Porter, G. (1991). RUSLE: Revised Universal Soil Loss Equation. *Journal of Soil and Water Conservation*, *46*(1), 30–33. DOI: 10.1080/00224561.1991.12456571

Shaheemath Suhara, K. K., Varughese, A., Sunny, A. C., & Krishna, A. P. R. (2023). Erosivity Factor of the Revised Universal Soil Loss Equation (RUSLE) - A Systematized Review. *Current World Environment*, *18*(2), 433–445. https://www.cwejournal.org/pdf/Vol18No2/CWE_Vol18_No2_p_433-445.pdf. DOI: 10.12944/CWE.18.2.02

SYGM. (2018). Ceyhan Havzasi Taşkın Yönetim Planı. https://www.tarimorman.gov.tr/SYGM/Belgeler/Taşkın Yönetim Planları 26.12.2022/CEYHAN HAVZASI TAŞKIN YONETIM PLANI.pdf

TOKB. (1987). Tarım, Orman ve Köyişleri Bakanlığı (1987) Türkiye'nin genel yönetim planlaması (toprak koruma ana planı).

Torri, D., Poesen, J., & Borselli, L. (1997). Predictability and Uncertainty of the Soil Erodibility Factor Using a Global Dataset. *Catena*, *31*(1–2), 1–22. https://doi.org/https://doi.org/10.1016/S0341-8162(97)00036-2. DOI: 10.1016/S0341-8162(97)00036-2

Wischmeier, W. H., & Smith, D. D. (1978). *Predicting Rainfall Erosion Losses: a Guide to Conservation Planning. Science and Education Administration*. US Department of Agriculture.

Chapter 6
Resilience in Action: Cambodia's Evolving Disaster Governance and Regional Cooperation in a Changing Climate

Sovanna Huot
http://orcid.org/0009-0006-8335-442X
Department of Political Science, University of Delhi, New Delhi, India

Sovannaroth Chheang
http://orcid.org/0009-0007-3382-7169
Department of Education, Sovanna Elton Foreign Language School, Cambodia

ABSTRACT

This chapter explores Cambodia's evolving disaster governance as a case of resilient transformation in the Global South. Once seen as highly vulnerable to climate-related disasters, Cambodia has become an active player in disaster risk reduction (DRR), climate adaptation, and regional humanitarian cooperation. The chapter highlights legal reforms, institutional shifts, local preparedness, and cross-border initiatives that reflect a shift from reactive to inclusive, forward-looking planning. It examines how Cambodia integrates DRR with development and climate politics, asserts leadership within ASEAN, and uses disaster governance to build state legitimacy and strengthen diplomacy. Key themes include multi-level governance, gender and youth inclusion, early warning systems, and community-based management. Cambodia's experience shows how resilience-building can drive regional engagement, urging greater recognition of Global South strategies and investment in digital infrastructure, civil society, and cross-border collaboration for lasting climate resilience.

DOI: 10.4018/979-8-3373-6274-8.ch006

I. INTRODUCTION

Disasters occur when natural or human-induced hazards exceed a society's capacity to cope, leading to loss of life, economic disruption, and social instability. Climate change intensifies these risks by altering temperature and precipitation patterns, raising sea levels, and increasing the frequency and severity of extreme weather events such as floods and droughts (Amsalu et al., 2026; Çelik, 2026). These impacts disproportionately affect developing countries, where climate-sensitive livelihoods and limited institutional capacity heighten vulnerability, making disaster risk a central challenge for governance, development, and long-term resilience.

Over the past two decades, Cambodia has undergone a noteworthy transformation in the realm of disaster governance. Once predominantly identified as a vulnerable, aid-dependent country in the face of recurring climate-induced hazards, Cambodia is now progressively recognized as a proactive regional contributor in shaping disaster risk reduction (DRR) frameworks and climate resilience strategies. According to certain studies, like Kelman (2017), countries that have been through a war often have a hard time getting rid of bad climate governance. Cambodia, on the other hand, has gone against this trend by making its institutions more consistent, investing in local skills, and working together with other countries, especially in Southeast Asia. This chapter claims that Cambodia has successfully altered its status from being a symbol of climate vulnerability to becoming a leader in regional resilience and new ways of running the government.

This development should be considered within the broader context of Southeast Asia region, where countries are facing increasing climatic challenges such as shifting monsoon patterns, rising sea levels, and more severe weather events (Hugo, 2011; IPCC, 2022). Cambodia shares a lot of environmental difficulties with its neighbors, like Vietnam, Thailand, and Laos, especially in the lower Mekong basin. According to regional experts (Yusuf & Francisco, 2009), the region's ecological concerns are connected, therefore responses need to be coordinated across borders. Cambodia is becoming more aware of this as it takes part in ASEAN mechanisms and climate cooperation with other countries.

Cambodia was quite vulnerable to climate shocks at first because it was often flooded, especially in lowland and deltaic areas, and because it relied on subsistence farming (Hoang et al., 2018; Huot, 2025a). But now the state has a more organized and thorough approach to dealing with disasters. Alexander (2020) and other policy experts suggest that the changes to Cambodia's institutions are not merely technical; they are also political. The government is connecting building resilience to state legitimacy, working with other countries, and wanting to prosper. There are two key research goals that this chapter talks about:

(1) To examine how Cambodia has included disaster management in its development plans at many levels, like national, sub-national, and community based.
(2) To see how Cambodia's disaster resilience strategy has changed because of engaging with other nations and areas, focusing on ASEAN frameworks, multilateral partnerships, and soft diplomacy.

As it investigates these goals, the chapter is directed by the following primary research questions:

(1) How has Cambodia included disaster management in its strategies for growth and climate change as a whole?
(2) What are the legal, institutional, and operational ways that make this integration easier?
(3) How has Cambodia used international and regional platforms to make its institutions better able to handle crises and learn more?

These issues are examined from multi perspectives, including political science, development studies, catastrophe risk research, and broader studies of Southeast Asia. The analysis incorporates major policy documents, like the Sub-Decree on Disaster Management from 2015 and the Cambodia Climate Change Strategic Plan from 2014 to 2023, as well as academic research and reviews from development organizations (ADPC, 2019; UNDP, 2023). This study is crucial because it could transform how people in the Global South communicate about climate vulnerability. People often think of Cambodia as a country that needs help, but it is becoming a leader in new policies and resilience. Researchers like Gaillard and Mercer (2013) underline how crucial it is to focus on disaster governance models that emanate from inside a community, especially in the Global South. Cambodia is an example of this kind of idea since it integrates national political will, locally risk management, and international cooperation into one plan for resilience (Huot, 2025a). This situation also helps us understand how small, climate-vulnerable countries may run their own regions while yet being part of global disaster management systems. The Cambodian situation is particularly relevant and useful since more people want global disaster policy frameworks that include everyone. O'Brien et al. (2011) suggest that resilience should not only be understood as a technology goal, but also as a process of change that is built on making governance, fairness, and social learning better. Cambodia's success in these areas suggests that it is on a steadier path to prosperity. This can help other countries with similar issues make substantial adjustments.

In short, this chapter reveals that Cambodia is not only an example of how to adapt, but it is also becoming a pioneer in managing risk in a region that is at the forefront of climate change. The study adds to the expanding corpus of research on the politics of resilience and supports policy discussions about how to construct a developmental state, work together regionally, and come up with innovative ideas in the face of disaster risk.

II. NATIONAL CONTEXT: CLIMATE VULNERABILITY AND GOVERNANCE FOUNDATIONS

Cambodia's developing approach to controlling disaster risk is heavily influenced by the country's historical issues and its expanding institutional ambition. People have long considered that Cambodia is one of the countries in Southeast Asia that is most vulnerable to climate change. It has had to deal with an environment that is unstable for plants and animals, weak for businesses, and has few strong buildings. But it has not stayed locked in this lame plot. Cambodia has changed from a country that was always in danger to one that is actively building resilience through governance reform, cross-sectoral planning, and partnerships between global and local organizations (Huot, 2025a). This shift has taken place during the previous 20 years. This part talks about four connected and supporting factors that make up the foundation of Cambodia's climate and disaster governance structure: (1) the country's geographic and socioeconomic exposure, (2) how past disasters can cause new ones, (3) how the National Committee for Disaster Management (NCDM) makes governance mechanisms more official, and (4) how disaster risk reduction (DRR) is included in national development and climate policy frameworks.

Cambodia's geography and how it has grown make it very vulnerable to climate change and other environmental shocks. The Tonle Sap system, which is vital for the ecosystem but not particularly reliable when it comes to water flow, is in Cambodia. The country is in the floodplains of the Lower Mekong Basin. Because of this, the country often has really bad weather, like floods, droughts, and more rain that comes out of nowhere. Yusuf and Francisco (2009) who study development geography, have always said that Cambodia is one of the Southeast Asian countries that is most likely to be affected by climate change. They believe this is due to things that happen in nature and things that people do. Moreover, agricultural employing seventy per cent of Cambodia's workforce work in agriculture, which is the most important part of the country's economy (FAO, 2013). It depends a lot on the monsoon seasons and the flow of rivers. Changes in temperature, rainfall, and water availability affect the lives of millions of Cambodians since they depend on these things.

Cambodia's socio-economic profile, which includes rural poverty, few insurance alternatives, and poorly established infrastructure, makes the hazards posed by its environment much worse. Nuorteva et al. (2010) and other environmental scientists say that building dams upstream, modifying the climate, and changing land use patterns make populations downstream even more vulnerable, especially those that rely on fishing and subsistence farming. However, it is important to note that policymakers have begun to consider these exposures as not just hazards, but also as opportunities to improve adaptive ability. Vulnerability assessments are now integrated into national climate strategies, and the Ministry of Environment (2014) has taken steps to incorporate risk mapping into subnational development plans. Rather than viewing climate hazards as external shocks to be endured, Cambodian institutions have begun interpreting them as governance challenges to be managed, anticipated, and learned from, a significant epistemological shift in national planning discourse.

Cambodia's catastrophe management has changed a lot over the years, and some things that happened in the past were highly crucial for making these changes happen. The worst of these were the floods in 2000 and 2011. Both had a huge effect on politics, the economy, and society. The flood in 2000, which killed more than 300 individuals and affected more than 3.4 million people, revealed how poor Cambodia's disaster response infrastructure was and how much more susceptible rural and marginalized communities were (ADPC, 2019). The event not only caused problems for people, but it also demonstrated how bad the mechanisms are for coordinating across agencies, making infrastructure stronger, and getting ready for public health emergencies.

The flood in 2011, while not as fatal, had much bigger consequences on the economy. It affected 18 provinces and cost more than US$521 million, which made governments and development partners think hard about what they were doing (UNDP, 2014). Researchers that study disasters, including Aldrich (2012) and Birkland (2006), claim that significant catastrophes are often focusing events or critical junctures that make it possible for policies to change. These disasters speed up the pace of altering institutions in Cambodia. They made early warning systems better, emergency communication infrastructure better, and community-based response training started. Hoang et al. (2018) and Huot (2025a) argue that such disasters have significantly shifted the approach to risk management, moving from reactive responses to anticipatory risk governance. Cambodia's delayed addition of disaster management to its larger development plans is an example of this transition.

The National Committee for Disaster Management (NCDM) is an important aspect of Cambodia's disaster management system. It was created in 1995 by Sub-Decree No. 30 ANK. First, the NCDM was supposed to be the main group that would organize emergency reactions. Now, it is a governing body that can do many things. Alexander (2020) adds that this transformation is part of a bigger trend in countries that have been through a war. They are going from meeting immediate needs to planning for long-term stability.

The NCDM oversees a lot of things right now, such as getting ready for disasters, reducing their effects, helping people recover quickly, and making sure that all levels of government follow the same rules. It has gotten aid with its work from several international development groups, including UNDP, JICA, and ASEAN. Reports from the Asian Disaster Preparedness Center (2019) show how the NCDM has used modern techniques including geographic mapping, simulation modeling, and vulnerability indexing to help them make decisions. It has also made risk management more local and made disaster preparations a part of everyday life by establishing up Provincial and Commune Disaster Management Committees. The 2015 Sub-Decree on Disaster Management made it official what different government agencies were supposed to do. This linked together response systems at the local, national, and regional levels (World Bank, 2016). Parry et al. (2018), legal experts suggest that such legal frameworks are essential for transforming from fragmented catastrophe risk management systems to a more inclusive and universally effective approach. Cambodia has been doing this as its institutions get stronger.

Cambodia's government has changed a lot, but one of the most important things is that it has made lowering the risk of disasters a core aspect of its national development and climate plan. The Strategic National Action Plan for Disaster Risk Reduction (SNAP-DRR) 2008–2013 was one of the first steps toward combining planning for social and economic issues with planning for disaster risk reduction. The Ministry of Environment (2014) carried on this momentum with the Cambodia Climate Change Strategic Plan (CCCSP) 2014–2023. It sets several goals that are all connected and recognizes that being able to deal with disasters is an important part of national sustainability. These goals link DRR directly to key fields such as education, agriculture, water management, and public health. This shows that the whole system needs to be more resilient, not only in certain scenarios. The National Strategic Development Plan (NSDP) 2019–2023 and Cambodia's Rectangular Strategy Phase IV make these ideas more formal. They indicate that the government is serious about conserving the environment, making sure that everyone can benefit from development, and making sure that growth is sustainable. RGC (2019) says that climate resilience is not just an afterthought in its development strategy; it is a crucial aspect of its long-term plan to alter the country's economy and society. Cambodia is an active participant in regional disaster diplomacy because it is a member of the ASEAN Agreement on Disaster Management and Emergency Response (AADMER). Caballero-Anthony (2015) says that smaller countries generally get additional authority through regional frameworks. Cambodia is a strong example of this since it uses ASEAN forums to share technical knowledge, come up with policies that are in line with those of other nations, and find ways to work together to respond, as seen in Figure 1.

Figure 1. Cambodia's pathway from climate vulnerability to institutional resilience, illustrating key stages from environmental exposure to policy integration in disaster risk governance

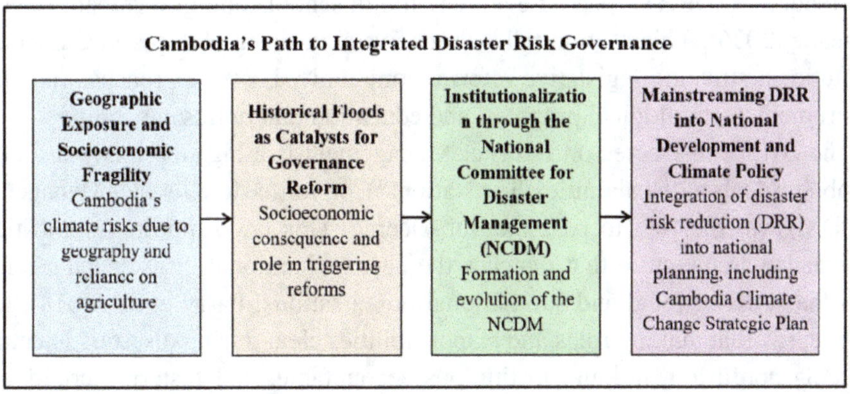

Cambodia's national position is a good example of how geographic exposure, historical learning, institutional innovation, and policy integration all function together in a changing fashion. Cambodia is neither just a passive recipient of foreign help nor an example of how to continually be weak. Instead, it has gained more power over establishing a strong, future-looking catastrophe governance model that includes everyone (Huot, 2025a). Resilience is not only a technique to manage risk; it is also a deliberate way to bring about change in development through planned reform, local empowerment, and global alignment. Cambodia's example can help other nations in the Global South learn how to deal with climate change and strengthen their institutions at the same time. It reveals that true resilience comes not from avoiding being vulnerable, but from tackling it with a clear plan, good communication, and a long-term investment.

III. INSTITUTIONAL INNOVATIONS AND LOCALIZED PREPAREDNESS

Cambodia's ideas about and how it handles disaster governance have changed a lot in the previous twenty years. The country has stopped depending on centralized government officials to deal with calamities after they happen and has instead moved toward a more proactive, decentralized, and community-driven strategy of managing risk (Huot, 2026a). This transformation is happening because of things happening all around the world, such the Sendai Framework for Disaster Risk Reduction, and because the country has learnt from a series of major floods, droughts, and climate-related shocks. A big element of this transition is the desire to make resilience more local. This means making choices and being ready for disasters closer to communities, giving women and youth more power, and making disaster risk reduction (DRR) a part of everyday life in schools and government (Huot & Chheang, 2026). This section talks about five significant changes in Cambodia's institutions: strategic legislative reform, community-based approaches, inclusive government, technological progress, and education that builds capability.

The 2015 Sub-Decree on Disaster Management is a big step forward for how Cambodia deals with calamities. The National Committee for Disaster Management (NCDM) was the first to pass the Sub-Decree. This law supports a multi-tiered governance architecture that matches the needs of national disaster management with the skills of local and sub-national organizations. Parry et al. (2018), legal experts, say that making roles and responsibilities clear during disasters is vital for reducing confusion, holding institutions accountable, and fostering coordinated actions. The Sub-Decree makes it clear that authorities at the province, district, and commune levels have the power to look at risks, create backup plans, and put in place risk-reduction measures that are right for their area. There is legal support for decentralization, and there are also efforts to make collaboration between different sectors more effective. The Asian Disaster Preparedness Center (2019) claims that the Sub-Decree makes it easier for agencies to work together by linking disaster preparedness efforts with public service areas like water management, agriculture, education, and healthcare. It also says that local disaster committees must be set up, which means that making the government better able to handle disasters is now part of its normal duties. This legal framework has allowed people at the sub-national level to not only carry out Cambodia's disaster policy but also help make it (Huot & Chheang, 2026). This is a very important change for making resilient systems that work in the local area.

Community-Based Disaster Risk Management (CBDRM) is now a big element of how Cambodia deals with risk. Because they are based on the concept that communities are best suited to discover and deal with the threats they face, CBDRM initiatives have been frequently implemented in rural and peri-urban areas. Some of the local NGOs that help with these programs are the Cambodian Red Cross, Save the Children, and ActionAid. The government, international aid, and other groups help them. Backed by development partners including JICA and UNDP, CBDRM has enabled local communities to map vulnerabilities, train emergency volunteers, and establish Village Disaster Risk Reduction Plans. As Andrew (2018) and Gaillard and Mercer (2013) assert, genuine resilience requires that communities be given not only tools but also agency in disaster planning and response. In Cambodia, this idea comes to life through community workshops, risk assessments done by local councils, and the incorporation of community plans in Commune Development

Plans. This vertical alignment makes sure that people at the top know about problems in their area while still preserving control at the bottom. CBDRM is significant in Cambodia since it does more than just deal with short-term concerns. It also helps bring people together, offers power to groups that are often left out, and inspires people to be creative when things are unclear.

Cambodia's disaster management has put more focus on inclusiveness, especially getting women and young people involved, because they know that social structures determine how vulnerable and resilient people are. Many gender scholars, including Enarson and Chakrabarti (2009), have emphasized for a long time that resilience must take into consideration the diverse effects and abilities of different groups of people. The National Action Plan for Disaster Risk Reduction is one of the national frameworks that Cambodian officials have used to build strategies. These plans call for programming that takes gender into account, maternal health during evacuations, and women leading disaster committees (MoWA, 2018). More young people are getting involved, especially through school-based simulations, youth-led risk assessments, and peer education programs supported by NGOs like Plan International and World Vision Cambodia. These programs not only teach kids practical skills, but they also help them become good communicators in their homes and communities. The Sendai Framework's focus on all-of society engagement works well with this plan, which encompasses everyone. It makes sure that resilience is built from the ground up, with different people setting priorities and reactions.

Cambodia is moving toward anticipatory governance by putting more money into using technology to control disaster risk. Cambodia has considerably enhanced its ability to anticipate, monitor, and communicate risk by introducing multi-hazard early warning systems (Huot, 2025a). This was made possible by working with UNESCAP, the World Meteorological Organization (WMO), and ADPC. These platforms now use weather, water, and geography data to assist consumers make choices and give out alerts in real time. One of the most important uses has been making SMS alert systems for mobile phones. These systems have been tested in areas that are sensitive to climate change, like Battambang, Kampong Thom, and Prey Veng. These methods let officials get in touch with individuals quickly, especially those who do not always have internet access. They tell people crucial things including where to locate shelter, flood alerts, and evacuation routes (WEF, 2020). Some examples of technology improvements that have made disaster response planning more accurate for policymakers are putting in place autonomous weather stations, flood gauges, and data visualization dashboards. Mechler et al. (2019) and other climate governance experts suggest that mechanisms to transition from reacting to losses to avoiding risks are essential, and Cambodia is quickly adopting this approach.

More individuals are coming to understand that education is an important part of being able to handle disasters. Cambodia has come a long way in making disaster risk education a common part of school. The Ministry of Education, Youth, and Sport have partnered with Save the Children, UNICEF, and the Japan International Cooperation Agency (JICA) to produce disaster curricula for elementary and secondary schools. These lectures cover a lot of information, like how to evacuate and offer first aid, as well as how climate change will influence the threats in your area. Ronan and Johnston (2005), who are experts in education, say that kids are highly crucial for sharing information about disasters with their families, which makes the community stronger. Cambodian pilot projects in provinces like Kandal and Siem Reap have demonstrated that students not only absorb preparedness knowledge but also actively share it with parents, neighbours, and peers. Encouraged by these successes, the Ministry now aims to institutionalize disaster education nationwide, viewing it as a strategic investment in long-term, generational resilience, as featured in Figure 2.

Figure 2. Cambodia's localized disaster governance model, highlighting the interlinked innovations that support community resilience

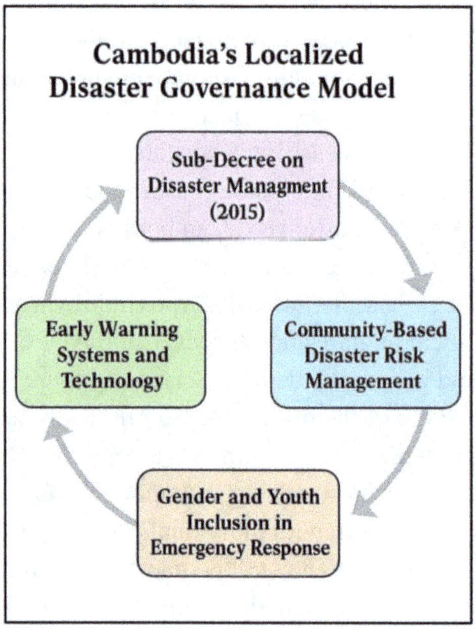

The Last but not least, Cambodia's new institutions and localized disaster preparedness plans show that disaster management may be very different in the future. The government has transformed how it handles risks from a centralized and reactionary way to one that is participative and adaptable. This was done by adjusting the law that looks ahead, giving communities more control, making sure that everyone is included in programs, using technology wisely, and reaching out to schools. These changes show that resilience is no longer merely a technological problem; it is now also a project for social, institutional, and developmental growth (Huot, 2025b). Cambodia's example provides other countries in the Global South that are at risk of climate change important lessons. For example, it shows that good crisis management comes from the community and is aware of what is going on, and that creating long-term resilience starts at the bottom.

IV. CAMBODIA'S REGIONAL AND INTERNATIONAL COOPERATION

Cambodia's disaster management has changed not just because it has changed its own policies, but also because it has worked closely with allies in the area and around the world. Cambodia has learnt how important it is to be part of networks that help people share information, plan together, and make sure that institutions are working together when climate-related disasters transcend borders (Huot, 2025b). It has improved both its technical skills and its diplomatic stature in Southeast Asia and abroad by doing this. This part talks about how Cambodia is a member of the ASEAN Agreement on Disaster Management and Emergency Response (AADMER), how it works with other countries and groups to prepare for and practice disasters, and how it uses disaster governance to build trust and do soft diplomacy in the area.

Cambodia has been an active member of ASEAN's disaster management system since the early 2000s. Cambodia's participation in the ASEAN Agreement on Disaster Management and Emergency Response (AADMER), which was signed in 2005 and went into force in 2009, is the main reason for this. Caballero-Anthony (2015) and other experts in regional governance think that AADMER is the first regional catastrophe pact in the world that is legally obligatory. It enables the countries in ASEAN to work together in a planned way, coordinate their efforts, and help each other. Cambodia has always made sure that its national plans follow the AADMER Work Program (2021–2025), especially when it comes to working with other stakeholders, analyzing risks, offering early warnings, and building capacity. The ASEAN Secretariat (2021) says that Cambodia has hosted and taken part in many regional policy discussions, technical workshops, and consultative gatherings. These have helped the National ASEAN Committee for Disaster Management (NCDM) in Cambodia and ASEAN's disaster governance bodies, such as the ASEAN Coordinating Centre for Humanitarian Assistance on disaster management (AHA Centre), operate more closely together. Another sign that Cambodia is becoming a more important leader is that it is organizing regional catastrophe simulation exercises like the ASEAN Regional catastrophe Emergency Response Simulation Exercise

(ARDEX). Cambodia has been able to improve its Standard Operating Procedures (SOPs), see how effectively they work with those of neighboring nations, and see how ready it is for scenarios that happen across borders (AHA Centre, 2023). These items not only prepare Cambodia for action, but they also indicate that Cambodia can be trusted to work with other countries in the area.

Cambodia's bilateral partnerships with its ASEAN neighbors, especially Thailand, Singapore, and Indonesia, make its cooperation in the area even stronger. These links have made it easier to share technical knowledge, plan for emergencies in an organized way, and undertake collaborative training sessions. For example, Cambodia has worked with the Singapore Civil Defence Force (SCDF) on intensive courses on urban search and rescue, managing logistics during disasters, and incident command systems (SCDF, 2020). Cambodian disaster officials also learn from each other a lot through peer-learning exchanges with Thailand's Department of Disaster Prevention and Mitigation (DDPM). Most of these talks are about how to get ready for disasters in your town, how to teach people about them, and how to let people know about floods early. It is clear how people in border areas get along with each other. Flooding in the Mekong River and forest fires, for example, are problems that span borders and need rules and agreements that everyone can use to aid each other. Disaster diplomacy experts like Kelman (2011) have argued that this kind of cooperation not only makes it simpler for diverse areas to work together, but it also builds trust between them, especially when matters are politically delicate.

Cambodia has gotten a lot better at dealing with disasters because it has worked with development partners from many countries, such as the United Nations Development Program (UNDP), the Food and Agriculture Organization (FAO), the Japan International Cooperation Agency (JICA), and the Asian Disaster Preparedness Center (ADPC). The UNDP has helped a lot to keep Cambodia's institutions strong. They did this by creating the Cambodia Climate Change Strategic Plan (CCCSP) and assisting the NCDM in developing national contingency mechanisms (UNDP, 2023). FAO-funded programs have helped Cambodia make its farming systems smarter about climate change and less likely to be affected by catastrophes. This has made it tougher for floods and droughts to damage the people who live in rural regions (FAO, 2013; Çelik et al., 2026). JICA's participation has had a particularly significant influence on sub-national capacity building and infrastructure risk management. This involves developing resilient rural roads and school structures, improving flood forecasting algorithms, and mapping hazards in areas that are vulnerable to flooding (JICA, 2021). Modern risk analytics, simulation tools, and training modules in early warning, emergency operations center design, and school-based DRR instruction have all been made more accessible to Cambodia thanks to design. Cambodia is ranked among the ASEAN nations with the best localized risk management systems and training outreach, according to assessments conducted by ADPC (2019).

Cambodia's regional and international collaboration in disaster governance has significant political value in addition to its operational and technical advantages. Disaster collaboration provides a non-political forum for communication, confidence-building, and the projection of soft power in an area historically characterized by interstate conflicts. Non-traditional security concerns, such disaster management, are strategically valuable in boosting ASEAN centrality and cohesion, according to political analysts like Emmers (2017) and Haacke (2013). Cambodia's continuous participation in donor coordination platforms, cross-border exercises, and AADMER demonstrates its dedication to these regional standards and strengthens its reputation as a cooperative and positive actor in international affairs. Furthermore, Cambodia can project national success stories, draw in climate finance, and advocate for the needs of least-developed and climate-vulnerable states thanks to its prominence in international fora, such as its involvement in the COP climate negotiations and the Global Platform for Disaster Risk Reduction. As such, disaster governance has become not only a tool of resilience but a vehicle for international legitimacy and regional leadership, as shown in Table 1.

Cambodia has a strong grasp of both government and diplomacy, which is seen in its approach to regional and international disaster cooperation. Cambodia has improved both its national readiness for disasters and its reputation as a dependable, forward-thinking regional actor by actively participating in ASEAN frameworks, conducting bilateral training initiatives, and forming strategic alliances with international organizations (Huot, 2026b). Its use of disaster governance as a forum for institutional learning, soft diplomacy, and regional collaboration demonstrates how technical capability and global participation can strategically merge. Other Global South nations looking to transform risk governance into a source of diplomatic capital and resilience might learn a lot from Cambodia's experience.

Table 1. Cambodia's Regional and International Disaster Cooperation

Cooperation Area	Partners/Institutions	Key Activities and Achievements
ASEAN Frameworks	ASEAN Secretariat, AHA Centre	Participation in AADMER; hosting ARDEX simulations; aligning national SOPs with ASEAN protocols
Bilateral Training	Thailand, Singapore, Indonesia	Joint training in emergency response, search and rescue; operational coordination in border provinces
Multilateral Partnerships	UNDP, FAO, JICA, ADPC	Technical support in planning, risk mapping, resilient infrastructure, and school-based DRR
Disaster Diplomacy	ASEAN Member States, Global Platforms	Use of disaster governance for regional trust-building, international visibility, and soft power diplomacy

V. GOVERNANCE, LEGITIMACY, AND THE POLITICS OF RESILIENCE

In Cambodia, disaster governance has evolved beyond technocratic coordination, playing a central role in the construction of state legitimacy, the functioning of multi-level governance, and the projection of resilience as a diplomatic and developmental narrative. As the country faces increasingly complex and transboundary hazards, effective disaster management has evolved into not only a practical but also a political and symbolic act. This section examines how Cambodia's disaster management strengthens intergovernmental cooperation, improves the country's reputation abroad, and advances an adaptive governance model that is in line with long-term resilience.

In fragile and post-conflict environments such as Cambodia, the state's ability to provide emergency services has a substantial impact on perceptions of government legitimacy (Mcloughlin, 2015; Slater, 2010). Historically, Cambodia's rural populations have had limited access to public services, particularly in distant districts hit by floods, droughts, and storms. Disaster response in these regions turns into an outward display of the state's power and presence. Cambodia's increased investment in community-based preparedness, rapid relief delivery, and disaster recovery has helped to close the public trust gap between urban and rural areas. Field evaluations by the Asian Disaster Preparedness Center (2019) and the UNDP (2023) demonstrate that timely government action following the 2013 floods in Kampong Thom and the 2020 flash floods in Banteay Meanchey increased public trust in local authorities. When citizens see that local communes are receiving timely aid, such as road repairs, food supply, and school rehabilitations, the legitimacy of the state is reinforced via responsive action rather than theory. According to scholars like Lund (2006), legitimacy in emerging nations frequently arises from crisis management performance rather than legislative requirements. Cambodia's disaster response mechanisms, particularly those led by the National Committee for Disaster Management (NCDM), put this theory into effect, transforming humanitarian intervention into a state-building tool.

The key to Cambodia's success in disaster administration is its move toward multi-level governance, in which province and commune disaster committees carry out national objectives with assistance from outside partners and regional coordination. Multi-level governance, according to political scientists like Hooghe and Marks (2003), is the distribution of power among several jurisdictions, improving both horizontal accountability and vertical coordination. The 2015 Sub-Decree on Disaster Management, which gave local disaster management committees legal authority, and the incorporation of Village Disaster Risk Reduction Plans into Commune Development Plans serve as examples of this system in Cambodia. Contextualized planning that takes into account local knowledge and vulnerabilities, effective resource deployment, and prompt early warning communication have all been made possible by this system. External partnerships, such as JICA's technical support for flood forecasting at the provincial level and UNDP's facilitation of community simulations and drills, help to build the synergy between levels of government. The vertical coherence of Cambodia's risk governance allows for accountability both upward (to national ministries) and downward (to people and civil society), which is consistent with governance models espoused by Mechler et al. (2019) and Ostrom (2012). Furthermore, because sub-national authorities can adjust plans to new hazards while keeping with national aims and ASEAN guidelines, this layered approach offers more adaptive flexibility. In order to respond to compound disasters, like COVID-19 that coincides with seasonal floods, in a way that preserves both public health and food security, such adaptability has proven essential.

Beyond its use in domestic governance, resilience is strategically and symbolically important to Cambodia's foreign policy and development narrative. Cambodia has been able to transform itself from a post-conflict aid recipient to a regional player in climate and catastrophe diplomacy by positioning resilience as a national strength. Cambodia regularly emphasizes how it has included disaster risk reduction into national development plans in multilateral forums like the Global Platform for Disaster Risk Reduction and ASEAN ministerial meetings, establishing itself as a responsible global citizen and policy innovator. Such regional participation enables smaller governments to develop normative power by providing ideas rather than merely resources, according to scholars like Acharya (2009). Additionally, the focus on resilience makes Cambodia more attractive to donor organizations, technical partnerships, and climate finance instruments. Its proactive disaster education initiatives and gender-responsive emergency planning, for example, have garnered favorable assessments from UNICEF, GIZ, and the Green Climate Fund, enhancing its reputation as a progressive, reform-focused state. In Southeast Asia, where practical collaboration and norm alignment are frequently the foundations of regional credibility, this symbolic capital is particularly significant. Cambodia has successfully employed resilience as a strategy of developmental branding and constructive diplomacy by fusing disaster governance with diplomatic message.

Cambodia's dedication to adaptive governance, the capacity to not only respond to previous catastrophes but also foresee and prepare for future, systemic risks, may be its most significant contribution to the conversation around disasters. Adaptive governance is a kind of governance that is defined by learning, flexibility, and cooperation across institutional boundaries, according to adaptation theorists such as Folke et al. (2005) and Pelling (2010). These ideas are becoming increasingly reflected in Cambodia's national strategies. The Cambodia Climate Change Strategic Plan (CCCSP) places a strong emphasis on multi-stakeholder collaborations, scientific knowledge integration, and iterative policy learning. Regular post-disaster assessments are carried out to guide future planning, and commune-level discussions are used to incorporate public input. Programs that combine disaster risk education, ecosystem-based techniques, and infrastructure resilience demonstrate Cambodia's move away from merely reactive response and toward proactive resilience-building. Additionally, its partnership with the Asian Disaster Preparedness Center (ADPC) on scenario modeling and risk data platforms represents an investment in governance systems that are focused on the future. By instituting adaptive thinking, Cambodia presents itself as a climate-ready nation that can handle complexity, change, and uncertainty in addition to being disaster-resilient.

The accomplishments of Cambodia in disaster governance highlight a larger shift in institutional accountability, state capability, and diplomatic maturity. State legitimacy has increased because of disaster management capabilities, particularly among rural communities' most vulnerable residents. A more integrated and contextually aware risk management system has resulted from the incorporation of multi-level governance. Resilience has become a strategic narrative as well as a policy objective, influencing Cambodia's interactions with foreign partners and raising its stature in the region. Above importantly, Cambodia's dedication to socially and structurally adaptive governance shows that resilience is a continuous process of change rather than a result. In this regard, Cambodia provides other climate-vulnerable countries looking to negotiate the politics of risk with realism, inclusivity, and foresight with important lessons.

VI. CONCLUSION

An exciting and educational example of how a post-conflict, climate-vulnerable nation may move toward institutional strength, local readiness, and regional cooperation is Cambodia's experience with disaster governance. In addition to lowering the risk of disasters, Cambodia has developed a resilience architecture that improves state legitimacy, developmental inclusiveness, and regional participation through persistent efforts in legal reform, decentralized governance, community-based projects, and international collaborations. This chapter has established that Cambodia's disaster management method should be understood as a transformative governance pathway rather than a technological apparatus. Cambodia has steadily developed a reputation for being capable, well-coordinated, and forward-thinking, despite its early image as a fragile and aid-dependent country dealing with regular floods and droughts. Today, it is an active contributor to ASEAN's regional risk frameworks, a collaborator in cross-border disaster exercises, and a policy innovator in inclusive, adaptive governance. Disaster governance in Cambodia is increasingly understood as developmental infrastructure, a mechanism through which the state not only responds to hazards but also invests in systems of education, health, agriculture, and digital capacity. As so, Cambodia demonstrates how catastrophe resilience may be both a development success story and a vehicle for regional integration, particularly given Southeast Asia's shared ecological and geopolitical backdrop.

As Cambodia enters a new stage of resilience-building, several proactive measures can support the consolidation and expansion of its achievements:

1. Expand digital Infrastructure for Real-Time Risk Alerts: Cambodia should increase its investment in digital connectivity, mobile platforms, and data-sharing technologies to improve real-time risk communication. This involves increasing mobile SMS alert systems, weather monitoring networks, and interoperable GIS platforms that connect national agencies to subnational responders. Collaboration with partners like the World Meteorological Organization, UNESCAP, and private telecom carriers might ensure that even isolated rural people receive timely and effective warnings.
2. Strengthening Civil Society Collaborations in Disaster Education and Monitoring: Local NGOs, women's organizations, faith-based networks, and youth organizations are vital in fostering resilience at the grassroots level. The government should formalize community-based risk monitoring tools, expand disaster teaching in schools, and enable civil society actors to co-create information and implement early warning strategies. These actions would not only boost local ownership but also promote social trust and responsibility.

3. Promote Transboundary Cooperation on Water and Climate Resilience: Given Cambodia's reliance on the Mekong River and shared ecological zones with Laos, Vietnam, and Thailand, transboundary water governance must become a fundamental component of its resilience strategy. Cambodia should actively participate in river basin organizations, climate adaptation forums, and collaborative water management programs to reduce upstream risks and increase data transparency. In addition to strengthening regional integration and trust, such cooperation would aid in managing common vulnerabilities.

A more comprehensive change in international disaster discussions that acknowledges, incorporates, and benefits from the ideas of Global South nations is necessary given Cambodia's trajectory. All too frequently, conceptions of resilience that are Northern-centric ignore the institutional experimentation, political agency, and adaptive capacities occurring in nations like Cambodia. International venues, such as the Global Platform for Disaster Risk Reduction, COP climate discussions, and UNDRR policy consultations, must prioritize the knowledge systems and lived experiences of Global South nations. The paradigm used in Cambodia shows that institutional coherence, social inclusion, and community-driven governance are just as important to resilience as cutting-edge technology or extensive funding. As a result, incorporating nations like Cambodia more meaningfully into global catastrophe governance frameworks is more than just representation; it is also about knowledge justice, ensuring that resilience strategies are anchored in diverse, context-specific realities.

The resilient turn of Cambodia has both symbolic meaning and useful lessons. It is a tale of change: from vulnerability to proactive readiness, from brittleness to foresight, and from discrete reactions to coordinated, regional approaches. Cambodia's governance style is a shining example of inclusive, cooperative, and adaptable resilience as climate change worsens, and disasters become more complicated. Because Cambodia is not only surviving in the face of climate uncertainty, but also setting the standard, the country's experience challenges the international community to reconsider how resilience is defined, which models are given priority, and how local triumphs might influence global norms.

REFERENCES

Acharya, A. (2009). *Constructing a security community in Southeast Asia: ASEAN and the problem of regional order*. Routledge., DOI: 10.4324/9780203939239

ADPC. (2019). *Disaster risk reduction in Cambodia*. Bangkok: Asian Disaster Preparedness Center. https://www.preventionweb.net/files/68249_682301cambodiaupdaed16oct2019.pdf

Aldrich, D. P. (2012). *Building resilience: Social capital in post-disaster recovery*. University of Chicago Press. DOI: 10.7208/chicago/9780226012896.001.0001

Alexander, D. (2020). Disaster risk reduction. In Devuyst, Y. (Ed.), *The European Union's community method: Foundations and evolutionlocked*. Oxford University Press., DOI: 10.1093/acrefore/9780190228637.013.1538

Amsalu, A., Garedew, W., Melka, G. A., Çelik, M. A., & Gemeda, D. O. (2026). Trends in climate extreme indices and implications for coffee farming in The Kafa Biosphere Reserve, southwest Ethiopia. *Climate Services*, *42*, 100637. DOI: 10.1016/j.cliser.2026.100637

Andrew, M. (2018). Revisiting community-based disaster risk management. In I. Davis, S. Bender, F. Krimgold, & F. McDonald (Eds.), *Reducing disaster risks: Progress and challenges in the caribbean region* (pp. 42–52). Routledge. DOI: 10.4324/9781315067261-5

Birkland, T. A. (2006). *Lessons of disaster: Policy change after catastrophic events*. Georgetown University Press.

Caballero-Anthony, M. (2015). *An introduction to non-traditional security studies: A transnational approach*. Sage Publications.

Çelik, M. A. (2026). Agro-climatic classification of Türkiye's Agricultural Basins using long-term ERA5-Land temperature data. *Physics and Chemistry of the Earth Parts A/B/C*, *142*, 104307. DOI: 10.1016/j.pce.2026.104307

Çelik, M. A., Kıraç, A., & Gouveia, C. M. (2026). North Atlantic Oscillation modulation of drought and vegetation activity in Türkiye. *Theoretical and Applied Climatology*, *157*(2), 74. Advance online publication. DOI: 10.1007/s00704-025-05971-5

Centre, A. H. A. (2023). *AHA centre annual report 2022*. Jakarta: ASEAN Coordinating Centre for Humanitarian Assistance on disaster management (AHA Centre). https://ahacentre.org/wp-content/uploads/2023/09/AHA-Centre-Annual-Report-2022-FA-Digital.pdf

Emmers, R. (2017). *Non-traditional security in Asia: Dilemmas in securitization*. Routledge. DOI: 10.4324/9781315247878

Enarson, E., & Chakrabarti, P. D. (2009). *Women, gender and disaster: Global issues and initiatives*. SAGE Publications India Pvt Ltd.

FAO. (2013). *Climate-smart agriculture sourcebook: Agriculture, forestry and fisheries*. Rome: Food and Agriculture Organization of the United Nations. https://www.fao.org/3/i3325e/i3325e.pdf

Folke, C., Hahn, T., Olsson, P., & Norberg, J. (2005). Adaptive governance of social-ecological systems. *Annual Review of Environment and Resources*, *30*(1), 441–473. DOI: 10.1146/annurev.energy.30.050504.144511

Gaillard, J.-C., & Mercer, J. (2013). From knowledge to action: Bridging gaps in disaster risk reduction. *Progress in Human Geography*, *37*(1), 93–114. https://doi.org/OI. DOI: 10.1177/0309132512446717

Haacke, J. (2013). *ASEAN's diplomatic and security culture: Origins, development and prospects*. Routledge. DOI: 10.4324/9780203037539

Hoang, L. P., Biesbroek, R., Tri, V. P. D., Kummu, M., van Vliet, M. T., Leemans, R., Kabat, P., & Ludwig, F. (2018). Managing flood risks in the Mekong Delta: How to address emerging challenges under climate change and socioeconomic developments. *Ambio*, *47*(6), 635–649. DOI: 10.1007/s13280-017-1009-4 PMID: 29478182

Hooghe, L., & Marks, G. (2003). Unraveling the central state, but how? Types of multi-level governance. *The American Political Science Review*, *97*(2), 233–243. https://garymarks.web.unc.edu/wp-content/uploads/sites/13018/2016/09/hooghe.marks_.unravelingcentralstate.apsr_.2003.pdf

Hugo, G. (2011). Future demographic change and its interactions with migration and climate change. *Global Environmental Change*, *21*, S21–S33. DOI: 10.1016/j.gloenvcha.2011.09.008

Huot, S. (2025a). Cultural diplomacy across the Mekong subregion: Cambodia's soft power diplomacy in Southeast Asia. *Literary Enigma: The International Journal of English Language. Literature and Culture*, *1*(6), 80–90. DOI: 10.5281/zenodo.15803843

Huot, S. (2025b). Digital tutors in the kingdom: Exploring ChatGPT's influence on Cambodian students' academic and cognitive growth. In Pandya, J. D., & Patel, K. C. (Eds.), *Interdisciplinary insights: Exploring multiple perspectives* (Vol. 2, pp. 36–54). Meadow Publication.

Huot, S. (2026a). Digital diplomacy in Cambodia: Embracing artificial intelligence for strategic engagement and national development. In Turan, A., & Ucuzal, H. (Eds.), *Diplomacy in the age of artificial intelligence* (pp. 307–330). IGI Global Scientific Publishing., DOI: 10.4018/979-8-3373-4937-4.ch013

Huot, S. (2026b). Emerging voices from Cambodia: International student journeys and educational aspirations in a globalized academic landscape. In Irham, I., Huda, M., & Yaqin, L. N. (Eds.), *International student narratives in globalized higher education* (pp. 365–400). IGI Global Scientific Publishing., DOI: 10.4018/979-8-3373-4212-2.ch014

Huot, S., & Chheang, S. (2026). Artificial intelligence and human rights in Cambodia: Pathways to social justice and positive developmental progress. In Pasupuleti, S., & Yadav, U. (Eds.), *Safeguarding social justice and human rights in the age of AI* (pp. 355–386). IGI Global Scientific Publishing., DOI: 10.4018/979-8-3373-6598-5.ch016

IPCC. (2022). *Climate change 2022: Impacts, adaptation and vulnerability.* Geneva: Intergovernmental Panel on Climate Change (IPCC). https://www.ipcc.ch/report/ar6/wg2/

JICA. (2021). *Building disaster and climate resilient cities in ASEAN.* Tokyo: Japan International Cooperation Agency. https://openjicareport.jica.go.jp/pdf/12321154_01.pdf

Kelman, I. (2011). *Disaster diplomacy: How disasters affect peace and conflict.* Routledge., DOI: 10.4324/9780203806210

Kelman, I. (2017). Linking disaster risk reduction, climate change, and the sustainable development goals. *Disaster Prevention and Management*, *26*(3), 254–258. DOI: 10.1108/DPM-02-2017-0043

Lund, C. (2006). Twilight institutions: Public authority and local politics in Africa. *Development and Change*, *37*(4), 685–705. DOI: 10.1111/j.1467-7660.2006.00497.x

Mcloughlin, C. (2015). When does service delivery improve the legitimacy of a fragile or conflict-affected state? *Governance: An International Journal of Policy, Administration and Institutions*, *28*(3), 341–356. DOI: 10.1111/gove.12091

Mechler, R., Bouwer, L. M., Schinko, T., Surminski, S., & Linnerooth-Bayer, J. (2019). *Loss and damage from climate change: Concepts, methods and policy options*. Springer Nature., DOI: 10.1007/978-3-319-72026-5

Ministry of Environment. (2014). *Cambodia climate change strategic plan 2014-2023.* Phnom Penh: Royal Government of Cambodia. https://www.undp.org/cambodia/publications/cambodia-climate-change-strategic-plan-2014-2023

MoWA. (2018). *Master plan on gender and climate change (2018-2030).* Phnom Penh: Ministry of Women's Affairs-Royal Government of Cambodia. https://ncsd.moe.gov.kh/sites/default/files/2019-06/MASTER%20PLAN%20ON%20GENDER%20AND%20CLIMATE%20CHANGE.pdf

Nuorteva, P., Keskinen, M., & Varis, O. (2010). Water, livelihoods and climate change adaptation in the Tonle Sap Lake area, Cambodia: Learning from the past to understand the future. *Journal of Water and Climate Change*, *1*(1), 87–101. DOI: 10.2166/wcc.2010.010

O'Brien, K., Eriksen, S., Nygaard, L. P., & Schjolden, A. (2011). Why different interpretations of vulnerability matter in climate change discourses. *Climate Policy*, 7(1), 73–88. DOI: 10.1080/14693062.2007.9685639

Ostrom, E. (2012). Polycentric systems for coping with collective action and global environmental change. In Barry, C. (Ed.), *Global justice* (pp. 423–430). Routledge., DOI: 10.4324/9781315254210

Parry, L., Davies, G., Almeida, O., Frausin, G., de Moraés, A., Rivero, S., Filizola, N., & Torres, P. (2018). Social vulnerability to climatic shocks is shaped by urban accessibility. *Annals of the American Association of Geographers*, *108*(1), 125–143. DOI: 10.1080/24694452.2017.1325726

Pelling, M. (2010). *Adaptation to climate change: From resilience to transformation.* Routledge., DOI: 10.4324/9780203889046

RGC. (2019). *National srategic development plan.* Phnom Penh. http://cdc-crdb.gov.kh/en/strategy/documents/nsdp-2019-2023_en.pdf

Ronan, K. R., & Johnston, D. M. (2005). Community resilience to disasters: Introduction to theory and review of research. In Ronan, K. R., & Johnston, D. M. (Eds.), *Promoting community resilience in disasters: The role for schools, youth, and families* (pp. 9–48). Springer., DOI: 10.1007/0-387-23821-2_2

SCDF. (2020). *Annual report FY2019-2020.* Singapore Civil Defence Force. https://www.scdf.gov.sg/home/about-scdf/media-room/publications/annual-report/FY2019-2020

Secretariat, A. S. E. A. N. (2021). *AADMER work programme 2021-2025.* ASEAN Secretariat., https://asean.org/wp-content/uploads/2021/08/AADMER-Work-Programme-2021-2025.pdf

Slater, D. (2010). *Ordering power: Contentious politics and authoritarian leviathans in Southeast Asia.* Cambridge University Press. DOI: 10.1017/CBO9780511760891

UNDP. (2014). *Cambodia post-flood early recovery needs assessment report 2013*. Royal Government of Cambodia-United Nations Development Programme.

UNDP. (2023). *Disaster financial preparedness analysis report for Cambodia*. Phnom Penh: United Nations Development Programme. https://www.undp.org/sites/g/files/zskgke326/files/2023-05/01_Cambodia_Financial%20Preparedness%20Plan_Final_May2023.pdf

WEF. (2020). *Cambodia Annual Country Report 2020*. Phnom Penh: World Food Programme. https://wfp.tind.io/record/130856/files/ACR%20Cambodia%202020.pdf

World Bank. (2016). *Building regulation for resilience: Toward high performance building regulatory systems for disaster risk reduction* Washington DC: World Bank. https://documents1.worldbank.org/curated/en/326581468337788007/pdf/ACS15966-WP-PUBLIC-BRR-report-002.pdf

Yusuf, A. A., & Francisco, H. (2009). *Climate change vulnerability mapping for Southeast Asia*. Economy and Environment Program for Southeast Asia (EEPSEA). https://www.preventionweb.net/files/7865_12324196651MappingReport1.pdf

KEY TERMS AND DEFINITIONS

Cambodia: In this chapter, Cambodia is portrayed as a post-conflict, climate-vulnerable state that has deliberately shifted from reactive disaster response to proactive, inclusive risk governance. Through legal reforms (e.g., the 2015 Sub-Decree), stronger national and sub-national institutions (NCDM and local committees), and active engagement in ASEAN frameworks, Cambodia uses disaster policy to bolster state legitimacy, enable local preparedness, and position itself as a constructive regional actor.

Disaster Risk Reduction: DRR is the term used to describe Cambodia's comprehensive set of laws and procedures that attempt to prevent or decrease the effects of disasters. These procedures and regulations include risk mapping, early warning systems, community disaster plans, and programs that are inclusive of women and youth. Risk management is ingrained in various sectors, including education, health, agriculture, and water, and is supported by national policies (SNAP-DRR, NSDP), as well as the Sendai Framework and AADMER.

Climate Resilience: The ability of Cambodia to predict, absorb, adapt to, and recover from climatic shocks (such as floods, droughts, and shifting monsoons) while preserving development benefits is referred to in the chapter as climate resilience. It is built through ecosystem and infrastructure measures, early digital warnings, climate-smart agriculture, and continuous learning loops that connect local knowledge to national policy and regional cooperation, turning vulnerability into a platform for long-term, systemic adaptation.

Governance: Governance denotes the multi-level, cross-sectoral arrangements that steer Cambodia's disaster and climate actions. Clear legal mandates, coordination between ministries and governmental levels, civil society involvement, and collaboration with foreign partners are all components of governance. The chapter places a strong emphasis on governance as both technical and political, connecting improved state legitimacy and global credibility to efficient service delivery and accountability.

Adaptive Capacity: The ability of Cambodia to modify its institutions, laws, and procedures in response to changing dangers and lessons discovered is known as adaptive capacity. It shows up in decentralized decision-making, community-based planning, iterative policy revisions, school-based DRR education, and the uptake of technology (forecasting, SMS alerts). Cambodia increases its flexibility to manage uncertainty by allowing provinces and communes to customize activities to local settings while adhering to ASEAN and national norms.

Humanitarian Diplomacy: Humanitarian Diplomacy: In this case, humanitarian diplomacy encapsulates how Cambodia uses disaster governance to foster collaboration, influence, and trust both locally and internationally. In order to project a resilient image, draw in resources and expertise, and strengthen ASEAN cohesion in the face of transboundary climate risks, Cambodia transforms operational competence into soft power through AADMER participation, ARDEX exercises, bilateral trainings, and multilateral partnerships (UNDP, JICA, FAO, ADPC).

Chapter 7
Public Perceptions of Earthquake Risk and Preparedness in Istanbul:
A Mixed-Methods Analysis of YouTube Comments

Volkan Duran
http://orcid.org/0000-0003-0692-0265
Iğdır University, Turkey

ABSTRACT

This study examines public perceptions and awareness of earthquake risks in Istanbul through a mixed-methods qualitative analysis of 765 YouTube comments posted under a video featuring seismologist Professor Naci Görür, recorded as of June 05, 2025. Employing thematic analysis, frequency statistics, and Latent Dirichlet Allocation (LDA) topic modeling, the research identifies seven disaster awareness subdimensions: Risk Perception, Knowledge/Information Level, Preparedness/Preventive Behaviors, Response Awareness, Post-Disaster Recovery Awareness, Social/Societal Awareness, and Fatalism vs. Agency. Findings reveal a dominant focus on Knowledge/Information Level (590 comments) and Preparedness/Preventive Behaviors (135 comments), reflecting a strong public appetite for scientific understanding and proactive measures, driven by Istanbul's proximity to the North Anatolian Fault.

1. INTRODUCTION

Disaster awareness is essential to community resiliency, allowing people or communities to reduce risks, respond, and recover from natural or human-made disasters.

DOI: 10.4018/979-8-3373-6274-8.ch007

Disasters i.e. earthquakes, hurricanes, droughts, extreme meteorological events, floods, and wildfires, are significant threats to life, property, and our environment (Amsalu et al., 2026; Çelik, 2026; Çelik et al., 2026). Disaster awareness - increase awareness of threats, potential consequences, and preparedness - can help people substantially mitigate their vulnerabilities and improve safety. In this chapter we will summarize the importance of disaster awareness, preparedness strategies, and the necessary application of awareness and education together that build community resiliency, supported by academic literature.

Disaster awareness is knowing the hazards that may impact your community, the possible consequences of those hazards, and what actions to take to prepare and mitigate their impact. The United Nations Office for Disaster Reduction (UNDRR) indicates that effectively reducing disaster risk (DRR), requires communities to be aware of hazards in order to anticipate or respond to them (UNDRR, 2020). Awareness is fundamental to proactive and informed actions, which allow individuals to remain conscious in decision-making during a crisis situation.

Disasters can be broadly categorized into two types:

- **Natural Disasters**: These include geophysical events (e.g., earthquakes, volcanic eruptions), meteorological events (e.g., hurricanes, tornadoes), hydrological events (e.g., floods, tsunamis), and biological events (e.g., pandemics).
- **Human-Induced Disasters**: These encompass industrial accidents, terrorist attacks, and technological failures.

Each type presents unique challenges and requires tailored preparedness measures.

The effects of disasters are numerous, including impacts on physical infrastructure, economics, and mental health. For example, when the Haiti earthquake occurred in 2010, it resulted in over 200,000 fatalities, and displacement was estimated in the millions. The earthquake illustrates that lack of preparedness has negative consequences for those live in vulnerable areas, and the more vulnerable area, the more devastation (Bilham, 2010). Hurricane Katrina resulted in an economic loss of over $125 billion (Knabb et al (2005). In terms of preventative awareness, it should be noted that awareness of a disaster and their mitigations combined with education are all encompassing when it comes to knowing when and how to proactively take disaster awareness to mitigated actions.

Disaster awareness is a major determining factor of preparedness and resilience. Awareness has many measures that include range of influences and is linked to more than a dozen explanatory variables. Various research studies reveal the strong connection between disaster awareness and preparedness behaviors. Patel and Kermanshachi (2020) suggest an economic status including education and income is a reasonable predictor of disaster awareness and readiness. In the educational context,

Park and Yang (2017) studied nursing students showed if students had high disaster awareness had better competencies on disaster response. Gerdan (2014) emphasized measuring awareness and attitudes about disaster preparedness will help identify where best to apply resources for university-based programs.

Community level data collected in Indonesia showed that increased awareness contributes to better levels of preparedness, particularly in vulnerable communities (Hargono et al., 2023). Socio-economic status and gender also affect seeking information, and low-income women had high levels of disaster awareness (Teo et al., 2018). High levels of awareness was also associated with quality of psychological coping, which was vital for reducing stress and anxiety during disaster occurrences (Lee & Lee, 2019). For the purpose of systematic measurement of awareness and planning interventions to stimulate awareness and improve preparedness, Dikmenli et al., (2018) developed a valid and robust disaster awareness scale whilst Ozkazanc and Yuksel (2015) measured levels of awareness and sensitivity in university undergraduates. Evidence from longitudinal community based disaster training research conducted in Turkey indicated that community disaster training can change cognition and behaviour long after the training has concluded (Karanci et al., 2005). Shaw et al., (2009) provide an evidence based plan of action for government policy: the role of public education and public awareness campaigns in developing community and societal resilience is emphasized. When consideration of these studies are placed together, the results demonstrate that disaster awareness is effectual in reducing risks associated with disaster, in particular the constructs of education, community and psychological resilience.

Disaster awareness and preparedness have changed due to technology. Today, mobile applications, social media, and geographic information systems (GIS) support real-time information and risk maps. For example, social media platforms as was the case with "X" that were used to share updates during disasters, allowed communities to stay informed and responsible during an emergency disaster and share community outreach (Alexander, 2014). IT applications are also being used to help first responders and government agencies share real-time data, helping with quick, accurate, and appropriate decision making (Murayama, 2021; Son et al., 2008).

Educating the public and training officals is an important step to developing disaster knowledge. It has been shown (Durmuş et al., 2024; Türksever, 2021; Turan et al., 2017) that disaster training increased self-reported knowledge and behaviour in both students and educators, and a first-aid and preparedness program in the Middle East reported statistically significant increases in knowledge level and satisfaction with the program (Elsaeidy et al., 2016, p. 15). Awareness of a disaster risk or disaster programs does not necessarily lead to a case of awareness and compliance with the disaster programs (Asio, 2021). While urban populations may be aware of the potential risks associated with a disaster, their level of compliance with disaster preparedness programs seems to be only in the moderate range. Even though nurses and staff typically had some basic awareness, there was a lack of practical preparedness (Sonopant, 2015; Sasmoko et al., 2020). After a study was conducted in a hospital pharmacy on disaster drill awareness of natural disasters, the drills increased awareness about disaster risks, but awareness did not seem to lead to behaviours associated with preparedness afterwards (Kajihara et al., 2023).

Awareness programs can address knowledge gaps for vulnerable populations, and can be useful in ensuring that marginalized populations are not going without access to such critical knowledge (Wisner et al., 2014). Awareness programs are equally important to allowing information to access different audiences to help make disaster preparedness a culture.

The advancement of disaster awareness and understanding has grown over the last decade but still faces immeasurable and unprecedented setbacks. With the emergence of socioeconomic inequities, limited access to education or digital literacy, language barriers or education system disparities as examples, access to critical knowledge can be unstable. In conjunction, misinformation during a crisis can cause confusion and slow response efforts (Veil et al., 2011). These previously identified obstacles involved in disaster awareness must be addressed in inclusionary approaches to considering at-risk populations, accessibility, and existing communication approaches.

Disaster awareness and understanding are crucial pillars of resiliency. This process enables a person to prepare for, act and respond to hazards, and recover from events in an innovative manner. By integrating opportunities to teach and learn about hazards through education, early warning systems, community engagement and available technologies, societies can diminish the effects of hazards and save lives. These approaches can act as an indication of intent when combined with the agency of local policy-makers, educators, and communities. Together, local stakeholders can assist in creating and nurturing the culture for disaster preparedness and risk reduction by providing the knowledge and tools required to address disasters as they arise.

2. METHODOLOGY

This study employs a mixed-methods qualitative approach to analyze public perceptions and awareness of earthquake risk, preparedness, and related subdimensions through YouTube comments. The methodology combines thematic analysis, frequency statistics, and topic modeling to provide a nuanced understanding of disaster awareness discourse in a high-risk urban context.

Thematic Coding and Subdimension Assignment: The first step was to perform a deductive thematic analysis on the basis of well-cited dimensions of disaster awareness in the literature: Risk Perception, Knowledge/Information Level, Preparedness/ Preventive Behaviors, Response Awareness, Post-Disaster Recovery Awareness, Social/Societal Awareness, and Fatalism vs. Agency. Coding schemes were designed and deployed by experts, with each dimension operationalized through lists of keywords and phrases in Turkish and English (e.g., "deprem," "risk," "hazırlık," "sigorta," "kader," "fay hattı," etc.). Comments were assigned to subdimensions through keyword matching, and overlapping was facilitated if more than one theme went into the making of a comment.

Frequency and Engagement Analysis:

For every subdimension, the aggregate number of comments corresponding was counted to show the distribution of public discussion. The like counts were added to determine which subdimensions and specific comments generated the most engagement. Therefore, quantitative comparisons between subdimensions could be made while allowing for qualitative reviews of the kinds of comments that tended to revolve most strongly around viewers.

Illustrative Quotation Selection:

Mixture quotations, representative of each subdimension, were selected from the pool of comments, preferably one with higher engagement or entries that implied clear articulation of one relevant theme. Each quotation was provided with a translation into English to ensure clarity and facilitate international scholarly comparison.

Timeline and Temporal Analysis:

To study temporal trends, the timestamp of each comment was then categorized into discrete intervals such as "1 month ago", "2 weeks ago", "11 days ago" and so on. The frequency of comments for each subdimension was plotted across these time points, thus providing clues of the waxing and waning of public interest in disaster awareness themes against the firm backdrop of current events or media attention.

Topic Modeling:

For complementing deductive coding with induction, LDA topic modeling was applied to the entire comment corpus. Preprocessing entailed removal of duplicates, tokenization, stopword removal, and vectorization by means of a CountVectorizer. The model was run for five topics-the number finally decided upon after exploratory runs and consideration in terms of interpretability. Top words and representative comments from each topic were extracted, and then the topics were mapped post hoc to disaster awareness subdimensions. Emergent themes such as trust in experts, debates among scientists, skepticism of the public, and fatalism could then be explored-aspects that maybe do not find a full representation in the predefined codes.

Trustworthiness and Limitations:

In order to enhance trustworthiness, the coding scheme was reviewed by two coders independently, and upon discrepancies, decisions were discussed and resolved. Then again, some limitations must be taken into account: comments may show self-selection bias only to relate to the more vocal users being internet-active; automated mapping of keywords may fail to represent sarcasm or complex argument scenes; Turkish stopwords at the LDA stage deserve further polishing by refinement. The analysis, however, provides a fairly solid overview of the topical earthquake awareness scenes among Istanbul's digitally speaking public.

Ethical Issues:

All data examined in this study were open to the public; no personal identifying information other than anonymized user names and aggregated engagement statistics was kept. The analysis abides by applicable ethical standards for research using publicly accessible social media data and platform terms of service.

3. FINDINGS

3.1. Thematic Analysis of the Comments

3.1.1. Risk Perception

Risk perception encompasses how or what people and communities discern about the likelihood, intensity and immediacy of earthquake threats depending on what they know and have experienced in their local environment. The variety of ways that individuals and communities respond to the risks they perceive will influence their attitudes and behaviours. There are many ways that risk can be perceived, depending on whether a person is experiencing anxiety about every shudder than can be perceived. In Istanbul's case, the ever-present awareness of the risk of natural disaster can lead to not only anxiousness but also even denial, for if an earthquake will definitely occur, then determining when and how and whether it will threaten lives is not part of risk awareness. The public's perception of risk will be affected by many factors including the science of earthquake prediction, media coverage of the earthquake threat, previous experience with earthquakes, and the prevailing attitudes which the community has adopted towards natural disasters.

"Istanbul will definitely have a big earthquake. It's not a question of If, but when."

This statement expresses a prevailing view among expert opinion within the geological and seismological community, that Istanbul will pose a significant and permanent threat to earthquake activity because of the knowledge related to the North Anatolian Fault and the high probability that a high magnitude earthquake will occur within decades to come. Studies of the North Anatolian Fault have established that the fault has exhibited considerable seismic activity, including the well documented and oft emphasized during earthquake preparedness planning the 1999 İzmit Earthquake which caused significant life lost and structural damage.

The concluding certainty which considers the fate of Istanbul relative to earthquake susceptibility generates considerable worry from the public and contributing to an acceptance of fatalism towards the earthquake without appropriate assuming management control responsibilities.

"I always have a level of anxiety about a disaster here because the likelihood is so high."

Living in constant suspense as to when the earthquake will come is chronically stressful, as many residents fret over. Much of this anxiety is magnified by media representations, documented evidence from scientists or experts, and concrete reminders like visibly cracked buildings or media reporting on even minor earthquakes. For some people, the anxiety leads to preparedness; for others, it can become a stay away or a denial response.

"No one can promise our safety in this city."

This reveals the weakness being felt in Istanbul, as rapid urbanization, particularly the degradation of infrastructure and lax protection measures has become heightened risk. Even a newly built building, educated people can not rely on the practice of enforcing seismic codes as a reason to trust them. People often have little influence over circumstances, and an experience of helplessness is created.

"The fault line is in our neighborhood. Everybody knows it exists but acts like it doesn't."

This is an expression of a common but human psychological response called cognitive dissonance. In other words, we acknowledge the risk but, at the same time, decide not to engage it in our daily lives. In neighborhoods like Avcılar or Zeytinburnu that have fault lines, residents normalize the threat to some extent and shield their daily life. This might mean neither talking about hazards nor engaging any measures to reduce discomfort.

3.1.2. Knowledge/Information Level

This category is specific to the acquisition and sharing of scientific, technical, or practical information concerning earthquakes, which is important for making informed decisions. Access for the majority of the public (in Istanbul) to ideas about seismic risk is widely variable. Some people seek out scientific knowledge related to earthquakes, while others have their understanding informed by a misunderstanding of the science, or by lagging historical views. Our educational activities, through a lecture, documentary, or social media activities, play an important role in bridging the gap and improving an informed public.

"Professor Naci explained the North Anatolian Fault so simply, now I understand the risk."

Experts, prominent Turkish seismologist, have helped raise public awareness of a potential major earthquake on the North Anatolian Fault. His simple explanations of tectonic movements and the chronology of earthquakes can help make complex scientific ideas more accessible to the public and make confusing things more concrete for them.

"We need more scientific broadcasts like this, people should learn the facts."

The public support for clear and reliable information on seismic risk helps indicate the need for ongoing and accessible educational campaigns in the public domain. Broadcasts on television, YouTube, or through social media platforms like X, would help address myths (such as "earthquakes only happen at night") while also providing people practical knowledge about the risk and how to prepare for earthquakes.

"The Richter scale is too limited to measure damage - the quality of soil is equally as important."

Some residents are beginning to understand that the impact of seismic occurrences can vary - and that is based on more than earthquake magnitude. For example, in Istanbul, there are places of alluvial soils, and other places solid as bedrock, as well as everywhere in between, and this will amplify or dampen shaking. In places like Bakırköy, which are prone to liquefaction in even moderate earthquakes, the danger is markedly increased post event.

"What we see on the Richter scale isn't really the issue - the issue is how prepared our buildings are. That's what the experts tell us."

The other major player in the outcome of damage post earthquake is structural resilience. Building and building code laws is important, particularly in older sections of Istanbul. Struck poorly or not retrofitted, can be far more dangerous than the magnitude of the quake itself - emphasizing the need for retrofitting programs and regulating building work.

"After watching this, I understand the fundamental difference between an epicenter and a focal point."

Public education can demystify the terminology with which people feel increasingly anxious, like knowing the difference between epicenter (the point on the Earth's surface located above the place of origin of the quake) and focal point (the actual origin of the quake under the ground). When a person identifies and understands these elements, they can then examine seismic reports safely without panic, and assess their risk in their own area.

3.1.3. Preparedness/Preventive Behaviors

Preparedness entails taking proactive steps to reduce earthquake risks, whether at the individual level or community level. In Istanbul, where millions of people live in urban and densely packed cities, it is essential to use preparedness, but it is often inconsistent because of cost, awareness, or cultural issues. Preparedness can include personal actions (e.g., securing furniture), community strategies, and policies to help ensure safer buildings.

"Everyone should have a go-bag ready and a meeting to plan where the family will meet in safe place."

A go-bag filled with necessities (water, food, first-aid kit, flashlight, important documents, etc.) and a pre-planned family meeting place are basic but essential. These preparations ensure individuals can evacuate quickly and be re-united with loved ones after an emergency and can reduce the chaos from a disaster.

"I keep telling my parents to secure the furniture, but they think it is unnecessary."

It is common to experience resistance when attempting low-cost forms of preventative measures- such as securing heavy furniture or heavy appliances- because it costs money, is too much effort, or individuals underestimate the risks. The previous examples highlight the need for awareness campaigns which can promote low-cost, high-impact measures which can help prevent injuries from falling objects.

"Why are there still so many unsafe buildings in Istanbul?"

Although there has been some development of urban renewal, many of the buildings in Istanbul considered as older - the buildings that were built before seismic codes were created - are still vulnerable. Corruption, red tape in the bureaucracy, and financial barriers have impeded progress in retrofitting these buildings, forcing millions to live in substandard conditions and at risk.

"We should have drills in schools and workplaces for earthquakes, an annual requirement."

Being exposed to drills in a repeatable pattern creates muscle memory for what people should do in an earthquake. For example, in Japan there are regular earthquakes drills that take place every year in both schools and workplaces. Istanbul needs to establish a model which normalizes preparedness and provides exercises to reduce panic when it does happen.

"If my building does not hold up in an earthquake, I will just move out - no excuses."

It does show that some people are increasing their awareness of the vulnerabilities around housing in seismically safe buildings. However, many people do not have the economic option to simply move. Buildings that are safer in Istanbul tend to be from a more expensive price point, compared to older, unsafe buildings.

"I will check my apartment's earthquake insurance right after this video."

Earthquake insurance, including DASK, Turkey's mandatory earthquake insurance, is an insurance that is a financial safety net. However, many people...either collapse on their own risk to save money, or for other reasons do not know about or trust that the insurance system in Turkey will be "there" for them after they had an incident. They are willing to live without coverage, all to face potential ruin, after a disaster.

3.1.4. Response Awareness

Response awareness includes the recommendations for doing the right thing during and immediately after an earthquake, enhancing safety (during an earthquake, response could involve several things, as necessary). Misinformation increases risk, and a famous dangerous mistake is outdated advice to stand in a doorway. Close to half of respondents stated they could take safe action without an earthquake. The practice of educating the public about evidence-based response is needed in a city like Istanbul, particularly with the possibility of millions affected by a major earthquake.

"We need to drop to the ground, take cover from moving objects below, and hold on, we should not run out the door."

The approach of "drop, cover, and hold on" which refers to (1) dropping to the ground, (2) covering oneself under a strong piece of furniture, and (3) holding on until the shaking stops, is the recommended action globally. In the past there was good advice to run to the door, if available, but that is not true as the protective aspects of doorframes are no longer standing or do not withstand the forces of collapsing buildings. Most stairways are safest because (1) elevators may not work or get stuck during an aftershock, (2) the stairs give a better opportunity to evacuate, and (3) the fastest route down.

"Do not use elevators after an earthquake. Always use the stairs for evacuating."

Because elevators malfunction (or stick) in aftershocks the stairs allow public occupants (with others) the opportunity to evacuate. Many public buildings do not assist the public to learn the proper way to evacuate or leave ground, especially within buildings that lack signage of particular stairs through identification on diagrams in emergency plans or walking exercises in emergency drills. Seeing signage and in turn practicing the methodology may reduce confusion during emergencies.

"Everyone should know emergency numbers, and children should be taught how to use them."

It is important for everyone in Turkey to learn the important emergency numbers, especially 112 as general emergency numbers and AFAD's disaster emergency lines. The reason for this is so children can learn these numbers as well. Children commonly become separated from their adults in emergencies, and it is incredibly beneficial for them to know how to contact someone to help them in an emergency situation.

"After an earthquake, avoid using your phone except for an emergency call to prevent network congestion."

Network congestion occurs regularly after an earthquake. This was clearly experienced following the 1999 earthquake when cellular and other telephone communications were completely overloaded after the quake. If people make emergency calls using the network, EVERYONE can get their emergency calls in. Non-emergency callers should use text messaging instead, as sending a text message uses a significantly less amount of bandwidth than making an emergency call.

"Should I shelter beside the strong sofa, or under the table? Experts are divided, though both agree to keep away from the windows."

Especially in "sheltering" advice, the distinction between "beside" and "under" furniture furnished through the theory "the triangle of life" is confused, although this has generated considerable controversy in the serviced action community. Consensus about avoiding windows and glass must be emphasized as unexplained injuries are likely to occur during severe earthquakes from shattered glass and various surfaces.

3.1.5. Post-Disaster Recovery Awareness

The theme centers on the long-term requirements facing earthquake survivors, such as psychosocial support, housing, economic recovery and support for community rebuilding. In large urban centres, like Istanbul, where an earthquake could displace several million people, the success of recovery from an earthquake is just as important as a well-planned response. Knowledge of survivor recovery needs can improve policy and community support systems.

"Some interventions after the disaster require psychological support by skilled people to assist survivors and victims besides physical assistance or support."

It is normal for survivors of natural disasters, including earthquakes, to experience post-traumatic stress disorder (PTSD). Community based temporary support can assist in this area of socially and incarceration deprived support, such as the programs many non-government organizations (NGOs) or the Disaster and Emergency Authority of Turkey (AFAD) offer; however scaling a service for a city like Istanbul with a population of 16 million people is not practical.

"The government should organize not only planning for search and rescue, but all long-term support for victims."

The efforts of search and rescue after an earthquake disaster event must lead to sustained planning for temporary housing, employment opportunities for survivors, and community infrastructure rebuilding plans. The earthquake in Kahramanmaraş in which many survivors struggled months after the event illustrated many gaps in the actual long-term recovery effort.

"We need more social housing when people are left homeless after the earthquake hammers the spine of our shallow or deep realities."

Many survivors experience evacuation during and after an earthquake, and displacement is a serious or major problem in urban earthquakes. The combined issues of high property costs, as well as limited public/social housing for the displaced in Istanbul reinforces the necessity for pro-active intervention directed by housing solutions by governments whereby displaced persons and/or survivors can find affordable seismic-safe housing options in a recovering community.

"People move on from earthquake victims in a week, but their difficulties span years."

Media and public interest usually vanish rapidly, leaving survivors to negotiate healing on their own. Local support networks among other community-led projects help to keep focus on long-term needs and support aid projects.

"Is there any support for those who lost their jobs after the disaster?"

An important problem following an earthquake is economic upheaval. Jobs lost as a result of relocated or damaged workplaces can keep families in poverty. Government or non-government organization programs providing retraining or financial assistance could help to reduce this, although such programs in Turkey are now underdeveloped.

3.1.6. Social/Societal Awareness

This category emphasizes collective responsibility, community solidarity and institutional accountability for earthquake preparedness and response. Local community culture, in Istanbul, is critical where individual activism alone is not sufficient, and collective responsibility, proactivity and collaborative decision-making is important.

"The solidarity of neighbors is vital - let's prepare together."

Grassroots initiatives, like neighborhood earthquake committees, can develop local drills, share resources, and deepen social trust. In tight community networks, solidarity can be the difference between life and death in a disaster that calls for mutual aid or collective response.

"The municipality should educate and inform the public, and proactively check buildings."

Municipalities have a pivotal role in oversight of building codes and public awareness. The municipality of Bağcılar has commenced a retrofit program, but in districts inconsistent checks exist.

"I really wish we had more community spirit, everybody is only looking out for themselves."

Urbanization and self-concern diminish collective action, which for an emergency situation mitigates collective preparedness. Events like neighborhood meetings, community gatherings or disaster preparedness fairs could embolden a change in their sense of collective responsibility.

"Every district should engage civil society organizations to plan awareness events."

Community-based agencies and organizations can help alleviate government responses in either education-based events, providing preparedness kits, or about organizing for vulnerable populations, e.g. low-income or elderly residents.

"Only if everyone accepts responsibility can we reduce the loss."

Combining individual preparedness, community planning, and institutional responsibility, collective action can greatly lessen the effects of earthquakes. This calls for a cultural change toward shared responsibility rather than reliance on top-down fixes.

"We want the authorities to act, but we also need to support one another."

Although governmental action is essential, citizens must also be proactive. Volunteering with groups like AKUT—a Turkish search-and-rescue nonprofit, for instance—can enable people to help to create community resilience.

3.1.7. Fatalism vs. Agency

This section addresses the interplay between culturally or religiously influenced fatalistic attitude and more proactive calls for agency and responsibility. In the case of Turkey, where fatalism is sometimes fostered by cultural modes of thought, the transition to empowerment is vital for earthquake preparedness.

"There was nothing we could do. If it is meant to be, we have to accept it."

The mindset of fatalism can lead to inaction because of the view that risk is inevitable. In this way, some residents view earthquakes as some kind of fateful acts of nature, thereby diminishing their motivation to prepare or act for change.

"I trust God to decide everything; we cannot change fate."

This statement demonstrates a common attitude in the Turkish context, where religious beliefs may influence a conception of disasters as divine will. It may ease the anxiety of people, but reduce actions to put preventive measures or structure in place.

"Enough blaming fate! We have to act and save ourselves."

This resistance to fatalism creates an idea of individual and collective agency. Those that advocate this approach strive to frame the earthquake as a natural event, but can advocate that damage from the earthquake may be mitigated through preparation, education and policy.

"People employ fate as justification for laziness and irresponsibility."

Critics of fatalism contend it offers a simple explanation for avoiding challenging or expensive activities including community readiness efforts and retrofitting houses.

"Though we should still be prepared, worrying about something we cannot manage makes no sense."

While urging realistic preparedness, this balanced perspective accepts the unpredictability of earthquakes. It is a pragmatic approach combining proactive conduct with acceptance.

"Let's do what needs to be done right now; let's not hide behind fate!"

Encapsulating the move toward empowerment, this rallying cry encourages people and groups to take tangible action—including securing homes, studying safety procedures, and advocating for improved policies—to lessen earthquake risk.

3.2. Number of Comments per Subdimension (with Detailed Quotes)

Thematic mapping of opinions on earthquake awareness in Istanbul shows a clear difference in public debate spread over seven subdimensions: Risk Perception, Knowledge/Information Level, Preparedness/Preventive Behaviors, Response Awareness, PostDisaster Recovery Awareness, Social/Societal Awareness, and Fatalism versus Agency. With 765 comments studied, the overwhelming emphasis on Knowledge/Information Level (590 comments) and Preparedness/Preventive Behaviors (135 comments) contrasts dramatically with the little representation of Response Awareness (3 comments), PostDisaster Recovery Awareness (2 comments), Social/Societal Awareness (2 comments), and Fatalism versus Agency (1 comment). Based on its nearness to the North Anatolian Fault, Istanbul is a highly threatened city; this distribution helps to reveal public priorities, cultural beliefs, and deficiencies in earthquake preparedness.

3.2.1. Dominance of Knowledge/Information Level

The 590 remarks under Knowledge/Information Level—including "I finally understood why the Kumburgaz fault is so hazardous" and "What matters is not the magnitude but how strong our buildings are. That's what the experts say"—show great public interest in scientific and technical knowledge on earthquakes. The prevalence of this subdimension results from several reasons:

1. Public understanding is molded mostly by people, a well-known Turkish seismologist. A point of focus is the Kumburgaz section of the North Anatolian Fault, determined to be a high-risk zone for a possible magnitude 7+ earthquake. Reliance on expert voices implies trust in scientific authority but also points up a need on accessible, clear communication to span the gap between complicated seismology and lay knowledge.
2. Istanbul, a megacity of more than 16 million people, has a dense population and extensive infrastructure that emphasizes the need of seismic risk awareness. Comments highlighting developing resilience over size highlight a rising understanding that local elements—like soil quality or construction standards—greatly affect earthquake results. This concentration might come from recent education initiatives or media coverage of earlier calamities, such as the 1999

İzmit earthquake, which killed over 17,000 people and highlighted flaws in Turkish urban planning.

3. High comment count indicates Istanbul people are purposefully looking for information to lower uncertainty. This matches psychological ideas of risk perception, where knowledge acquisition helps to handle anxiety about uncontrollable events like earthquakes. The great number of remarks under this area, meanwhile, might point to a gap in general, accessible public knowledge, which would drive people to depend on intermittent sources like media interviews or social media sites like X.

3.2.2. Emphasis on Preparedness/Preventive Behaviors

At 135 remarks, Preparedness/Preventive Behaviors is the second most represented subdimension, shown by comments such "Why are there still so many unsafe buildings in Istanbul?" and "We must have earthquake drills in schools and workplaces daily." This emphasis both exposes systematic issues and shows that among some inhabitants there is a proactive attitude:

1. The statement on dangerous structures brings up a major problem in Istanbul, where fast urbanization and weak implementation of seismic standards in the past have left many constructions susceptible. Despite urban renewal efforts, progress is uneven and public dissatisfaction is clear. This implies a gap between public expectations and policy execution, particularly in vulnerable areas like Avcılar or Zeytinburnu.
2. The focus on drills in schools and businesses reveals a want for institutionalized readiness akin to models in earthquake-prone nations like Japan. But the relatively little number of comments as opposed to Knowledge/Information Level indicates that translating information into actual actions is still difficult even if some people are driven to act. Cost, time, denial, or other obstacles could restrict preparedness initiatives.
3. The comments capture a conflict between personal acts (e.g., getting gobags ready) and systemic failures (e.g., hazardous structures). This duplicity stresses the need for both local projects and more government intervention to guarantee seismic safety throughout Istanbul's varied districts.

3.2.3. Underrepresentation of Response Awareness

Answer With just three comments like "Take your precautions before New Year's; I'm an expert; I never compromise on what I know" and "Be careful; evacuate using stairs; never elevators after a quake," awareness is remarkably underrepresented. This points to a number of fundamental problems:

1. Concentrate on Pre-Disaster Issues: The small number of comments fits with the note that unless an event is current, people seem to focus on risk and readiness. Without a rapid earthquake, people might give preparation or learning about hazards top priority above rote answer memorizing. This is worrisome since proper actions during an earthquake (e.g., "drop, cover, and hold on") can greatly minimize injuries.
2. The dearth of comments may point to a dearth of public understanding of quick response protocols. Without consistent training or campaigns, false information—such as obsolete advise to stand in doorways—could linger. Istanbul's dense metropolitan setting, featuring highrise buildings and crowded public areas, makes response awareness especially important in avoiding panic-driven accidents.
3. Cultural Elements: The expert's authoritarian tone in the comment implies a topdown method to dispersing response knowledge, which might not appeal to all inhabitants. Campaigns customized culturally utilizing familiar language or community leaders may boost participation with this subdimension.

3.2.4. Neglect of Post-Disaster Recovery Awareness

PostDisaster Recovery only has two comments: "After the earthquake, there must be longterm support for survivors, not just rescue operations" and "I have witnessed for 26 years that none of this man's predictions about Marmara have come true." There is little awareness. This underrepresentation has major consequences:

1. The emphasis on current hazards and readiness conceals long-term recovery strategy. Particularly in a city where millions could be displaced by a major quake, the comment about long-term support underlines a dire need for policies targeting mental health, economic recovery, and housing. Though public debate in Istanbul appears to lag in this field, the 2023 Kahramanmaraş earthquake revealed faults in ongoing support.
2. The second remark expresses skepticism about expert predictions, perhaps resulting from continuous cautions about a Marmara earthquake that has yet not materialised. This weariness could make residents wonder about the need of getting ready for a remote or ambiguous future, therefore diverting their attention from recovery planning.
3. The absence of conversation on recovery implies that government and institutional messages emphasize prevention and reaction, therefore ignoring the longrun

consequences of catastrophes. Istanbul could be unprepared for the social and economic problems of post-earthquake rehabilitation.

3.2.5. Minimal Social/Societal Awareness

Social/Societal Awareness, also with 2 comments like, "Why is there no-one to organize a public debate between experts? Does no-one in this country have that kind of initiative?" indicates a lack of focus on shared responsibility and institutional accountability:

1. Individualism vs. Community: The statement about public debates implies frustration over the lack of public channels for collective conversation. Urban sprawl and diversity in Istanbul could reinforce individualism at the expense of community cohesion and limit grassroots initiatives or neighborhood-based cooperation.
2. Distrust in Institutions: The call for initiative insinuates a general distrust in authorities or civil society to facilitate genuine conversations. Lingering anti-state notions in Turkey, along with potential bureaucratic inefficiencies or perceptions of corruption, could contribute to a lack of confidence in institutions to adequately prepare for disasters.
3. Need for Civic Engagement: The relatively low number of comments hints that discussion on social solidarity and collective action is not yet a vital leitmotiv in public life. One approach is to seek ways to promote community-led events, such as awareness fairs or community preparedness groups, that could raise the profile of this subdimension.

3.2.6. Rare Expressions of Fatalism vs. Agency

Fatalism vs Agency, with only 1 comment: "Those who are saying that there are faults when there are not, are betraying this country. This is a ploy to remove Turks from Istanbul. What a shameful thing, we run commercials with such people on TV," is the underrepresented subdimension:

1. Conspiracy and Denial: This comment contained a rare but strong combination of fatalism and conspiracy thinking, and at the same time denied scientific evidence of the context and location of the fault lines. While views like this were rare in the sample, it is indicative of a cultural proclivity in some contexts in Turkey to frame disasters as the result of external actors or fate, instead of the natural causes the science shows. It is possible that this could spur some

members of the public to ignore preparedness messages and be dismissive of scientists warnings.

2. Cultural context: For some, fatalism is likely culturally or religiously determined in Turkey (e.g. "it is God's will"), and while it is likely underrepresented in this sample for various reasons, it is likely that it is the predominant discourse. Fatalism as demonstrated in the single comment, or lack of action, can also run opposite to agency, and hence be a form of resistance to the scientific stories being told.
3. Need for empowerment: The dearth of comments representing agency, in other words, comments suggesting or calling to "act now" show how effectively campaigning can implicitly proffer a kind of mindset on how to feel empowered as an individual or part of a collective. Rather than fatalism, campaigns focused on agency and showing how an individual or collective can exert control over an outcome can counteract the tendency of fatalism.

3.2.7. Statistics: Number of Comments per Subdimension

Some subdimensions (like "Knowledge/Information Level") are highly represented, while others ("Recovery", "Response Awareness", etc.) appear far less in the sample.

Table 1. The results of the thematic mapping.

Subdimension	Number of Comments	Example Comments (English, Translated from Turkish)
Risk Perception	32	- "He openly says there's a chance we might die, that's why the risk triggers our brain so much." - "The earthquake map clearly shows which regions are at higher risk. The government should consider relocating strategic investments."
Knowledge/Information Level	590	- "I finally understood why the Kumburgaz fault is so dangerous." - "What matters is not the magnitude but how strong our buildings are. That's what the experts say."
Preparedness/Preventive Behaviors	135	- "Why are there still so many unsafe buildings in Istanbul?" - "We must have earthquake drills in schools and workplaces regularly."
Response Awareness	3	- "Take your precautions before New Year's, I'm an expert, I never compromise on what I know." - "Be careful, evacuate using stairs, never elevators after a quake."

continued on following page

Table 1. Continued

Subdimension	Number of Comments	Example Comments (English, Translated from Turkish)
Post-Disaster Recovery Awareness	2	- "After the earthquake, there must be long-term support for survivors, not just rescue operations." - "I have witnessed for 26 years that none of this man's predictions about Marmara have come true."
Social/Societal Awareness	2	- "Why is there no one to organize a public debate between experts? Does nobody in this country have that initiative?"
Fatalism vs. Agency	1	- "Those who claim there is a fault where there isn't one are betraying this country. This is a game to get Turks out of Istanbul. Shame on those who air such people on TV."

Note: The corresponding counts for "Knowledge/Information Level" and "Preparedness/Preventive Behaviors" exceed all the other subdimensions. There was an underrepresentation of "Response," "Recovery," and "Social/Societal Awareness," which is a recognized occurrence in comment analysis involving disaster scenarios (comment activity is primarily risk and preparedness focused versus response and recovery until such time that a disaster is taking place).

Overall the comments suggest a public discourse dominating focused on risk identification and risk preparation with considerably less emphasis placed on planning for response, recovery, or social collaboration. This focus on preparedness versus response and recovery matches global trends where there is a predominance of pre-disaster commentary with little thought given to crisis, response, or recovery, unless an impending crisis is imminent or arose. Factors prompting this focus on risk and preparedness over response and recovery are possible in Istanbul's post-earthquake context, which may include:

- High-Risk Awareness: İstanbul's exposure to the North Anatolian Fault, in conjunction with its history (e.g., 1999 İzmit) of serious urban earthquakes, affords risk perception and risk knowledge a privileged position. Consistent media coverage as well as warnings and accounts of experts or authorities is in turn a source of acknowledgment and hence preserves risk awareness.
- Urban Constraints: A large urban population, infrastructural decay, and socioeconomically diverse community, amongst other factors, presents urgency, but also complexity for our understandings of preparedness. Comments about unsafe buildings, for example, reflect a broader sense of discontent with politics and institutions.
- Cultural Considerations: A culture of fatalism, collective individualism and/or mistrust of institutions may serve to stifle conversations about response, recovery, and accountability as a society. Education campaigns that acknowledge the cultural underpinnings of these beliefs and recognize a collective return may lessen the public discontent.

- Educational Gaps: Low levels of awareness of response and recovery suggest that public education and campaigns are not tackling these two areas as effectively as we would expect, or people's own practices may undermine this knowledge and awareness. Everyday and regularly scheduled drills, known protocols of practice, and immediate to long term recovery discussions could strengthen our understandings in both subdimensions.

3.3. Frequency of Disaster Awareness Subdimensions Over Time

The timeline plot illustrates the frequency of disaster awareness subdimensions—Risk Perception, Knowledge/Information Level, Preparedness/Preventive Behaviors, Response Awareness, Post-Disaster Recovery Awareness, Social/Societal Awareness, and Fatalism vs. Agency—across various time periods (e.g., “1 month ago” to “9 days ago”). The analysis highlights consistent engagement patterns and underscores the YouTube audience's focus areas, with notable implications for earthquake preparedness in Istanbul.

Figure 1. Frequency of disaster awareness subdimensions over time

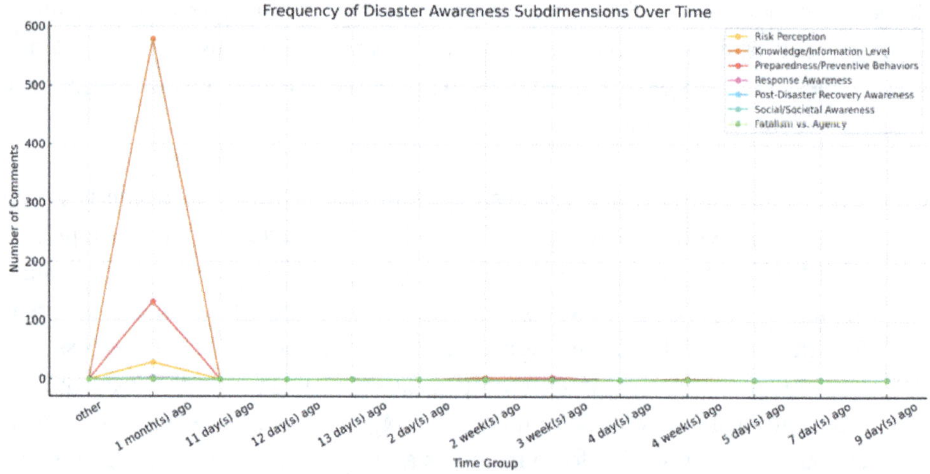

3.3.1. Consistent High Engagement with Knowledge/ Information Level and Preparedness/Preventive Behaviors

The consistently high rates of Knowledge/Information Level and Preparedness/ Preventive Behaviors across all time periods, particularly when “11 days ago” for

Knowledge/Information Level collections yielded over 500 comments and almost 100 or 150 comments for Preparedness/Preventive Behaviors, depicts a sustained interest from the public to learn about seismic risks and act upon this knowledge through preparedness efforts. This difference sheds light on the following:

- The demand for education is ongoing: Knowledge/Information Level always appears at the top of the comments and reveals a continuing desire for scientific information. This demand is likely a result of the input from experts and the media covering Istanbul's vulnerability to earthquakes. The audience has engaged on complementary subjects on the relative strength of buildings compared with magnitude of earthquakes—indicating they are becoming more sophisticated in their assessment of risk.
- The audience has a proactive mindset: The ongoing rate of comments calling for earthquake drills or concerns about unsafe buildings indicated that there are some viewers who are willing to take action. This could be related to the ongoing reminder of the 1999 İzmit earthquake, or the recent debates around urban renewal in Turkey, which made this subject relevant to many viewers even if an earthquake is not near at medium or long term.
- There is temporal consistency: The lack of significant fluctuations in the data resemble ongoing debate rather than comments related to an event.

3.3.2. Sporadic Risk Perception

Risk Perception follows a pattern when tracked over time. We see an irregular sample in the data presentation in that Risk Perception seems to pop up without a lot of consistency with the recent benchmark around the "11 days" (50-100 comments approximately) and trail off fairly quickly. Based on the data to this point identifies that there are probably two critical features of the data:

- Event-driven attention: The short term spike is likely aligned with some event (trivial tremor, expert pointing to issue, media referencing Kumburgaz fault). Spikes in anxiety around earthquakes are extremely common. In these situations, when people are informed they can root their anxiety and emotion to new information, over time they typically fall back to sensing a "certain level of risk".
- Lower sustained attention: Unlike Knowledge/Information Level, Risk Perception does not have a high sustained level of attention data to indicate they there is much focus. The pattern tends to reflect that the risk acknowledgment (ex: "Istanbul is going to have a big earthquake") aligns to somewhat what is occurring, but residents are likely hunting around for some un-

derstanding, or solution, or they'll probably distance from the emotion of that risk.

3.3.3. Underrepresentation of Response, Recovery, Social/ Societal Awareness, and Fatalism vs. Agency

The subdimensions of Response Awareness, Post-Disaster Recovery Awareness, Social/Societal Awareness, and Fatalism vs. Agency continue to be low and steady, with frequencies generally less than 10-20 comments in each time period across all categories. This under-representation, in validation of the previous findings, indicates:

- Limited Attention to Practical Response: The low and flat trend of Response Awareness indicates that little attention is given to direct actions (e.g., "drop, cover, and hold on") likely because of inadequate approaches to education campaigns and/or the belief that knowing how to respond is less important than being prepared. This gap is vital for an urban setting like Istanbul where rapid response can prevent casualties.
- Absence of Planning for Recovery: The low reference to Post-Disaster Recovery Awareness highlights the focus on pre-disaster activities, with limited consideration of longer-term needs such as housing or mental health support. This could be due to a human tendency of avoidance in consideration of post-disaster conditions or could suggest that there is not visible consideration of recovery planning by the public.
- Limited Attention to Community: The low prevalence of Social/Societal Awareness implies that concepts of collective responsibility and accountability of institutions are not at the forefront of the audience's consciousness. This could suggest an urban style of individual coping in Istanbul or diminished trust in authorities to enact community-level or governmental efforts.
- Rare Fatalistic Views: Fatalism vs. Agency's near-zero representation indicates that conspiracy theories or fatalistic attitudes (e.g., "it's God's will") are not prominent in this sample. This may suggest that the YouTube audience is more engaged with rational discourse, though it could also reflect self-selection bias toward science-oriented content.

3.3.4. Audience Engagement Trends and Implications

The trends as a whole indicate that this YouTube audience engages most with learning, science talk, and preparedness... and implies that there is a knowledge-rich community with certain blind spots.

- Educational Opportunity: The strong and steady engagement with Knowledge/ Information Level and Preparedness/Preventive Behaviors presents an opportunity for expanding awareness. Content creators can engage with people's existing interest by incorporating protocols for response, planning for recovery, and social engagement into their education videos, eventually moving toward under-represented topics.
- Event Sensitivity: The erratic engagement with Risk Perception suggests that it may be possible to start public health practices or expert-led live sessions when natural seismic events occur to keep the collective consciousness and sustain it into action.
- Cultural, Urban Experience: The low rate of engagement with the response, recovery, and social aspects may indicate the urban experience of Istanbul: dense people and construction, ageing infrastructure, and socio-economic inequities could accelerate the consideration of personal survival (i.e., securing the home) at the expense of a more communal or post-disaster perspective. Other cultural factors, such as fatalism and reliance on government, may inhibit awareness of response and recovery discussions.
- Preparedness Deficit: The disparity in the topic registries suggests that, although the audience are knowledgeable and motivated to prepare, they may not have access to the practical tools or community resources to act when they are responding during or after an earthquake. This gap could exacerbate losses in a future seismic event, given Istanbul's high-risk status.

The timeline plot reveals a YouTube audience in Istanbul deeply invested in learning about and preparing for earthquakes, with Knowledge/Information Level and Preparedness/Preventive Behaviors as the cornerstones of engagement. However, the persistent underrepresentation of response, recovery, and social dimensions highlights a need for broader education and community focus. By building on the audience's existing interest, tailored content can create a more balanced and resilient disaster awareness culture, critical for mitigating the impacts of a potential major quake along the North Anatolian Fault.

3.4. Interpretation of YouTube Comments on Earthquake Awareness

The data from the YouTube comments analysis indicates a varied public engagement with earthquake awareness. Comment counts greatly differed across subdimensions, and vote distributions helped identify what ultimately received audience approval. This interpretation of their patterns will discuss the importance of both dimensions of data as it relates to Istanbul's risk associated with the North

Anatolian Fault, and how these comments may have implications for a more resilient community.

3.4.1. Comments Associated With Knowledge/Information Level and Preparedness/Preventive Behaviors

The subdimensions Knowledge/Information Level (590 comments) and Preventative/Preventive Behaviors comments (135) encompass the majority of the comments and the most liked comment—"It seemed very ill-intentioned to me that Professor Naci was not invited to the meeting" received 480 votes across these two subdimensions. This overlap may mean:

- Expert Influence and Trust: The high comment count and vote count of a comment that criticized removing Professor Naci Görür - an esteemed seismologist - demonstrates public trust in expert voices. The 480 votes reflect either a level of public shared frustration or aggregation (which could be based on a current event like a meeting or media date around June 2025), which indicates the audience appreciates a scientific voice and considers not getting one an affront. Additionally, it could help enhance involvement in knowledge based and preparedness planning.
- Demand for Accountability: Comment intent being focused on intent ("ill-intentioned"), hints at a level of distrust in institutional decision-making, which could provide motivation for the number of comments. They are presumably wanting information that would dictate some practical means for preparedness, (i.e. building safety/drills), which can be seen from the 135 comments with this subdimension.
- Engagement Discrepancy: A comment which nets the comment count of 480 (out of a maximum possible which reflects the sample size) indicates a number of people left comments, but only a few users scored broadly across the sample. This concentration of voting suggests a polarized or focused audience, where certain issues (expert exclusion) resonated more than diverse perspectives presented.

3.4.2. Moderate Engagement With Risk Perception

Risk Perception, with 32 comments and the best-rated comment—"Professor Şener does not assert the risk is over; he says there may be a 6.2 event; however, nothing greater than 7, therefore an earthquake will still occur"—had 6 votes, the responses showed a moderate level of interest. This indicates:

- Intermediate Concern: The commenter's precision and specificity regarding the eventuality of a specified earthquake (6.2) and the existing acknowledgement of an inevitable quake indicates a public that is attuned to expert evaluations of the North Anatolian fault being active. The 6 votes indicate limited but nonetheless significant approval from these individuals that the discussion of risk is more engaging when it speaks to some predicted eventuality rather than speaking to fears or wondering about an unknowable.
- Noteworthy in Second Place: The number of comments and votes here is lower than the knowledge and preparedness subdimensions of public perceptions of risk, suggesting a focus on risk that bracketing before venturing into opinion. The public accepts risk; being prepared and understanding how to act is a priority that coupled with risk does not leave room for fear; the public cannot avoid recognizing the spectre of a major quake but out of necessity must take action to prioritize and operationalize risk management as a coping strategy against anxiety in a city where the looming odds of a significant quake are sundry.

3.4.3. Underrepresentation of Response, Recovery, and Other Subdimensions

Subdimensions such as Response Awareness (3 comments, 0 votes), Post-Disaster Recovery Awareness (2 comments, 0 votes), Social/Societal Awareness (2 comments, 1 vote), and Fatalism vs. Agency (1 comment, 0 votes) are glaringly absent with no voter engagement. This pattern is indicative of the following:

- Lack of Practical Explicitness: The zero votes for Response Awareness comments, which reflect academic discussions versus actionable efforts (e.g., "drop, cover, and hold on") indicate a disconnection between theoretical discussion and what is being asked of the public. This is troubling because a well-thought-out response plan can help mitigate mortality and injury in a dense urban area like Istanbul.
- Lack of Concern for Recovery Planning: Post-Disaster Recovery Awareness has likewise drawn zero votes, and is simply another academic debate rather than long-term scenario planning. This too may be a a result of psychological avoidance to think about anything that may occur after the disaster, or it may be a conceptual predisposition to assume recovery will be undertaken by the government, preventing the public from identifying the issues they will encounter after a potentially traumatic recovery phase post-quake.
- Weak Community Sentiment: The only vote for Social/Societal Awareness—"It's really sad for a knowledgeable person to appeal to others for validation,

if he believes in his knowledge, he'll make a debate about it? Or does this country not anyone who can set it up?"— suggests very weak interaction with collective action. The comment expresses disappointment, indicating irritation toward institutional or expert inaction, but very few votes suggest limited communal interest in action-oriented solutions.

- Marginal Fatalism: The only Fatalism vs. Agency comment—"Those who claim there are faults where they are not are a betrayal of this country.... This country did not acquire this land through a lottery; they acquired it through blood and misery...."—was the only comment with 0 votes, suggesting that conspiratorial or fatalistic perspectives are rare in this sample. This might signal that the YouTube audience is generally biased toward rational discussion, although the nationalistic tone of the comment suggests that such a disparaging and fatalistic view is a small, albeit uncommonly held, viewpoint that can make an emerge in larger discursive spaces.

3.4.4. Implications of Vote Distribution

The comment vote distribution numbers—full scale of 480 for the Knowledge/ Information Level and Preparedness/Preventive Behaviors overlap, and all others with six or fewer—suggest that the clear audience preference is around:

- Polarization Around Experts: The 480 votes split between 1 comment indicates some indication of polarization around the inclusion of experts, indicating public admiration from and trust of individuals and a public skepticism toward authorities. This polarization, if addressed, can inform educational outreach or, if a resolution is not made, can contribute to increased reluctance.
- Disengagement with Practicality: The zero votes for Response and Recovery subdimensions suggest that academic or theoretical type material simply do not engage viewers, possibly as the lack of relatable or actionable advice reduces viewer interest. This emphasizes the need for material to translate the science into something relatable to everyday safety.
- Low Collective Voice: The low votes for Social/Societal Awareness and Fatalism vs. Agency, suggests that community action or alternate perspectives are not available to motivate the audience yet. This may reflect Istanbul's urban individualism or the audience may be focused on individual survival measures versus collective survival measures.

Table 2. Subdimension Statistics & Sample Comments

Subdimension	Comment Count	Most Liked Comment (English, Translated)	Votes
Risk Perception	32	"Professor Şener does not say the earthquake risk is over; he says there might be a 6.2 magnitude quake, but nothing over 7. So, an earthquake will still happen."	6
Knowledge/Information Level	590	"It seemed very ill-intentioned to me that Professor Naci was not invited to the meeting."	480
Preparedness/Preventive Behaviors	135	"It seemed very ill-intentioned to me that Professor Naci was not invited to the meeting." (overlaps with above, reflects expert exclusion in planning)	480
Response Awareness	3	(Academic debate, not a practical tip: discusses earthquake prediction models and publications.)	0
Post-Disaster Recovery Awareness	2	(Same academic debate, not a recovery plan.)	0
Social/Societal Awareness	2	"It's really sad for a knowledgeable person to seek the approval of others. If he trusts his knowledge, why doesn't he ask for a debate? Or is there no one in this country who can organize such a thing?"	1
Fatalism vs. Agency	1	"Those who claim there are faults where there aren't any are betraying this country. ... This nation did not win this land by lottery, it won it with blood and sacrifice..."	0

In the case of Istanbul, where the North Anatolian Fault is always there, the trends here mirror previous studies which indicate a knowledge-oriented disproportionate discourse. That there was a high amount of engagement with Knowledge/ Information Level and Preparedness/Preventive Behaviors shows a public willing to engage with learning and taking action. This engagement is likely within the context of the past earthquakes that occurred in or impacted Istanbul, such as the 1999 İzmit earthquake, as well as current discussions regarding urban renewal projects that are very locally debated and informed by a public that is largely uninformed about seismic risks. Lack of engagement with the response, recovery, and social dimensions of earthquake preparedness could stem from cultural factors, such as the public's expectation of government intervention in all aspects of a disaster response situation, or the possibility that, in a context like Istanbul, much of the urban population has some fatalistic tendencies, as outside of this sample, the majority are in circumstances where a potential catastrophic earthquake is less of a remote possibility. Similarly, we are in an urban context with more than millions of people with a logistics burden of coordination.

3.5. The Topic Modeling Results

The Topic Modeling Results from Latent Dirichlet Allocation (LDA) with five topics gives us the most complete views into the topical structure of earthquake awareness, based on YouTube comments, relevant to earthquakes generally, but specifically within the context of Istanbul's seismic risk, as of 06:12 PM +03,

Thursday June 05, 2025. By looking at the tendency of the top words of each topic and interpreting the topics-related representative sample comments and mapping the disaster awareness subdimensions to those comments, we can identify what the dominant narrative, public perception, and the concentration of content was present among viewers' comments. This interpretation looks at the next phase of understanding public engagement and supporting earthquake preparedness.

The first topic appears to be related to science, and relevant keywords include "ve"(and), "bu"(this), "bir"(a/one), "bilim"(science), "gibi"(like/as), "da"(also), "hocam"(professor/teacher), "daha"(more), "naci"(referring to Naci Görür), and "değil"(not). This topic is a likely mixture of trust and distrust of authority, and frequently referenced Naci Görür, a well-known seismologist from Turkey, while "hocam" references or respects or promotes authority. Sample comments include "Hocam, bilim gibi düşünülmeli, yoksa yanlış sonuçlara varılır" (Prof., we must think scientifically, or we will reach wrong conclusions) and "Naci hocam, sizin bilimsel açıklamalarınızı önemsiyoruz, lütfen devam edin" (Prof. Naci, we value your scientific explanations, so please continue). These comments reveal a like-minded community who value scientific reasoning, albeit, perhaps doubting its use. This behaviour matches the Knowledge/Information Level subdimension; the public appears to seek clarity and reliability from experts, all while acknowledging and/or feeling complexity of seismic science.

In the second topic, which included the keywords of "deprem", "bu", "bir", "ve", "sonra", "olacak", "naci", "da", "stıstanbul", and "oldu", were all around earthquake prediction and risk perception. The mention of timing ("sonra, then") and the references to Istanbul and Naci Görür seem to imply a story about when and where the next quake is going to happen. Sample comments like "Deprem İstanbul'da olacak, bunu herkes biliyor ama kimse önlem almıyor" (An earthquake will happen in Istanbul, everyone knows it but no one takes precautions) and "Naci Hoca yine depremden sonra olacakları söyledi" (Naci again spoke about what will happen after the earthquake) reflect a higher level of awareness of risk and have a sense of frustration about (lack) of action. This topic maps precisely to the Risk Perception subdimension, highlighting the public's obsession with the inevitability of a seismic event along the North Anatolian Fault.

The third theme, with prominent keywords "bir" (a/one), "naci" (Naci Görür), "bu" (this), "hoca" (professor/teacher), "şener" (a reference to another expert, presumably Şener Üşümezsoy), "ve" (and), "da" (also), "deprem" (earthquake), "diyor" (says), and "ama" (but), concerns debates and comparisons between experts. The specific contrasting expert names and the qualifiers such as "says, but" signal public uneasiness, confusion, or critical scrutiny toward differing opinions. For example, when the concerned individual disclaims objections, comments such as "Bir Naci diyor bir Şener diyor, hangisi doğru?" (One says Naci, the other Şener, which is true?), or “Hocalar arasında bu kadar fark olmamalı, biri diyor olacak, diğeri diyor olmayacak” (There shouldn't be such a difference between Hocalar, one says it will happen, the other says it won't), highlight discord in conflicting predictive messages, and exhibit awareness of cultural norms related to expectations of informed certainty from experts. This theme is situated within the Knowledge/Information Level and Risk Perception subdimensions, both of which are indicated at Special Level by experts developing their own familiarity with knowledge expanded and their risk assessment of that knowledge compared to other seismic experts.

The fourth topic has some phrases related to a facetious and informal tone of voice signaled by key terms such as: “bu” (this), “ne” (what), “bi” (a abbreviated form of bir), “bir” (a/one), “bilim” (science), “ya” (an interjection), “sen” (you), “naci” (Naci Görür), “adam” (man), “kadar” (as much as), and conveys an informal and perhaps derisive tone. For example, the question “bu ne” (what is this?) and the words “you” and “man” suggest a layperson's challenge to the establishment or a scepticism about scientific truth and expert communication. Comments such as “Bu ne biçim bilim, kimse anlamıyor, herkes kendine göre konuşuyor.” (What kind of science is this, no one understands, everyone speaks for themselves.) and “Adam gibi açıklayan yok, sadece laf kalabalığı” (There's no one explaining properly, just a bunch of talk.) indicate doubts or confusion as future building blocks of knowledge based on one's beliefs in modern science can lead to frustration with perceived ambiguity or inconsistency. This topic describes elements that sit within the Social/Societal Awareness and Fatalism vs. Agency sub-dimensions of capability as it conveys the potential risk of sceptical critique belonging to the public about authorities, and suggests inquirers are concerned about questioning scientific legitimacy and tend to experience either a sense of futility or fatalistic resignation.

The fifth topic, which has keywords such as “deprem” (earthquake), “bir” (a/one), “bu” (this), “ve” (and), “ne” (what), “çok” (very), “ama” (but), “god” (allah), “da” (also), and “yok” (no/none), concerns fatalism and religious resignation. The existence of “god” and the “very” and “no” qualifiers all indicate some form of helplessness or frustration relating to the phenomenon of earthquakes, which could stem from the view that the phenomenon is completely out of human control. The comments “Allah'tan başka kimse depremi engelleyemez” (No one but God can prevent an earthquake) and “Ne yaparsak yapalım, deprem olacaksa olur” (No matter what we do, if an earthquake is going to happen, it is going to happen) affirm a fatalistic worldview. This topic is obvious mapped to the Fatalism vs. Agency subdimension because this portion of the audience is adopting a position that the decisions people make do not really matter, because outcomes are determined by divine forces rather than humanizing and agency.

Taken as a whole, the topic modeling demonstrates a multivalent discussion around earthquake awareness. Topics 1 and 2's relative prominence (science and risk prediction) reflects their very high comment counts in Knowledge/Information Level and Risk Perception from the earlier analysis, further reinforcing the public's desire to know and predict seismicity. Topic 3's expert debate further enriched the knowledge segment, while Topics 4 and 5 reflected a contrasting current of skepticism and fatalism that may interrogate the scientific narratives toward which public discourse typically gravitates. In combination with the competing elements of knowledge-seeking and uncertainty, we have a community that embraces knowledge, but sometimes resigns itself to fatalism. In Istanbul, where people are not only asking whether they will experience a significant earthquake, but what impact urban policy is having in the meantime, these themes are directly related to a call for more education- interventions aimed at helping the public help themselves in the face of fatalism by promoting education about the concrete steps people can take in light of their knowledge about earthquakes (and where there is scientific consensus).

4. DISCUSSION

This study reveals a central paradox in Istanbul's earthquake-awareness discourse on YouTube: a strong appetite for scientific knowledge coexists with weak, uneven engagement with action-oriented preparedness. Viewers actively seek explanations about fault dynamics, probability, and expert credibility, yet they rarely discuss what to do during shaking, how to coordinate collectively, or how to plan for long-term recovery. The primary analytical problem, therefore, is not whether the public is aware of the earthquake threat; it is why a knowledge-rich conversation remains operationally thin in a megacity where response competence, recovery readiness, and community coordination are decisive for survival and resilience.

4.1. A Knowledge-Forward Discourse and the Appeal of Expert Translation

The thematic distribution is dominated by Knowledge/Information Level (590 comments) and, to a lesser extent, Preparedness/Preventive Behaviors (135 comments), indicating that YouTube comments function as an informal "public classroom." Many users treat seismologists—particularly Professor Naci Görür—as interpreters of complex geology, and the LDA topics similarly cluster around scientific explanation, risk prediction, and debates among experts. This knowledge-forward pattern is understandable in Istanbul's high-risk context (North Anatolian Fault proximity), where uncertainty produces anxiety and information-seeking can operate as a coping strategy. In this sense, the public's cognitive engagement should be interpreted as an asset: it signals receptivity to evidence-based communication and sustained interest beyond short-lived media spikes.

At the same time, the public's reliance on expert figures also highlights a vulnerability: epistemic dependence can outpace capability-building. When the dominant "rewarded" content in the comment space is explanation (what the risk is, who is right, what magnitude matters), discussion may stall at diagnosis—particularly when competing expert narratives invite continual comparison rather than conversion into standardized routines.

4.2. Preparedness Talk as a Blend of Individual Intent and Systemic Frustration

Preparedness/Preventive Behaviors emerges as the second most salient subdimension, frequently expressed through calls for household measures (e.g., go-bags, securing furniture) and institutional practices (e.g., drills), alongside intense frustration about unsafe buildings and uneven enforcement. This duality matters: some preparedness discourse reflects personal agency, yet much of it reflects structural constraints—housing affordability, retrofit costs, and distrust in oversight. In a city where many residents cannot easily relocate to safer housing, preparedness becomes both a private responsibility and a contested political issue, shaping the tone of the discourse toward grievance and accountability rather than toward practical rehearsal.

4.3. Analysis of Absence: Silence as a Substantive Finding

A key contribution of this study is that the striking scarcity of several subdimensions should not be treated as a null result. The underrepresentation of Response Awareness (3 comments), Post-Disaster Recovery Awareness (2 comments), and Social/Societal Awareness (2 comments) constitutes an "analysis of absence": the silence itself is empirical evidence of a procedural and collective-preparedness deficit. In a setting where an earthquake could simultaneously affect millions, limited attention to response routines, recovery planning, and mutual aid suggests that public earthquake awareness is cognitively developed but operationally incomplete.

This absence plausibly reflects broader societal failures. If response protocols and recovery planning are not visible in everyday civic life—through routine drills, accessible guidance, neighborhood-level organizing, and institutional messaging—then it is predictable that they will not appear organically in a comment corpus shaped by a knowledge-oriented video. Put differently, the absence of these themes is consistent with a preparedness ecosystem that emphasizes risk talk and structural vulnerability more than capability rehearsal and post-event continuity planning.

4.4. Socio-Cultural Mechanisms That Suppress Recovery and Collective Planning

Several socio-cultural factors can help explain why recovery planning and collective action remain marginal even in a high-risk city:

Culture of state dependency. In Turkey's disaster governance context, many residents expect that large-scale response, sheltering, and recovery will be state-led. This expectation can coexist with institutional distrust: individuals may doubt effectiveness yet still assume that recovery is fundamentally "someone else's job." The result is reduced public incentive to discuss recovery logistics, psychosocial support, employment disruption, or community-led continuity planning—even though these domains strongly influence long-term outcomes.

Urban individualism and weakened neighborhood ties. Istanbul's megacity dynamics—mobility, density, heterogeneity, and fragmented social networks—can erode the everyday foundations of collective preparedness. When neighbor-to-neighbor trust is limited, community-based disaster planning feels less feasible and less salient, and preparedness remains framed as an individual household problem rather than a cooperative capacity.

Credibility fractures and expert polarization. Topic modeling and high-engagement comments suggest strong attention to which experts are trusted and how institutions treat them. When the public sphere becomes organized around expert contestation and perceived exclusion, attention can shift from "what we should do" to "who should be believed." This can crowd out procedural talk, especially if people interpret disagreement as evidence that action guidance is uncertain or politically mediated.

4.5. Cognitive Biases That Inhibit Engagement with Recovery and Response

Beyond socio-cultural factors, well-established cognitive tendencies can suppress engagement with response and recovery planning:

Normalcy bias leads individuals to assume that future conditions will resemble the present, dampening motivation to imagine prolonged disruption and recovery needs.

Optimism bias ("it won't happen to me," "my building will be fine") reduces perceived benefit of rehearsing response behaviors and preparing for displacement.

Present bias / temporal discounting makes long-term recovery planning (insurance, livelihood continuity, mental health support, housing contingencies) feel too distant and costly relative to immediate concerns.

Availability effects mean that unless a recent earthquake experience is vivid, people prioritize informational debate over procedural rehearsal; conversely, risk-perception spikes tend to be event-driven and short-lived.

Cognitive overload—especially under economic stress and housing insecurity—can push residents toward passive information consumption rather than effortful planning for complex post-disaster scenarios.

These biases help explain why even a knowledge-motivated audience may not spontaneously discuss recovery or response: the cognitive load of planning for a severe disruption is high, and avoidance becomes psychologically adaptive in the short term, despite long-term costs.

4.6. Platform and Discourse Effects: Why YouTube Comments Favor "Knowing" Over "Doing"

The YouTube comment environment also shapes what is likely to be expressed. Comments tend to reward short, reactive statements rather than structured contingency planning. Moreover, the source video's framing—earthquake "reality," expert interpretation, and risk explanation—likely primes knowledge-seeking participation. In that sense, the platform and the content jointly encourage an epistemic mode ("explain the risk") and weakly scaffold a procedural mode ("practice and coordinate"). This does not diminish the importance of the findings; it clarifies that the observed imbalance is produced by both social context and communication affordances.

4.7. Implications and Targeted Recommendations Derived from the Gap Structure

The practical implication is that Istanbul's public discourse—at least in this corpus—is knowledge-forward but capability-light. Bridging this gap requires interventions that deliberately convert knowledge appetite into procedural competence and collective readiness. Targeted recommendations follow directly from the missing subdimensions:

1. Response Awareness (procedural micro-skills): Embed short, repeatable "micro-protocols" (e.g., drop–cover–hold on; post-shaking stairs not elevators; family communication rules) into the same expert-facing formats that already attract engagement. Normalize drills as routine rather than exceptional by linking them to schools, workplaces, and building management.
2. Recovery Awareness (continuity literacy): Treat recovery as a teachable domain—psychosocial support, temporary housing pathways, employment disruption, documentation, and insurance—using practical checklists and scenario-based communication. Make recovery resources visible before disasters occur so that "recovery talk" becomes socially legitimate rather than avoidable.

3. Social/Societal Awareness (collective capacity): Create neighborhood- and building-level preparedness templates (roles, contact chains, support for vulnerable residents) that reduce the coordination burden. Where urban individualism is high, low-friction collective mechanisms (building WhatsApp groups, simple meeting points, shared supply plans) can serve as entry points.
4. Trust and credibility fractures (consensus messaging): Reduce dependence on polarizing expert disputes by emphasizing areas of professional consensus and communicating uncertainty transparently. Joint messaging by universities, municipalities, and disaster agencies can shift attention from "who is right" toward "what to do regardless."

In sum, this study suggests that Istanbul's digital earthquake discourse reflects substantial cognitive engagement with risk but insufficient integration of response routines, recovery planning, and collective coordination. The observed silence in these domains is best interpreted as an indicator of systemic preparedness shortcomings shaped by socio-cultural expectations, institutional credibility gaps, cognitive biases, and platform dynamics. Addressing the paradox requires not more information alone, but communication and civic design that turns understanding into practiced capability and shared resilience.

5. CONCLUSION

The results indicate that the YouTube audience in Istanbul is highly involved with learning and preparing for earthquakes as a result of relatively high rates of awareness regarding the earthquake risks of the city (given the North Anatolian Fault) and when including a historic context like the 1999 İzmit earthquake. However, the involvement is unbalanced because while the audience is heavily favored on the Knowledge / Information Level and Preparedness / Preventive Behaviors, there are miles of unmet need in categories like Response Awareness, Post-Disaster Recovery Awareness, and Social / Societal Awareness. The imbalance in disaster awareness may leave the public vulnerable to both the immediate and long-term impacts of a future quake because while knowledge of one's risks and the intent to prepare is certainly a positive step, there are no practical response skills, recovery plans or collaboration with social networks. The culture (individualism, institutional mistrust, occasional fatalism, etc.) geographical academics and demographics (the social/economics of urban convenience and aging infrastructure) plays a role in the gaps that exist. Resolving these gaps is essential for promoting and establishing a competent disaster awareness culture that has the ability to mitigate the negative impacts of a significant seismic event.

6. RECOMMENDATIONS

In an effort to improve earthquake readiness and to close previously identified gaps, the following recommendations are suggested:

1.Enhance Response and Recovery Education: conduct grass roots campaigns to boost Response Awareness, including key evidence-based practices such as, "drop, cover, and hold on", or what to do after the quake - not to use elevators, etc. These could be integrated into existing knowledge content such as YouTube clips, and perhaps systematically embed drills into the school system, or the workplace. In this same vein, encourage Post-disaster Recovery Awareness, but focus on comparative long-term needs such as psychological support, housing, economic support, etc. to help frame the narrative, and sharing, or listening to experts help the public contemplate the needs of the public in the aftermath of the quake.

2.Engage and build Community: When trying to build community remain mindful of counter urban individualism, promote the awareness of Social/ Societal Awareness through grassroots initiatives. By engaging municipalities and many NGOs, seek to grow district awareness campaigns such as preparedness fairs, or district drills to elicit solidarity. Platforms like X can be engaged in their own concepts to start district conversation about collective duty, thus limiting dependence on top down efforts.

3. Identify Cultural Barriers and Fatalism. Develop culturally aware/culturally-sensitive campaigns to chart a journey from fatalism to agency by portraying preparedness as a way to mitigate the impacts of the natural (i.e., earthquakes are natural, but their impacts can be mitigated). Incorporate leaders of the community (like religious leaders) in framing preparedness as a moral good and to link preparedness to their cultural values, where acceptance of taking proactive measures will be possible.
4. Improve systemic accountability and the built environment. Given the public frustration around unsafe buildings, urban renewal programs should be accelerated, transparency should be ensured around retrofitting and enforcing seismic codes on buildings. Municipal governments should be open with their residents about the building assessments as had been suggested in the comments section when rebuilding trust to actively engage the individual to prepare.
5. Utilize expertise and utilize media. Continue to feature trusted experts like Naci Görür, ensuring they are presented on public forums to safeguard credibility and to diffuse frustration about their inability to be included. Utilise "media spikes" in Risk Perception (for example, around minor tremors) to

promote preparedness strategies, ensuring that a level of public attention will exist beyond a period of transient anxiety.

By implementing these recommendations, stakeholders—government, civil society, and media—can build on the public's existing knowledge and motivation to create a more balanced and resilient disaster awareness culture in Istanbul, ultimately reducing the human and economic toll of a future earthquake.

REFERENCES

Alexander, D. E. (2014). Social media in disaster risk reduction and crisis management. *Science and Engineering Ethics*, *20*(3), 717–733. DOI: 10.1007/s11948-013-9502-z PMID: 24306994

Amsalu, A., Garedew, W., Melka, G. A., Çelik, M. A., & Gemeda, D. O. (2026). Trends in climate extreme indices and implications for coffee farming in The Kafa Biosphere Reserve, southwest Ethiopia. *Climate Services*, *42*, 100637. DOI: 10.1016/j.cliser.2026.100637

Asio, J. (2021). *Disaster awareness and level of compliance to disaster programs in a highly urbanized city*. Aquademia., DOI: 10.21601/aquademia/9581

Bilham, R. (2010). Lessons from the Haiti earthquake. *Nature*, *463*(7283), 878–879. DOI: 10.1038/463878a PMID: 20164905

Çelik, M. A. (2026). Agro-climatic classification of Türkiye's Agricultural Basins using long-term ERA5-Land temperature data. *Physics and Chemistry of the Earth Parts A/B/C*, *142*, 104307. DOI: 10.1016/j.pce.2026.104307

Çelik, M. A., Kıraç, A., & Gouveia, C. M. (2026). North Atlantic Oscillation modulation of drought and vegetation activity in Türkiye. *Theoretical and Applied Climatology*, *157*(2), 74. Advance online publication. DOI: 10.1007/s00704-025-05971-5

Dikmenli, Y., Yakar, H., & Konca, A. S. (2018). Development of a disaster awareness scale. *Review of International Geographical Education Online*, *8*(2), 206–220.

Durmuş, E., Kinaci, M., & Kuruyer, D. (2024). *Examining the disaster awareness levels of secondary school students*. Uluslararası Türk Eğitim Bilimleri Dergisi., DOI: 10.46778/goputeb.1494675

Elsaeidy, A., Saboor, A., Hassan, A., & Redwan, A. (2016). Disaster-resistant communities, public awareness program. *Journal of Emergency Medicine. Trauma & Acute Care*, *137*(2). Advance online publication. DOI: 10.5339/jemtac.2016.icepq.137

Gerdan, S. (2014). Determination of Disaster Awareness, Attitude Levels and Individual Priorities at Kocaeli University. *Dergipark. Eurasian Journal of Educational Research*, *2014*(55), 159–176. DOI: 10.14689/ejer.2014.55.10

Hargono, A., Artanti, K. D., Astutik, E., Widodo, P. P., Trisnawati, A. N., Wardani, D. K., & Lioni, E. (2023). Relationship between disaster awareness and disaster preparedness: Online survey of the community in Indonesia. *Journal of Public Health in Africa*, *14*(9), 2376. DOI: 10.4081/jphia.2023.2376 PMID: 37927360

Kajihara, C., Kashiwabara, M., Yamao, M., Aoki, Y., Yabe, K., Kaneko, M., Sano, M., & Munechika, M. (2023). Evaluation of the degree of crisis awareness and behavioral change through disaster exercise: A case study of a disaster response exercise in a pharmacy department. *Prehospital and Disaster Medicine, 38*(S1), s22–s22. DOI: 10.1017/S1049023X23000997

Karanci, A. N., Aksit, B., & Dirik, G. (2005). Impact of a community disaster awareness training program in Turkey: Does it influence hazard-related cognitions and preparedness behaviors. *Social Behavior and Personality, 33*(3), 243–258. DOI: 10.2224/sbp.2005.33.3.243

Knabb, R. D., Rhome, J. R., & Brown, D. P. (2005). *Tropical cyclone report: Hurricane Katrina*. National Hurricane Center.

Lee, E., & Lee, H. (2019). Disaster awareness and coping: Impact on stress, anxiety, and depression. *Perspectives in Psychiatric Care, 55*(2), 311–318. DOI: 10.1111/ppc.12351 PMID: 30648274

Murayama, Y. (2021). Use of IT for situation awareness for disaster risk reduction. In *Disaster risk reduction* (pp. 145–160). Springer., DOI: 10.1007/978-981-16-0360-0_9

Özkazanç, S., & Yüksel, Ü. (2015). Evaluation of disaster awareness and sensitivity level of higher education students. *Procedia: Social and Behavioral Sciences, 197*, 745–753. DOI: 10.1016/j.sbspro.2015.07.168

Park, A.-S., & Yang, J.-H. (2017). The effects of disaster awareness, attitude, and preparedness on the basic competencies of disaster nursing among nursing students. *International Journal of Advanced Nursing Education and Research, 2*(1), 59–64. DOI: 10.21742/IJANER.2017.2.1.11

Patel, R. K., & Kermanshachi, S. (2020). *A socioeconomic-based analysis of disaster preparedness, awareness and education. ResearchGate PDF.* Retrieved from. [ResearchGate]

Shaw, R., Takeuchi, Y., & Rouhban, B. (2009). Education and public awareness for disaster reduction. In *Landslides – Disaster Risk Reduction.* Springer., DOI: 10.1007/978-3-540-69970-5_26

Son, J., Aziz, Z., & Peña-Mora, F. (2008). Supporting disaster response and recovery through improved situation awareness. *Structural Survey, 26*(5), 411–425. DOI: 10.1108/02630800810922757

Sonopant, J. (2015). Nurses' awareness and perceptions towards disaster situations. *Nursing Journal of India, 106*(4), 147–150. DOI: 10.48029/NJI.2015.CVI404 PMID: 30650932

Teo, M., Goonetilleke, A., Ahankoob, A., Deilami, K., & Lawie, M. (2018). Disaster awareness and information seeking behaviour among residents from low socio-economic backgrounds. *International Journal of Disaster Risk Reduction, 31*, 1121–1131. DOI: 10.1016/j.ijdrr.2018.09.008

Turan, M., Bulut, Y., Öztürk, G., & Göktekin, Z. (2017). The knowledge and behavior levels of the students taking disaster awareness training: The example of Tekirda province, Turkey. *Nepal Journal of Environmental Science, 5*, 57–60. DOI: 10.3126/njes.v5i0.22716

Türksever, Ö. (2021). Analysis of disaster awareness perception levels of students in social studies teaching undergraduate program. *OSF Preprints*. https://doi.org/ DOI: 10.31219/osf.io/4zq8v

United Nations Office for Disaster Risk Reduction (UNDRR). (2020). *Global assessment report on disaster risk reduction*. UNDRR.

Veil, S. R., Buehner, T., & Palenchar, M. J. (2011). A work-in-process literature review: Incorporating social media in risk and crisis communication. *Journal of Contingencies and Crisis Management, 19*(2), 110–122. DOI: 10.1111/j.1468-5973.2011.00639.x

Wisner, B., Blaikie, P., Cannon, T., & Davis, I. (2014). *At risk: Natural hazards, people's vulnerability and disasters* (2nd ed.). Routledge.

Chapter 8
Landslides and Mass Movements:
Natural and Human Triggered

Pardeep Kumar
http://orcid.org/0000-0003-0308-7634
Lovely Professional University, India

Sanjeev Kumar
http://orcid.org/0000-0002-7375-7341
Lovely Professional University, India

ABSTRACT

Natural or man-made landslides and mass movements are important geophysical processes that may cause disastrous effects on the environment, as well as social and economic directions. Natural factors include excessive rain, earthquakes, volcanic activities and fast melting of snows, whereas the human aspect includes clearing forests, mining and infrastructure construction, as also the poor land-use methods. Such convulsions cause soil erosion culminating in infrastructure damages and loss of lives. Disaster prevention and mitigation requires knowledge on the reasons behind the disaster, risk assessment and early warning systems. This chapter discusses the forces behind landslides and mass movements, the impact of climate change, and how effective mitigation measures such as slope stabilization, afforestation and the community awareness programs are in alleviating them. The main objective of using interdisciplinary methods is to bring better resilience and policy options on sustainable management of land.

DOI: 10.4018/979-8-3373-6274-8.ch008

1. INTRODUCTION TO LANDSLIDES AND MASS MOVEMENTS

Landslides and mass movements: These are natural geo-physical phenomena, which entail the down-slope movement of soil, rocks and debris under the impact of gravity. They frequently happen in mountainous and hilly regions even though they can take place anywhere whose slopes are not stable. Landslides and mass movements are also major natural hazards which may lead to unimaginable harm to life, property and the environment (Fayaz et al., 2022). They occur because of natural and man-made conditions. Natural causes are heavy rainfall, earthquakes, volcanic eruptions, and rapid snowmelt whereas man-made factors such as deforestation, mining, construction activities, and alteration of land use are anthropogenic causes. These movements are categorized broadly on the basis of the nature of movement and the material moved over. Falls, slides, flows, creeps and topples are considered to be the main types of mass movements. Falls are rock movements related to sudden dislodging of rock and falling down the steep slopes (Mollamahmutoğlu & Acar, 2020). Slides happen when a coherent mass of the earth or rock moves on a well defined surface and it often produces rotational or translational slide. Flows Materials that react as a fluid take the form of flows such as mudflows and debris flows most often in events that follow or accompany high amounts of rainfall when the soil is completely wet. Creeps are gradual, undetected movement of soil and rocks that occur over a long time and topple are forward rotation and the movement of big blocks of rock.

The history of mankind is full of landslides and mass movements that in many cases can be devastating. The Vajont Dam in Italy 1963 is one of the most devastating of incidents. A huge landslide fell into the reservoir creating a wave that rose above the dam and washed away a waterfall, killing some 2,000 people in the downstream. This is a reminder that a geological risk assessment in large complex projects is important. The second significant event was the debris avalanche in Yungay, Peru in the year 1970 caused by a fierce earthquake (Strupler et al., 2018). The landslide engulfed the whole town and led to massive casualties of more than 20,000 people. The Oso mudslide in Washington state in 2014 also exemplified how unstable slopes represent constant risks since the engulfment of an avalanche swept off houses and the loss of lives (43 people were killed). These disasters have more than just direct human deaths. Long-term consequences of landslides include permanent settlement displacement, agricultural land being rendered unproductive and infrastructural destructions such as roads, bridges and other infrastructural facilities (Damm & Klose, 2015). Vegetation is destroyed, levels of sedimentation in rivers and streams are enhanced and eco systems are disrupted. At an economic level, the expenses involved in the recovery, rehabilitation and recreation are very high and this is more difficult in developing countries which do not have the essential resources and planning procedures (Parkash, 2023).

To counter these challenges, there has been an attempt to learn how landslides operate and the reasons behind them, so that their impacts could be curtailed. Our capacity to observe and forecast such events has been boosted by the technological progress made in remote sensing, satellite images, and geographic information systems (GIS). Measures like slope stabilization methods, adequate drainage system and retaining walls have been set up in the prone places. Also community-based disaster preparedness programs and awareness campaigns have been effective in the reduction of the risks and promoting the resilience. In general, there must be an in-depth knowledge of landslides and mass movements in order to mitigate them. This is through scientific research, policy interventions, and community participations (Mundhe & Jaybhaye, 2023). Climate change may lead to a greater (or more extreme) frequency and magnitude of weather events (Amsalu et al., 2026; Çelik et al., 2026); as such, this could result in an increase in the likelihood of landslides happening in at-risk regions. As such, it needs an interdisciplinary strategy prior to devising sustainable and effective means of managing the risk of landslides; one that employs a combination of environmental science, engineering, urban planning, and public policy (Joshi et al., 2019).

2. GEOLOGICAL AND ENVIRONMENTAL STIMULI

Landslides and mass movements are mainly triggered by geological and environmental causes such as the destabilization of the slope that causes the movement. Such triggers are seismic activity, heavy rainfall, hydrologic changes and volcanic eruptions. All these agents may either work alone or in concert or enhance the susceptibility of the slopes, particularly in the sites which already possess a weak condition (Alonso, 2021). Among the natural causes of landslides, earthquakes count among the most important ones. Seismic shaking on the ground may interfere with the stability of slopes through decreasing shear strength of materials, opening cracks and promoting internal displacement. Earthquake magnitude, depth and duration and geologic aspects of the area are critical contributing factors to landslide susceptibility. As an example, during the Sichuan earthquake in China in 2008, more than 60,000 landslides were initiated causing loss of life and property on a large scale. Steep area and poor surfaces of rock, loose or fracture are especially susceptible (Klose, 2015). Aftershocks themselves are another cause of slope failures particularly when the slopes have already been weakened by the main shock (Bottelin et al., 2014).

Landslides are also associated with rain as one of the highest probable environmental causes especially in subtropical and tropical areas. Heavy precipitation or persistent precipitation will saturate the ground, raise water pressure in the pores and lower the friction that particles have on each other. This action reduces the slope stability and can cause the severe mass movement termed as mudslides and debris flows. Hydrological determinants such as the changing levels of groundwater, ineffective drainage and deposition of surface water further intensify the danger (Bozzolan et al., 2020). As an illustration, landslides usually occur in hilly and mountainous terrain following monsoon seasons in south Asia and typhoons in East Asia. Modification by humans through poor management of water through drainage as well as the cutting down of trees by human can be a major factor of increase in susceptibility since the waterline can easily be damaged depending on how well the slants are cushioned (McColl, 2022).

Volcanic activity adds to landslides in a number of ways. The pressure exerted on the steep slopes of volcanic debris, including ash and pyroclastic streams, can turn unstable especially when these materials are precipitated by overflow of water. The explosive eruptions are capable of causing enormous ground slumps referred to as volcanic landslides or sector collapses by cracking the volcanic cones or destabilizing the slopes due to the earthquakes (Aversa et al., 2018). Furthermore, sudden glacier or snowcap melting in the course of eruptions may cause lahars-volcanic mud flows that move very fast and wash away everything in their path. Great records are the eruption of Mount St. Helens in the United States in 1980, which resulted in a tremendous land sliding event in the area and another eruption that happened in Mount Pinatubo in the Philippines in 1991, that produced extensive lahars. The danger poses a particular threat in volcanic areas that receive large amounts of precipitation and have unstable geologic formations. These geological and environmental triggers are key issues that should be acquired to make effective landslide prediction, monitoring and risk mitigation. Holistic hazard evaluation, along with monitoring technologies in real time and community-based preparedness, form critical approaches to reducing the devastating effects of these natural processes (Vrba, 2007).

3. HUMAN-INDUCED FACTORS

Although natural conditions play the critical role as the cause of landslides, in most cases, people can enhance the odds and rate of landslides. Deforestation, mining and fast urbanization disturb natural slope balance and may severely compromise slope stability. Anthropogenic processes include deforestation as a great cause of landslides. The vegetation acts very important in the slope stability by holding the soil together with roots and slowing down the rate of surface run off (Stefanescu et al., 2004). When forests are cut down and converted to be used as land crops, cut down as timber or even under development, the vegetation cover is cleared to expose the soil which becomes more susceptible to erosion and displacement. Surface water Inputs: Changes in land use involve converting forests into farmlands or land dedicated to urbanization activities, affecting the normal drainage. This enhances large amounts of water on the surface. The extra water gets into the ground and exerts pressure on the pore water which lowers the cohesive strength of the pore (Ehrlich et al., 2017). Such changes, in most instances, are made without due environmental evaluation and make the terrain more prone to landslides. Natural geological formations get disrupted by mining activities and the mass movements can be started as a result of the extensive excavations. Open-pit mining, quarrying and tunneling tend to undermine the structural stability of the slopes, which becomes unstable. More risk is added with the process of the removal of overburden and steep walls of the mines. Furthermore, vibrations that occur as a result of blasting may induce failure of slopes, particularly the regions of where there are predisposed cracks (Si-Tayeb et al., 2010). Poor spillage of mine tailings and waste material on or close to slopes maybe a cause of landslides as well. Abandoned mines that are no longer operated and whose slopes are not monitored remain a threat several years after closure in some cases.

The most significant reason causing human-made landslides include rapid urbanization and haphazard construction of infrastructure. During construction there is the cutting of the slopes, disruption of the natural drainage patterns and saturating of the slopes by construction works, such as buildings and roads. These interventions may upset the level of stress in the soil and the rock, which decreases stability. Poor engineering works only add to the problem with poor slope design, failure to include retaining structures as well as lacking provided drainages. The roads, rail, and tunnels built in the hilly terrains without keeping geological conditions under consideration are prone to landslides. There is also the urban runoff on the impervious surfaces that causes the increased infiltration of water in slopes which enhances the erosion and consequent slope failures. Knowledge of causes that are created by humans is essential in terms of planning and establishment of effective strategies of land management and risk management. The human role in landslide hazards can be curbed to a large extent by adopting sustainable development practices, enforcing environmental regulations and creating awareness among the stakeholders (Servetti et al., 2025).

Human-Induced Factors Contributing to Landslides and Mass Movements shown in table1

Table 1. Human-Induced Factors Contributing to Landslides and Mass Movements

Factor	Description	Examples	Impact
Deforestation	Removal of trees reduces root strength and soil cohesion	Logging in Himalayan regions, Amazon deforestation	Increased surface runoff, slope instability, and erosion
Land-Use Changes	Alteration of natural terrain for agriculture or construction	Hill slope agriculture in Northeastern India	Destabilization of slopes, altered drainage patterns
Mining and Quarrying	Blasting and excavation weaken rock formations and natural slopes	Open-cast mining in Jharkhand, sand mining in riverbeds	Subsurface voids, vibrations, increased risk of slope failure
Infrastructure Development	Road construction, tunneling, and dam building disturb geological stability	Road building in Western Ghats, hydropower projects in Uttarakhand	Increased erosion, landslides due to undercutting and vibration
Urban Expansion	Unregulated construction on steep slopes and flood-prone zones	Expanding cities like Shimla, Darjeeling	Overloaded slopes, disrupted drainage, poor slope management

Source: Author Compilation

4. MECHANICS AND DYNAMICS OF MASS MOVEMENTS

Mass movements, otherwise known as mass wasting, are the downslope movement of soil, rocks and debris induced by gravity. The traits of such movements are different and they run at various speeds as well as being induced by different things but collectively these movements shape the landscape and cause environmental hazards. Rockfalls, slumps and debris flows are among the main types of mass movements, and their mechanisms of action and effects are different (Irschik & Belyaev, 2014). Rockfalls take place when a single rock or a big chunk becomes dislodged due to the steep slope or cliffs and falls down at high speeds. These usually occur abruptly either as a result of weathering, freeze-thaw processes or by an earthquake and are especially hazardous in mountainous or cliff prone areas. Slumps Slumps In contrast, slumps involve downward, outward rotational surface movement of material along a concave slip (Pradhan & Siddique, 2019). They are slower, less violent than rockfalls, but may cause major damage in the long run, particularly in regions of weak, water-saturated soils. Debris flows are a fast and messy process of a water, soil, rock, and organic matter combination downslope. They are mostly caused by heavy rains or quick melting snow and can move quite fast overtaking all the objects in their way and affecting a mass destruction in valleys or lower lands (Schuster & Highland, 2007).

To study the mechanics of these movements we need to have a clear analysis of soil and slope stability. Whether slop is stable or not depends mainly on the driving forces (examples are gravity and water pressure) versus resisting (mainly the shear strength of the material). In case of a situation of driving forces overcoming the resisting forces, slope failure will occur. The nature and composition of the soil or rock type, angle on this slope, vegetation cover, moisture condition, as well as external forces such as construction or earthquakes play their role in such a balance. durable changeling Models and methods exist to analyze slope stability and these consist of the limit equilibrium method and finite element analysis, among others used by engineers and geologists. Mass Movements and their Types shown in Figure 1.

Figure 1. Mass Movements and their Types

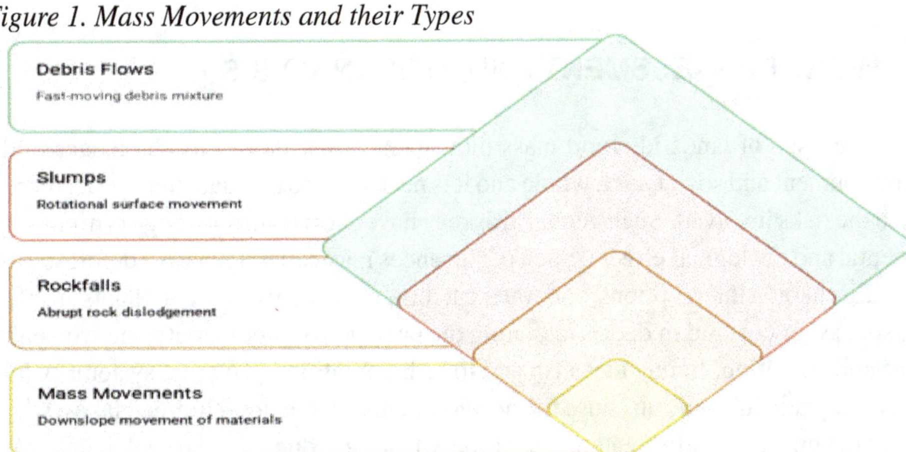

Source: Author Compilation

Mass movements are also affected by the process of erosion and sediment transport. Erosion means lifting and displacement of soil and other rocks as a result of a working agent like wind, water, and ice. Erosion may also cause slopes to grow steeper, undermine stabilizing layers and strip slopes of their vegetation, making slope failure increasingly possible over time. Conversely, the element of sediment transport entails the detachment of the eroded matter and the relocation of the matter into other areas, which mostly causes downstream or lowland challenges. The rivers, specifically, contribute to a large extent of landslide and other mass movement derived sediments transportation (Scheffers et al., 2015). This deposition may change river channels, augment the flood risk and disturb the ecosystems. In addition, the processes of erosion and transportation of the sediment belong to a feedback loop:

the erosion process is promoted by mass movements; the latter, in their turn, are promoted by erosion. Such natural activities are precipitated by human activities like deforestation, building of roads and misuse of lands to destabilize the slopes and raise the sedimentation material in water courses. Summarizing the presented above, it could be stated that the processes and phenomena of mass movements are impacted by an extremely diverse combination of nature and human activity factors. Knowledge of the nature of the mass movements helps add important information to the assessment and mitigation of the risks. Also, erosion and sediment transport management is the key to maintaining the integrity of the landscape and reduction in the destructive forces of such events (Hradecký & Pánek, 2008). Human development is infiltrating into vulnerable terrains more than ever before, making the investigation of the dynamics of mass movement even more important as a sustainable strategy of planning land-use and mitigating disasters.

5. IMPACT ASSESSMENT AND RISK ANALYSIS

The issue of landslides and mass movement poses enormous challenges to the environment and society as a whole and it is necessary to evaluate the impact thereof and the risks involved. Such geohazards may have short term and long-term environmental and ecological effects. Such occurrences may cause a massive deforestation, wildlife habitat interruptions, and water catchments systems changes. Plants are often displaced or covered in debris lessening the presence of biodiversity and worsening the soil condition. Landslides also add to sedimentation in riverine systems, which affect aquatic life and can cause the development of landslide-dammed lakes, which become unstable and threaten catastrophic flooding due to failure (Zeleňáková & Zvijáková, 2017). Moreover, erosion is facilitated by the destruction of natural cushions such as forests, and this also decreases the capacity of the land to withstand any upcoming landslides. The changes in socioeconomic aspects due to landslides are also tremendous. The infrastructure like the road, bridges, and communications are commonly damaged or destroyed and usually isolate the communities and impede the rescue and relief operations. Intensive cultivation leads to the destruction of arable land and to the interruption of irrigation systems, which directly bears consequences on food security and local economies. Landslides may cause displacement of the groups of people leading to psychological trauma in the long term, making them poor, and being displaced to other regions. The disaster response, rehabilitation and reconstruction costs are also a burden to the governmental and non-governmental capabilities (Jones, 2001). In tourist-dependent areas, tourism is the biggest revenue earning activity and in the cases where landslides occur, this can discourage tourists hence a loss to the locals. Further, frequent occurrences of landslides further diminish the confidence that people had in the safety of infrastructures as well as its governance, particularly in an instance where preventive systems are weak.

Hazard mapping and Vulnerability mapping are extremely significant in planning and risk alleviation. Vulnerability analysis incorporates an evaluation of the exposure and adjustment capacity of people, infrastructure and environments to a hazard of landslides. This is affected by a number of issues such as population density, the state of the economy and land use practices as well as emergency preparedness. Conversely, there is hazard mapping that is concerned with the harnessing of the dangers areas prone to landslides depending on such factors as slope grades (Şahin & Kurum, 2002). Geographic Information Systems (GIS), remote sensing and machine learning algorithms have been the modern tools that have significantly improved the accuracy of such maps. The tools enable the policymakers and planners to find high-risk areas, allocate and prioritize resources to be used to carry out land-use regulations or set up early warning (Gallina et al., 2016). The task of impact assessment and risk analysis should be interdisciplinary and include such sciences as geotechnical, environmental and social. Government and research agencies together with local communities can be important in providing locale-based solutions. Local resilience can be facilitated a great deal by the use of public awareness campaigns, community-based monitoring systems, and integration of indigenous knowledge. Moreover, hazard assessment should also consider projections of climate change because altered precipitation and extreme weather behaviors will likely amplify the rate and magnitude of mass movements (Eduljee, 2000). To sum up, it is necessary to learn more about the various impacts landslides and mass movements on people, infrastructure, ecosystems, and invest in correct risk investigation instruments to protect human life, infrastructure, and ecosystems. Societies can only develop appropriate resilience to this now expanding geohazard through an integrated and participatory approach.

6. MONITORING AND EARLY WARNING SYSTEMS

Early warning systems and monitoring are very crucial in the elimination of landslide and mass movement risks because of the provision of correct and timely information. They can also inform on the possible slope instabilities and help in reducing deaths, destruction of infrastructure, and economic losses by detecting the problem earlier in these systems. Thanks to technological changes, there have been significant transformations in the methods of monitoring, which now provide more accurate forms of data collection and assessment towards disaster risk reduction (Pecoraro et al., 2019). Modern monitoring systems are based on remote sensing technologies. Such technologies are satellite images, LiDAR (Light Detection and

Ranging), InSAR (Interferometric Synthetic Aperture Radar), and drones, by which the scientists are able to track land surface transformations throughout time. Satellite sensors are able to measure ground deformation, changes in vegetation cover and the extent to which the land is saturated, which are important criteria that can give rise to potential landslides. LiDAR gives topographic maps in high resolution form with particular applicability in mapping of slopes and identifying the slight movements when major failures are about to occur. InSAR, however, is able to sense ground displacement at millimeter level and so becomes invaluable in tracking slow landslides (Balis et al., 2017). Key Components of Monitoring and Early Warning Systems for Disaster Risk Reduction shown in table 2

Table 2. Key Components of Monitoring and Early Warning Systems for Disaster Risk Reduction

Component	Description	Examples / Applications
Remote Sensing Technologies	Use of satellite imagery and aerial sensors to monitor terrain changes, moisture levels, and slopes	Landsat, Sentinel, LiDAR, InSAR
Geographic Information Systems (GIS)	Tools for spatial data analysis and mapping of risk zones	Landslide hazard zonation maps, slope instability modeling
Community-Based Warning Systems	Involving local communities in hazard monitoring, using low-cost tools and local knowledge	Sirens, mobile alerts, local response teams in hilly and flood-prone areas
Ground-Based Instrumentation	Installation of sensors and monitoring equipment on slopes	Extensometers, tiltmeters, rain gauges, ground vibration sensors
Data Integration Platforms	Systems that compile data from various sources to assess risk and issue alerts	National Disaster Management dashboards, real-time monitoring systems
Early Warning Communication	Mechanisms for disseminating alerts and warnings to authorities and the public	SMS alerts, radio announcements, social media updates
Risk Threshold Parameters	Predefined indicators for triggering alerts	Rainfall thresholds, slope displacement rates, seismic activity levels

Source: Author Compilation

Geographic Information Systems (GIS) are quite important in handling, examining and mapping geographical information regarding landslide prone areas. GIS combines several layers of data (geology, slope gradient, rainfall patterns and land use), allowing researchers to create zonation maps of different hazards with a high level of detail and perform multi-criteria risk analysis. By use of GIS, the authorities are able to pinpoint areas of high risks, platform various mass movement possibilities and develop evacuation pathways and measures to take. In addition, GIS is also useful in long-range planning giving the ability to measure long-term accumulating environmental effects as well as aiding sustainable land use plans. The use of community-based warning systems supplements the use of technology as it approximates local populations in surveillance (Yang & Wang, 2017). Such systems are based on the expertise and experience of local communities living in high-hazard zones, and may also combine basic mechanisms, such as rain gauges, crack meters, and hand-powered slope surveys. Awareness and training programs make residents identify early stages of slope failure, and reveal abnormalities. Community participation will support swifter response time and enhance resilience, especially in the remote location where the access to high technology could be restricted (Kyriacos et al., 2011). Comprehensively, with its community participation, integration of high-tech monitoring systems has given a solid basis of both early warning and disaster preparedness. Preemptive solutions, like the creation of real-time information networks and the use of upfront modeling would be critical to alleviating the negative effects of landslides and mass movements

7. MITIGATION AND CONTROL STRATEGIES

Landslides and mass movements should be mitigated and managed through the efforts of engineering, ecological and policy practices. The control of the movement of soil and rock and frequency stabilization of slopes engineering applications are also used in terms of the first line of defense. An example is having retaining walls which are built to hold back the unstable soil at a physical level and does not allow collapsing or slipping (Elsaid et al., 2020). These are concrete-, steel-, or geosynthetic structures, which are developed, depending on a particular geotechnical evaluation of the slope-condition. Another equally important feature is the drainage systems which control the flow of water on the surface as well as underground. As water infiltration is one of the significant causes of landslides, proper drainage structure (culverts, surface drains and sub-drains) can minimize the pore water pressure and increase slope stability. Strategies for Landslide Mitigation shown in figure 2.

Figure 2. Strategies for Landslide Mitigation

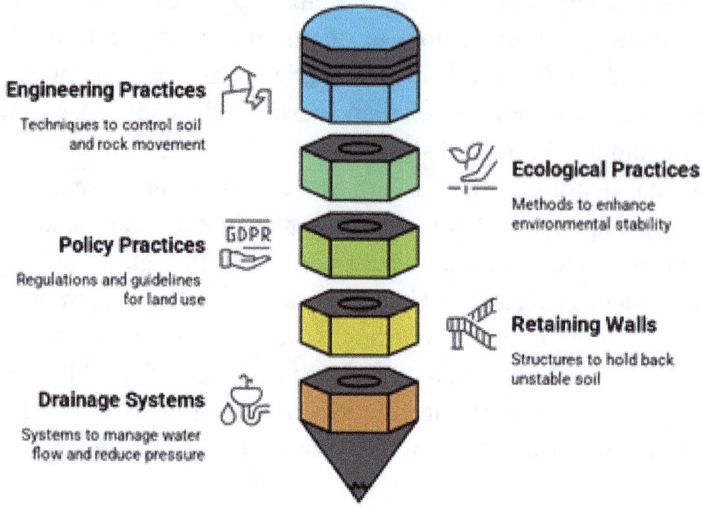

Source: Author Compilation

Besides engineering solutions, there is a natural and sustainable way of controlling landslides through the reforestation and vegetative stabilization. The presence of vegetation, more so the deep-rooted tree and shrubs contribute to this soil, by increasing its cohesion and minimizing erosion, they keep their surface stable with their roots (Fawzy et al., 2020). Rain provides good soil absorption as well because of forest cover, reducing surface runoff. It is more economical and eco-friendly to replenish native vegetation in spots exposed to landslides because of logging or farming. Besides, bioengineering methods where plant materials are used in conjunction to reinforce physical structures to support slopes are also gaining use.

It is also vital to have strong policy and regulatory structures. Land-use planning regulations that can limit the activities of construction and development on risky areas need to be pushed by governments and local authorities. The building codes must also include a condition of slope stabilization and especially those regions that have generally been identified as geologically unstable. Prior to development of major infrastructure projects or mining, the environmental orientations ought to be accompanied by an environmental impact assessment (EIA) necessitate in the environmental regulations. In addition, policies should also adopt community involvement and arrangements of public awareness campaigns on disaster risk management, and policies that connect to long term resilience (Perrings, 2005).

8. ROLE OF CLIMATE CHANGE IN MASS MOVEMENTS

Climate change is emerging in the rate, frequency, and distribution of the mass movements around the world (Çelik, 2026). These days, the changing climate destabilizes natural processes and changes geophysical conditions, for which reason slopes are increasingly vulnerable to collapse and destabilization. This chapter discusses the multiple ways climate change is considered as a precondition to landslides and other mass movements through the lens of shifting weather patterns, glacier melting, and the degradation of permafrost (Stoffel et al., 2014). Change in precipitation is one of the most direct effects of climate change. High occurrence and rates of rain events, especially very short periods of high rain intensity, considerably heighten the chances of the slopes failing. Water penetrates into soil strata, raises the pressure of the pores, weakens the soil and finally causes landslides. On the other hand, extended periods of dry spells may cause soils to dry and loosen causing soil structures to become vulnerable to collapse once heavy rains occur (Stoffel & Huggel, 2012). The randomness and instability of weather phenomena as a result of climate change thus presents critical issues to slope stability and risk control. A second key factor is the advancing thawing of permafrost and melting of glaciers in highlands and polar zones. Valley walls which had previously been glacier buttresses are receding causing adjacent slopes to become unstable. The missing ice mass initiates the redistribution of stress in the Earth crust, which jumps into rockfalls, debris flows, and outbursts floods of glacier lakes (GLOFs). Thawing permafrost also serves as destabilization, since thawing of formerly solid frozen ground to a soft water-saturated layer can no longer sustain overburden material, causing slope failure. It is estimated that such trends will only continue in the future, especially where land is high in both altitude and latitude (Smith, 2007). The temperatures will increase with the high occurrence of the extreme weather condition, accelerating the possibility of the extension of landslides-prone areas and escalating the magnitude of the mass movements. To this effect, there is need to embrace adaptation measures such as infrastructure planning that is climate resilient, slope stabilization in prone locations.

9. CASE STUDIES AND REGIONAL ANALYSIS

Mass movements are common all over the world but also varied in the causes, effects, and actions. Case studies are important in explaining the unique geological, environmental and anthropogenic conditions and the circumstances that induce such incidents. The chapter explores three key areas: Himalayan landslides, coastal erosion in Southeast Asia and urban landslides in Latin America and how diverse geographic and socio-economic conditions predispose areas to mass movement hazards. Himalaya The Himalaya region is one of the most landslide-prone areas of the world, mainly because of its young and frail geology, high seismicity, and extensive rainfall during monsoons. Its high topography, geological processes and massive rainfall provide conducive environment to cause regular landslides especially when the monsoon season comes. Risks are further increased by human activities such as road constructions, deforestation, and unplanned urbanization. One notable accident is the Kedarnath tragedy in 2013, Uttarakhand India, which occurred due to a combination of extreme rainfalls leading to large landslides and flash floods with a devastating human and material loss of life and properties. Such events keep reoccurring in Nepal, Bhutan and northern India leading to blockage of roads, displacement of communities and destruction of infrastructure etc. Such incidents emphasize how terrain-specific planning, hazard zoning, and even climate-adaptive infrastructure is required in the Himalayan zone. Mass movements in the form of coastal erosion and shoreline backflows are also on the rise in Southeast Asia, especially in those low-lying deltas such as the Mekong Delta in Vietnam, the Indonesian stretch, and the Philippines. Sediment flows and therefore the coastal stability has been considerably changed through sea-level rise and increased storm power and human activities like sand mining, mangrove deforestation, and construction of dams upstream. In Vietnam, 50 percent of the Mekong Shore is eroding with thousands of people displaced every year and agricultural livelihoods at risk. What adds to the problem is subsidence through over-pumping of ground water. Coastal erosion in this case is not merely a geological issue but socio-economic issue and requires some form of management in integrated coastal zone management (ICZM), mangrove restoration. In Rio de Janeiro, in 2011, hundreds were killed and thousands displaced during landslides in the mountainous areas in the city that were as a result of heavy downpours over an extended period of time. These cities are weak in terms of uncontrollable constructed houses, bad drainages, cutting down trees and there is a shortage of early warning systems. Such incidents demonstrate that a combination of geotechnical assessment, education of the people and investment in drainage infrastructure and slope stabilization, should be considered in urban planning considerations. The problems presented by the tectonically active Himalayas and the sediment-deprived Southeast Asian and Latin American coasts individual house building systems as well as the shanty towns of Latin America are just some of the challenges which have very specialized solutions concerning risk mitigation. The local conditions can be better understood based on case studies that can not only contribute to preparing and resiliency but can also be used to create the global frameworks of disaster risk reduction and sustainable land management.

10. EMERGENCY RESPONSE AND RECOVERY PLANNING

The planning undertaken to address the recovery process and emergencies is crucial in the extreme mitigation of the effects associated with mass movements like landslides, debris flow, and rockfalls. These disasters are usually un-foreseen and may cause a lot of loss to lives, infrastructure as well as cause matriculation of citizens. Through an organized and proactive characteristic the affected populations will be assisted at the right time and the reconstruction process is effective and sustainable (Perry & Lindell, 2003). Search and rescue (SAR) operation is the first and most important stage of operations that start after a mass movement event. These activities are intended to search and rescue the people who might be trapped under rubbles or in risky locations. The team of SAR that can commonly include emergency services of the locality, military units, and special volunteers needs to have the tools to operate in dangerous terrain effectively, including thermal imaging cameras, drones, sniffer dogs, heavy machinery. The key point is time because survival rates drops drastically past the initial 72 hours. Agency coordination, quick deployment, and sharing real-time information are critical success factors of SAR missions. Moreover, with pre-existing command centers and trained local working forces, response time and efficiency can be highly increased (Fischer, 1998). After the repressive phase, the emphasis is put on the rehabilitation and restoration process. Some of the actions of this stage include rehabilitating key facilities such as electricity, water, and transport, and repairing homes, schools, and other forms of infrastructure. Displaced people have to be accommodated in temporary homes, and also those who have witnessed such traumatic experiences need prompt psychology counseling. The rebuilding should integrate the lessons learned during the disaster including avoidance of reconstruction on high risk areas, enhancement of better construction standards, and adoption of methods of slope stabilization. It must aim at restoring what has been lost in addition to making communities less susceptible in the future by advancing planning on proper land use, engineering measures and environmental rehabilitation (Iloglu & Albert, 2018).

The recovery process requires the involvement and empowerment of the communities that have been affected in the long term. Development of resilience presupposes the awareness of the dangers of mass movements among the residents and training them in preparing, responding to disasters and first aid. Modern patterns of disaster management can be complemented by local knowledge and traditional patterns. As part of this, community-based disaster risk reduction (CBDRR) initiatives have become viewed as an important part of establishing a safety culture. These programs will promote local involvement in the process of planning, mapping hazards, and generating evacuation plans and emergency communication strategies. Moreover, the use of drills and simulations on a regular basis will assist with the community preparedness. Economic resilience can also be promoted through access to micro-insurance, financial literacy, and livelihood diversification schemes to promote communities better equipped to endure and recover after such events. Emergency response and recovery planning is a complex undertaking that involves coordination in multiple areas including (but not limited to) the government, non-governmental organizations, and the local communities and the international partnerships. It has to respond to short term needs and long term development agenda (Goltz et al., 2001). The combination of SAR activities, efficient rehabilitation plans, and community-based resilience building will make the affected areas due to mass movement to recuperate faster and be more prepared to respond to related incidents down the line. Investing in such spheres in advance is not only life-saving, but also leads to sustainable development and improvement of disaster risks on a larger scale.

11. TECHNOLOGICAL INNOVATIONS AND RESEARCH ADVANCEMENTS

New technology developments have helped greatly in gaining knowledge, tracking and controlling mass movements, e.g., landslides, debris flows, and soil erosions. The development of advanced simulation models and the derivation of their application has been one of the most paradigmatic developments. Such models further enable researchers and engineers to make forecasts about the instability of any slope provided certain environmental conditions are given. They replicate the physical events that result in mass movements such as rainfall infiltration, saturation and build-up in pore-water pressure along with gravity forces. These simulations allow a detailed analysis of possible failure mechanism of slopes by integrating both hydrological and geological data along with non-hydrological loadings such as geological/mechanical loading. Some of the most common techniques include finite element models; limit equilibrium analyses, coupled hydromechanical models among others. These tools also aid in conducting academic research as well as practical decision-making on matters concerning infrastructure development, disaster preparedness, and land-use planning in vulnerable regions across the globe. The combination of Artificial Intelligence (AI) and Machine Learning (ML), in geohazard studies, has also paved way to further breakthroughs in risk assessment and early warning. These technologies are capable of analyzing very big and complicated data sets- satellite pictures, weather data, soil moisture sensors readings, and land topography among others- in an extremely quick and accurate manner. Machine learning models can detect the patterns and unevenness that can precondition an oncoming landslide, like over-rapid ground change or bizarre precipitation conditions. It is also possible to predict with predictive models using past evidence on the possibility of the future events happen that the authorities can send timely warnings and exercise counter-action measures. Moreover, remote sensors can provide input through the use of AI-driven platforms, where the input would be processed and can be monitored in the real-time risk exposure domain. Specifically, convolutional neural networks (CNNs) and the support vector machine (SVM) have been found useful in the classification of terrain types and the identification of slope instabilities using visual information.

The other prominent change is and use of the satellite-based monitoring/drone-based systems which provides continuous high-resolution measurements to land surfaces of large and inaccessible areas. Technologies Using satellite remote sensing, such as Interferometric Synthetic Aperture Radar (InSAR) offers an early warning capability that is sensitive to millimeter-scale ground movements over time and thus offers early evidence of slope failure. These factors can be identified through multispectral, hyperspectral imagery to monitor vegetation stress and soil moisture fluctuations which could add to or reflect higher chances of landslides. In the meantime, use of localized mapping and inspection is rapidly done through drones, which have cameras, LiDAR, and thermal sensors installed. They offer 3D simulations of geography where the volume of landslide material can be calculated with precision, the slope angle evaluated and they can be monitored in real time during emergency situations. The main advantage of drone technology is that it is particularly useful during the aftermath phase of a disaster when field-based surveys are either unsafe or not feasible. Pooled together, such advancements in technology are transforming the terrain of mass movement research and management. They play a pivotal role in helping to make more accurate predictions, respond quicker, and make more informed policy decisions that lead to strengthening community resilience in preventing the human and financial cost of landslide disasters. In addition, continuous research and cooperation among geoscientists, engineers, data scientists and policymakers would be crucial in exploiting the true potential of these technologies. These can be invaluable tools in proactive and adaptive disaster risk reduction plans as we continue towards a more uncertain environmental future presented to us by climate change and urbanization as mentioned in the table 03.

Table 3. Technological Innovations and Research Advancements; Source: Author Compilation

Innovation/Advancement	Field	Key Impact	Example Application
Artificial Intelligence (AI)	Computer Science	Automates tasks, enhances decision-making	AI-powered medical diagnostics
Quantum Computing	Physics/Computer Science	Solves complex problems faster than classical computers	Drug discovery simulations
CRISPR Gene Editing	Biotechnology	Precise genetic modifications	Treatment of genetic disorders
Renewable Energy Tech	Environmental Science	Reduces carbon footprint	Solar panels, wind turbines
Nanotechnology	Materials Science	Creates stronger, lighter, and smarter materials	Targeted drug delivery
5G & Beyond Networks	Telecommunications	Ultra-fast connectivity	Smart cities, IoT devices
Space Exploration Tech	Aerospace Engineering	Expands human reach beyond Earth	Mars rovers, reusable rockets
Brain-Computer Interfaces	Neuroscience	Direct communication between brain and machines	Assistive devices for paralysis
Robotics & Automation	Engineering	Increases efficiency and safety	Autonomous manufacturing systems
Synthetic Biology	Life Sciences	Designs new biological systems	Biofuels, lab-grown meat

12. CONCLUSION AND FUTURE DIRECTIONS

Mass movements (landslides, debris flows, rockfalls, etc.) are a continuous threat to people and ecosystems. In the course of this discussion, we have looked at the complex nature of both causes and effects of such geological hazards taking note of the natural causes and the man-made causes. With increased insights in understanding these events, it is becoming ever more evident that to managing and minimizing their effects, a multidisciplinary, proactive approach is important. The complexity of the factors that are implicated in mass movements and how these factors are connected comes out as one of the biggest realizations that have emerged. These processes seldom occur due to one event or condition; they could be through the geological structures and slope stability and through the human encroachments of the area as well as the changes that are caused by the climate. An example is the expanding urban area that causes deforestation leading to landslides due to loss of vegetative anchors, altering soil composition and the water runoff. This has caused numerous and disastrous slope failures especially in the mountainous areas of the world such as the Himalayas. Also, the sphere of community sensitization and early warning can never be underestimated. Most of the disasters in the past might have been managed had there been proper preparedness, mapping of risks, and strategies of managing disasters. The integration of indigenous-based knowledge and scientific models make the disaster risks management systems stronger to take place. Governments should consider integrative land use planning and should not allow construction on high-hazard zones.

All infrastructure projects in the vulnerable zones should be obligatory to Environmental Impact Assessment (EIAs). Moreover, there should be policies that promote reforestation, sustainable agriculture and proper drainage mechanisms so that slope destabilization can be reduced. A bigger collaboration, both national and international is also required to commit some funds to research and technology that would help in the monitoring and prevention of landslides. To exchange data, enhance the modification of hazards and train expertise on disaster risk reduction (DRR), institutions need to cooperate. Resilience measures such as sustainable land and water management would also be encouraged by subsidizing or giving tax breaks to green infrastructure and community-based resilience programs. This involves replenishing natural forests, the development of agroforests in the countryside, and the creation of engineered work such as terracing and retention of walls only when the given need is inevitable and when it proves to be environmentally sound. The most suitable watershed management and slope stabilization methods should applied depending on geology in the region and climatic conditions. Moreover, the inclusion of local communities in the planning and decision-making process can enhance the effectiveness of planning and decision-making on the one hand but create a preparedness culture and stewardship on the other. With the changing climatic conditions giving rise to more frequent and more severe mass movement hazards, it is critical to move to a more anticipatory posture as opposed to a reactive one. The integration of new technologies such as remote sensing and AI into the combination of the community involvement and policy facilitation creates a complete avenue of minimizing risks. It is only with long-term cooperation across industries and levels that we can work towards a more resilient, safer future of vulnerable landscapes and populations.

REFERENCES

Alonso, E. E. (2021). Triggering and motion of landslides. *Geotechnique*, *71*(1), 3–59. DOI: 10.1680/jgeot.20.RL.001

Amsalu, A., Garedew, W., Melka, G. A., Çelik, M. A., & Gemeda, D. O. (2026). Trends in climate extreme indices and implications for coffee farming in The Kafa Biosphere Reserve, southwest Ethiopia. *Climate Services*, *42*, 100637. DOI: 10.1016/j.cliser.2026.100637

Aversa, S., Cascini, L., Picarelli, L., & Scavia, C. (Eds.). (2018). *Landslides and Engineered Slopes. Experience, Theory and Practice* (1st ed.). CRC Press., DOI: 10.1201/9781315375007

Balis, B., Bubak, M., Harezlak, D., Nowakowski, P., Pawlik, M., & Wilk, B. (2017). Towards an operational database for real-time environmental monitoring and early warning systems. *Procedia Computer Science*, *108*, 2250–2259. DOI: 10.1016/j.procs.2017.05.193

Bottelin, P., Jongmans, D., Daudon, D., Mathy, A., Helmstetter, A., Bonilla-Sierra, V., Cadet, H., Amitrano, D., Richefeu, V., Lorier, L., Baillet, L., Villard, P., & Donzé, F. (2014). Seismic and mechanical studies of the artificially triggered rockfall at Mount Néron (French Alps, December 2011). *Natural Hazards and Earth System Sciences*, *14*(12), 3175–3193. DOI: 10.5194/nhess-14-3175-2014

Bozzolan, E., Holcombe, E., Pianosi, F., & Wagener, T. (2020). Including informal housing in slope stability analysis – an application to a data-scarce location in the humid tropics. *Natural Hazards and Earth System Sciences*, *20*(11), 3161–3177. DOI: 10.5194/nhess-20-3161-2020

Çelik, M. A. (2026). Agro-climatic classification of Türkiye's Agricultural Basins using long-term ERA5-Land temperature data. *Physics and Chemistry of the Earth Parts A/B/C*, *142*, 104307. DOI: 10.1016/j.pce.2026.104307

Çelik, M. A., Kıraç, A., & Gouveia, C. M. (2026). North Atlantic Oscillation modulation of drought and vegetation activity in Türkiye. *Theoretical and Applied Climatology*, *157*(2), 74. Advance online publication. DOI: 10.1007/s00704-025-05971-5

Damm, B., & Klose, M. (2015). The landslide database for Germany: Closing the gap at national level. *Geomorphology*, *249*, 82–93. DOI: 10.1016/j.geomorph.2015.03.021

Eduljee, G. H. (2000). Trends in risk assessment and risk management. *The Science of the Total Environment*, *249*(1–3), 13–23. DOI: 10.1016/S0048-9697(99)00507-0 PMID: 10813443

Ehrlich, M., Mozafari, S., Glatza, M., Starost, L., Velychko, S., Hallmann, A.-L., Cui, Q.-L., Schambach, A., Kim, K.-P., Bachelin, C., Marteyn, A., Hargus, G., Johnson, R. M., Antel, J., Sterneckert, J., Zaehres, H., Schöler, H. R., Baron-Van Evercooren, A., & Kuhlmann, T. (2017). Rapid and efficient generation of oligodendrocytes from human induced pluripotent stem cells using transcription factors. *Proceedings of the National Academy of Sciences of the United States of America*, *114*(11). Advance online publication. DOI: 10.1073/pnas.1614412114 PMID: 28246330

Elsaid, K., Sayed, E. T., Abdelkareem, M. A., Baroutaji, A., & Olabi, A. G. (2020). Environmental impact of desalination processes: Mitigation and control strategies. *The Science of the Total Environment*, *740*, 140125. DOI: 10.1016/j.scitotenv.2020.140125 PMID: 32927546

Fawzy, S., Osman, A. I., Doran, J., & Rooney, D. W. (2020). Strategies for mitigation of climate change: A review. *Environmental Chemistry Letters*, *18*(6), 2069–2094. DOI: 10.1007/s10311-020-01059-w

Fayaz, M., Meraj, G., Khader, S. A., & Farooq, M. (2022). ARIMA and SPSS statistics based assessment of landslide occurrence in western Himalayas. *Environmental Challenges*, *9*, 100624. DOI: 10.1016/j.envc.2022.100624

Fischer, H. W. (1998). The role of the new information technologies in emergency mitigation, planning, response and recovery. *Disaster Prevention and Management*, *7*(1), 28–37. DOI: 10.1108/09653569810206262

Gallina, V., Torresan, S., Critto, A., Sperotto, A., Glade, T., & Marcomini, A. (2016). A review of multi-risk methodologies for natural hazards: Consequences and challenges for a climate change impact assessment. *Journal of Environmental Management*, *168*, 123–132. DOI: 10.1016/j.jenvman.2015.11.011 PMID: 26704454

Goltz, J. D., Flores, P. J., Chang, S. E., & Atsumi, T. (2001). Emergency Response and Early Recovery. *Earthquake Spectra*, *17*(1, suppl), 173–183. DOI: 10.1193/1.1586198

Hradecký, J., & Pánek, T. (2008). Deep-seated gravitational slope deformations and their influence on consequent mass movements (case studies from the highest part of the Czech Carpathians). *Natural Hazards*, *45*(2), 235–253. DOI: 10.1007/s11069-007-9157-7

Iloglu, S., & Albert, L. A. (2018). An integrated network design and scheduling problem for network recovery and emergency response. *Operations Research Perspectives*, *5*, 218–231. DOI: 10.1016/j.orp.2018.08.001

Irschik, H., & Belyaev, A. K. (Eds.). (2014). *Dynamics of Mechanical Systems with Variable Mass* (Vol. 557). Springer Vienna., DOI: 10.1007/978-3-7091-1809-2

Jones, R. N. (2001). An Environmental Risk Assessment/Management Framework for Climate Change Impact Assessments. *Natural Hazards, 23*(2–3), 197–230. DOI: 10.1023/A:1011148019213

Joshi, A., Grover, J., Kanungo, D. P., & Panigrahi, R. K. (2019). Edge assisted Reliable Landslide Early Warning System. *2019 IEEE 16th India Council International Conference (INDICON)*, 1–4. DOI: 10.1109/INDICON47234.2019.9028898

Klose, M. (2015). Landslide Databases—State of Research and the Case of Germany. In M. Klose, *Landslide Databases as Tools for Integrated Assessment of Landslide Risk* (pp. 25–43). Springer International Publishing. DOI: 10.1007/978-3-319-20403-1_2

Kyriacos, U., Jelsma, J., & Jordan, S. (2011). Monitoring vital signs using early warning scoring systems: A review of the literature: Monitoring vital signs - early warning scoring systems. *Journal of Nursing Management, 19*(3), 311–330. DOI: 10.1111/j.1365-2834.2011.01246.x PMID: 21507102

McColl, S. T. (2022). Landslide causes and triggers. In *Landslide Hazards, Risks, and Disasters* (pp. 13–41). Elsevier. DOI: 10.1016/B978-0-12-818464-6.00011-1

Mollamahmutoğlu, M., & Acar, Y. (2020). Human-Induced Landslide Risk Analysis at Durhasanlı Reservoir Site. In Shehata, H., Brandl, H., Bouassida, M., & Sorour, T. (Eds.), *Sustainable Thoughts in Ground Improvement and Soil Stability* (pp. 104–115). Springer International Publishing., DOI: 10.1007/978-3-030-34184-8_7

Mundhe, N. N., & Jaybhaye, R. G. (2023). Land Suitability Analysis for In Situ Slum Redevelopment of Pune City Using AHP, Remote Sensing and GIS Techniques. *Photonirvachak (Dehra Dun), 51*(8), 1777–1795. DOI: 10.1007/s12524-023-01731-y

Parkash, S. (2023). Lessons Learned from Landslides of Socio-economic and Environmental Significance in India. In I. Alcántara-Ayala, Ž. Arbanas, D. Huntley, K. Konagai, M. Mikoš, K. Sassa, S. Sassa, H. Tang, & B. Tiwari (Eds.), *Progress in Landslide Research and Technology, Volume 1 Issue 2, 2022* (pp. 309–315). Springer International Publishing. DOI: 10.1007/978-3-031-18471-0_23

Pecoraro, G., Calvello, M., & Piciullo, L. (2019). Monitoring strategies for local landslide early warning systems. *Landslides, 16*(2), 213–231. DOI: 10.1007/s10346-018-1068-z

Perrings, C. (2005). Mitigation and adaptation strategies for the control of biological invasions. *Ecological Economics, 52*(3), 315–325. DOI: 10.1016/j.ecolecon.2004.08.001

Perry, R. W., & Lindell, M. K. (2003). Preparedness for Emergency Response: Guidelines for the Emergency Planning Process. *Disasters*, *27*(4), 336–350. DOI: 10.1111/j.0361-3666.2003.00237.x PMID: 14725091

Pradhan, S. P., & Siddique, T. (2019). Mass Wasting: An Overview. In Pradhan, S. P., Vishal, V., & Singh, T. N. (Eds.), *Landslides: Theory, Practice and Modelling* (Vol. 50, pp. 3–20). Springer International Publishing., DOI: 10.1007/978-3-319-77377-3_1

Şahin, Ş., & Kurum, E. (2002). Erosion risk analysis by GIS in environmental impact assessments: A case study—Seyhan Köprü Dam construction. *Journal of Environmental Management*, *66*(3), 239–247. DOI: 10.1016/S0301-4797(02)90574-8 PMID: 12448403

Scheffers, A. M., May, S. M., & Kelletat, D. H. (2015). Mass Movements: Landforms Shaped Under the Force of Gravity. In A. M. Scheffers, S. M. May, & D. H. Kelletat, *Landforms of the World with Google Earth* (pp. 163–182). Springer Netherlands. DOI: 10.1007/978-94-017-9713-9_8

Schuster, R. L., & Highland, L. M. (2007). Overview of the Effects of Mass Wasting on the Natural Environment. *Environmental & Engineering Geoscience*, *13*(1), 25–44. DOI: 10.2113/gseegeosci.13.1.25

Servetti, M., Caramia, M., Parodi, G., Loiacono, F., Nano, E., Biddau, G., Ferrando, L., Morinelli, L., Valente, P., Martinoia, S., Escelsior, A., Serafini, G., Tamburro, S., Baldassari, S., Fassio, A., Benfenati, F., Corradi, A., & Sterlini, B. (2025). Optimization of Transcription Factor-Driven Neuronal Differentiation from Human Induced Pluripotent Stem Cells for Disease Modelling and Drug Screening. *Stem Cell Reviews and Reports*, *21*(3), 816–833. DOI: 10.1007/s12015-025-10845-4 PMID: 39888571

Si-Tayeb, K., Noto, F. K., Sepac, A., Sedlic, F., Bosnjak, Z. J., Lough, J. W., & Duncan, S. A. (2010). Generation of human induced pluripotent stem cells by simple transient transfection of plasmid DNA encoding reprogramming factors. *BMC Developmental Biology*, *10*(1), 81. DOI: 10.1186/1471-213X-10-81 PMID: 20682060

Smith, P. J. (2007). Climate Change, Mass Migration and the Military Response. *Orbis*, *51*(4), 617–633. DOI: 10.1016/j.orbis.2007.08.006

Stefanescu, C., Herrando, S., & Páramo, F. (2004). Butterfly species richness in the north-west Mediterranean Basin: The role of natural and human-induced factors. *Journal of Biogeography*, *31*(6), 905–915. DOI: 10.1111/j.1365-2699.2004.01088.x

Stoffel, M., & Huggel, C. (2012). Effects of climate change on mass movements in mountain environments. *Progress in Physical Geography, 36*(3), 421–439. DOI: 10.1177/0309133312441010

Stoffel, M., Tiranti, D., & Huggel, C. (2014). Climate change impacts on mass movements—Case studies from the European Alps. *The Science of the Total Environment, 493*, 1255–1266. DOI: 10.1016/j.scitotenv.2014.02.102 PMID: 24630951

Strupler, M., Hilbe, M., Kremer, K., Danciu, L., Anselmetti, F. S., Strasser, M., & Wiemer, S. (2018). Subaqueous landslide-triggered tsunami hazard for Lake Zurich, Switzerland. *Swiss Journal of Geosciences, 111*(1–2), 353–371. DOI: 10.1007/s00015-018-0308-5

Vrba, E. S. (2007). 4 Role of Environmental Stimuli in Hominid Origins. In W. Henke & I. Tattersall, *Handbook of Paleoanthropology* (pp. 1441–1481). Springer Berlin Heidelberg. DOI: 10.1007/978-3-540-33761-4_47

Yang, Z., & Wang, J. (2017). A new air quality monitoring and early warning system: Air quality assessment and air pollutant concentration prediction. *Environmental Research, 158*, 105–117. DOI: 10.1016/j.envres.2017.06.002 PMID: 28623745

Zeleňáková, M., & Zvijáková, L. (2017). Risk analysis within environmental impact assessment of proposed construction activity. *Environmental Impact Assessment Review, 62*, 76–89. DOI: 10.1016/j.eiar.2016.10.003

Chapter 9

Integrating GIS and Remote Sensing in Early Warning Systems for Real-Time Disaster Preparedness

Sindy Rahmahsari
http://orcid.org/0009-0000-8491-3242
Universitas Esa Unggul, Indonesia

Binastya Anggara Sekti
http://orcid.org/0000-0001-5489-4888
Universitas Esa Unggul, Indonesia

ABSTRACT

This chapter per the authors explores how Geographic Information Systems (GIS) and Remote Sensing (RS) are transforming early warning systems for hydrometeorological disasters through real-time, data-driven approaches. It examines the integration of multi-source datasets topographic, climatic, and socio-economic to assess flood and drought risks with high spatial and temporal precision. The discussion highlights how cloud-based platforms such as Google Earth Engine and Copernicus Open Access Hub, combined with artificial intelligence and Internet of Things (IoT) networks, enable dynamic monitoring and predictive modeling of environmental hazards. Methodological frameworks including multi-criteria decision analysis, machine learning, and digital twin technologies are reviewed for their capacity to

DOI: 10.4018/979-8-3373-6274-8.ch009

support adaptive management and policy decision-making. The chapter concludes by emphasizing the role of open data, interoperability, and ethical governance in advancing global disaster preparedness and climate resilience.

1. INTRODUCTION

Figure 1. Conceptual Integration of GIS and Remote Sensing for Flood and Drought Early Warning Systems

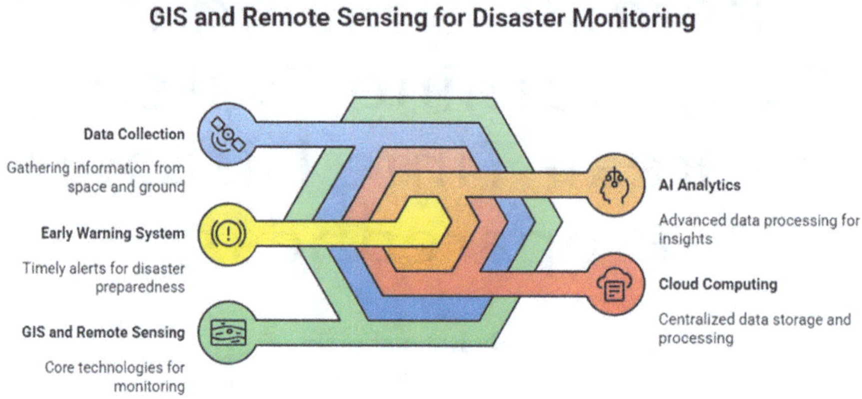

Two of the most destructive hydrometeorological disasters in the world, flooding and droughts impact millions of people annually and result in significant socioeconomic and environmental damage. Climate unpredictability, changes in land use, and growing human demands on natural ecosystems are the main causes of these catastrophic events. Global climate change has increased the frequency and severity of these disasters in recent decades, especially in tropical and subtropical countries where prolonged dry spells and heavy rains are becoming more frequent. Creating successful mitigation and adaptation plans requires an understanding of the temporal and spatial dynamics of these hazards. (G. Liu et al., 2026). The foundation for integrating diverse datasets, such as digital elevation models (DEM), land use, soil characteristics, rainfall, and socioeconomic indicators, is GIS. Researchers can assess exposure, measure susceptibility, and spatially depict flood and drought threats using GIS in a logical manner. On the other hand, remote sensing offers real-time observations from satellites and aerial sensors, obtaining vital environmental data like the amount of surface water, vegetation health, soil moisture, and rainfall intensity (Abdelkrim & Eslamian, 2026). These technologies work in concert to create a special synergy that turns static data into knowledge that decision-makers may use. The potential of GIS and RS in environmental research has been greatly increased in recent years by the growth of cloud computing, open-access satellite data, and artificial intelligence (AI) (Rohanadeera & Hewagamage, 2026). Near-real-time study of flood inundation, drought severity, and land degradation at global scales is now possible thanks to platforms like Google Earth Engine (GEE), NASA's EarthData, and Copernicus Open Access Hub, which have democratized access to huge Earth Observation datasets. This digital revolution is not just technical; it is also epistemological, changing the way environmental risk is conceptualized, modeled, and managed by academics. This chapter's main goal is to give a thorough examination of the ways that GIS and RS support flood and drought risk assessment in a worldwide setting. By combining geospatial modeling techniques, applications, and advancements, the chapter emphasizes the transition from descriptive mapping to predictive, data-driven, and decision-support systems. By doing this, it also establishes a connection between scholarly research and technical system development, offering guidance on how to construct integrated platforms for early warning and adaptive management. (Warlina & Afredi, 2026).

2. CONCEPTUAL FRAMEWORK OF GIS AND REMOTE SENSING APPLICATIONS

Remote sensing (RS) and geographic information systems (GIS) technologies offer complimentary capabilities for precisely comprehending, mapping, and forecasting natural hazards over time and space (Çelik, 2026). Three fundamental elements are conceptually needed for flood and drought risk assessments: hazard, exposure, and vulnerability. The hazard component uses RS-based indices like the Normalized Difference Water Index (NDWI) or Standardized Precipitation Index (SPI) to determine the physical attributes and likelihood of the natural event, such as the extent of flooding, the intensity of rainfall, or the length of drought (Çelik et al., 2026). The geographic distribution of people, infrastructure, and land use within hazard-prone areas is described by the exposure component, which is often represented by GIS-based demographic and land cover data. Last but not least, vulnerability assesses how resilient ecosystems or populations are to such events, sometimes integrating socioeconomic and environmental aspects. (Mwinjuma et al., 2026). These elements can be operationalized into a thorough risk framework by integrating GIS and RS. While GIS enables the organization, display, and geographic analysis of multi-layered datasets, remote sensing offers current, synoptic observations of environmental factors. For instance, satellite imagery can be used to identify areas that have been flooded or that are under vegetation stress in order to simulate spatial risk patterns. Then, layers of population density, elevation, and infrastructure are layered on top of these locations. Analytical approaches such as weighted overlay, multi-criteria decision analysis (MCDA), or machine learning-based hazard mapping further enhance the prediction capabilities of GIS-RS systems. (Nesanır et al., 2026).

Table 1. Core components of GIS and RS-based disaster-risk assessment

Component	Description	Typical GIS/RS Inputs	Key Analytical Techniques
Hazard	Probability and intensity of flood or drought occurrence	Precipitation (TRMM, GPM), soil moisture (SMAP), NDVI, land-use maps, DEM	Hydrological modeling, change detection, time-series analysis
Exposure	Spatial distribution of elements at risk (population, infrastructure, agriculture)	Census data, land-parcel maps, urban footprint (Sentinel-2, Landsat-9)	Overlay analysis, spatial interpolation, buffer modeling
Vulnerability	Susceptibility of exposed elements to hazard impacts	Socio-economic indices, building materials, crop type, water demand	Multi-criteria evaluation (AHP, Fuzzy logic), regression analysis
Component	Description	Typical GIS/RS Inputs	Key Analytical Techniques

Flood risk assessment has evolved from simple hydrological estimations into complex, spatially explicit modeling systems that integrate multiple environmental, climatic, and anthropogenic variables. At its core, flood risk is generally modeled by synthesizing rainfall intensity, catchment topography, drainage density, land-use and land-cover patterns, and soil infiltration capacity each representing a crucial component of hydrological response. Using digital terrain inputs and precipitation time series, more complex models like HEC-RAS, SWAT, or LISFLOOD further simulate hydrodynamic processes and provide dynamic estimates of flood depth and extent under various rainfall scenarios (Song et al., 2025). Conversely, drought risk mapping requires a distinct yet equally integrative approach, focusing on the prolonged deficit of water availability rather than its excess. Drought risk models typically rely on vegetation indices, temperature anomalies, evapotranspiration rates, and water-balance parameters derived from multi-temporal Remote Sensing datasets. Vegetation indices such as the Normalized Difference Vegetation Index (NDVI) or Vegetation Condition Index (VCI) provide indirect measures of plant health and moisture stress, while thermal infrared data from sensors like Landsat TIRS or ASTER quantify surface temperature anomalies that signal evaporative stress. Evapotranspiration, an essential indicator of hydrological imbalance, can be retrieved from MODIS or Sentinel-derived products using energy balance algorithms such as SEBAL (Surface Energy Balance Algorithm for Land) or METRIC (Mapping Evapotranspiration at High Resolution with Internalized Calibration). Integrating these indicators with precipitation anomalies (from CHIRPS or TRMM) and soil moisture observations (from SMAP or Sentinel-1 SAR) enables the construction of comprehensive drought severity indices such as the Standardized Precipitation Evapotranspiration Index (SPEI) or the Vegetation Health Index (VHI). These indices, when mapped in GIS, reveal spatial and temporal variations in drought intensity, duration, and frequency critical for agricultural planning and water resource management. Remote Sensing serves as the backbone of these analytical processes by providing consistent, high-frequency, and wide-coverage environmental observations. Optical sensors such as MODIS and Sentinel-2 MSI capture spectral reflectance across visible and near-infrared wavelengths, essential for analyzing vegetation vigor, soil exposure, and water bodies. Land surface temperature is recorded by thermal sensors such as Landsat TIRS, ASTER, and MODIS LST, which makes it possible to estimate energy fluxes and evapotranspiration, two crucial variables for diagnosing drought. In the meanwhile, vital day-and-night, all-weather studies of soil moisture and surface roughness are made possible by microwave sensors like Sentinel-1 Synthetic Aperture Radar (SAR), RADARSAT, and SMAP (Soil Moisture Active Passive) (Taylor & Salvucci, 2025). SAR data is extremely useful for flood detection during monsoon or storm events because, in contrast to optical images, it can penetrate cloud cover and is sensitive to the presence of surface water. These multi-sensor datasets enable both synoptic (regional to global) and granular (local) hazard assessments when incorporated into GIS, bridging geographical scales and guaranteeing that decision-makers can more accurately identify hydrological threats (Song et al., 2025).

The GIS platform functions as the analytical and integrative environment where these multi-source datasets are fused into coherent spatial models. Through spatial overlay operations, raster algebra, and geostatistical analysis, GIS enables the computation of composite hazard indices that account for physical, environmental, and socio-economic dimensions of risk. Statistical techniques such as logistic regression, principal component analysis (PCA), and multi-linear correlation have traditionally been used to model relationships between hazard determinants and observed impacts. By detecting nonlinear interactions and enhancing prediction accuracy, machine learning methods such as Random Forest (RF), Support Vector Machine (SVM), Gradient Boosting, and Artificial Neural Networks (ANN) have transformed hazard mapping in more recent times (Vlasenko et al., 2024). These algorithms are particularly effective when trained on large datasets combining Remote Sensing variables with field observations or historical disaster records. For example, flood susceptibility mapping using Random Forest models trained on rainfall, slope, land cover, and distance to rivers has achieved predictive accuracies exceeding 90% in various case studies across Southeast Asia. Similarly, ANN-based drought models integrating NDVI, Land Surface Temperature (LST), and precipitation anomalies have been successfully used to predict vegetation stress weeks in advance, offering actionable lead time for agricultural intervention.

The integration of AI-driven algorithms within these cloud platforms further enhances analytical automation and predictive intelligence. Deep learning architectures, such as Convolutional Neural Networks (CNNs) for spatial feature extraction and Recurrent Neural Networks (RNNs) for temporal pattern recognition, enable the creation of dynamic hazard models that continuously learn from new data inputs. When embedded into GIS systems, these AI models can perform tasks ranging from automatic flood extent delineation to drought anomaly prediction and water resource optimization (Saleh et al., 2025). Additionally, the use of transfer learning allows pre-trained models to be adapted across different geographic regions, improving performance in data-scarce environments. This combination of cloud computing, AI, and GIS-RS integration is paving the way for global-scale environmental intelligence systems, capable of continuously monitoring hydrological extremes, updating hazard assessments in real time, and providing actionable information for decision-makers at every level from local municipalities to international disaster response agencies. Collectively, these developments signify a paradigm shift from reactive hazard assessment to proactive, data-driven, and adaptive environmental management. Flood and drought risk are no longer static entities to be mapped periodically; they are dynamic processes that can now be observed, modeled, and forecasted continuously (Pineda et al., 2026). Through the seamless integration of Remote Sensing, GIS, and Artificial Intelligence within cloud infrastructures, the modern geospatial ecosystem enables the transition toward anticipatory climate governance where risk assessment is not an endpoint but an ongoing cycle of observation, learning, and response.

3. METHODOLOGICAL APPROACHES TO FLOOD AND DROUGHT RISK ASSESSMENT

Data collection, pre-processing, spatial modeling, and validation are all integrated into a methodical workflow that is part of the GIS and remote sensing-based risk assessment approach. GIS-RS approaches use distributed datasets and probabilistic modeling, which enable flexible application across many geographic settings, in contrast to classic deterministic hydrological modeling, which depends on localized characteristics (Liao et al., 2024). Geometric correction aligns images to a common coordinate system, whereas atmospheric and radiometric corrections eliminate distortions brought on by atmospheric interference or sensor noise (Mariani et al., 2025). Data are resampled to uniform spatial resolutions, often between 10 and 30 meters, to enable accurate overlay within GIS environments. For optical imagery, cloud masking is applied using algorithms such as Fmask or QA60, while radar data require speckle filtering and terrain correction. Once the datasets are harmonized, researchers derive thematic indices that quantify environmental conditions. The Normalized Difference Water Index (NDWI) identifies water bodies and flood extents; NDVI and VHI detect vegetation stress; SPEI integrates temperature and precipitation anomalies to gauge drought intensity. These indices are statistically normalized to create comparable scales across time and space. The next stage involves modeling and analysis, which can be deterministic, empirical, or data-driven. Deterministic hydrological models such as SWAT or HEC-RAS simulate surface runoff and streamflow based on physical parameters, while empirical models correlate environmental variables using statistical relationships. More recently, machine-learning techniques Random Forest, Gradient Boosting, and Convolutional Neural Networks have been applied to capture complex nonlinear interactions between environmental predictors and hazard occurrences. Risk is then computed through multi-criteria integration. In GIS, this is achieved by assigning weights to different variables based on their relative importance to flooding or drought susceptibility. The Analytic Hierarchy Process (AHP) is still a widely used technique for expert judgment in calculating these weights. Zones of low, moderate, and high risk are spatially categorized using the derived composite index (Gibson & Sokolov, 2026). Validation is a critical component of methodological rigor. For flood models, validation may involve comparing simulated flood extents with SAR-derived inundation maps from Sentinel-1, while drought models are validated against agricultural yield data or ground-based meteorological indices. Metrics such as the Area Under the Curve (AUC) from Receiver Operating Characteristic (ROC) analysis or the Kappa coefficient assess the agreement between modeled and observed outcomes (Li et al., 2025). Ultimately, GIS and RS methodologies are iterative rather than linear. Each phase data collection, modeling, validation, and visualization informs the next, creating a feedback loop that improves predictive accuracy. This methodological adaptability explains why GIS-RS integration has become a cornerstone of modern environmental risk science.

4. APPLICATIONS OF GIS AND REMOTE SENSING IN FLOOD RISK ASSESSMENT

Figure 2. GIS and Remote Sensing Workflow for Flood Hazard Mapping and Early Warning

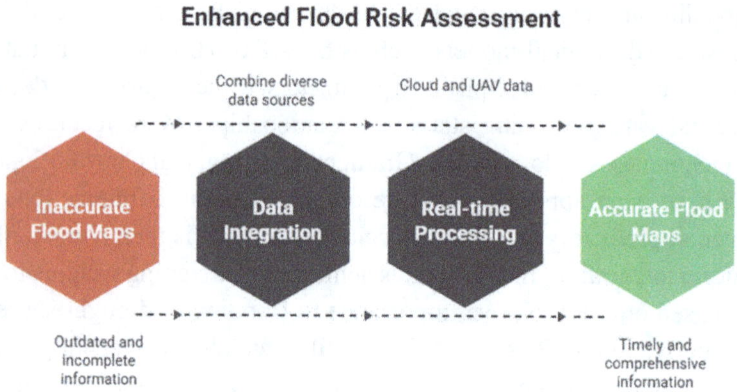

Floods represent one of the most pervasive natural hazards globally, exerting significant socio-economic, environmental, and infrastructural impacts that are exacerbated by urbanization, deforestation, and climate change. Between 1990 and 2023, floods have accounted for nearly 45% of all recorded natural disasters, causing economic losses exceeding USD 600 billion and affecting more than 1.8 billion people worldwide. The increasing complexity and magnitude of flood disasters necessitate advanced geospatial tools capable of integrating diverse datasets, capturing hydrological dynamics across multiple spatial scales, and translating raw environmental data into actionable risk information. Because they provide a comprehensive platform for spatial modeling, temporal analysis, and decision-support applications, Geographic Information Systems (GIS) and Remote Sensing (RS) have become essential facilitators of flood hazard assessment in this setting (Cerrada Ayllón et al., 2023). GIS enables the visualization and quantification of flood-prone zones by combining multiple spatial layers topography, hydrology, rainfall, land use, and population density while Remote Sensing provides the empirical basis for observing precipitation trends, water surface dynamics, and land-cover transformations through time. The combination of these technologies allows researchers and planners to not only delineate areas susceptible to flooding but also assess the evolving exposure and vulnerability patterns that determine disaster outcomes. The capacity of GIS and RS to produce high-resolution flood danger maps that take into account both physical and socioeconomic factors is one of their most potent uses in flood assessment. Delineating floodplain boundaries and simulating water flow accumulation require the use of Digital Elevation Models (DEMs), especially those obtained from the Shuttle Radar Topography Mission (SRTM), the Advanced Spaceborne Thermal Emission and Reflection Radiometer (ASTER), and Light Detection and Ranging (LiDAR) (Zhou et al., 2023). Another major strength of GIS-RS integration is in vulnerability and exposure analysis. Flood risk is not solely a function of hydrological intensity but also of human and infrastructural susceptibility. GIS provides a robust analytical environment for overlaying hazard layers with socio-economic datasets such as population density, building footprints, road networks, and land-use zoning. The assessment of exposure indices, which measure the possible number of individuals, properties, or agricultural regions that may be impacted under different flood scenarios, is made easier by this geographical integration. Furthermore, to assess the relative significance of various parameters like slope, rainfall intensity, drainage density, and soil permeability in determining flood susceptibility, GIS-based multi-criteria decision analysis (MCDA) techniques are used, such as fuzzy logic models and the Analytic Hierarchy Process (AHP) (Mulesa & Bohdan, 2024). These models enable a systematic and reproducible approach to spatial risk classification, producing maps that inform both urban planning and disaster mitigation strategies. In countries like Vietnam, Nigeria, and India, such GIS-based frameworks have been institutionalized for flood zoning, insurance modeling, and infrastructure planning.

As satellite and cloud computer technologies have advanced, so too has the operational application of RS in flood monitoring and early warning. Automated flood extent extraction and near-real-time computation of dynamic indices like the Normalized Difference Water Index (NDWI) or Modified Normalized Difference Water Index (MNDWI) are made possible by platforms such as Google Earth Engine (GEE) (Bournez et al., 2024). This capability has transformed how agencies such as NASA, the European Space Agency (ESA), and the United Nations World Food Programme (WFP) conduct global flood surveillance. The Global Flood Monitoring System (GFMS), for example, uses RS-based precipitation and soil-moisture data integrated within GIS frameworks to issue alerts for over 150 countries. The scalability of such systems lies in their ability to incorporate global datasets while maintaining flexibility for regional customization. Furthermore, Unmanned Aerial Vehicles (UAVs) are being used more and more for mapping floods at the local level and assessing damage since they offer centimeter-level spatial resolution that enhances satellite data (Kierzkowski et al., 2024). The integration of UAV imagery into GIS databases supports rapid damage estimation, particularly in urban environments where satellite data may be obstructed by cloud cover or urban canyons.

5. INTEGRATING FLOOD AND DROUGHT ASSESSMENTS FOR COMPARATIVE INSIGHTS

Floods and droughts, though contrasting in manifestation, are two sides of the same hydrological coin products of extreme variability within the Earth's water and energy cycles. Both are governed by the complex interplay of atmospheric circulation, land-surface processes, and anthropogenic factors that alter watershed hydrodynamics. The global trend toward intensified hydrological extremes, driven by climate change and land-use transformation, has prompted scientists to reconsider the traditional compartmentalization of these phenomena. Instead of being viewed as discrete events, floods and droughts are increasingly understood as interconnected states within a continuous hydroclimatic system an oscillation between surplus and deficit modulated by climatic anomalies such as El Niño Southern Oscillation (ENSO), Pacific Decadal Oscillation (PDO), and Indian Ocean Dipole (IOD). This shift in perspective has given rise to Integrated Flood and Drought Risk Management (IFDRM), a paradigm emphasizing the need for cross-hazard understanding, joint monitoring, and coordinated mitigation. Within this emerging framework, Geographic Information Systems (GIS) and Remote Sensing (RS) have become indispensable, offering the analytical capacity to synthesize spatial, temporal, and thematic data across both extremes. The integration of GIS and RS facilitates the simultaneous observation of parameters critical to both floods and droughts, such as rainfall variability, soil moisture dynamics, vegetation stress, and surface water extent. Unlike traditional observation systems that operate in isolation, satellite-based platforms such as NASA's Global Precipitation Measurement (GPM), the European Space Agency's Sentinel constellation, and the Gravity Recovery and Climate Experiment (GRACE) allow near-continuous global monitoring of water distribution and storage. These datasets can be processed within GIS environments to derive indices that quantify the spatiotemporal transitions between wet and dry states. For instance, combining the Normalized Difference Water Index (NDWI) and the Normalized Difference Vegetation Index (NDVI) provides a multidimensional picture of how excess water from floods may evolve into vegetation stress and eventual drought conditions. Similarly, GRACE-based terrestrial water storage anomalies, when integrated with rainfall and evapotranspiration data in GIS, can highlight basins where hydrological surpluses deplete rapidly, signaling a shift from flood recovery to drought onset. This integrative approach moves beyond single-event analysis toward systemic monitoring of water-cycle dynamics.

Case studies across the globe illustrate the practical importance of this integrated perspective. In the Indo-Gangetic Basin, where monsoon variability governs both flooding and water scarcity, GIS-RS analyses have revealed a distinct pattern of "hydrological inversion": heavy monsoon floods are followed by groundwater depletion and agricultural drought due to excessive extraction and reduced recharge (Xu et al., 2024). Using GRACE-FO and CHIRPS datasets, researchers have quantified that nearly 60% of the basin's post-monsoon groundwater loss is linked to insufficient infiltration following major flood events. Similarly, in East Africa, the interplay between ENSO-induced rainfall variability and land degradation has led to alternating cycles of drought and flash flooding. These examples underscore how GIS-RS integration enables the diagnosis of complex hydroclimatic linkages that traditional models, focused on isolated hazards, fail to capture. From a modeling standpoint, the integrated analysis of flood and drought requires reconciling differences in temporal and spatial resolution. Floods are typically fast-onset events that demand hourly or daily monitoring at fine spatial scales, while droughts unfold gradually, requiring monthly or seasonal analyses over broader regions. GIS provides the computational infrastructure to harmonize these disparate datasets through multi-resolution modeling and temporal resampling techniques.

Table 2. Comparative Overview of GIS and RS Applications for Flood and Drought Risk Assessment

Analytical Component	Flood Assessment	Drought Assessment	Shared GIS-RS Techniques
Key Variables	Rainfall intensity, runoff, topography	Precipitation deficit, evapotranspiration, vegetation stress	Time-series raster analysis, spatial interpolation
Satellite Data Sources	GPM, Sentinel-1, SRTM, Landsat	MODIS, SMAP, GRACE, Sentinel-2	Multi-sensor data fusion, cloud-based analytics
Modeling Tools	HEC-RAS, SWAT, AHP, Hydrodynamic Simulation	SPEI, VHI, Random Forest, LSTM Networks	Multi-criteria analysis, machine learning integration
Primary Outputs	Inundation maps, hazard zones, exposure layers	Drought severity, soil-moisture anomaly, vulnerability maps	Integrated hazard dashboards, water-balance visualization

Beyond scientific analysis, the integration of flood and drought assessment through GIS and RS carries critical policy and operational implications. Governments, development agencies, and water management institutions increasingly recognize that resilience to one extreme cannot be achieved without considering the other. For instance, flood mitigation infrastructure such as dams and levees must also be evaluated for their long-term effects on downstream drought vulnerability, as excessive flood retention can disrupt groundwater recharge and ecosystem balance. GIS-based scenario modeling can simulate such trade-offs, allowing decision-makers to optimize water storage operations under varying climate conditions. In the Mekong River Basin, for example, GIS-RS-integrated decision-support systems developed under the Mekong Water Data Initiative are used to balance flood control, agricultural irrigation, and hydropower generation, helping riparian countries coordinate adaptive water governance. Similar systems have been deployed in California's Central Valley, where RS-based soil moisture monitoring informs both flood forecasting and groundwater management policies, exemplifying how integrated geospatial tools can bridge scientific modeling with real-world decision-making. Moreover, integrated flood drought assessment contributes directly to achieving global sustainability agendas such as the United Nations Sustainable Development Goals (SDGs) specifically SDG 6 (Clean Water and Sanitation) and SDG 13 (Climate Action). By facilitating the monitoring of water availability, land-use change, and climate extremes, GIS-RS frameworks provide the empirical foundation for policy evaluation and adaptive planning (Amsalu, 2026). These technologies enable the quantification of resilience indicators, support transboundary water diplomacy through shared datasets, and promote equitable adaptation by identifying vulnerable populations affected by both flood and drought hazards. However, achieving this vision requires not only technical innovation but also institutional coordination and data-sharing mechanisms across jurisdictions. While the science of integration is advancing rapidly, governance structures and funding mechanisms often remain siloed, hindering the full realization of IFDRM potential. More than just a technical improvement, the incorporation of flood and drought evaluations into GIS and RS platforms offers a new paradigm in environmental risk management based on spatial intelligence and systems thinking (Güngör & Serttaş, 2024). By linking the spatial footprints of hydrological extremes to socio-economic processes and ecosystem dynamics, GIS and Remote Sensing allow for the creation of predictive, adaptive, and resilient management frameworks. As climate change continues to amplify hydroclimatic volatility, the ability to monitor, model, and manage both ends of the water spectrum within a single geospatial framework will become indispensable for safeguarding human welfare, ecosystems, and sustainable development worldwide.

6. TECHNOLOGICAL INNOVATIONS AND SYSTEM DEVELOPMENT

The rapid evolution of geospatial technologies over the past two decades has revolutionized how floods and droughts are monitored, modeled, and managed. The convergence of Geographic Information Systems (GIS), Remote Sensing (RS), artificial intelligence (AI), and cloud computing has enabled a paradigm shift from static mapping toward dynamic, data-driven, and real-time environmental intelligence. This transformation reflects not only advances in hardware and software but also a broader epistemological shift toward automation, interoperability, and open-access analytics in environmental science. The ability to collect, process, and understand multi-source hydrological data has significantly increased with the advent of new technologies including satellite constellations, Internet of Things (IoT) networks, machine learning algorithms, and geospatial web services (Jeyaraj et al., 2023). As a result, flood and drought risk assessment is no longer confined to post-event analysis; it now encompasses predictive modeling, automated alerts, and adaptive system development capable of guiding decision-making across spatial and temporal scales. One of the most significant innovations reshaping GIS and RS applications is the advent of cloud-based geospatial computing. Traditional geospatial analysis required high-performance local hardware and extensive data storage, limiting accessibility for researchers in data-scarce or resource-limited regions. Cloud platforms such as Google Earth Engine (GEE), Amazon Web Services (AWS), Microsoft Planetary Computer, and Copernicus Data Space Ecosystem have fundamentally changed this landscape. These platforms host petabytes of satellite data ranging from Landsat and Sentinel imagery to MODIS and CHIRPS climate datasets and allow users to perform analyses directly in the cloud using parallelized computation. The implications for hydrological research are profound: massive time-series datasets can be processed within minutes, enabling near-real-time flood mapping, drought index computation, and change detection at regional or global scales. Another revolutionary step is the combination of machine learning (ML) and artificial intelligence (AI) (Sekti, 2025). Traditional statistical models frequently fail to identify complicated, nonlinear interactions between hydrological, climatic, and land-surface factors; AI-driven analytics make this possible. In flood assessment, convolutional neural networks (CNNs) have been trained on multi-spectral and radar imagery to automatically classify inundated areas with remarkable precision, even under cloud-covered or low-light conditions. Deep learning architectures such as U-Net and ResNet have further enhanced the segmentation of flood extent from Sentinel-1 SAR data, allowing automated mapping of flood boundaries across large regions. In drought research, Recurrent Neural Networks (RNNs) and Long Short-Term Memory (LSTM) models have proven particularly effective in capturing temporal dependencies within long-term climatic and vegetation datasets. When integrated into GIS workflows, these models can predict drought onset, duration, and severity based on historical trends and near-future climate forecasts. A study by Kumar et al. (2025), for example, demonstrated that an LSTM-based model trained on 20 years of MODIS NDVI and ERA5 climate reanalysis data could forecast drought severity in South Asia with over 87% accuracy outperforming traditional regression-based approaches by nearly 20%. These developments underscore how AI not only enhances analytical precision but also transforms GIS-RS systems into intelligent, predictive infrastructures capable of autonomous learning and adaptation. The rise of the Internet of Things (IoT) has introduced a new dimension of granularity to geospatial monitoring. IoT-based hydrological stations equipped with sensors for rainfall, soil moisture, river discharge, and groundwater levels can transmit data in real time to centralized GIS platforms via wireless networks. This continuous data flow bridges the gap between ground-based measurements and satellite observations, improving model calibration and validation. For instance, IoT-enabled sensor networks deployed along flood-prone rivers in the Netherlands and Japan have been integrated with real-time satellite rainfall data, enabling automated flood forecasting and early warning dissemination through web-GIS dashboards. Similarly, in drought-prone regions of Sub-Saharan Africa, IoT soil-moisture sensors connected to mobile-based GIS applications support precision irrigation and crop stress monitoring (Anwar et al., 2025). When combined with RS-derived indicators, these systems enhance decision-making for farmers, local authorities, and water resource managers. The synergy between IoT and GIS thus transforms environmental monitoring from a periodic, data-scarce process into a continuous, adaptive, and participatory system.

Another pivotal advancement is the rise of open-source geospatial ecosystems and interoperable system architectures. Tools such as QGIS, GRASS GIS, and OpenLayers, together with open data standards promoted by the Open Geospatial Consortium (OGC), have democratized access to advanced geospatial capabilities. The open-data movement, championed by organizations like NASA, ESA, and the Group on Earth Observations (GEO), ensures that global datasets from land cover to atmospheric composition are freely available to researchers and policymakers. This has catalyzed the development of participatory and collaborative geospatial systems, where users across sectors can access, share, and analyze data in near-real time. The Global Flood Monitoring System (GFMS) and the Famine Early Warning Systems Network (FEWS NET) exemplify this approach. Both platforms integrate satellite observations, hydrological models, and socio-economic data within web-GIS environments, providing dynamic dashboards that support early warning and humanitarian response worldwide. The use of APIs and interoperable services allows these systems to interface seamlessly with national disaster management platforms, ensuring that scientific outputs translate into actionable insights. Furthermore, the integration of Digital Twin technology and 3D geospatial modeling marks a frontier in environmental system development. A Digital Twin represents a virtual replica of a physical environment that continuously assimilates real-world data to simulate and forecast system behavior. In the context of flood and drought management, Digital Twins of river basins, urban drainage systems, or agricultural landscapes allow real-time scenario testing, where different management interventions such as reservoir operations, land-use changes, or irrigation policies can be simulated to evaluate their impacts on hydrological extremes. For instance, the European Union's Destination Earth (DestinE) initiative aims to create a comprehensive Digital Twin of the planet, integrating satellite data, AI models, and GIS-based simulations to support climate adaptation policy. Such developments demonstrate the expanding horizon of GIS and RS from static analysis tools to dynamic, intelligent systems that continuously learn from the environment. The convergence of these technologies also introduces visualization and decision-support innovations that make complex data accessible to non-specialist audiences. Interactive web-GIS dashboards, augmented reality (AR) visualizations, and immersive 3D mapping applications enable policymakers, community leaders, and the general public to explore risk scenarios intuitively. However, the rapid technological evolution also raises challenges related to data governance, interoperability, and sustainability. The integration of diverse datasets from satellites, IoT devices, and social media introduces issues of data standardization, privacy, and long-term maintenance. Moreover, while open-access platforms expand opportunities, they also demand significant computational literacy and analytical skills, which remain unevenly distributed across regions. Ensuring that developing countries can fully leverage these innovations requires targeted investments in capacity building, education, and institutional collaboration. As technological frontiers expand, the future of GIS and RS in flood and drought management will depend not only on innovation itself but on the creation of equitable systems that make those innovations globally accessible. In essence, technological innovation has transformed GIS and Remote Sensing from static, descriptive tools into dynamic, intelligent ecosystems. The fusion of AI, IoT, cloud computing, and digital twin technologies enables continuous environmental monitoring, predictive risk modeling, and participatory decision-making at unprecedented scales. As climate-induced hydrological extremes become more frequent and complex, the development of these integrated, adaptive systems will define the next generation of environmental governance one grounded in spatial intelligence, data democratization, and global collaboration.

7. CHALLENGES AND LIMITATIONS

Despite their transformative potential, the integration of Geographic Information Systems (GIS) and Remote Sensing (RS) into flood and drought risk assessment remains a multifaceted challenge that extends far beyond technological capability. Although advances in satellite observation, cloud computing, and artificial intelligence have dramatically improved data availability, analytical precision, and automation, the effective realization of these tools for environmental management continues to face substantial barriers. The proliferation of global datasets and analytical platforms does not automatically translate into accuracy or usability, as inconsistencies in spatial resolution, sensor calibration, and temporal coverage often compromise the reliability of derived models. Moreover, methodological complexities such as uncertainty propagation in multi-sensor data fusion, the limitations of empirical assumptions in nonlinear hydrological systems, and the opacity of machine-learning algorithms frequently constrain interpretability and policy adoption. Institutional fragmentation further compounds these issues: many national agencies lack the capacity, coordination, and long-term funding to sustain geospatial operations beyond short-term projects, particularly in developing regions where vulnerability to climate extremes is highest. In addition, socio-political factors ranging from data sovereignty disputes and restricted information sharing to the ethical implications of surveillance and digital inequality undermine global collaboration and equitable access to spatial intelligence. Collectively, these interrelated challenges underscore that while GIS and RS provide unprecedented analytical power, their true impact depends on the quality of data, methodological rigor, human expertise, and governance frameworks that ensure the translation of scientific insight into actionable climate resilience and sustainable development strategies.

7.1. Data Availability, Accuracy, and Resolution

The most fundamental challenge lies in the accuracy, consistency, and resolution of spatial data. Remote Sensing, by nature, provides indirect measurements of surface and atmospheric parameters, which are subject to uncertainties arising from sensor characteristics, atmospheric interference, calibration errors, and retrieval algorithms. Satellite-based rainfall products, for instance, such as CHIRPS, TRMM, or GPM, while invaluable for large-scale hydrological modeling, often underestimate extreme precipitation events or fail to capture localized convective storms phenomena critical for flash flood prediction. Similarly, optical imagery from satellites like Landsat or Sentinel is frequently obstructed by cloud cover during monsoon seasons, precisely when flood mapping is most needed. In contrast, radar sensors such as Sentinel-1 and RADARSAT can penetrate clouds, but their interpretation requires complex pre-processing (speckle filtering, radiometric calibration) that demands technical expertise. These inconsistencies can propagate errors through subsequent GIS analyses, leading to uncertainty in hazard delineation, exposure estimation, and vulnerability assessment.

Spatial and temporal resolution also pose trade-offs that affect analytical reliability. High-resolution imagery (<10 m) provides detailed information suitable for urban flood modeling but is computationally demanding and often expensive for continuous monitoring. Conversely, coarse-resolution datasets like MODIS or GRACE, while freely available and globally consistent, may obscure localized hydrological variations essential for small-watershed drought assessment. The challenge thus lies in balancing resolution with scalability an issue partially addressed through data fusion and multi-sensor integration, yet still limited by computational costs and data harmonization complexities. The absence of standardized global protocols for data preprocessing, fusion, and validation further exacerbates discrepancies across studies, making cross-regional comparisons difficult and often inconsistent.

7.2. Methodological and Modeling Constraints

Beyond data availability, methodological challenges constrain the robustness and reproducibility of GIS-RS-based assessments. Hydrological models, while increasingly sophisticated, are still abstractions of complex natural systems. Parameters such as infiltration rates, soil hydraulic conductivity, and evapotranspiration coefficients vary spatially and temporally, often in ways that cannot be fully captured by remote sensing proxies. Moreover, many GIS-RS models rely on assumptions of stationarity implying that historical climate and land-use patterns remain valid predictors of future conditions an assumption increasingly invalidated under accelerating climate change.

Uncertainties also emerge from model coupling and calibration. Flood modeling tools such as HEC-RAS or MIKE FLOOD often require integration with precipitation and soil data derived from RS sources, yet mismatched spatial grids or temporal scales can introduce systematic biases. Similarly, drought indices like SPEI or NDVI are highly sensitive to input variable accuracy, algorithmic design, and local calibration. The lack of standardized validation against ground observations further limits the reliability of derived hazard maps. This issue is particularly acute in developing regions where hydrometeorological station density is low, leading to overreliance on global datasets without sufficient local adjustment. Additionally, machine learning models though powerful in predictive accuracy often function as "black boxes," providing limited interpretability. This opacity challenges their adoption in policy environments that demand transparency and explainability for decision-making.

7.3. Socio-Political and Ethical Considerations

The adoption of advanced geospatial technologies, while transformative for environmental monitoring and disaster management, simultaneously introduces a complex landscape of ethical, legal, and socio-political challenges that must be critically examined alongside technical advancement. The increasing availability of open-access satellite imagery and big geospatial data has undoubtedly democratized information, empowering researchers, institutions, and even local communities to engage in evidence-based decision-making. However, this democratization is accompanied by growing concerns regarding data sovereignty, privacy, and surveillance ethics. High-resolution imagery from satellites such as WorldView, GeoEye, or PlanetScope can capture details down to the level of individual buildings and property boundaries, raising questions about consent, ownership, and the potential misuse of sensitive spatial information. In politically volatile regions, the publication of detailed land-use or infrastructure maps can inadvertently expose military installations, refugee settlements, or resource-rich territories, leading to geopolitical tensions or security threats. Moreover, the increasing reliance on Artificial Intelligence (AI) and automated data processing introduces subtler yet equally significant risks of algorithmic bias and data asymmetry. When machine-learning models are trained predominantly on datasets from technologically advanced regions often in the Global North they may internalize spatial and socio-economic biases that distort predictions for less-represented contexts, such as rural Africa or Southeast Asia. This not only compromises scientific validity but also reinforces existing hierarchies in knowledge production, where the capacity to define "risk" or "vulnerability" becomes concentrated in a few technologically dominant institutions. Addressing such imbalances requires a paradigm shift toward ethical geospatial intelligence, emphasizing transparency in algorithmic design, inclusion of local data contexts in model training, and community consent in data collection processes.

Another equally pressing issue concerns the digital divide and inequitable access to geospatial technology, which manifests as both a technological and epistemological gap between nations and communities. While the proliferation of open-access platforms like Google Earth Engine (GEE), NASA EarthData, and Copernicus Hub has lowered financial barriers to data acquisition, effective utilization still requires stable internet connectivity, advanced computational infrastructure, and skilled human capital resources that remain scarce in many developing regions. Consequently, the benefits of GIS and RS technologies are disproportionately concentrated in countries and institutions with strong digital ecosystems, leaving vulnerable populations with limited capacity to harness these tools for disaster preparedness or climate adaptation. This asymmetry risks reproducing a new form of "digital environmental inequality", wherein those most exposed to hydrometeorological hazards such as drought-prone regions in sub-Saharan Africa or flood-prone deltas in South Asia are the least equipped to benefit from geospatial innovation. Furthermore, issues of data colonialism emerge when global institutions or private corporations collect, process, and monetize spatial data from the Global South without equitable data-sharing agreements or local capacity development. Such dynamics perpetuate dependency and marginalization, undermining the very goals of resilience and sustainability. To confront these challenges, future geospatial governance must evolve beyond technological determinism toward inclusive, participatory, and transparent frameworks that guarantee equitable access, ethical standards, and shared ownership of data. Policies that promote open data with local custodianship, enforce ethical use of AI in environmental analytics, and strengthen digital capacity across all levels of society are essential for ensuring that the geospatial revolution contributes not only to scientific progress but also to social justice and global climate equity.

8. FUTURE DIRECTIONS AND RESEARCH OPPORTUNITIES

The trajectory of Geographic Information Systems (GIS) and Remote Sensing (RS) in flood and drought risk assessment is poised to undergo a profound transformation driven by advances in artificial intelligence, multi-sensor fusion, and participatory geospatial technologies. As the pace of climate change accelerates, the demand for more accurate, predictive, and actionable hydrological intelligence will intensify, necessitating a shift from descriptive mapping toward real-time, adaptive, and self-learning systems. The next generation of research and development in this domain will be characterized by deeper integration across technological, methodological, and socio-institutional dimensions linking global data infrastructures with local decision-making, combining physics-based models with machine intelligence, and embedding human participation into the very architecture of geospatial systems. To put it simply, the goal of GIS and RS in this area is to develop what is known as cognitive environmental analytics systems that process geographical data while simultaneously anticipating change, learning from feedback, and facilitating on-going adaptation (D. Liu et al., 2025). One major frontier in this transformation is multi-sensor and multi-source data fusion, a process that seeks to overcome the limitations of individual datasets by integrating complementary information from diverse sensors, platforms, and domains. Traditional flood and drought studies often rely on either optical or radar data; however, each has inherent weaknesses. Optical imagery, though valuable for vegetation and land cover analysis, is constrained by cloud interference during critical wet periods, while radar data, though cloud-penetrating, are limited in spectral detail. Thc fusion of multi-modal datasets such as integrating Sentinel-1 SAR data with Sentinel-2 optical imagery or combining MODIS thermal observations with GRACE terrestrial water storage data enables the generation of richer, multi-dimensional insights into hydrological processes. Moreover, integrating these satellite observations with in-situ IoT sensor data, citizen-generated observations, and climate model outputs within GIS environments can create near-continuous representations of water cycle dynamics. Research into data assimilation frameworks for instance, coupling RS data with numerical models like SWAT or LISFLOOD will play a crucial role in bridging observation and simulation. This fusion-driven paradigm will allow for the creation of "hyper-resolution" environmental models capable of tracking floods and droughts simultaneously across temporal scales ranging from hours to decades.

Equally transformative will be the continued integration of Artificial Intelligence (AI) and Machine Learning (ML) into geospatial analytics. The application of AI has already improved classification accuracy and prediction capabilities in flood and drought mapping, but its future lies in creating autonomous, adaptive systems that continuously learn from new data streams. Deep learning architectures such as Convolutional Neural Networks (CNNs), Recurrent Neural Networks (RNNs), and Transformer-based models are increasingly capable of identifying nonlinear patterns within massive, multi-temporal datasets that traditional models cannot capture. For instance, hybrid deep learning systems combining physical hydrological models with neural networks can simulate both the physical behavior of water and its data-driven anomalies, offering a dual perspective that improves prediction robustness. Research in this direction is expected to move toward Physics-Informed AI (PIAI), where machine learning models are constrained by physical laws of hydrology, ensuring that outputs are scientifically consistent rather than purely statistical. These AI-driven systems will not only enhance early warning capabilities but also allow predictive "what-if" scenario testing for climate adaptation policies. Integrating such models within cloud-based GIS infrastructures especially those leveraging edge computing and distributed data architectures will enable near-real-time hazard forecasting at global scales. Another major opportunity lies in the development of Digital Twins of hydrological systems, which represent virtual replicas of physical environments capable of continuous synchronization with real-world data. Unlike static models, digital twins assimilate satellite observations, IoT sensor readings, and social data in real time to simulate evolving flood and drought scenarios. This approach enables policymakers and water managers to visualize multiple scenarios such as changes in rainfall patterns, reservoir releases, or land-use transitions and to test intervention strategies before implementation. For instance, digital twins of river basins can forecast how reservoir operations during monsoon seasons influence downstream drought risk, allowing for adaptive water governance. Europe's Destination Earth (DestinE) project has already demonstrated the feasibility of global-scale environmental twins, but future research will focus on translating this concept into basin-level and local applications, especially in regions where hydrological variability is both high and under-monitored. The integration of Digital Twins within web-based GIS systems will redefine decision support frameworks, making them predictive, interactive, and continuously updated. Parallel to technological innovation, the future of GIS and RS research must also address interoperability, transparency, and ethical governance. As the volume and diversity of geospatial data expand, ensuring that systems can communicate seamlessly and that models are interpretable becomes increasingly vital. The emergence of FAIR (Findable, Accessible, Interoperable, and Reusable) data principles and open-source geospatial ecosystems provides a foundation for greater collaboration and reproducibility. However, the growing reliance on AI and automated decision systems introduces risks related to bias, accountability, and data privacy. Future research must therefore develop frameworks for explainable AI (XAI) in environmental analytics, ensuring that model outputs are transparent and traceable. Ethical guidelines for data use, particularly in contexts involving population exposure or land ownership, will be critical for maintaining public trust and policy legitimacy. Furthermore, enhancing interoperability between global data platforms (NASA EarthData, Copernicus Hub, GEO) and regional systems will require standardized metadata conventions and APIs that allow fluid data exchange across institutional and geographic boundaries.

Capacity building and knowledge co-creation must become integral components of future research agendas. While technological innovations proliferate, their impact depends on human capability to operationalize them. Collaborative research networks, open educational resources, and regional centers of excellence such as SERVIR (a joint initiative by NASA and USAID) should be expanded to ensure equitable access to geospatial tools and training. Building local analytical expertise will not only sustain the use of GIS and RS systems but also foster innovation that reflects local contexts and priorities. Bridging the gap between technological capacity and social equity will be the defining challenge and opportunity of the next decade. In conclusion, the future of GIS and Remote Sensing in flood and drought risk assessment will be characterized by convergence of data, models, disciplines, and people. The integration of AI, multi-sensor fusion, and participatory technologies will redefine environmental monitoring as a continuous, intelligent, and inclusive process. However, to achieve this goal, geospatial science must be rethought as a dynamic ecosystem of knowledge, ethics, and cooperation rather than as a collection of instruments. In order to ensure that the potential of geospatial intelligence advances not only science but also planetary resilience, environmental justice, and the sustainability of life itself, the next generation of researchers must strike a balance between technological innovation and societal requirements (Liang & Fernandez, 2026).

9. CONCLUSION

The integration of Geographic Information Systems (GIS) and Remote Sensing (RS) in flood and drought risk assessment represents one of the most significant paradigm shifts in the history of environmental science and disaster management. Once constrained by sparse ground measurements and reactive policies, the global community now possesses the capacity to observe, model, and forecast hydrological extremes with unprecedented precision and timeliness. The convergence of multi-sensor Earth observation, spatial analytics, and computational intelligence has transformed environmental monitoring from a fragmented, retrospective practice into a unified, predictive, and adaptive system. At its core, GIS provides the analytical and spatial framework for integrating heterogeneous datasets topography, hydrology, land use, and socio-economic variables while RS supplies the dynamic observations needed to monitor environmental change continuously. Together, they have established a new epistemology of environmental knowledge one that is inherently spatial, data-driven, and systemic. This integration has not only enhanced scientific understanding of hydrometeorological hazards but has also laid the groundwork for proactive risk governance, sustainable resource management, and climate resilience at both local and global scales.

At a scientific level, the synthesis of GIS and RS has redefined the methodological frontiers of hydrology, climatology, and spatial modeling. The capacity to combine real-time satellite data with physics-based hydrological models allows researchers to capture both the complexity and the dynamism of flood and drought processes. Unlike traditional empirical approaches, which rely heavily on historical averages, modern geospatial systems can now detect evolving hazard patterns, simulate multiple climate scenarios, and quantify cascading impacts across sectors. In particular, the integration of AI and cloud-based computation has enabled global-scale modeling that was unimaginable even a decade ago. From a practical standpoint, the operationalization of GIS and RS has redefined how governments, organizations, and communities manage water-related risks. Early warning systems that once relied on manual gauge readings now function as fully automated geospatial infrastructures, integrating near-real-time satellite data, IoT sensor networks, and predictive algorithms. Decision-makers can visualize live flood inundations, forecast drought stress, and access vulnerability maps through interactive web-GIS dashboards that translate scientific models into intuitive policy tools. Such systems have already proven transformative in various regions from the Global Flood Monitoring System (GFMS) that tracks riverine floods worldwide, to the Famine Early Warning Systems Network (FEWS NET) that monitors drought and food insecurity across Africa and Asia. More importantly, GIS and RS technologies have enabled anticipatory governance, allowing policymakers not only to respond to disasters but to anticipate them shifting disaster management from reaction to prevention, and from crisis response to long-term resilience planning. This transformation is critical in a century where climate-induced hydrological volatility threatens to destabilize economies, ecosystems, and societies alike. Building capacity through training, fostering open-data collaborations, and embedding geospatial literacy into education systems are essential steps toward ensuring that the advantages of GIS and RS are equitably shared. The vision of a climate-resilient planet will remain incomplete unless every country regardless of its level of development can access, utilize, and benefit from the same spatial intelligence that drives global decision-making. The integration of GIS and RS also underscores the need for a new ethical and governance framework in the age of data abundance. As satellite constellations, drones, and IoT devices generate vast quantities of environmental data, concerns over data sovereignty, privacy, and algorithmic bias become increasingly relevant. Future geospatial systems must therefore adhere to the principles of transparency, accountability, and interoperability. Open standards and FAIR data practices should become the norm, ensuring that scientific progress is coupled with ethical responsibility. This balance between technological innovation and governance integrity will define the credibility and sustainability of geospatial science in the coming decades.

REFERENCES

Abdelkrim, Z., & Eslamian, S. (2026). An evaluation of spatial and temporal variations of land surface temperature in Biskra, Algeria using remote sensing and GIS. *Journal of African Earth Sciences*, *235*, 105942. DOI: 10.1016/j.jafrearsci.2025.105942

Amsalu, A., Garedew, W., Melka, G. A., Çelik, M. A., & Gemeda, D. O. (2026). Trends in climate extreme indices and implications for coffee farming in The Kafa Biosphere Reserve, southwest Ethiopia. *Climate Services*, *42*, 100637. DOI: 10.1016/j.cliser.2026.100637

Anwar, N., Rahaman, R., Widodo, A. M., Sekti, B. A., Erzed, N., Santiago Tangkudung, R. R., Muchlis, M., Yuhefizar, Y., & Budhisantosa, N. (2025). Sustainable cybersecurity in the AI era: Innovations in green computing and security. In *Sustainable Information Security in the Age of AI and Green Computing* (pp. 603–620). DOI: 10.4018/979-8-3693-8034-5.ch030

Bournez, P., Salzinger, J., Cella, M., Vultaggio, F., d'Apolito, F., & Fanta-Jende, P. (2024). Pushing the limit to near real-time indoor LiDAR-based semantic segmentation. *International Archives of the Photogrammetry, Remote Sensing and Spatial Information Sciences - ISPRS Archives, 48*(2/W8-2024), 45–52. DOI: 10.5194/isprs-archives-XLVIII-2-W8-2024-45-2024

Çelik, M. A. (2026). Agro-climatic classification of Türkiye's Agricultural Basins using long-term ERA5-Land temperature data. *Physics and Chemistry of the Earth Parts A/B/C*, *142*, 104307. DOI: 10.1016/j.pce.2026.104307

Çelik, M. A., Kıraç, A., & Gouveia, C. M. (2026). North Atlantic Oscillation modulation of drought and vegetation activity in Türkiye. *Theoretical and Applied Climatology*, *157*(2), 74. Advance online publication. DOI: 10.1007/s00704-025-05971-5

Cerrada Ayllón, A. M., Cocero Matesanz, D., Cerrillo Cuenca, E., & Jordá Pardo, J. F. (2023). SPATIAL STUDY BY GEOGRAPHICAL INFORMATION SYSTEMS OF THE MEGALITHS IN THE REGION TIERRA DE LEDESMA (SALAMANCA, SPAIN). *GeoFocus*, *2023*(32), 5–26. DOI: 10.21138/GF.826

Gibson, A. L., & Sokolov, D. (2026). A Modular Cognitive Architecture for Collective Intelligence Systems. *Lecture Notes in Computer Science, 16057 LNAI*, 181–191. DOI: 10.1007/978-3-032-00686-8_17

Güngör, E., & Serttaş, S. (2024). EOG Region Detection using Fuzzy k-NN for Virtual Reality Applications. *Sakarya University Journal of Science*, *28*(6), 1146–1157. DOI: 10.16984/saufenbilder.1386568

Jeyaraj, R., Balasubramaniam, A., Kumara, M. A. A., Guizani, N., & Paul, A. (2023). Resource Management in Cloud and Cloud-influenced Technologies for Internet of Things Applications. *ACM Computing Surveys*, *55*(12). Advance online publication. DOI: 10.1145/3571729

Kierzkowski, A., Kisiel, T., Milewski, M., Török, Á., Stosiak, M., & Wróbel, J. (2024). COMPUTER-AIDED SIMULATION OF UNMANNED AERIAL VEHICLE COMPOSITE STRUCTURE DYNAMICS. *Transport*, *39*(4), 302–312. DOI: 10.3846/transport.2024.23159

Li, Y., Zhong, X., Dai, P., Chen, B., Zhou, F., Fan, Y., Wu, J., & He, L. (2025). Construction and validation of a risk predictive model for fear of disease progression in patients after percutaneous coronary intervention. *Scientific Reports*, *16*(1), 1444. DOI: 10.1038/s41598-025-31099-7 PMID: 41353494

Liang, Y., & Fernandez, D. (2026). A bibliometric analysis of RPA and AI research: Current trends and future directions. *Multidisciplinary Science Journal*, *8*(3), 2026145. Advance online publication. DOI: 10.31893/multiscience.2026145

Liao, J., Xie, Y., Zhao, P., Xia, K., Xu, B., Wang, H., Li, C., Li, C., & Liu, H. (2024). Probabilistic assessment of the thermal performance of low-enthalpy geothermal system under impact of spatially correlated heterogeneity by using XGBoost algorithms. *Energy*, *313*, 133947. Advance online publication. DOI: 10.1016/j.energy.2024.133947

Liu, D., Zhao, Y., Ma, Z., & Ding, Q. (2025). Intelligent target detection and encrypted transmission system based on FPGA. *Computers & Electrical Engineering*, *128*, 110653. Advance online publication. DOI: 10.1016/j.compeleceng.2025.110653

Liu, G., Qiu, J., Huang, J., & Yuan, Y. (2026). GLGF-CR: A Gated Local-Global Fusion approach for cloud removal in real-world remote sensing. *Pattern Recognition*, *172*, 112319. Advance online publication. DOI: 10.1016/j.patcog.2025.112319

Mariani, S., Zeighami, F., Marzani, A., & Palermo, A. (2025). PSAFE: A semi-analytical finite element formulation for spatially periodic structural systems. *Journal of Sound and Vibration*, *619*, 119391. Advance online publication. DOI: 10.1016/j.jsv.2025.119391

Mulesa, O., & Bohdan, Y. (2024). DEVELOPMENT OF A FUZZY PRODUCTION MODEL FOR ASSESSING THE DEGREE OF INFORMATION SECURITY IN INTERNATIONAL COOPERATION. *Technology Audit and Production Reserves*, *6*(2), 6–10. DOI: 10.15587/2706-5448.2024.318446

Mwinjuma, M., Wang, R., Mtupili, M., & Twaha, M. (2026). Comparisons of SPI and SPEI in capturing drought dynamics: A Global assessment across arid and humid regions. *Atmospheric Research, 329*, 108475. Advance online publication. DOI: 10.1016/j.atmosres.2025.108475

Nesanır, M. O., Zenezini, G., Senvar, O., & Ottaviani, F. M. (2026). A Literature Review on the Integration of System Dynamics and Machine Learning/Artificial Intelligence in Addressing Environmental Matters. *IFIP Advances in Information and Communication Technology, 764 IFIPAI*, 370–384. DOI: 10.1007/978-3-032-03515-8_26

Pineda, V. G., Valencia-Arias, A., Giraldo, F. E. L., & Zapata-Ochoa, E. A. (2026). Integrating artificial intelligence and quantum computing: A systematic literature review of features and applications. *International Journal of Cognitive Computing in Engineering*, *7*, 26–39. DOI: 10.1016/j.ijcce.2025.08.002

Rohanadeera, S. L., & Hewagamage, K. P. (2026). Generative-AI-Powered TA BOT for Enhanced Personalized Support in Software Engineering Education. *Communications in Computer and Information Science, 2380 CCIS*, 197–204. DOI: 10.1007/978-981-96-6291-3_16

Saleh, K., Mabrouk, W. M., & Metwally, A. M. (2025). Convolutional neural network approach for automated well zonation in the Lower Bahariya member north Western Desert Egypt. *Scientific Reports*, *16*(1), 625. DOI: 10.1038/s41598-025-32119-2 PMID: 41453993

Sekti, B. A. (2025). Machine learning and artificial intelligence (AI) in manufacturing. In *Integrating Digital Innovation and Integrated Frameworks in Manufacturing* (pp. 255–292). DOI: 10.4018/979-8-3373-1082-4.ch011

Song, X., Zhang, L., Lu, Z., Liang, H., & Ren, W. (2025). Optimizing DEM error mitigation in multi-temporal InSAR: A detection-and-estimation strategy based on phase gradient direction consistency. *Remote Sensing of Environment*, *331*, 115028. Advance online publication. DOI: 10.1016/j.rse.2025.115028

Taylor, J., & Salvucci, G. (2025). Estimating contrasting soil moisture-precipitation feedbacks across global landmass using data from the Soil Moisture Active Passive satellite mission. *Science of Remote Sensing*, *12*, 100247. Advance online publication. DOI: 10.1016/j.srs.2025.100247

Vlasenko, T., Vlasovets, V., Kovalyshyn, O., & Bilovod, O. (2024). ANALYSIS OF THE PROJECT OF A MARKETING CAMPAIGN TO PROMOTE ROBOTIC SOLUTIONS USING RANDOM FOREST CLASSIFICATION. *Technology Audit and Production Reserves*, *6*(4), 43–50. DOI: 10.15587/2706-5448.2024.320367

Warlina, L., & Afredi, A. (2026). Geographic Information System (GIS) Application to Identify the Distribution of Tourist Attractions and Supporting Facilities Related to the Sustainable Development Goals (SDGs). *Indonesian Journal of Science and Technology*, *11*(2), 341–356. DOI: 10.17509/ijost.v11i2.89803

Xu, T., Xiao, S., Reuschen, S., Wildt, N., Hendricks Franssen, H.-J., & Nowak, W. (2024). Towards a community-wide effort for benchmarking in subsurface hydrological inversion: Benchmarking cases, high-fidelity reference solutions, procedure, and first comparison. *Hydrology and Earth System Sciences*, *28*(24), 5375–5400. DOI: 10.5194/hess-28-5375-2024

Zhou, M., Li, C., Li, Z., Yu, Z., & Zhou, X. (2023). Stratification of vertical canopy structure to improve estimation of forest inventory attributes using airborne LiDAR data in a large subtropical region of China. *Annals of Forest Research*, *66*(2), 101–120. DOI: 10.15287/afr.2023.3361

Chapter 10

A Special Focus on Natural Disasters in the Field of Earth Sciences: What Do the Reports Tell Us?

Yaver Batu
http://orcid.org/0009-0008-5586-3264
Igdir University, Turkey

Mehmet Ali Çelik
http://orcid.org/0000-0002-7729-6650
Igdir University, Turkey

Muhittin Kulak
http://orcid.org/0000-0003-3673-9221
Igdir University, Turkey

ABSTRACT

This study analyzes 9,116 scientific reports to explore research trends in natural disasters within Earth sciences. Keyword network and clustering analyses reveal that "natural disasters" and "climate change" are the most central themes, highlighting their strong interconnection. Technologies such as GIS, remote sensing, and artificial intelligence (deep learning, neural networks) play crucial roles in disaster prediction, risk assessment, and damage evaluation. Tsunami-related research exemplifies advancements in early warning and management systems, particularly in Japan, Indonesia, and Turkey. The study also emphasizes societal resilience and recovery processes after disasters. Based on the World Risk Index, India and Indonesia show the highest vulnerability, while the U.S. and Germany

DOI: 10.4018/979-8-3373-6274-8.ch010

demonstrate strong adaptive capacities. Turkey's moderate index underlines the need for improved preparedness. Bibliometric data identify China, the U.S., and India as leading contributors to disaster-related research.

INTRODUCTION

It is possible to define a disaster as the damage caused by an ordinary movement of nature to humans, other living beings, and settlements. Disasters, which may be of technological or natural origin, lead to significant economic and social losses (Ergünay, 2007; Erkal et al., 2009). To avoid being affected by disasters or to minimize their damage, it is essential to establish a Disaster Management Plan and to form well-trained and equipped disaster response teams. Effective disaster management depends on the proper implementation of the stages of *mitigation, preparedness, rescue, first aid, recovery, and reconstruction* (Varol & Gültekin, 2016). Holding nature responsible for losses caused by human mistakes-such as poor site selection, the use of substandard materials, or the lack of a disaster management plan-is unreasonable, since such activities are a natural part of the Earth's dynamics. Following the Industrial Revolution, the demand for labour led to increased employment and urbanization. Consequently, the rise in production caused greenhouse gases and chemical pollutants released from various industrial facilities to trigger untimely and abnormal natural phenomena in the Earth's climate and components (Bilke, 2018).

While technological advancement undeniably enhances human comfort, humanity also experiences the cost of disrupting nature's balance through exposure to disasters. Hydrological disasters resulting from geological movements, particularly in coastal cities, have become increasingly destructive. In recent years, emerging technologies have been used to reduce or even prevent such destructive impacts. Highly efficient early warning systems have been developed to detect potential disasters, along with software such as *VOSviewer* and *GIS* that enable the collection and analysis of disaster data. Through education and public awareness, it is possible to transform natural events into manageable phenomena and minimize their damage.

Recent years have seen a notable expansion of bibliometric and science-mapping studies that quantify how disaster scholarship is evolving across hazards, regions, and methodological toolkits. New bibliometric evidence (2023–2025) highlights accelerating growth in topic clusters that integrate climate change, compound risk, and technology-enabled assessment (e.g., remote sensing and AI) with governance and resilience outcomes. For example, bibliometric syntheses have recently examined earthquake-focused research surges following major events, as well as hazard-specific domains such as flood mitigation and tsunami vulnerability modelling, showing increasing reliance on Scopus- indexed datasets and network-based indicators (e.g., co-authorship, co-citation, keyword co-occurrence) to characterize global research fronts and gaps.

RESEARCH QUESTIONS

a. Which types of natural disasters are the most frequently studied?

b. Are Early Warning Systems truly effective in providing timely alerts?

c. What are the mechanisms of formation and the extent of damage caused by rare disasters such as tsunamis?

d. Is it possible to develop a single, universal guide for disaster management applicable to all types of disasters?

IMPORTANCE OF THE STUDY

The frequency with which a topic is addressed is closely related to the magnitude of its effects and the number of elements it influences. Throughout human history, natural disasters have remained in memory as extraordinary events-both in ancient times when their causes were unknown and in modern periods when science has revealed their origins (Türk, 2024; Sağlam et al., 2021). This study is significant because it examines and quantitatively analyzes research conducted on natural disasters, which-beyond causing economic losses to people and their possessions-often result in the loss of human life.

NOVELTY OF THE STUDY

Natural disasters have been a serious subject throughout history in terms of both their origins and consequences. In today's world, where population density and urbanization have increased, their social and economic impacts have become even more profound. Thus, disasters continue to represent not only vital threats but also costly realities (Kuş, 2018). This research examines the number of studies focused on natural disasters and the specific types of disasters most frequently investigated. It presents information and analyses regarding the formation, types, frequency, and impacts of various disasters.

AIMS AND OBJECTIVES OF THE STUDY

The aim of this study is to investigate, through quantitative analysis, the types and characteristics of natural phenomena that cause suffering to humanity and are referred to as "disasters," as well as to examine the findings of scientific studies conducted on these events.

METHODOLOGY

Bibliometric and Science Mapping Design

This study employed a bibliometric and science mapping approach to systematically analyze global research trends in natural disasters within the field of Earth sciences. Unlike a traditional narrative literature review, bibliometric analysis enables quantitative evaluation of scientific production, collaboration networks, and thematic evolution through structured database mining and visualization techniques.

Data Sources and Search Strategy

Scientific records were retrieved from internationally recognized citation database (Scopus) which provides comprehensive coverage of peer-reviewed journals and conference proceedings. The search was conducted using combinations of the following keywords: ("natural disaster*" OR "disaster risk reduction" OR "hazard*" OR "tsunami" OR "earthquake" OR "flood*" OR "climate-related disaster*")

The search was limited to: Subject areas: Earth sciences, environmental sciences, geography, disaster management; Document types: Articles and reviews; Language: English]. This process yielded 9,116 records. Regarding screening and pre-processing; we ensured data quality. In this context, I) duplicates were removed. II) Non-relevant subject areas were excluded; III) Keywords were standardized (e.g., "GIS" and "Geographic Information Systems" merged); IV) Country and institution names were harmonized; V) Records with missing bibliographic information were discarded, and VI) The cleaned dataset was exported in CSV format for analysis. We finally performed the bibliometric and network analysis using VOSviewer (v1.6.x). The following analyses were conducted: I) Keyword co-occurrence networks; II) Country collaboration networks, III) Citation and publication counts, IV) Cluster detection based on association strength. The following parameters were set: I) Counting method: Full counting, II) Minimum keyword occurrence: 5, III) Normalization: Association strength, IV) Clustering resolution: Default, and V)

Visualization: Network and density maps. These procedures enabled identification of dominant themes, research clusters, and global collaboration structures.

RESULTS AND DISCUSSION

In the study, to strengthen the interpretation of bibliometric patterns beyond descriptive mapping, we situate the study within two widely used perspectives in disaster research. First, the Pressure and Release (PAR) model explains disaster risk as the intersection of hazards with socially produced vulnerability, progressing from root causes to dynamic pressures and unsafe conditions. This lens helps interpret why certain regions and topics dominate research attention: publication intensity is not only associated with exposure, but also with vulnerability drivers and institutional capacity (Wisner and Wisner, 2004; Blaikie et al., 2014). We here secondly discussed the findings using Resilience Theory, which frames recovery and adaptation as capacities of systems and communities to absorb shocks, reorganize, and learn, emphasizing coping and adaptive mechanisms that reduce long-term losses. Using this lens, our keyword clusters related to recovery, risk reduction, and preparedness can be interpreted as signals of a broader research shift from response-oriented studies toward resilience-building agendas and governance-relevant outcomes (Hu et al., 2024; Du et al., 2025). Along with the study, we further connected the scientific trends to current disaster governance; we align our findings with the global framework for disaster risk reduction. The Sendai Framework for Disaster Risk Reduction (2015–2030) is operationalized through the Sendai Framework Monitor, which supports country reporting and uses indicators aligned with the Sustainable Development Goals-particularly SDGs 1, 11, and 13. We therefore interpret the growing prominence of risk assessment, early warning, and resilience/recovery concepts as consistent with the shift toward risk-informed development and measurable DRR targets. We also reference recent UN system progress reporting on DRR for resilience, which underscores the continued importance of integrated DRR–climate adaptation–development approaches (Sebesvari et al., 2019; Chmutina et al., 2021; Green et al., 2019; Ningsih et al., 2022).

The DNA of Scientific Reports: Keywords

The analysis of the most frequently emphasized keywords in the 9,116 scientific reports examined within the scope of this study is presented in Figure 1. This figure provides important insights into the relationships among keywords and helps to understand research trends in the fields of natural disasters and earth sciences. According to the figure analysis, the keywords "Natural disasters" and "Climate change" occupy central positions. This indicates a strong interconnection between natural disasters and climate change in the scientific literature. As highlighted in many scientific reports, most natural events are known to be linked to climatic changes (Morrissey & Reser, 2007; Benevolenza & DeRigne, 2019). Furthermore, the figure shows that technological tools such as "GIS (Geographic Information Systems)" and "Remote sensing" play a critical role in monitoring climate changes and analyzing the impacts of natural disasters (Cüce, 2023). These technologies are used in several crucial processes, including predicting natural disasters, mapping their effects, and conducting risk assessments.

Figure 1. Co-occurrences of the keywords

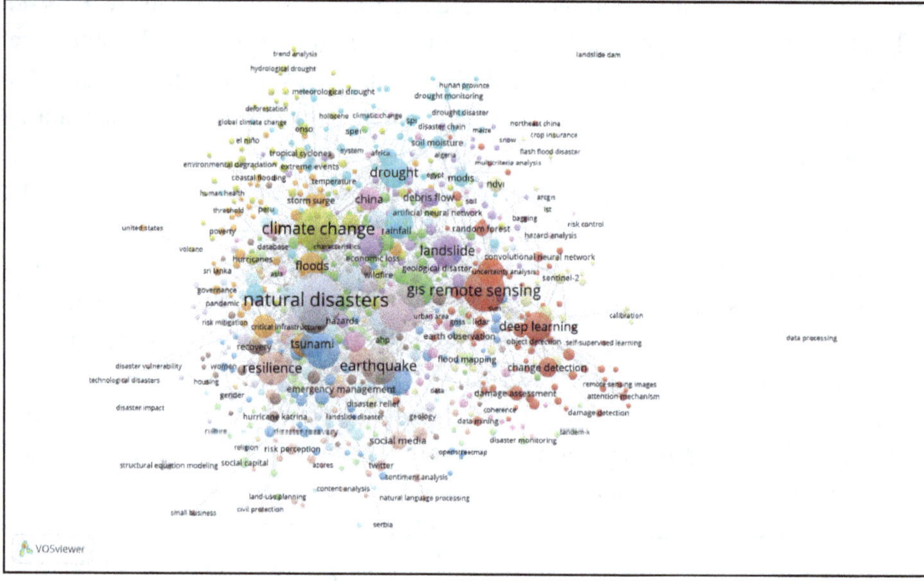

When the connections (co-occurrence network) and cluster formations (clustering) among the keywords are evaluated, it is observed that specific types of disasters-such as "Floods," "Tsunami," and "Earthquake"-form distinct clusters in different regions of the network. These findings indicate that the studies include disaster-type-specific and in-depth analyses (Figure 2). The emergence of thematic clusters in the analyzed reports clearly demonstrates that scientific studies are focused on specific types of disasters and that a separate body of knowledge has been developed for each disaster category. In this context, the prominence of keywords such as "Disaster management," "Disaster preparedness," and "Risk reduction" suggests that the research largely emphasizes solution-oriented approaches to disaster response and mitigation.

Figure 2. Risk assessment and related terms

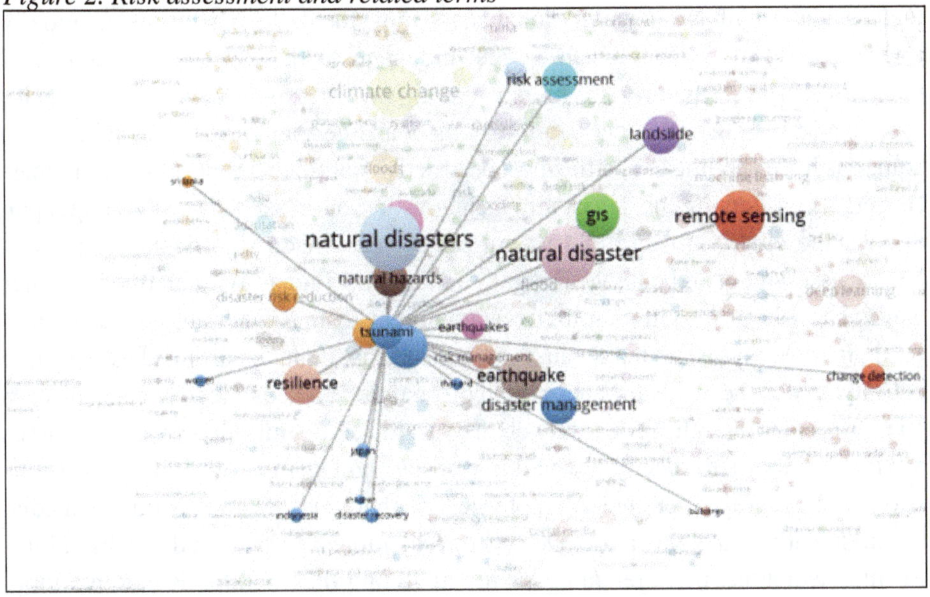

The frequent emphasis on the concept of “Risk assessment” highlights once again the importance of studies focused on identifying and reducing risks. Relevant studies (Shi et al., 2013) have indicated that topics such as evaluating the likelihood of disasters, minimizing potential damages, and developing disaster management strategies hold a significant place in the scientific literature. It is also evident that modern approaches and advanced technologies are increasingly being used in the analysis of disasters. In this context, artificial intelligence-based technologies such as “Deep learning” and “Artificial neural networks” have come to the forefront in disaster research. These technologies enable more accurate predictions of disasters and facilitate rapid analysis of their impacts by processing large datasets. In parallel, the keywords “Change detection” and “Damage assessment” emphasize the critical role of remote sensing technologies in post-disaster evaluation studies. These concepts contribute significantly to post-disaster response and recovery processes by enabling the rapid and effective identification of physical and environmental changes caused by natural disasters (Peña-Mora et al., 2008).

The Anatomy of a Tsunami: The Catastrophe Behind the Rising Wave

Tsunamis are generally clusters of water waves occurring in oceans or large bodies of water. A tsunami is generated as a result of the sudden displacement of a water column due to various natural and triggering factors such as earthquakes, underwater or subaerial gravitational mass movements, and volcanism. Initially referred to as Oshio (meaning "high tide" in Japanese), the phenomenon later came to be known as Tsunami (tsu: harbor, nami: wave) (Röbke & Vätt, 2017). With urbanization, industrial concentration along coastlines, and increasing population, tsunamis have become one of the most destructive natural disasters of modern times (Figure 3). Following the devastating tsunamis in Southeast Asia (2004) and Japan (2011), the emerging field of tsunami science has made significant progress in hazard assessment and damage mitigation planning. Tsunamis are defined as water waves capable of crossing entire oceans without losing much energy, depending on their wave type. The main causes of such events include earthquakes, volcanic activity, atmospheric disturbances (meteotsunamis), and cosmic impacts (Röbke & Vätt, 2017). Although it is a relatively new field, "Tsunami science" encompasses a wide range of sub-disciplines within disaster management, including earth sciences, oceanography, engineering, physics, mathematics, as well as policy, media, communication, and education. Despite considerable efforts from these disciplines to study tsunamis, there remain gaps in integrating these subfields within a more interdisciplinary framework (Röbke & Vätt, 2017).

The core research areas of tsunami science involve numerical and experimental modeling of tsunami generation, propagation, and inundation processes; field studies investigating the geomorphological impacts of tsunamis; and post-tsunami mass movement assessments. Examples include monitoring tsunami waves along ocean coasts through tools such as Deep-ocean Assessment and Reporting of Tsunamis (DART) buoys, satellites, and other instruments. In recent years, significant progress has been made in this field, and although current systems remain insufficient to prevent tsunami impacts, they have provided valuable early warning capabilities. One notable example is the German-Indonesian Tsunami Early Warning System (GITEWS), established after the 2004 Indian Ocean tsunami, which utilizes several seismometers in combination with GPS and measurement stations (Röbke & Vätt, 2017).

Figure 3. Tsunami and its effects (URL-1-4, respectively)

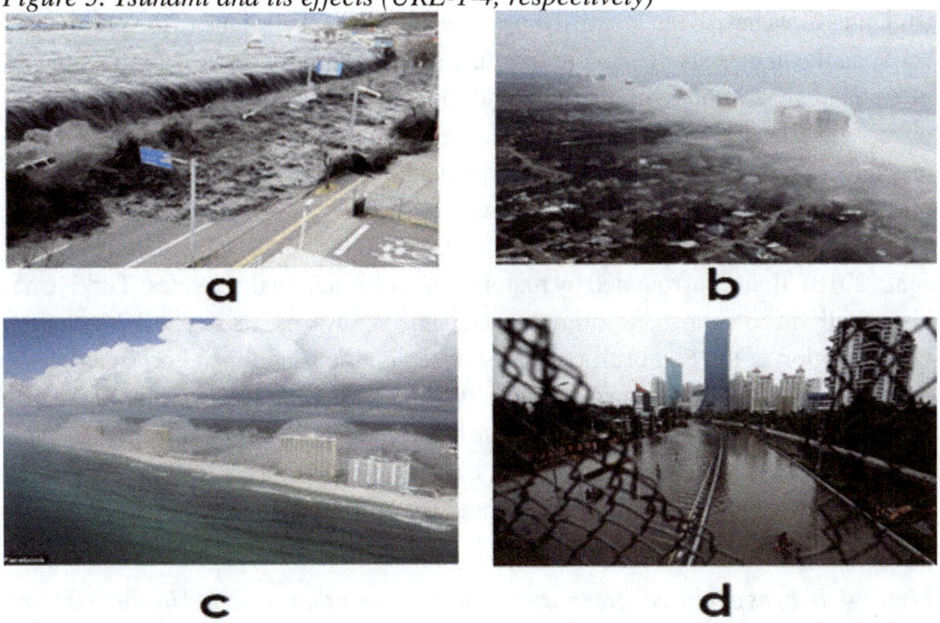

In the earthquake and subsequent tsunami disasters that occurred in Japan in 2011, it was reported that more than 13,000 people lost their lives and 14,600 people went missing. This event was the most devastating natural disaster to affect Japan since the 1923 earthquake and, with an estimated cost of 309 billion USD, became the most expensive disaster in world history (URL-5). The 2004 tsunami disaster that struck Sumatra Island, Indonesia, serves as one of the most significant examples of the potential scale of destruction caused by such events (Figure 4a). The tsunami, triggered by a magnitude 9.1 earthquake in December 2004, claimed the lives of 283,100 people. This disaster was recognized as the deadliest earthquake of the century up to that date. The tsunami waves exceeded 10 meters in some regions, and according to the NOAA Tsunami Catalogue (National Oceanic & Atmospheric Administration), the maximum recorded run-up height reached 33 meters (Güler et al., 2018).

The History of Tsunamis in Türkiye

An examination of Türkiye's tsunami history reveals that numerous tsunamis have occurred as a result of seismic activity, submarine landslides, and volcanism. A considerable number of tsunami events have been recorded in the Marmara Sea and Eastern Mediterranean regions. Over the past 2,000 years, 35 tsunamis have been documented in the Marmara Sea, while in the Eastern Mediterranean, 120 tsunami events have been identified over the last 3,500 years. Due to the presence of active faults in these areas, potential future seismic activity poses a significant risk of tsunami generation, with the potential for severe material and human losses. As observed in the 2011 Japan Earthquake, the possibility of an earthquake exceeding predicted magnitudes is scientifically plausible. Based on this information, it is of great importance to implement preventive measures against potential tsunamis along Türkiye's coasts and to evaluate the vulnerability of coastal structures (Güler et al., 2018). Being surrounded by major bodies of water on three sides, Türkiye has historically been exposed to numerous seismic sea waves. From the 17th century BCE eruption of the Santorini volcano-located on the island then known as Thera, north of Crete in the Aegean Sea-to the wave surges recorded in southern Aegean waters in 1956, many tsunami events have been observed. Historical records provide extensive information about these events, as noted in studies by Ambraseys (1962), Altınok and Ersoy (2000), and Papadopoulos (2000) (Yalçıner et al., 2005).

Figure 4. Images captured after major tsunami events that occurred in various parts of the world: a) The 2004 Sumatra earthquake and tsunami (URL-6), b) The 2011 tsunami in Japan (URL-7), c) The 2011 tsunami and the town of Yamada (URL-8), d) A boat stranded on top of a building following the 2011 Japan tsunami (URL-9).

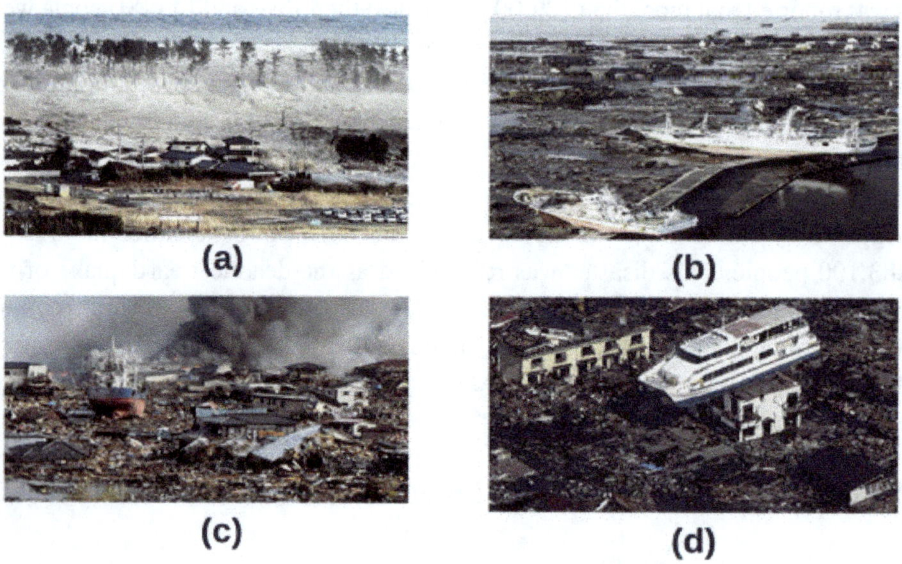

(a) (b) (c) (d)

The Final Scene of Disasters: How Societies Are Affected — Resilience and Recovery

When assessing social and regional factors, it becomes evident that concepts such as "Resilience" and "Recovery" come to the forefront. Research has increasingly focused on the social impacts of disasters and how societies recover in their aftermath (İnternet Kaynaklar Islam & Walkerden, 2014). These studies aim to understand how disasters affect social structures, the processes through which communities recover, and the challenges encountered during these periods. However, the consequences of disasters are multidimensional and cannot always be analyzed from every perspective. As emphasized by Arcaya et al. (2020), while disasters often follow recognizable patterns, it is impossible to predict precisely which individuals will be affected and when. Disasters can be seen as external shocks that expose survivors to trauma, difficulties in adapting to new environments, and significant changes in the sources of risk that are meaningful to them. Disasters have the power to negatively impact individuals' physical and psychological health. Psychological issues such as post-traumatic stress disorder (PTSD), depression, and anxiety are commonly observed among survivors (Arcaya et al., 2020). In this context, the level of social support individuals receive is a crucial factor in coping with PTSD. Research findings indicate that individuals with strong social support networks tend to achieve better psychological outcomes (Bakic & Ajdukovic, 2021). Moreover, since societies are the collective reflection of individuals, understanding how communities respond collectively and when the stressor is mitigated are also important research topics (Özkan, 2024).

According to research reports, disasters are extraordinary and unpredictable events that can disrupt the social fabric of communities. The nature of post-disaster interventions and the reestablishment of critical services mark the beginning of a long recovery journey once public attention fades. A well-established fact is that some communities are more severely affected than others, and their recovery takes longer. Researchers from the National Center for Disaster Preparedness (NCDP) have intensely investigated what makes a community resilient in the face of disaster-why some individuals and places recover faster than others-a question that remains complex and multidisciplinary (URL-10). Inequalities among communities can significantly influence recovery processes after disasters. Factors such as socio-economic status, ethnicity, and age groups affect the capacity of communities to recover (National Research Council, 2006). The impacts of disasters extend beyond individuals and societies, inflicting significant damage at the national level as well. As many research reports highlight, disasters negatively affect local economies. For instance, small businesses and local institutions often cannot absorb the financial losses following a disaster, making economic recovery difficult for communities.

Furthermore, disasters damage local infrastructure and essential services, thereby affecting both economic activity and the overall quality of life (Lindell, 2013).

Post-Disaster Recovery Processes of Societies

Post-disaster recovery processes are examined under three main categories: short-term, medium-term, and long-term recovery.

Short-Term: Emergency Response and Rescue

In the short term following a disaster, communities generally focus on emergency response activities, including search and rescue operations, temporary shelter, and medical assistance for victims. Emergency response encompasses a series of actions aimed at protecting lives and property and managing the social disruption caused by disasters. Examples include clearing debris from channels to prevent flood threats, controlling fires or hazardous material leaks triggered by earthquakes before they cause additional damage, and implementing population protection measures such as warnings, evacuations, rescue operations, emergency medical care, and shelter provision. Raising awareness and providing pre-disaster first-response training are essential responsibilities for individuals, institutions, and governments alike. Among all stages of disaster and risk management, initial emergency response is the most crucial, as it directly affects the extent of loss and the success of subsequent recovery efforts (National Research Council, 2006).

Medium- and Long-Term: Physical and Social Reconstruction

During the medium and long term, communities concentrate on rebuilding physical infrastructure and revitalizing economic activities. This phase involves repairing damaged buildings, reconstructing roads and utilities, and supporting the recovery of local businesses to ensure the continuity of infrastructure, superstructure, and economic systems (Smith, 2014). Strengthening social capital and rebuilding social support networks are also vital in the recovery process. Efforts may include enhancing community leadership, education, and health services to promote collective resilience and stability (National Research Council, 2006).

Long-Term: Resilience and Sustainability

In the long term, recovery focuses on building resilience and ensuring sustainability. This includes developing and implementing risk reduction strategies to create more disaster-resistant societies (Mayer, 2019; URL-11). Effective recovery requires collaboration among local governments, civil society organizations, and community members, ensuring that resources are distributed efficiently and community needs are adequately met. Ultimately, the social impacts of disasters and the recovery processes of affected communities are complex and multifaceted, encompassing psychological and social dimensions, economic consequences, and long-term adaptation strategies at both individual and societal levels.

Do the Regions We Live in Shape Our Lives? The Effects of Geography

Regional examples such as Japan, Indonesia, and Thailand highlight that these are areas with a high risk of natural disasters. Among these, Japan-an island nation located in the Pacific Ocean and consisting of 6,852 islands-covers an area of 377,864 km^2. The country is situated in the Pacific Ring of Fire, a volcanic zone where about 75% of the world's volcanic activity and 90% of global earthquakes occur. Natural disasters in Japan have profoundly affected both economic development and social life. The most frequent disasters Japan experiences include earthquakes, tsunamis, typhoons, and volcanic eruptions (Biricik, 2001). Over the last century, Japan has been struck by ten major natural disasters and has experienced about 20% of the world's most severe earthquakes. The most notable examples include the 1995 Kobe earthquake and the 2011 Sendai earthquake. The Hanshin–Awaji Earthquake, which occurred in January 1995 with a magnitude of 6.9, struck the city center of Kobe, Japan's sixth-largest city and the capital of the Hyogo Prefecture. The disaster resulted in 5,493 deaths, 41,521 injuries, and 600,000 displaced residents, with an estimated cost of 200 billion USD.

Sendai, with a population of over one million and the capital of Miyagi Prefecture in the Tohoku region, was hit by a 9.0-magnitude earthquake on March 12, 2011, known as the Great East Japan Earthquake. It was recorded as the third strongest earthquake in history. About 30 minutes after the quake, a tsunami reaching up to 30 meters high formed, pushing seawater as far as 10 kilometers inland. The Fukushima Daiichi Nuclear Power Plant was also severely affected, resulting in a radioactive leak that contaminated water sources, food supplies, and the Pacific Ocean. The International Atomic Energy Agency (IAEA) rated this incident as Level 7 on the International Nuclear Event Scale-the highest level, denoting a major radioactive release with widespread health and environmental effects. The total estimated economic cost of the disaster was 360 billion USD. In Japan, where tremors occur every five minutes on average and around 2,000 are felt each year, awareness of disaster risks has led to extensive preventive measures. The country has developed advanced infrastructure, established early warning systems across the Pacific and Indian Oceans, and led regional cooperation efforts. Following the 2011 disaster,

Japan adhered strictly to its disaster management plans, successfully clearing all debris and rebuilding infrastructure within two years (Yalçin, 2020).

Figure 5. The images showing Japanese cities before and after the tsunami (URL-12): a) Sendai Airport, before the tsunami b) Sendai Airport, after the tsunami c) Arahama area (Sendai), before the tsunami d) Arahama area (Sendai), after the tsunami

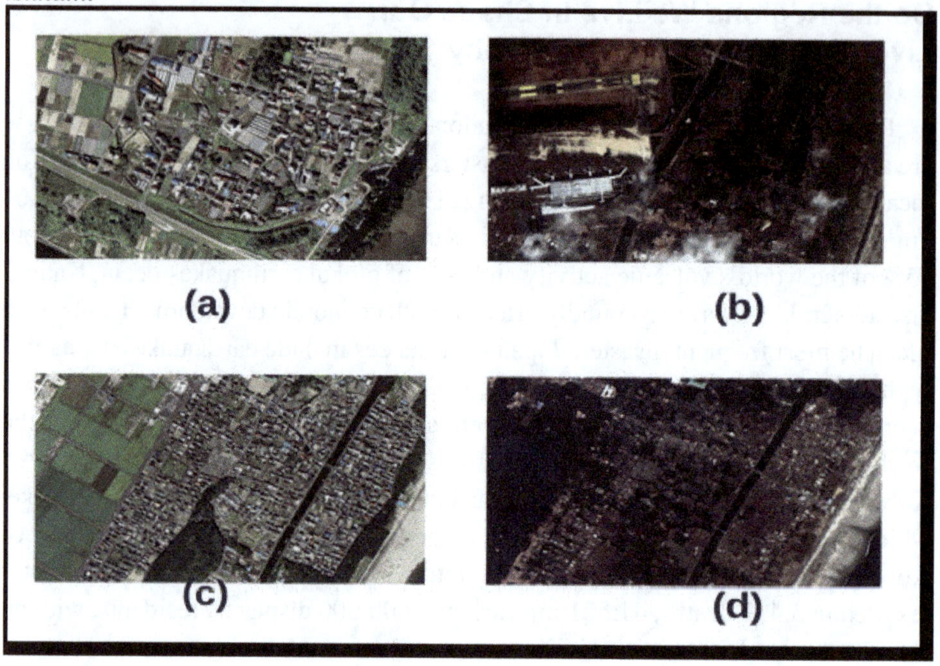

Indonesia: The Epicenter of Natural Disasters in the Pacific Ring of Fire

Located along the Equatorial Belt, Indonesia lies within the Pacific Ring of Fire, one of the most active earthquake and volcanic zones in the world. As a result, Indonesia is among the countries most frequently affected by natural disasters globally. Between 2009 and 2019, various natural disasters-including floods, landslides, earthquakes, tsunamis, and volcanic eruptions-caused the deaths of 12,449 people in Indonesia (Table 1). Over the past decade, the country has experienced 21,000 natural disasters, resulting in 75,000 injuries and affecting approximately 35 million people. Indonesia records an average of 5,000 earthquakes annually, hundreds of which have magnitudes above 5.0, making it highly vulnerable to natural hazards. In

2018, a 7.5-magnitude earthquake and subsequent tsunami struck Palu Bay, killing 4,340 people. The same year, a tsunami triggered by volcanic activity from Anak Krakatau in the Sunda Strait claimed 437 lives. Between October and April, intense rainfall leads to frequent floods (Figure 3d) and landslides. During this period, 7,378 floods and 4,898 landslides occurred, resulting in the deaths of 4,117 people. In 2010 alone, flood and landslide disasters claimed 886 lives. Indonesia hosts 130 active volcanoes, of which 104 erupted during the decade, causing 432 deaths. The 2010 eruption of Mount Merapi alone resulted in 347 fatalities. Due to natural disasters, Indonesia suffers over 20 billion USD in direct annual economic losses, and in the past 10 years, approximately 1.5 million buildings have been damaged. In response, the Indonesian government has prioritized disaster risk awareness and education, planning to include disaster preparedness courses in school curricula and strengthening risk reduction programs as part of its 2020–2024 disaster policy framework. However, many Indonesians have criticized the government for the insufficient early warning systems and inadequate emergency response infrastructure (Anadolu Agency, 2020).

Table 1. Total Number of Disasters in Indonesia Between 2009–2019 (URL-13)

Date	Type of disasters	Number of disasters	Losses	Number of injured	The population affected	The structures affected	Losses (billion $)
2009/ 2019	Earthquake-Tsunami, Flood-Landslide, and Volcanic Eruption	21.000	12.449	75.000	35 million	1.5 million	20

Thailand: Current Disaster Sensitivity and 2024 Impacts

Thailand remains highly vulnerable to natural disasters, and the events of 2024 illustrate the country's increasing exposure to climatic and meteorological hazards. Since the beginning of the year, extreme heatwaves have resulted in 61 fatalities, while intense rainfall and Typhoon Yagi triggered widespread floods and landslides, causing 43 additional deaths. Furthermore, persistent flooding across 11 southern provinces has affected approximately 762,000 households (Table 2). The 2024 disasters have caused both human and economic losses, prompting extensive government and emergency responses to drain floodwaters and rescue affected communities. However, despite these efforts, experts emphasize the urgent need for enhanced preventive measures and stronger disaster preparedness systems. Thailand's experience in 2024 demonstrates that effective risk management, early warning systems, and infrastructure resilience are essential to reduce future disaster impacts and safeguard communities across the country.

Table 2. Total Number of Disasters in Thailand in 2024 (URL-14)

Date	Disaster types	Losses	Number of affected states	Number of household affected
2024	Extreme temperature	61	11	762.000
	Heavy Rainfall-Typhoons, Floods, and Landslides	43		

Thailand: A Country Highly Vulnerable to Natural Disasters

Thailand is a country highly sensitive to natural disasters, and in 2024, several catastrophic events were recorded, resulting in significant loss of life and property. Since the beginning of 2024, 61 people have died due to extreme heat, while severe rainfall and Typhoon Yagi caused floods and landslides across the country, leading to an additional 43 fatalities. The ongoing flood disaster in the southern provinces has affected approximately 762,000 households (Table 2). The natural disasters that struck Thailand in 2024 led to both human casualties and economic damage. Although the government and disaster response teams worked to drain floodwaters and rescue affected residents, experts emphasize the urgent need for greater preventive measures and stronger disaster preparedness strategies to mitigate future risks.

Spatial Distribution of Publications: Is There a Relationship Between the Number of Studies and Disasters?

The World Risk Index (WRI) evaluates global risk levels and their components across different classifications-very low, low, medium, high, and very high. The index considers multiple dimensions of disaster risk, including exposure, vulnerability, susceptibility, lack of coping capacities, and lack of adaptive capacities, as shown in Table 3. Each category reflects a country's level of risk to natural disasters and crises. The index clearly demonstrates that risk does not arise from a single factor, but rather from the interaction of multiple variables-specifically, exposure, vulnerability, sensitivity, and the inadequacy of coping and adaptive capacities. This multidimensional and integrated approach allows for a comprehensive assessment of risks, providing valuable insights for both researchers and policy or decision-makers. It underscores the importance of accurately analyzing risks and developing context-specific strategies tailored to each country's needs. In this sense, the number and spatial distribution of scientific publications often correlate with regions categorized as high-risk or very high-risk, indicating that research activity tends to increase in areas more prone to natural disasters.

Figure 6. World Risk Index and Its Components: Exposure, Vulnerability, and Capacity Deficiencies

Risk Classification Thresholds

Classification	WorldRiskIndex	Exposure	Vulnerability	Susceptibility	Lack of Coping Capacities	Lack of Adaptive Capacities
Very low	0.00 – 1.84	0.00 – 0.17	0.00 – 9.90	0.00 – 7.17	0.00 – 3.47	0.00 – 25.28
Low	1.85 – 3.20	0.18 – 0.56	9.91 – 15.87	7.18 – 11.85	3.48 – 10.01	25.29 – 37.47
Medium	3.21 – 5.87	0.57 – 1.76	15.88 – 24.43	11.86 – 19.31	10.02 – 12.64	37.48 – 48.04
High	5.88 – 12.88	1.77 – 7.78	24.44 – 33.01	19.32 – 34.16	12.65 – 39.05	48.05 – 59.00
Very high	12.89 – 100.00	7.79 – 100.00	33.02 – 100.00	34.17 – 100.00	39.06 – 100.00	59.01 – 100.00

Global Risk Profiles According to the World Risk Index: India, Indonesia, China, Turkey, and Comparative Insights

Among the countries evaluated, India and Indonesia rank among those with the highest global disaster risk levels, according to the World Risk Index (WRI). Both countries' elevated risk levels are closely linked to their geographical exposure, social vulnerability, and limited coping and adaptive capacities.

India (Hindistan): High Exposure and Weak Coping Capacity

India stands out as one of the world's most disaster-prone countries, with a World Risk Index value of 42.31, placing it in the very high-risk category. I) Exposure: 35.99; II) Vulnerability: 49.75; III) Susceptibility: 39.50, and IV) Lack of Coping Capacity: 55.38. These values demonstrate that India's high risk is primarily due to its extensive exposure to natural hazards-such as floods, cyclones, and earthquakes-combined with social and economic fragility. The high vulnerability and weak coping capacity indicate that post-disaster interventions are not sufficiently effective, largely due to infrastructure limitations, inadequate coordination, and resource constraints. Although the available data do not allow for a detailed analysis of all causal factors, it is evident that deficiencies in emergency response and resilience infrastructure significantly contribute to India's overall risk.

Indonesia (Endonezya): Geographical Exposure and Adaptive Deficiencies

Indonesia, another country within the Pacific Ring of Fire, exhibits a World Risk Index value of 41.46, also placing it among the highest-risk nations. I) Exposure: 39.89 and II) Lack of Adaptive Capacity: 47.19 The high exposure level clearly reflects Indonesia's geological and climatic conditions, including its vulnerability to earthquakes, tsunamis, volcanic eruptions, and floods. However, the high lack of adaptive capacity reveals the urgent need to improve long-term resilience measures, such as better urban planning, stronger infrastructure, and community-based disaster preparedness programs.

China (Çin): High Exposure but Strong Resilience Mechanisms

China has a World Risk Index value of 28.70, placing it near the high-risk group, but it demonstrates stronger coping and adaptive mechanisms compared to India and Indonesia. I) Exposure: 64.59; II) Lack of Coping Capacity: 10.84, and III) Lack of Adaptive Capacity: 11.84. While China's exposure value is very high-reflecting its vast and diverse geography prone to earthquakes, floods, and typhoons-its relatively low vulnerability can be attributed to robust economic capacity, technological development, and well-established disaster management systems (Zhou et al., 2015). Low coping and adaptive capacity deficiency values indicate that China has a strong institutional response system and long-term adaptation strategies, enabling rapid intervention and effective mitigation in the face of disasters.

Moderate- and Low-Risk Countries: Turkey, Iran, and Western Nations

Countries such as the United States, Germany, and Canada fall within the low-risk category, exhibiting high adaptive capacity and effective risk reduction strategies. Their strong governance, advanced infrastructure, and public awareness contribute to their resilience against natural disasters. In contrast, Turkey and Iran represent countries in the medium-risk group, where exposure and vulnerability remain significant concerns. I) Turkey's World Risk Index: 16.23; II) Vulnerability: 29.85, and III) Lack of Adaptive Capacity: 35.85. These figures suggest that, while Turkey's overall risk is moderate, social and economic vulnerabilities heighten disaster sensitivity. The relatively high lack of adaptive capacity highlights the necessity for long-term resilience-building strategies-including investment in infrastructure, education, and institutional preparedness. Similar to Turkey, Iran demonstrates a comparable pattern, facing challenges in both coping and adaptation. Overall, the World Risk Index underscores the fact that high exposure alone does not determine total disaster risk-the ability to cope and adapt plays an equally critical role. We can deduce that: I) India and Indonesia: High exposure and weak institutional capacity elevate overall risk. II) China: High exposure but low vulnerability, supported by effective governance and infrastructure. III) Turkey and Iran: Moderate risk, with clear potential for resilience improvement.IV) Western nations: Low risk due to strong social, economic, and technological capacity.

Overall, the data reveal that resilience depends not only on geography but also on governance, preparedness, and socioeconomic development—factors that determine how societies confront and recover from disasters (Table 4).

Figure 7. World Risk Index and Its Components by Country (Exposure, Vulnerability, and Capacity Deficiencies)

Rank	Country	WorldRiskIndex	Exposure	Vulnerability	Susceptibility	Lack of Coping Capacities	Lack of Adaptive Capacities
1.	Philippines	46.82	39.99	54.81	51.35	57.81	55.48
2.	India	42.31	35.99	49.75	39.50	55.38	56.29
3.	Indonesia	41.46	39.89	43.10	33.48	50.67	47.19
4.	Colombia	38.37	31.54	46.69	47.84	48.23	44.11
5.	Mexico	37.55	50.08	28.16	37.26	12.09	49.55
6.	Myanmar	35.49	22.43	56.14	53.39	58.85	56.30
7.	Mozambique	34.37	18.10	65.28	64.57	64.54	66.76
8.	China	28.70	64.59	12.75	15.78	12.11	10.84
9.	Bangladesh	27.90	16.57	46.97	36.81	59.18	47.58
10.	Pakistan	26.75	13.11	54.58	41.42	60.96	64.41
11.	Russian Federation	26.54	28.35	24.85	11.22	39.19	34.91
12.	Viet Nam	25.85	26.73	25.00	26.54	12.98	45.38
13.	Peru	25.41	16.65	38.79	26.38	48.09	46.00
14.	Somalia	25.07	8.55	73.49	65.01	79.09	77.20
15.	Yemen	24.26	9.12	64.52	60.66	68.05	65.06
16.	Papua New Guinea	24.10	18.84	30.82	30.86	14.12	67.21
17.	Madagascar	23.48	18.38	29.99	25.36	15.49	68.69
18.	United States of America	22.73	39.59	13.05	11.60	6.15	31.16
19.	Bolivarian Republic of Venezuela	22.45	19.52	25.82	22.87	12.30	61.16
20.	Ecuador	22.42	14.57	34.51	20.20	46.01	44.21
21.	Nicaragua	22.35	18.71	26.71	28.28	14.02	48.06
22.	Australia	21.36	31.21	14.62	8.48	13.66	26.96
23.	Thailand	20.91	14.32	30.53	15.87	48.68	36.83
24.	Egypt	20.65	10.74	39.71	27.10	47.91	48.24
25.	Canada	18.99	25.89	13.93	10.38	7.72	33.74
26.	Iran (Islamic Republic of)	18.48	12.49	27.34	20.73	56.65	17.41
27.	Panama	18.38	15.89	21.27	20.64	10.68	43.63
28.	Japan	17.03	43.67	6.64	7.65	5.15	7.44
29.	United Republic of Tanzania	16.38	5.49	48.85	34.12	55.69	61.35
30.	Turkey	16.23	8.90	29.58	15.73	45.85	35.89
31.	Honduras	16.00	8.82	29.02	29.30	14.43	57.81
32.	Argentina	15.61	11.54	21.12	17.92	10.84	48.49
33.	Solomon Islands	14.62	9.62	22.23	17.95	11.25	54.43
34.	El Salvador	14.37	7.30	28.27	39.20	12.30	46.85
35.	Malaysia	14.36	8.64	23.86	19.10	19.81	35.88
36.	Libyan Arab Jamahiriya	14.31	4.94	41.47	24.51	54.53	53.35
37.	Costa Rica	14.20	9.89	20.39	22.57	10.82	34.72
38.	Kenya	13.92	3.27	59.27	59.30	58.50	60.03
39.	Chile	13.84	12.86	14.89	10.05	8.29	39.61

Exposure

Sphere of exposure to earthquakes, tsunamis, coastal flooding, riverine flooding, cyclone, droughts and sea level rise

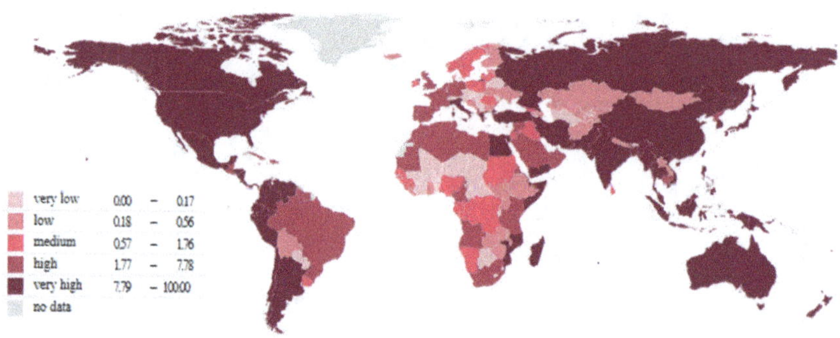

Distribution of Publications by Country and Their Scientific Impact

The country-wise distribution of publications within the scope of the study is presented in Table 5. The table also includes data on citation numbers and total link strength, which reflects the level of international academic collaboration among countries.

Publication Output by Country

The analysis reveals that the countries with the highest number of publications in the field are China, the United States (USA), and India. China stands out as the leading country, with 2,341 published articles, demonstrating its high research productivity and dominant position in scientific contributions related to disaster and risk studies. The United States ranks second, with 1,406 publications, maintaining a strong presence and providing significant contributions to the global research landscape. India follows as another major contributor, reflecting growing scientific engagement in disaster-related research across developing regions.

Citation Analysis

Based on the citation data, the United States holds a clear lead, with a total of 52,448 citations, highlighting the influence and impact of its research on the global scientific community. China ranks second, with 30,388 citations, underscoring its rapidly expanding academic visibility and contribution to the field. Notably, Switzerland, despite having fewer publications, exhibits a high citation count of 12,306, indicating that its research outputs are of high quality and global relevance.

Academic Collaboration (Total Link Strength)

The analysis of academic collaboration networks-measured by total link strength (TLS)-shows that the United States is at the center of the global collaboration network, with a TLS value of 576, representing the widest and strongest international partnerships in disaster-related research. China follows with a total link strength of 398, reflecting a broad and active scientific cooperation network with other countries. The United Kingdom ranks third, with a total link strength of 283, maintaining an influential role in cross-national scientific collaboration. Overall, in summary, China, the United States, and India are the leading nations in publication output, while the United States dominates in both citation impact and international collaboration strength. The case of Switzerland exemplifies how quality and influence can outweigh sheer publication volume. Together, these findings highlight the geographical concentration of scientific productivity and the importance of global research networks in advancing disaster risk and resilience studies.

Table 5. Number of Publications, Citation Counts, and Academic Collaboration Strength by Country

Country	Documents	Citation	Total link strength
China	2341	30388	398
United States	1406	52448	576
India	646	7184	160
Japan	505	8065	207
Indonesia	492	2688	99
United Kingdom	408	13028	283
Italy	356	11885	211
Germany	352	10535	241
Australia	313	10835	187
Canada	233	9634	150
Brazil	216	2459	47
Iran	203	5833	140
France	185	3045	148
Turkey	181	2521	62
Russian Federation	179	2133	55
South Korea	177	2046	92
Malaysia	160	2358	87
Taiwan	155	2903	65
Spain	141	3421	116
Netherlands	134	5984	134
Bangladesh	120	2589	98
Switzerland	101	12306	86

Academic Collaboration and Global Research Networks Among Countries

The international academic collaboration and scientific research networks are visualized in Figure 7. The figure clearly illustrates the density and frequency of scientific collaborations between countries through the interconnected links within the global research network.

Global Collaboration Hubs

According to the analysis results, the United States (USA) and China occupy the central positions in the global scientific collaboration network and are the countries with the highest number of connections. This finding highlights their leadership roles in global research and demonstrates their extensive international cooperation networks across multiple disciplines, particularly in disaster and risk-related studies. The central positioning of these two countries indicates not only their high publication and citation performance but also their ability to build and maintain strong transnational research partnerships, which significantly influence global scientific directions.

Regional Collaboration Dynamics

Other countries, such as India and Japan, also form important nodes within the network, establishing active collaboration clusters in the Asian region. These nations contribute significantly to both regional and global research efforts, engaging in joint publications and cross-institutional projects. Their collaborative engagement supports knowledge exchange, enhances scientific visibility, and strengthens Asia's role in the global research ecosystem.

Turkey's Position in the Global Network

The analysis shows that Turkey produced 181 scientific publications and received 2,521 citations, indicating a growing level of academic productivity and scientific contribution. Turkey's total link strength (TLS) is 120, reflecting its active participation in international research collaborations. However, the relatively moderate TLS value suggests that while Turkey is involved in global partnerships, there remains potential for expanding and intensifying these collaborations to move closer to the core of the international scientific network. Overall, the network visualization (Figure 6) underscores a hierarchical structure of global academic collaboration, where the USA and China lead as primary hubs, followed by India, Japan, and several European countries that act as regional bridges. Turkey, though actively engaged, would benefit from further strengthening its bilateral and multilateral research ties, particularly through participation in international projects, co-authorships, and academic consortia. Such strategic expansion would enhance Turkey's visibility, influence, and centrality within the global scientific community.

Figure 8. Co-occurrences of the countries

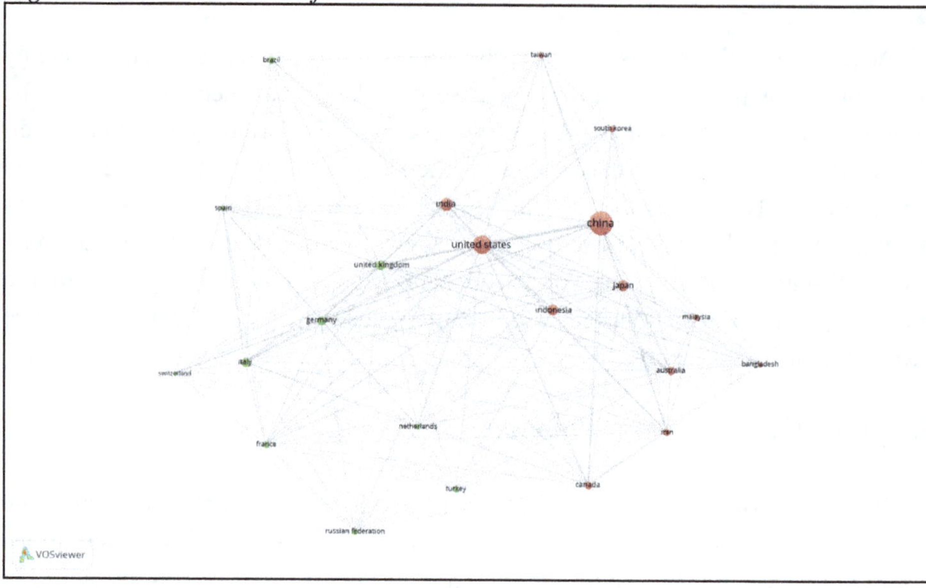

Integrated Evaluation: Relationship Between Publication Distribution, Exposure, and Risk Levels

When all data are evaluated together, it becomes evident that the spatial distribution of scientific publications is directly related to disaster exposure and risk levels. For instance, in high-risk countries such as India and Indonesia, the high concentration of publications may reflect greater scientific awareness and a stronger perception of disaster management importance. The frequent occurrence of disasters in these regions likely motivates researchers and institutions to focus more on hazard analysis, preparedness, and mitigation strategies. Conversely, low-risk countries such as the United States and Germany exhibit both high publication and citation counts, despite being less exposed to natural hazards. This indicates that these countries not only develop advanced disaster management systems for their own contexts but also serve as global models, contributing significantly to the scientific and methodological development of disaster risk reduction (DRR) worldwide. Countries in the medium-risk category, such as Turkey and Iran, demonstrate increasing research engagement, yet their data emphasize the need to strengthen preparedness and resilience capacities. Enhancing infrastructure, coordination, and adaptive capabilities in these nations would not only mitigate future risks but also support their integration into global scientific and disaster management networks.

Overall, the relationship between research productivity and disaster risk highlights how scientific attention aligns with exposure and vulnerability, shaping both policy development and international cooperation in disaster resilience.

Disaster Science in the Context of Climate Change Adaptation and Socio-Economic Resilience

Recent advances in disaster research increasingly emphasize that natural hazards cannot be evaluated independently of climate change processes. According to the IPCC risk framework, disaster risk emerges from the interaction of hazard, exposure, and vulnerability, all of which are intensified under changing climatic conditions. Rising temperatures, sea-level rise, extreme precipitation events, and prolonged droughts amplify both the frequency and severity of disasters, thereby transforming many hazards from isolated events into systemic risks that threaten long-term development.

In this context, disaster science has gradually shifted from a reactive, response-oriented approach toward a proactive framework centred on climate adaptation, risk reduction, and resilience building. International policy instruments such as the Sendai Framework for Disaster Risk Reduction (2015-2030) explicitly promote integrating early warning systems, risk assessment, and sustainable planning into national development strategies. Similarly, the Sustainable Development Goals-particularly SDGs 1 (No Poverty), 11 (Sustainable Cities and Communities), and 13 (Climate Action)-highlight the necessity of reducing disaster losses through adaptive governance and community preparedness. These frameworks demonstrate that disaster management is no longer solely an emergency issue but also a core component of climate adaptation and sustainable development policy.

Beyond physical impacts, disasters generate profound socio-economic consequences. Loss of livelihoods, displacement, infrastructure collapse, and psychological stress often persist long after the initial event. Vulnerable populations-including low-income households, elderly individuals, and marginalized communities-experience disproportionate losses, thereby deepening existing social inequalities. Consequently, community resilience has emerged as a critical concept, referring to the capacity of societies to absorb shocks, reorganize, and recover while maintaining essential functions. Strengthening social capital, local knowledge, education, and institutional coordination has proven to be as important as technological solutions in reducing long-term disaster vulnerability. Comparative evidence further illustrates the importance of socio-economic capacity in shaping outcomes. Developed countries such as the United States, Germany, and Japan typically exhibit stronger institutional preparedness, advanced infrastructure, and effective early warning systems, enabling faster recovery despite high exposure. In contrast, developing countries-including

India and Indonesia-face greater challenges due to limited coping capacity, rapid urbanization, and resource constraints, resulting in higher human and economic losses. Turkey represents an intermediate case, where moderate exposure combined with evolving institutional capacity highlights both progress and the need for further resilience investment. These contrasts indicate that disaster impacts are determined not only by geographical hazard exposure but also by governance quality, economic strength, and social organization. Taken together, these perspectives reinforce the importance of linking bibliometric findings with climate adaptation policies. The growing prominence of keywords such as "risk assessment," "resilience," "recovery," and "early warning systems" observed in our network analysis reflects this global transition toward integrated, climate-responsive disaster management. Thus, disaster science should be interpreted as part of a broader resilience agenda that combines environmental monitoring, socio-economic planning, and adaptive governance to reduce future risks.

From Bibliometric Trends to Policy and Practice: Implications and Future Directions in Disaster Science

While bibliometric analyses primarily reveal patterns of scientific productivity, collaboration, and thematic evolution, their value extends beyond academic mapping. The trends identified in this study have direct implications for practical disaster management and policy design. For example, the increasing prominence of keywords such as "risk assessment," "early warning systems," "GIS," "remote sensing," and "artificial intelligence" reflects a global shift from reactive disaster response toward proactive risk reduction and preparedness strategies. These developments indicate that scientific advances are increasingly being translated into operational tools for hazard monitoring, evacuation planning, and rapid damage assessment. Thus, bibliometric evidence not only describes research activity but also signals where technological and managerial solutions are being prioritized worldwide. Linking publication dynamics to real-world challenges further highlights important policy implications. Countries with high research output-such as the United States, China, and Japan-tend to demonstrate stronger disaster governance, institutional coordination, and infrastructure resilience. In contrast, regions with limited scientific production often face greater difficulties in implementing early warning systems, data-driven planning, and recovery mechanisms. This relationship suggests that research capacity itself may function as a component of adaptive capacity. Therefore, strengthening scientific collaboration and knowledge transfer should be considered an integral part of disaster risk reduction policy, particularly for developing and moderate-risk countries.

Despite recent progress, several emerging research gaps remain. First, although artificial intelligence and machine learning approaches are increasingly visible in the literature, their operational integration into real-time disaster prediction and decision-support systems is still limited. Many studies remain experimental or small-scale, and challenges related to data quality, model generalization, and ethical considerations persist. Second, resilience financing has received comparatively less attention within disaster science. Mechanisms such as climate risk insurance, catastrophe bonds, and disaster contingency funds are critical for reducing long-term socio-economic losses, yet they remain underrepresented in bibliometric networks. Expanding interdisciplinary research that connects earth sciences with economics, public policy, and finance would therefore enhance both preparedness and recovery capacities. Looking ahead, global disaster science is expected to follow several key directions. Future research will likely prioritize (i) AI-driven predictive modelling and real-time early warning systems, (ii) integration of big data and satellite-based monitoring platforms, (iii) community-centred resilience planning and social vulnerability reduction, and (iv) innovative financing strategies to support rapid recovery and adaptation. In addition, stronger international collaboration-particularly between developed and developing countries-will be essential for equitable knowledge sharing and capacity building. Overall, the transition observed in the bibliometric landscape-from hazard documentation toward prediction, resilience, and governance-indicates that disaster research is increasingly aligned with practical and policy-oriented objectives. Interpreting these trends through a management lens allows scientific outputs to inform not only academic debates but also evidence-based decision-making, ultimately contributing to safer and more sustainable societies.

CONCLUSION

This study provides a comprehensive analysis of 9,116 scientific reports focusing on natural disasters within the field of Earth sciences. The results clearly indicate that "Natural disasters" and "Climate change" are central, interrelated themes driving global scientific inquiry. The keyword network and clustering analyses reveal that disasters such as floods, earthquakes, and tsunamis are studied as distinct research clusters, reflecting the scientific community's commitment to exploring their mechanisms, impacts, and management strategies in depth.

Technological innovations-particularly GIS, remote sensing, deep learning, and artificial neural networks-are increasingly integrated into disaster analysis, risk assessment, and post-disaster management. These tools have enhanced the accuracy of predictions, facilitated rapid damage evaluation, and strengthened early warning systems. The case studies of Japan, Indonesia, and Turkey demonstrate how high-risk regions utilize scientific knowledge and technological capacity to mitigate disaster effects.

From a global risk perspective, findings based on the World Risk Index (WRI) show that India and Indonesia are among the most vulnerable nations, primarily due to high exposure and limited coping capacities. In contrast, China, the United States, and Germany exhibit stronger resilience through well-developed infrastructure and adaptive mechanisms. Turkey and Iran, situated in the moderate-risk category, demonstrate growing awareness but require further improvement in preparedness and institutional coordination.

Bibliometric analysis revealed that China, the United States, and India lead in publication output, while the United States stands out in citation impact and global collaboration strength. The international co-authorship network (Figure 7) confirms that these countries act as global hubs of scientific cooperation, forming the backbone of the disaster research ecosystem. Turkey, with 181 publications and total link strength of 120, has an active but still peripheral role, emphasizing the potential to strengthen international collaboration and enhance visibility in global research networks. Collectively, these findings demonstrate that the spatial distribution of research intensity correlates with exposure and risk levels-countries most vulnerable to disasters also exhibit higher research activity. However, low-risk yet high-output nations serve as global models in disaster resilience, contributing scientific expertise that transcends geographic boundaries.

Recommendations

Enhancement of Early Warning and Monitoring Systems:

Countries with high exposure and weak coping capacities-such as India, Indonesia, and Iran-should prioritize investment in advanced early warning systems integrating remote sensing, AI-driven forecasting, and community-based communication mechanisms.

Integration of Multidisciplinary Approaches:

Disaster management requires the combined input of earth sciences, engineering, social sciences, and policy studies. Strengthening interdisciplinary collaboration will facilitate more holistic disaster preparedness and recovery planning.

Strengthening International Cooperation:

Expanding global research partnerships-especially for developing countries-will foster technology transfer, capacity building, and shared learning. Turkey, in particular, should enhance participation in EU Horizon, UNDRR, and Bilateral Disaster Risk Reduction Programs.

Focus on Community-Based Resilience:

Building resilience at the local level requires integrating education, training, and social support networks into national disaster strategies. Community participation ensures sustainability and social cohesion during recovery.

Data Transparency and Accessibility:

Establishing open-access platforms for real-time data on disasters, hazard mapping, and vulnerability assessments would improve decision-making and support regional cooperation.

Future Perspectives

The future of disaster research lies in digital transformation and artificial intelligence. The integration of machine learning, big data analytics, and satellite-based observation systems will revolutionize risk assessment and real-time disaster response. Moreover, climate change modelling will increasingly inform adaptation policies and sustainable development strategies. For developing and moderate-risk countries such as Turkey, future research should focus on: Creating dynamic resilience indicators that integrate socioeconomic, infrastructural, and environmental variables. Developing AI-based predictive models tailored to local geological and climatic conditions. Establishing regional research consortia to link Mediterranean, Middle Eastern, and Asian scientific efforts under a unified disaster resilience framework. In conclusion, natural disaster research is transitioning from observation to prediction, from response to resilience. The integration of technological innovation, international cooperation, and social inclusion will define the next era of sustainable disaster management and climate-resilient societies.

REFERENCES

Arcaya, M., Raker, E. J., & Waters, M. C. (2020). The socialconsequences of disasters: Individual and communitychange. *Annual Review of Sociology*, *46*(1), 671–691. DOI: 10.1146/annurev-soc-121919-054827

Bakic, H., & Ajdukovic, D. (2021). Resilienceafternaturaldisasters: The process of harnessingresources in communitiesdifferentiallyexposed to a flood. *European Journal ofPsychotraumatology*, *12*(1), 1891733. DOI: 10.1080/20008198.2021.1891733 PMID: 34992751

Benevolenza, M. A., & DeRigne, L. (2019). The impact of climate change and natural disasters on vulnerable populations: A systematic review of literature. *Journal of Human Behavior in the Social Environment*, *29*(2), 266–281. DOI: 10.1080/10911359.2018.1527739

Biricik, A. S. (2001). YERYUVARI'NDA DOĞAL OLAYLAR VE AFETLER. *Marmara Coğrafya Dergisi*, (3), 7–26.

Blaikie, P., Cannon, T., Davis, I., & Wisner, B. (2014). *At risk: natural hazards, people's vulnerability and disasters*. Routledge. DOI: 10.4324/9780203714775

Chmutina, K., Von Meding, J., Sandoval, V., Boyland, M., Forino, G., Cheek, W., Williams, D. A., Gonzalez-Muzzio, C., Tomassi, I., Páez, H., & Marchezini, V. (2021). What we measure matters: The case of the missing development data in Sendai framework for disaster risk reduction monitoring. *International Journal of Disaster Risk Science*, *12*(6), 779–789. DOI: 10.1007/s13753-021-00382-2

Cüce, H. (2023). Climate Change-related natural disasters: Environmental effects and sustainable monitoring. [IJENT]. *International Journal of Environmental Trends*, *7*(2), 113–125.

Du, Y., Li, Y., Yang, M., Wang, Y., Lian, Q., & Hong, R. (2025). Bibliometric analysis of advances in heritage resilience at the district scale. npj. *Heritage Science*, *13*(1), 606.

Ergünay, O. (2007). Türkiye'nin afet profili. *TMMOB afet sempozyumu bildiriler kitabı*, *5*(7), 1-14.

Erkal, S., Şafak, Ş., Yertutan, C., & Coşkuner, S. (2009). Bir Hastanede Fiziksel Çevre ve Yönetimi (Kurum Ev İdaresi) Hizmetlerini Yürüten Personelin İş Bölümüne İlişkin Görüşlerinin İncelenmesi. *Hacettepe Sağlık İdaresi Dergisi*, *12*(2), 157–176.

Green, H. K., Lysaght, O., Saulnier, D. D., Blanchard, K., Humphrey, A., Fakhruddin, B., & Murray, V. (2019). Challenges with disaster mortality data and measuring progress towards the implementation of the Sendai framework. *International Journal of Disaster Risk Science, 10*(4), 449–461. DOI: 10.1007/s13753-019-00237-x

Güler, H. G., Sözdinler, C. Ö., Arikawa, T., & Yalçıner, A. C. (2018). Tsunami Afeti Sonrası Yapısal ve Yapısal Olmayan Önlemler ve Farkındalık Çalışmaları: Japonya Örneği. *Teknik Dergi, 8605*(8629), 514.

Hu, H., Yan, K., Zhu, Z., & Lv, T. (2024). Building forwards: Bibliometric analysis and framework construction for community resilience. *Journal of Contingencies and Crisis Management, 32*(3), e12604. DOI: 10.1111/1468-5973.12604

İnternet Kaynaklar Islam, R., & Walkerden, G. (2014). How bonding and bridging networks contribute to disaster resilience and recovery on the Bangladeshi coast. *International Journal of Disaster Risk Reduction, 10*, 281–291. DOI: 10.1016/j.ijdrr.2014.09.016

Lindell, M. K. (2013). Recovery and reconstructionafterdisaster. In *Encyclopedia of naturalhazards* (pp. 812–824). Springer.

Mayer, B. (2019). A review of the literature on community resilience and disaster recovery. *Current Environmental Health Reports, 6*(3), 167–173. DOI: 10.1007/s40572-019-00239-3 PMID: 31267349

Morrissey, S. A., & Reser, J. P. (2007). National Research Council, Division on Earth, Life Studies, Committee on Disaster Research in the Social Sciences, FutureChallenges, &Opportunities. (2006). Facinghazards and disasters: Understandinghumandimensions. NationalAcademiesPress. Natural disasters, climate change and mental health considerations for rural Australia. *The Australian Journal of Rural Health, 15*(2), 120–125. DOI: 10.1111/j.1440-1584.2007.00865.x PMID: 17441821

Ningsih, R. D., Ariyaningsih, A., Sukhwani, V., & Shaw, R. (2022). Sendai Framework and its implementation. In *Handbook on Climate Change and Disasters* (pp. 30–41). Edward Elgar Publishing.

Özkan, G. G. (2024). Deprem döneminde belirsizliğe tahammülsüzlük düzeyinin çalışanların iş stresi üzerindeki etkisinde psikolojik iyi oluşun aracılık rolü (Master's thesis, Hasan Kalyoncu Üniversitesi Lisansüstü Eğitim Enstitüsü İşletme Ana Bilim Dalı).

Peña-Mora, F., Aziz, Z., Chen, A. Y., Plans, A., & Foltz, S. (2008). Building assessment during disaster response and recovery. *Proceedings of the Institution of Civil Engineers. Urban Design and Planning*, *161*(4), 183–195. DOI: 10.1680/udap.2008.161.4.183

Röbke, B. R., & Vött, A. (2017). The tsunami phenomenon. *Progress in Oceanography*, *159*, 296–322. DOI: 10.1016/j.pocean.2017.09.003

Sağlam, R., Uyguner, Z., & Küp, F. (2021). Şanlıurfa Bölgesinde Farklı Yıllarda Çırçırlanmış Pamuğun Lif Kalitesine Etki Eden Faktörlerin Belirlenmesi. *Journal of the Institute of Science and Technology*, *11*(1), 709–721.

Sebesvari, Z., Woelki, J., Walz, Y., Sudmeier-Rieux, K., Sandholz, S., Tol, S., Ruíz García, V., Blackwood, K., & Renaud, F. G. (2019). Opportunities for considering green infrastructure and ecosystems in the Sendai Framework Monitor. *Progress in Disaster Science*, *2*, 100021. DOI: 10.1016/j.pdisas.2019.100021

Shi, X., Tan, J., Guo, Z., & Liu, Q. (2013). A review of risk assessment of storm surge disaster. *Diqiu Kexue Jinzhan*, *28*(8), 866.

Smith, P. G. (2014). *Healthy, Resilient, and SustainableCommunities After Disasters: Strategies*. Opportunities, and Planning for Recovery.

Varol, N., & Gültekin, T. (2016). Afet antropolojisi. *Elektronik Sosyal Bilimler Dergisi, 15*(59).

Wisner, B., & Wisner, B. (2004). *At risk: natural hazards, people's vulnerability and disasters*. Psychology Press.

Yalçin, G. (2020). *Doğal afetlerin etkilerine ve afet risk yönetimine toplumsal cinsiyet perspektifinden bakış: Türk ve Japon kadınların duruş noktasından doğal afetler*. Master'sthesis, Sosyal Bilimler Enstitüsü.

Yalçıner, A. C., Kuran, U., Mınoura, K., Imamura, F., Takahashı, T., Papadopoulos, G., & Ersoy, Ş. (2005). *Türkiye Mühendislik Haberleri Sayı 438 - 2005/4.Türkiye Kıyılarında Depreşim Dalgası (Tsunami)*. İzleri.

Zhou, Y., Liu, Y., Wu, W., & Li, N. (2015). Integrated risk assessment of multi-hazards in China. *Natural Hazards*, *78*(1), 257–280. DOI: 10.1007/s11069-015-1713-y

Chapter 11

Forest Fires and the Pandemic in Mediterranean Tourism: A MEREC-CoCoSo-Based Emergency Response State Budget

Çağlar Karamaşa
http://orcid.org/0009-0001-8034-270X
Anadolu Univerity, Turkey

Osman Pala
http://orcid.org/0000-0002-2634-2653
Karamanoğlu Mehmet Bey University, Turkey

Filiz Ekici
http://orcid.org/0000-0002-4904-3381
Iğdır University, Turkey

Melahat Batu Ağırkaya
http://orcid.org/0000-0002-8703-5622
Iğdır University, Turkey

Öner Gümüş
http://orcid.org/0000-0002-3231-3293
Kutahya Dumlupinar University, Turkey

ABSTRACT

Using the MEREC-CoCoSo method, this study evaluates the impact of pandemics and forest fires on the tourism sector in Greece, France, Spain, Turkey, and Italy,

DOI: 10.4018/979-8-3373-6274-8.ch011

proposing a Mediterranean Tourism Emergency Response Budget. It addresses the potential benefits of such a fund in preserving the tourism industry during disasters, particularly examining the impact of the 2020 pandemic and forest fires as well as explores whether pre-emptive creation of such a fund could allow these nations to safeguard their tourism industry and the extent of potential benefits. The study emphasizes the need for proportional financing based on the MEREC-CoCoSo method to ensure fairness and cooperation among participating countries, offering a potential model for other regions. The findings suggest that if implemented, the fund could notably benefit France, fostering regional cooperation and setting a precedent for sustainable tourism practices.

INTRODUCTION

Motivating Factors Behind the Research

The common preference of governments is to ensure a high standard of living for their citizens through stable economic growth and sustainability (Demir and Bahar, 2021). The volume of gross domestic product (GDP), used in measuring the standard of living, refers to the sum of the added value of households, the government, and other units operating in the economy measured at fixed prices. The contribution of sectors to economic growth is measured by the increase in added value in the current year compared to the specified base year (World Bank, 2024). When measuring the sectoral sizes, four main sectors are mainly addressed as being agriculture, industrial, production, and service sectors, according to the Organization for Economic Co-operation and Development (OECD, 2024). According to the data for 2021, the sector volumes were determined to be 4.16% for the agriculture sector, 19.20% for the industrial sector, including construction, 11.93% for the manufacturing sector, and 64.70% for the service sector. The tourism sector is a subcomponent of the service sector. Considering the OECD member countries, the tourism sector accounts for an average of 4.4% of the GDP and 6.9% of the employment, while it accounts for 21.5% of service exports (OECD, 2020). Tourism-based activities are considered one of the sources of growth, especially in countries experiencing economic difficulties (Nene and Taivan, 2017). The sector contributes to the economy by encouraging employment and the possibility of establishing business networks with the opportunities it provides (Cortes-Jimenez and Pulina, 2010). In addition, it allows the elimination of payment deficits by increasing foreign exchange reserves

(Kumar and Hussain, 2014) and provides an alternative form of export earnings through additional tax revenues (Davis et al., 1988).

In order to benefit from such beneficial effects, it is important for the sector's management to be balanced and for the sector to act in harmony with other sectors of economic activity (Khan et al., 2020). One of the factors unveiling this importance is the contribution of the tourism sector to the foreign exchange earnings obtained both through further economic expansion and exports by stimulating the economic activity in the host country (Li et al., 2013). The second factor is that sectoral activities create positive effects on the economy with positive externalities (Marin, 1992). The third factor is the ability to quickly turn the investments into results and create new employment areas due to the sector's low capital requirements (Ritchie and Goeldner, 1994). In addition, there are economic fields in which the sector offers direct and indirect contributions. The tourism sector has a direct positive effect on numerous fields such as accommodation, transportation, entertainment, food, and retail. It also has an indirect effect on fields such as printing and publishing, utilities, financial services, cleaning services, furniture and equipment supply, security services, car rental services, tourism promotion, resort development, glass products, and iron and steel (World Travel and Tourism Council, 2019).

Despite all these positive contributions, there are also points where the strength of the tourism sector is weak. Compared to other sectors, it is affected rapidly by epidemic disasters, such as COVID-19, and natural disasters, such as earthquakes, fires, and floods, and it gets a sensitive appearance through accelerated regression (Aslan, 2016; Gössling et al., 2021). The basis of its sensitive appearance lies in the fact that it arises directly from human mobility. In cases of epidemic diseases or natural disasters (Dann, 1977), human mobility decreases due to concerns about health and safety (Sánchez-Cañizares et al., 2021). Therefore, the shrinkage experienced by the sector increases its sphere of influence through its manifestation in economic fields that it indirectly affects (Gelashvili, 2022). Due to the effect of the COVID-19 pandemic, the sector's contribution to global GDP decreased from 10.3% to 5.3% between 2019 and 2020, and employment provided by the sector was also negatively affected by this, experiencing a decrease of 18.6%. Another circumstance that negatively affects the sector is forest fires. Almost 50% of the bookings in Greece in 2001 were cancelled due to fires experienced in the summer of 2000 (Scott and Lemieux, 2010). In 2007, fires in the Peloponnese damaged the archaeological site of Olympia, which is also the birthplace of the Olympic Games (Boustras and Boukas, 2013). According to Stevens (2023), in 2023, mass evacuations occurred as a result of forest fires in Greece, and 20,000 people, including tourists, were evacuated in the coastal villages of Kiotari, Gennadi, Pefki, Lindos, Lar Decos, and Kalathos. In Italy, Palermo airport was closed due to fires, and thus disruptions in public transport services occurred. Evacuations were also carried out around the cities of Catania, Trapani, Syracuse, and Messina due to fires. In addition, power and water failures were experienced in the referred regions. More than 2,000 tourists were evacuated in Puglia due to fires. Croatia, Spain, Portugal, and Turkey were among the countries affected by the forest fires in the Mediterranean basin. Although the negative effects were not permanent, the economy and a large number of people were negatively affected by the tourism sector's regression (Duro et al., 2021).

The Vulnerability of the Tourism Sector to Crises: Theoretical Foundations

The structural fragility of the tourism sector can be explained by the concept of 'adaptive capacity' proposed by Holling (1973). Adaptive capacity is a system's ability to adjust and correct itself in response to external shocks. The tourism sector, due to its direct dependence on human mobility, is affected more quickly and profoundly than other sectors by health shocks such as COVID-19 or natural disasters such as forest fires. This is because such events immediately restrict human mobility and rapidly reduce tourist demand. Miller and Ritchie (2003) explained this problem with tourism vulnerability. Tourism vulnerability is defined as the degree to which the tourism system is exposed to external shocks and its capacity to cope with these shocks. The authors demonstrated that this vulnerability is not only a theoretical but also a practical problem by showing how quickly tourism demand fell after the attacks on Australian tourism (9/11) and in the aftermath. Ritchie and Jiang (2019) have demonstrated that creating resilience in disaster and crisis management in the tourism sector can be achieved not only through private sector efforts, but also through public funding and regional cooperation mechanisms. Resilience is the ability of a system or community to return to its previous state or rebuild itself stronger after a shock.

However, contrary to the mistake made by most researchers, resilience cannot be achieved solely through the private sector's investment and innovation efforts. Instead, it is achieved through various public financing mechanisms, policy coordination, and regional cooperation institutions. The occurrence of the same shocks (pandemic, forest fire) in the Mediterranean basin does not mean that the impact is homogeneous across countries. The vulnerability of destinations varies according to differences in the level of dependence on tourism, the size of financial buffers, and, in particular, institutional/managerial capacity (crisis coordination, data production, rapid budget transfer, and accountability). Therefore, multiple crises may lead to deeper income and employment losses and longer recovery periods in some countries. The design of the proposed regional fund should be structured with criteria and transparent rules (contribution-benefit balance, triggers, damage indicators) that make this asymmetric impact visible. In particular, the fact that the tourism sector in geographically close countries (such as the Mediterranean region) is exposed to similar shocks (fires, pandemics) demonstrates the need to establish regional financing mechanisms. This study combines the tourism crisis management framework proposed by Faulkner (2001) with the concept of a regional emergency fund. The flexibility cycle proposed by Faulkner is explained through five stages of recovery after a crisis: (i) Preparedness, (ii) Crisis Response, (iii) Emergency Recovery, (iv) Long-Term Recovery, (v) Reconstruction and Adaptation. The proposed Mediterranean Tourism Emergency Fund could strengthen regional resilience by providing financial support at each stage of this cycle. The MEREC-CoCoSo method provides a mechanism for the fair and objective distribution of this fund.

The Gap and Limits in the Literature

The literature on crisis management and destination resilience in tourism shows that the most effective approach to crises is not only to intervene at the moment of crisis, but to institutionalize risk reduction and preparedness capacity. Nevertheless, in the face of multiple shocks such as pandemics and climate-related fires, public support for tourism is often designed at the national level in the form of compensatory packages implemented after the crisis; the financial liquidity necessary for destinations to recover quickly is sought post-event. However, in regions such as the Mediterranean basin, which share a common destination perception, have intense tourism activity, and are exposed to cross-border risks, a predefined regional fund mechanism is a rational policy tool for both risk sharing and rapid resource allocation. The fundamental point that remains limited in the literature is what criteria should underpin the design of such a regional tourism emergency fund and what analytical approach should be used to transparently determine the balance of contributions/benefits between countries. This study aims to contribute to this gap by (i) developing a regional fund proposal for the Mediterranean basin within the tourism policy–public finance axis and (ii) quantifying the potential benefit distribution of this fund using a MEREC-CoCoSo-based decision support approach.

The proposed regional tourism emergency fund approach is related to the tourism crisis literature as well as the disaster risk financing and climate resilience financing literature. This literature proposes mechanisms (e.g., reserve/contingency funds, conditional credit lines, parametric insurance, and similar instruments) that operate based on predefined triggers, multi-layered risk, and ex-ante financing logic, rather than post-crisis ad-hoc support. The financial protection frameworks of institutions such as the World Bank/GFDRR and the IMF's financial resilience approach for economies vulnerable to disasters provide a reference framework that theoretically strengthens the activation criteria, contribution-benefit balance, and accountability components of the Mediterranean Tourism Emergency Fund proposed in this study. This framework also strengthens the conceptual basis of the fund's institutional design when read in conjunction with regional solidarity instruments such as the EU Solidarity Fund and crisis financing mechanisms such as the IMF's Catastrophe Containment and Relief Trust (CCRT) (trigger–eligibility –expenditure monitoring) and UNDRR's multi-stakeholder risk governance approach, thereby strengthening the conceptual basis of the fund's institutional design. Therefore, the study positions the fund design not only within the context of tourism crisis management but also as a policy tool consistent with disaster risk financing principles (World Bank & GFDRR, 2014; IMF, 2019; UNDP, 2025).

Although there are many national or international institutions and organizations that have the power to direct the tourism sector worldwide, the organization that has data, publications, and research at an expert level on growth and development with sustainable policies in the sector is the World Tourism Organization (UNWTO) (Jafari, 2001). The organization has advocated the sector and has proposed to provide new incentives to the sector by publishing new action plans in order to rebuild tourism, take precautions against various future risks, and disseminate sustainable tourism policies, especially during the COVID-19 pandemic, as well as initiating a rapid sectoral recovery process with appropriate tax policies (UN Tourism, 2020). Even though many countries have adopted national measures as a matter of urgency, the measures taken couldn't yet preserve other related sectors from the impacts of the crises. The governments observe the tourism sector's impact on the entire economy (Mckibbin and Fernando, 2020). In particular, considering the contribution of the tourism sector to countries' foreign exchange income and balance of payments, the issue necessitates multifaceted perspectives and measures from an economic point of view (Çelik et al., 2013). The measures focused on are generally not preventive measures but compensatory measures. From this point of view, considering the economic dimensions of the issue, the establishment of emergency response funds through the cooperation of geographically proximate countries that use the same tourism basin and are affected by disasters to a similar extent will play an effective role. As a result of the problems encountered in world history, many funds were established for precautionary purposes (Boughton, 2002). However, there is no fund to protect the tourism construct of the Mediterranean basin against possible problems. In this sense, the present study addressed a regional emergency response fund proposal for the first time, and it filled the gap in the literature.

Contributions of the Present Research

The present study examines five countries located in the Mediterranean basin, where tourism is an important source of income, especially at times of forest fires and the COVID-19 pandemic, and revealed the benefits that the countries may acquire through a regional fund that is expected to be established on an international scale. The study dealt with the consequences of the disasters experienced in 2020 and aimed to fill a gap in the literature through the proposal of a "Mediterranean tourism emergency response budget" in the field of the tourism sector, considering how the countries are affected by the disasters they experience. In the study, the MEREC-CoCoSo method was used in the analyses conducted while seeking answers regarding how much the tourism sector in Greece, France, Spain, Turkey and Italy was affected by the pandemics and forest fires and to what extent the sector may benefit from a fund to be established as a solution.

This study aims to fill the gap in the literature by focusing on the answers to the following questions: ***(i)*** Can a regional fund be established against extraordinary circumstances such as forest fires and pandemics? ***(ii)*** By which method or methods may the benefits of this fund be revealed? ***(iii)*** Who will benefit from the fund in case of its establishment?

The Mediterranean Region Tourism Emergency Response Fund proposed in this section is a regional risk-sharing mechanism that aims to reduce the collapse in tourism demand, loss of income/employment, and the financing gap during the recovery process in the event of simultaneous multiple shocks such as a pandemic and forest fires. The fund design is divided into two decision problems: (i) determining annual contribution shares during normal periods and (ii) determining payment/allocation shares to countries after a crisis is triggered. The study uses MEREC to generate criterion weights and CoCoSo to generate country priority/benefit scores; these outputs are combined with payment capacity and crisis damage dimensions to translate into the fund's contribution and payment rules. The limitations are as follows:

- As the data of only five countries located in the Mediterranean basin could be completely acquired, modelling was conducted by these five countries in the present study.
- The proposed emergency response fund model requires agreement-based cooperation between the countries. It is important for each country to participate in this model and accept the proposed resources. In addition, it is necessary to establish the necessary administrative structures and mechanisms for the effective operation of this fund.
- The operability of the fund proposal, which is the main subject of the study, can be measured during extraordinary periods such as periods of natural disasters and pandemics. As a result, the study was limited to the years 2019 and 2020, when both the pandemic and intense fires were experienced.
- Fourth, this study is limited to a quantitative MCDM-based comparison. Future research can validate the proposed fund design through country-

specific case studies and stakeholder interviews (e.g., ministries, DMOs, tourism SMEs) to test institutional feasibility and implementation details.

These limitations will have an impact on the viability and success of the proposed emergency response fund model. However, the proposed fund can protect the tourism sector from the effects of natural disasters and pandemics, support national economies, and assist quick recovery under circumstances of crisis. Therefore, it is important to take these constraints into account and develop appropriate strategies.

The data will be made available on reasonable request. There is no conflict of interest between the authors of the study. This study did not receive any external funding. The study does not require ethical approval, as it does not involve any work with human or animal subjects. This article has been proofread and linguistically edited using GenAI tools to improve grammar and fluency. No part of the article, including data analysis, interpretations, or conclusions, was generated by AI tools. The authors take full responsibility for the content of this publication.

DATA SET AND METHODOLOGY

The multi-criteria decision-making framework used in this study is a decision-support approach aimed at addressing the problem of resource allocation and risk sharing in tourism under crisis/disaster conditions, rather than a direct method comparison. The selected criteria are designed to represent, on the one hand, the economic carrying capacity of countries (e.g., GDP, total employment) and, on the other hand, their exposure to tourism shocks and the magnitude of damage (e.g., COVID-19 cost, burned forest area, changes in tourist numbers). Thus, the MEREC-CoCoSo steps provide a numerical, transparent, and repeatable basis for policy discussions on the contribution and benefit distribution of the regional fund proposal. In this section, the method is positioned not as a 'method comparison' in itself, but as a decision support tool that transparently discusses the contribution-payment rule of the regional emergency tourism fund.

The methodology section is divided into two parts. While the criteria are prioritized using an objective weighting method, namely MEREC, alternatives are ranked by way of the CoCoSo method. The steps of these two methods are explained in the following parts.

In the present study, the five countries' data for the years 2019 and 2020 relevant to gross national product, tourism revenues, costs of COVID-19, number of incoming tourists, number of people employed in tourism sector, total employment, total forest area, and burnt forest area are used.

Since the study covers the 2019–2020 period, it is more meaningful in terms of policy implications to assess the impact of the crisis not only in absolute terms for 2020 but also in terms of change rates (shock magnitude) relative to 2019. Therefore, the following dynamic measures have been defined for indicators such as tourism revenues, number of incoming tourists, and tourism employment.

- Tourism revenue change rate: $\Delta TR_i = \frac{TR_{2020i} - TR_{2019i}}{TR_{2019i}}$
- Tourism revenue change rate ($\Delta C2_i$): ($C2_i$,2020-$C2_i$,2019)/$C2_i$,2019 x 100
- Percentage change in tourist arrivals: $\Delta TA_i = \frac{TA_{2020i} - TA_{2019i}}{TA_{2019i}}$
- Percentage change in tourist arrivals ($\Delta C4_i$): ($C4_i$,2020-$C4_i$,2019)/$C4_i$,2019 x 100
- Tourism employment change rate: $\Delta TE_i = \frac{TE_{2020i} - TE_{2019i}}{TE_{2019i}}$
- Tourism employment change rate ($\Delta C5_i$): ($C5_i$,2020-$C5_i$,2019)/$C5_i$,2019 x 100

These dynamic indicators can be used in two ways: (i) integrated as an additional robustness/scenario analysis alongside the level indicators in the main model to test the sensitivity of country rankings to the time dimension; (ii) used to generate a damage index in the fund's activation and payment mechanism during a crisis. Thus, the analysis results are more directly linked from ranking to rule design.

The data for the eight indicators used in this study were obtained from international official databases; where possible, cross-checks between sources and definition/unit consistency checks were performed. GDP is shown as C1, tourism revenues as C2, COVID-19 costs as C3, number of incoming tourists as C4, tourism employment as C5, total employment as C6, total forest area as C7, and burned forest area as C8.

The necessary indicators for the country-year combinations used in the analysis have been completed. The units of all variables have been defined individually. In cases where there were inconsistencies in the definitions of series from different sources, they have been re-expressed under the same definition. The positivity condition required in the MEREC steps has been controlled. The transformation rule to be applied in the event of zero or negative values is specified in the methodology section. Although differences between countries can be observed, particularly in fire indicators, these differences are treated as actual observations reflecting the magnitude of the crisis shock rather than as data errors.

The reasons for selecting France, Greece, Italy, Spain, and Turkey in this study are as follows:

- **Geographical proximity and shared risk basin:** Countries are located in the Mediterranean basin; they are exposed to climate-related risks such as forest fires during the summer season.
- **Tourism volume and sensitivity to tourism shocks:** These countries are Mediterranean destinations where the tourism sector plays a significant role in terms of economic activity and employment.
- **Joint shock period:** It provides a sample that allows for the simultaneous assessment of tourism shocks in the context of both COVID-19 and fire risks during the 2019–2020 period.
- **Data accessibility and comparability:** The availability of data for the selected countries on eight indicators for the same period and with consistent definitions has enhanced comparability in the analysis.

Although some countries in the Mediterranean basin are geographically suitable, they have not been included in the sample due to a lack of data integrity/definition consistency for the relevant period and indicators, or due to differences in risk profile and institutional comparability. Therefore, the following steps should be read as a decision-support framework that supports the discussion of resource allocation and risk sharing under multiple crises, rather than a comparison of methods.

Method Based on the Removal Effects of Criteria (MEREC)

Keshavarz-Ghorabaee et al (2021) propose a novel principle for obtaining objective weights related to criteria using MEREC. Apart from other weighting methods, MEREC handles the removal effects of each criterion on the aggregate performance of alternatives in order to compute the weights. In this method, when a criterion having greater weight is removed, it leads to higher effects on the aggregate performance of alternatives. The causality concept is the basis of this method. The aim of this method is to determine the criterion having the greatest impact on the overall performance of alternatives and assign the highest weight to this criterion. In the present study, a logarithmic function with equal weights was used to assess the aggregate performance of alternatives. In addition, the absolute deviation measure was considered for determining the effects of removing each criterion. The steps of the MEREC can be stated as below.

Step 1. Considering that there are m alternatives and n criteria, the initial decision matrix D is constructed as below:

$$D = \begin{bmatrix} d_{11} & \cdots & d_{1n} \\ \vdots & \ddots & \vdots \\ d_{m1} & \cdots & d_{mn} \end{bmatrix}, \quad \begin{matrix} i = 1,\ldots,m \\ j = 1,\ldots,n \end{matrix} \tag{1}$$

The elements of this matrix $\left(d_{ij}\right)$ need to be greater than zero $\left(d_{ij} > 0\right)$. If negative or zero values exist in the decision matrix, an appropriate transformation technique needs to be applied to obtain positive ones.

Step 2. The elements of the decision matrix are normalized by way of simple linear normalization as specified in Eq. (2).

$$e_{ij} = \begin{cases} \dfrac{\min_i d_{ij}}{d_{ij}}, & j \in B \\ \dfrac{\max_i d_{ij}}{d_{ij}}, & j \in C \end{cases} \tag{2}$$

where B denotes the set of benefit-based criteria, and C shows the set of cost-based criteria. Additionally, the elements of the normalized decision matrix are represented by $\left(e_{ij}\right)$.

Step 3. The overall performance of alternatives is acquired by way of a logarithmic measure with equal criteria weights, as can be seen in Eq. (3). A non-linear function is considered to form this measure. By taking the normalized values into account, it can be implied that the lower values of $\left(e_{ij}\right)$ lead to greater performance values for the considered alternatives $\left(S_i\right)$.

$$S_i = \ln\left(1 + \left(\frac{1}{n}\sum_j \left|\ln(e_{ij})\right|\right)\right) \tag{3}$$

Step 4. The performance of the alternatives in terms of removing different criteria at each step is computed by way of a logarithmic measure as can be seen in Eq. (4).

$$S'_{ij} = \ln\left(1 + \left(\frac{1}{n}\sum_{k,k\neq j} \left|\ln(e_{ij})\right|\right)\right) \tag{4}$$

According to Eq.(4), S'_{ij} shows the overall performance of ith alternative related to the removal of jth criterion.

Step 5. The removal effect of jth criterion is calculated according to Eq.(5) shown as below:

$$f_j = \sum_i \left|S'_{ij} - S_i\right| \tag{5}$$

where f_j represents the effect of removing j[th] criterion.

Step 6. The objective weight for each criterion (w_j) is computed by way of Eq.(6) by utilizing the f_j values obtained in the fifth step.

$$w_j = \frac{f_j}{\sum_k f_k} \tag{6}$$

The Combined Compromise Solution (CoCoSo) Method

Yazdani et al (2019) established the CoCoSo method, which includes the SAW (simple additive weighting) and EWP (exponentially weighted product) methods. The CoCoSo method aims to present balanced results by taking different methods and approaches into account.

The steps of the CoCoSo method can be summarized as below (Yazdani et al., 2019):

Step 1. Constructing the decision matrix: The decision matrix is created by considering Eq. (1).

Step 2. Normalization: The decision matrix is normalized according to Eq. (7).

$$y_{ij} = \begin{cases} \dfrac{d_{ij} - \min_i d_{ij}}{\max_i d_{ij} - \min_i d_{ij}}, & \textit{for benefit criteria} \\ \dfrac{\max_i d_{ij} - d_{ij}}{\max_i d_{ij} - \min_i d_{ij}}, & \textit{for cost criteria} \end{cases} \tag{7}$$

Step 3. Computing gray relational and multiplicative comparison values: S_i values are created with the gray relational approach denoted in Eq. (8).

$$S_i = \sum_{j=1}^{n} w_j y_{ij} \tag{8}$$

Multiplicative comparison values (P_i) are generated according to Eq. (9), as shown below:

$$P_i = \sum_{j=1}^{n} (y_{ij})^{w_j} \tag{9}$$

Step 4. Acquiring relative weights of alternatives: Eqs. (10-12) as three appraisal score strategies are used to obtain relative weights of alternatives according to different effect levels of S_i and P_i values.

$$k_{ia} = \frac{P_i + S_i}{\sum_{i=1}^{m}(P_i + S_i)} \tag{10}$$

$$k_{ib} = \frac{S_i}{\underset{i}{min}\, S_i} + \frac{P_i}{\underset{i}{min}\, P_i} \tag{11}$$

$$k_{ic} = \frac{\lambda S_i + (1 - \lambda)P_i}{\lambda\left(\underset{i}{max}\, S_i\right) + (1 - \lambda)\left(\underset{i}{max}\, P_i\right)}, 0 \leq \lambda \leq 1 \tag{12}$$

Step 5. Computing integrated assessment scores of alternatives: The integrated assessment scores of the alternatives are calculated by way of Eq. (13).

$$k_i = (k_{ia}k_{ib}k_{ic})^{\frac{1}{3}} + \frac{1}{3}(k_{ia} + k_{ib} + k_{ic}) \tag{13}$$

Alternatives are ranked in descending order based on their k_i values.

RESULTS AND DISCUSSION

This section illustrates the need for resource allocation and risk sharing in tourism under multiple shocks such as the pandemic and forest fires. The country ranking is not an absolute measure of success but a policy signal reflecting the relative vulnerability or capacity profile of countries under selected indicators. This signal provides input for the transparent establishment of contribution (capacity-based) and payment/allocation (damage-based) rules in fund design during normal and crisis periods, respectively.

Application

In the present study, eight criteria were considered, and they are shown in Table 1. As alternatives, five countries are considered in this study. These are France (A1), Greece (A2), Italy (A3), Spain (A4), and Turkey (A5). The details of the calculation steps are provided in the methodology section. To improve readability, rather than repeating the calculation of the intermediate matrices (Tables 2–9) here, we treat these tables as technical background; we present the composite scores and country rankings (Table 10) as the primary output, which are critical for the policy discussion.

The relative weights of the alternatives and their integrated assessment scores are obtained according to Eqs .(10-13), and they can be seen in Table 10 ($\lambda = 0.5$) . According to Table 10, France (A1) has the highest integrated assessment score, while Greece (A2) has the lowest score. These scores indicate the relative position of countries under the selected indicators; they should not be interpreted as an absolute judgment of success or failure. In general, the performance of the countries is ranked as A1>A3>A5>A4>A2. Table 10's composite ranking measures countries' simultaneous exposure to (i) crisis shocks (such as COVID-19 costs and burned areas), (ii) their vulnerability to a decline in tourism demand (such as the number of incoming tourists and employment indicators), and (iii) their recovery/financing capacity (such as GDP and total employment). Therefore, the rankings should not be read solely as an expression of success or failure; they should be interpreted as a policy signal that reveals the asymmetric pressures created by multiple shocks, such as the pandemic and climate-related fires in the Mediterranean basin. A low composite score indicates that destinations may be more vulnerable to demand shocks and have relatively limited financial buffers, while a high score points to a stronger capacity to contribute. In this context, linking contribution shares to capacity in fund design and linking payments during crises to damage/decline criteria produces a more consistent outcome in terms of destination resilience and regional solidarity.

The policy signal in question shows that multiple shocks (pandemic + wildfires) have produced asymmetric effects in the Mediterranean basin: while some destinations have experienced higher demand/employment losses, some countries have relatively stronger financial buffers and management capacity. Therefore, the logic of the fund is not merely compensation; it is a resilience tool that supports the stages of preparedness, rapid response, temporary recovery, and long-term reconstruction. In practice, part of the fund's resources should be directed in the short term to liquidity–employment protection and crisis communication; another part should be directed in the long term to destination adaptation investments (fire risk reduction, reforestation, infrastructure, and product diversification). Thus, the ranking output is not merely a technical list; it provides a basis for prioritization for regional coordination and resilience. These asymmetric impacts are concentrated not only on macro indicators but also on local tourism workers, small businesses, and rural/community-based economies. Therefore, the fund's emergency window should be designed to include not only business liquidity but also inclusive recovery outcomes such as wage/employment protection and local supply chain continuity.

The practical implications of these results can be made more visible with recent shock examples: for instance, the successive occurrence of large-scale fire seasons and pandemic-induced demand collapse in some destinations demonstrates that the fund must simultaneously pursue rapid liquidity/employment protection and medium-to-long-term adaptation/risk mitigation objectives. Another example is that events such as the 2021 fire season in Greece or the 2020 tourism demand shock in Turkey clearly demonstrate the need for the fund to operate both its rapid support and adaptation investment windows simultaneously. Therefore, the rankings should be interpreted not merely as a technical country list, but as a policy signal that initiates the calibration of fund windows (emergency support vs. recovery/adaptation investment) according to the type of crisis.

Sensitivity Analysis

In sensitivity analysis, the extent to which the ranking changed in response to potential measurement/calculation errors in the decision matrix values was tested. The detailed simulation procedure (random error generation and number of repetitions) is presented in simplified form; findings are reported in Table(s).

The fact that the ranking is largely preserved under error/deviation scenarios indicates that the proposed fund's contribution/payment rule is relatively resilient to measurement uncertainty and therefore increases its policy feasibility.

Table 1. Criteria considered in the analysis.

Refs.	Criterion Code	Criterion Name	Explanation	Optimization
Brezina (2012)	C1	Gross national product	A measure of the size of the economy	Max
(Qin et al 2018)	C2	Tourism income	Economic activity measurement indicator of the tourism sector	Max
Yamin (2020)	C3	Cost of COVID-19	The cost incurred by countries due to the COVID-19 pandemic	Min
Mou et al (2020)	C4	Number of inbound tourists	Number of tourists visiting the country in the relevant year	Max
Vanhove (1981)	C5	Number of person employed in tourism	Number of people working in the tourism sector in the relevant year	Max
Fields (1988)	C6	Total employment	Total number of people working in the relevant period	Max
Zhang et al (2006)	C7	Forestland (hectare)	Total forest area in the country	Max
Canadell et al (2021)	C8	Burnt forestland (hectare)	Burned forest area (hectares)	Min

Table 2. The initial decision matrix.

Country / Criteria	C1	C2	C3	C4	C5	C6	C7	C8
A1	2630317731455	35958000000	180000000000	117109000	1340078	30379167	17253000	17077
A2	188835201626	6193000000	23500000000	7406000	535801	4643796	3901800	8925
A3	1892574064222	20459000000	110000000000	38419000	2410000	25126337	9566100	31060
A4	1276962685648	18500000000	165631000000	36410000	2367560	22838137	18572170	66783
A5	720288811762	13771000000	70211219013	15971000	1851788	31643711	22220360	20971

Table 3. Normalized decision matrix.

Country / Criteria	C1	C2	C3	C4	C5	C6	C7	C8
A1	0.071791	0.172229	1.000000	0.063240	0.399828	0.152861	0.226152	0.255709
A2	1.000000	1.000000	0.130556	1.000000	1.000000	1.000000	1.000000	0.133642
A3	0.099776	0.302703	0.611111	0.192769	0.222324	0.184818	0.407878	0.465088
A4	0.147878	0.334757	0.920172	0.203406	0.226309	0.203335	0.210089	1.000000
A5	0.262166	0.449713	0.390062	0.463715	0.289343	0.146753	0.175596	0.314017

Table 4. Overall performance values of (S_i) the countries considered as alternatives.

Country	(S_i)
A1	0.955460538
A2	0.409502668
A3	0.837944188
A4	0.772445865
A5	0.805632103

Table 5. Overall performance $\left(S'_{ij}\right)$ of the countries considered as alternatives by removing different criterion at each step.

Country / Criteria	C1	C2	C3	C4	C5	C6	C7	C8
A1	0.820052	0.867100	0.955460	0.813046	0.910384	0.860815	0.881305	0.887645
A2	0.409502	0.409502	0.224401	0.409502	0.409502	0.409502	0.409502	0.226732
A3	0.704832	0.771142	0.810952	0.744708	0.753139	0.742204	0.788235	0.795667
A4	0.655515	0.707178	0.767631	0.675993	0.682755	0.675971	0.678047	0.772445
A5	0.727917	0.759972	0.751618	0.761763	0.733853	0.692264	0.703426	0.738752

Table 6. The f_j values and criteria weights.

	C1	C2	C3	C4	C5	C6	C7	C8
f_j	0.463164	0.266088	0.270920	0.375970	0.291349	0.400225	0.320468	0.359741
Weight	0.168550	0.096832	0.098590	0.136819	0.106025	0.145646	0.116621	0.130913

Table 7. Normalized decision matrix.

Country / Criteria	C1	C2	C3	C4	C5	C6	C7	C8
A1	1	1	0	1	0.429131	0.953164	0.728834	0.859103
A2	0	0	1	0	0	0	0	1
A3	0.697829	0.479287	0.447284	0.282699	1	0.758615	0.309210	0.617425
A4	0.445683	0.413472	0.091814	0.264386	0.977355	0.673866	0.800847	0
A5	0.217676	0.254594	0.701525	0.078074	0.702159	1	1	0.791800

Table 8. The weighted normalized decision matrix for gray relational S_i values.

Country / Criteria	C1	C2	C3	C4	C5	C6	C7	C8	S_i
A1	0.168550	0.096832	0	0.136819	0.045498	0.138824	0.084997	0.112468	0.783992
A2	0	0	0.098590	0	0	0	0	0.130913	0.229504
A3	0.117619	0.046410	0.044098	0.038678	0.106025	0.110489	0.036060	0.080829	0.580212
A4	0.075120	0.040037	0.009052	0.036173	0.103624	0.098146	0.093396	0	0.455550
A5	0.036689	0.024652	0.069163	0.010682	0.074446	0.145646	0.116621	0.103657	0.581561

Table 9. The weighted normalized decision matrix for multiplicative P_i values.

Country / Criteria	C1	C2	C3	C4	C5	C6	C7	C8	P_i
A1	1	1	0	1	0.914208	0.993038	0.963783	0.980314	6.851345
A2	0	0	1	0	0	0	0	1	2
A3	0.941160	0.931261	0.923742	0.841260	1	0.960562	0.872072	0.938825	7.408885
A4	0.872656	0.918035	0.790228	0.833587	0.997574	0.944131	0.974432	0	6.330646
A5	0.773370	0.875925	0.965653	0.705461	0.963204	1	1	0.969901	7.253517

Table 10. The relative weights and integrated assessment scores of the considered alternatives.

Country	k_a	Ranking (k_a)	k_b	Ranking (k_b)	k_c	Ranking (k_c)	k	Final ranking
A1	0.235113	3	6.841695	1	0.931948	3	3.814072	1
A2	0.068652	5	2	5	0.272127	5	1.114572	5
A3	0.246006	1	6.232549	2	0.975127	1	3.628029	2
A4	0.208965	4	5.150250	4	0.828304	4	3.024926	4
A5	0.241263	2	6.160743	3	0.956328	2	3.577151	3

The results of the ranking for each MCDM method using identical MEREC weights are shown in Table 11. According to the methods, while there is consensus on the best and worst considered alternatives as A1 and A2, respectively, the rank of the others varied according to the methods used. The CoCoSo, ARAS, COPRAS, and EDAS methods ranked A3 as the second best, while WSM and TOPSIS ranked A3 as the third. All the methods ranked A4 as the fourth best, with the exception of TOPSIS, which ranked A4 as the second best. This policy signal also highlights regional disparities: in countries with higher demand/employment shocks and more limited fiscal buffers, the principle of earlier/priority support during a crisis becomes justified. Furthermore, differences in administrative capacity between countries make the fund's role in developing data standardization and reporting capacity through its technical secretariat critical.

These findings should be read in conjunction with three trends that have become apparent in post-pandemic tourism: (i) the increasing necessity of sustainable/green tourism investments due to rising climate-related risks, (ii) a shift in demand behavior focused on security and flexibility, and (iii) the growing need for digitalization and product diversification in destination management. Therefore, the rankings should be interpreted not only as an indicator of vulnerability level but also as a policy learning input aimed at designing funding mechanisms that combine short-term shock absorption (liquidity–employment protection) with medium-to-long-term transformation (green/digital adaptation and diversification).

Table 11. The ranking of the considered alternatives with the MCDM methods using the MEREC weights.

	CoCoSo Score	WSM Score	TOPSIS Score	ARAS Score	COPRAS Score	EDAS Score
A1	3.8141 (1)	0.7840 (1)	0.6947 (1)	0.7604 (1)	100.0000 (1)	0.9284 (1)
A2	1.1146 (5)	0.2295 (5)	0.0000 (5)	0.3572 (5)	50.8380 (5)	0.1380 (5)
A3	3.6280 (2)	0.5802 (3)	0.5078 (3)	0.5170 (2)	68.8265 (2)	0.5580 (2)
A4	3.0249 (4)	0.4555 (4)	0.5669 (2)	0.4691 (4)	61.5558 (4)	0.3506 (4)
A5	3.5772 (3)	0.5816 (2)	0.3855 (4)	0.4850 (3)	64.0995 (3)	0.5053 (3)

The first sensitivity analysis performed in the study is the statistical approach proposed by Mukhametzyanov and Pamučar (2018), which tests the sensitivity of the methods to possible estimation errors in the decision matrix values. We used their statistical approach with a slight modification, as follows;

$$Re_{ij} = R_{ij} + (-1 + 2.rnd()).\gamma.R_{ij} (\forall i = 1,...,n \; and \; \forall i = 1,...m) \qquad (14)$$

where, R_{ij} is the initial decision matrix, *rnd*() provides a random number with a uniform distribution [0, 1] for each element of the decision matrix, and γ is the relative error of estimating the element of the decision matrix. We used the same error values (γ = 0.05, 0.10, 0.15, 0.20, 0.25) used by Mukhametzyanov and Pamučar (2018). We further re-performed all the evaluations, including the criteria weighing with MEREC and the alternative ranking with all six MCDM methods on this newly obtained Re_{ij} matrix.

We ran Eq. (14) 1000 times for each relative error value and obtained a ranking for each MCDM method based on one thousand Re_{ij}. The ranking results are presented in Tables 12-16, and while the error values are low, there is little or no rank change observed for the methods other than for the WSM. As the error values increased, the only methods that continued to resist this change are the TOPSIS, EDAS, and CoCoSo methods. Apart from these, the rank changes occurred only in the ARAS, EDAS, and COPRAS methods for A2, which are observed as the worst-considered alternatives by all the methods.

Table 12. The ranking results for each MCDM method, according to 1000 Re_{ij} (γ = 0.05).

CoCoSo	RANK	1	2	3	4	5	WSM	RANK	1	2	3	4	5
	A1	1000	0	0	0	0		A1	1000	0	0	0	0
	A2	0	0	0	0	1000		A2	0	0	0	0	1000
	A3	0	975	25	0	0		A3	0	433	567	0	0
	A4	0	0	0	1000	0		A4	0	0	0	1000	0
	A5	0	25	975	0	0		A5	0	567	433	0	0
TOPSIS	**RANK**	**1**	**2**	**3**	**4**	**5**	**ARAS**	**RANK**	**1**	**2**	**3**	**4**	**5**
	A1	1000	0	0	0	0		A1	1000	0	0	0	0
	A2	0	0	0	0	1000		A2	0	0	0	0	1000
	A3	0	0	1000	0	0		A3	0	1000	0	0	0
	A4	0	1000	0	0	0		A4	0	0	13	987	0
	A5	0	0	0	1000	0		A5	0	0	987	13	0
COPRAS	**RANK**	**1**	**2**	**3**	**4**	**5**	**EDAS**	**RANK**	**1**	**2**	**3**	**4**	**5**
	A1	1000	0	0	0	0		A1	1000	0	0	0	0
	A2	0	0	0	0	1000		A2	0	0	0	0	1000
	A3	0	1000	0	0	0		A3	0	1000	0	0	0
	A4	0	0	3	997	0		A4	0	0	0	1000	0
	A5	0	0	997	3	0		A5	0	0	1000	0	0

Table 13. The ranking results for each MCDM method, according to 1000 Re_{ij} (γ = 0.10).

CoCoSo	RANK	1	2	3	4	5	WSM	RANK	1	2	3	4	5
	A1	1000	0	0	0	0		A1	1000	0	0	0	0
	A2	0	0	0	0	1000		A2	0	0	0	0	1000
	A3	0	869	131	0	0		A3	0	490	510	0	0
	A4	0	0	0	1000	0		A4	0	0	0	1000	0
	A5	0	131	869	0	0		A5	0	510	490	0	0
TOPSIS	**RANK**	**1**	**2**	**3**	**4**	**5**	**ARAS**	**RANK**	**1**	**2**	**3**	**4**	**5**
	A1	1000	0	0	0	0		A1	1000	0	0	0	0
	A2	0	0	0	0	1000		A2	0	0	0	0	1000
	A3	0	0	1000	0	0		A3	0	974	26	0	0
	A4	0	1000	0	0	0		A4	0	2	171	827	0
	A5	0	0	0	1000	0		A5	0	24	803	173	0
COPRAS	**RANK**	**1**	**2**	**3**	**4**	**5**	**EDAS**	**RANK**	**1**	**2**	**3**	**4**	**5**

continued on following page

Table 13. Continued

CoCoSo	RANK	1	2	3	4	5	WSM	RANK	1	2	3	4	5
	A1	1000	0	0	0	0		A1	1000	0	0	0	0
	A2	0	0	0	0	1000		A2	0	0	0	0	1000
	A3	0	985	15	0	0		A3	0	981	19	0	0
	A4	0	0	112	888	0		A4	0	0	0	1000	0
	A5	0	15	873	112	0		A5	0	19	981	0	0

Table 14. The ranking results for each MCDM method, according to 1000 Re_{ij} ($\gamma = 0.15$).

CoCoSo	RANK	1	2	3	4	5	WSM	RANK	1	2	3	4	5
	A1	988	12	0	0	0		A1	1000	0	0	0	0
	A2	0	0	0	0	1000		A2	0	0	0	0	1000
	A3	12	758	230	0	0		A3	0	507	493	0	0
	A4	0	0	0	1000	0		A4	0	0	0	1000	0
	A5	0	230	770	0	0		A5	0	493	507	0	0
TOPSIS	**RANK**	**1**	**2**	**3**	**4**	**5**	**ARAS**	**RANK**	**1**	**2**	**3**	**4**	**5**
	A1	1000	0	0	0	0		A1	1000	0	0	0	0
	A2	0	0	0	0	1000		A2	0	0	0	0	1000
	A3	0	23	977	0	0		A3	0	900	90	10	0
	A4	0	977	23	0	0		A4	0	14	256	730	0
	A5	0	0	0	1000	0		A5	0	86	654	260	0
COPRAS	**RANK**	**1**	**2**	**3**	**4**	**5**	**EDAS**	**RANK**	**1**	**2**	**3**	**4**	**5**
	A1	1000	0	0	0	0		A1	1000	0	0	0	0
	A2	0	0	0	2	998		A2	0	0	0	0	1000
	A3	0	941	56	3	0		A3	0	929	71	0	0
	A4	0	4	213	781	2		A4	0	0	0	1000	0
	A5	0	55	731	214	0		A5	0	71	929	0	0

Table 15. The ranking results for each MCDM method, according to 1000 Re_{ij} ($\gamma = 0.20$).

CoCoSo	RANK	1	2	3	4	5	WSM	RANK	1	2	3	4	5
	A1	964	34	2	0	0		A1	1000	0	0	0	0
	A2	0	0	0	0	1000		A2	0	0	0	0	1000
	A3	33	713	254	0	0		A3	0	559	439	2	0
	A4	0	0	0	1000	0		A4	0	0	4	996	0
	A5	3	253	744	0	0		A5	0	441	557	2	0
TOPSIS	**RANK**	**1**	**2**	**3**	**4**	**5**	**ARAS**	**RANK**	**1**	**2**	**3**	**4**	**5**
	A1	984	16	0	0	0		A1	1000	0	0	0	0
	A2	0	0	0	0	1000		A2	0	0	0	1	999
	A3	0	47	953	0	0		A3	0	831	134	35	0
	A4	16	937	47	0	0		A4	0	39	295	665	1
	A5	0	0	0	1000	0		A5	0	130	571	299	0
COPRAS	**RANK**	**1**	**2**	**3**	**4**	**5**	**EDAS**	**RANK**	**1**	**2**	**3**	**4**	**5**
	A1	1000	0	0	0	0		A1	1000	0	0	0	0
	A2	0	0	2	12	986		A2	0	0	0	0	1000
	A3	0	862	112	26	0		A3	0	859	141	0	0
	A4	0	23	256	708	13		A4	0	0	2	998	0
	A5	0	115	630	254	1		A5	0	141	857	2	0

Table 16. The ranking results for each MCDM method, according to 1000 Re_{ij} ($\gamma = 0.25$).

CoCoSo	RANK	1	2	3	4	5	WSM	RANK	1	2	3	4	5
	A1	916	74	10	0	0		A1	1000	0	0	0	0
	A2	0	0	0	0	1000		A2	0	0	0	0	1000
	A3	71	611	318	0	0		A3	0	520	471	9	0
	A4	0	0	0	1000	0		A4	0	2	14	984	0
	A5	13	315	672	0	0		A5	0	478	515	7	0
TOPSIS	**RANK**	**1**	**2**	**3**	**4**	**5**	**ARAS**	**RANK**	**1**	**2**	**3**	**4**	**5**
	A1	948	52	0	0	0		A1	1000	0	0	0	0
	A2	0	0	0	0	1000		A2	0	0	0	4	996
	A3	0	115	882	3	0		A3	0	716	221	63	0
	A4	52	833	114	1	0		A4	0	72	298	626	4
	A5	0	0	4	996	0		A5	0	212	481	307	0

continued on following page

Table 16. Continued

CoCoSo	RANK	1	2	3	4	5	WSM	RANK	1	2	3	4	5
COPRAS	**RANK**	**1**	**2**	**3**	**4**	**5**	**EDAS**	**RANK**	**1**	**2**	**3**	**4**	**5**
	A1	1000	0	0	0	0		**A1**	1000	0	0	0	0
	A2	0	0	5	44	951		**A2**	0	0	0	1	999
	A3	0	748	208	44	0		**A3**	0	769	229	2	0
	A4	0	50	272	637	41		**A4**	0	2	9	988	1
	A5	0	202	515	275	8		**A5**	0	229	762	9	0

The overall percentages (for all error values) of maintaining the overall original ranking of all the considered alternatives for each method are presented in Fig. 1. If we are to rank the methods in terms of resistance to error values, it could be TOPSIS, EDAS, CoCoSo, COPRAS, ARAS, and WSM. In addition, TOPSIS, EDAS, and CoCoSo are ahead of the others and are the foremost methods in stable ranking. Furthermore, WSM is the most sensitive to error values due to its relatively unstable ranking.

Figure 1. A comparison of the methods based on maintaining the original ranking.

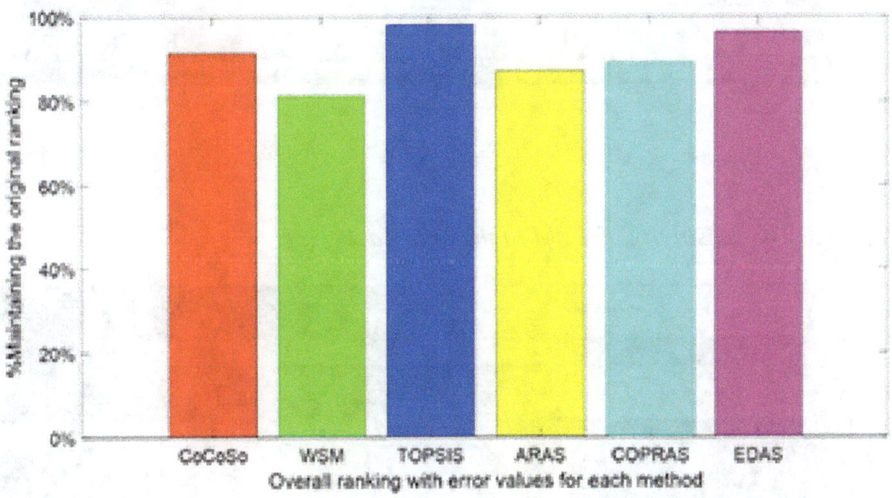

In the second sensitivity analysis, we examined the effect of the change in criteria importance values on the ranking of the methods and the sensitivity of the methods to this circumstance, as in Lee and Chang (2018).

A total of 168 new weight vectors is created by multiplying the importance values of each criterion (eight in total) by the values between 0 and 10 (twenty-one in total) and correcting the importance values of the other criteria so that the total importance values are 1. For each MCDM method, we employed these weight vectors and obtained alternative rankings. In Figs. 2-7, for each MCDM method, the light green color indicates that there is no change in rank according to the relative criterion importance change rate, while the change in rank is indicated in dark green. According to the results, in general, C2 and C8 are the most and least sensitive criteria, respectively. Accordingly, it is revealed that the TOPSIS, EDAS, and CoCoSo methods are less sensitive than other methods in terms of criteria weight change. While the WSM had the highest sensitivity, the COPRAS and ARAS followed. Fig. 8 indicates the percentages of unchanged rankings relative to the 168 weight vectors for each MCDM method. It can be seen that TOPSIS, EDAS, and CoCoSo were separated positively from the other methods.

Figure 2. The sensitivity of CoCoSo to the criteria weight.

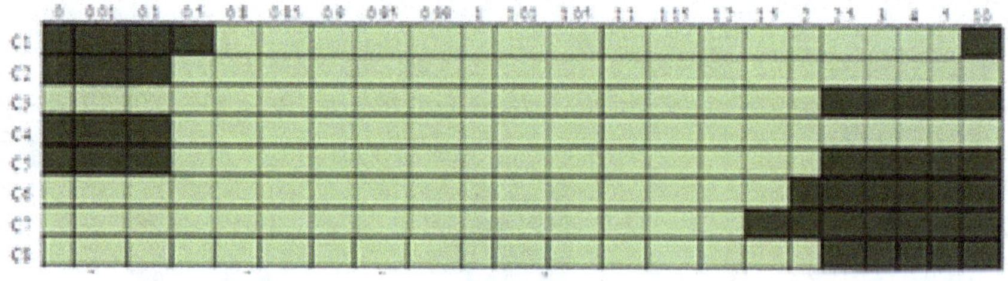

Figure 3. The sensitivity of WSM to the criteria weight.

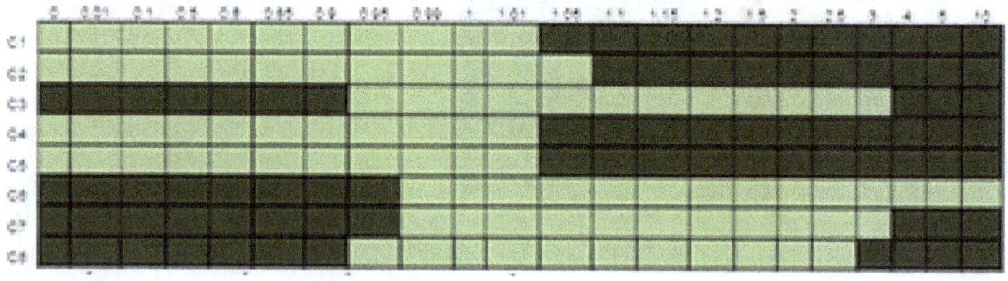

Figure 4. The sensitivity of TOPSIS to the criteria weight.

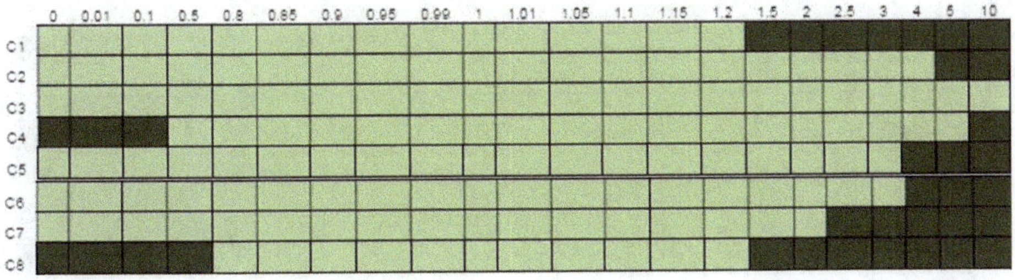

Figure 5. The sensitivity of ARAS to the criteria weight.

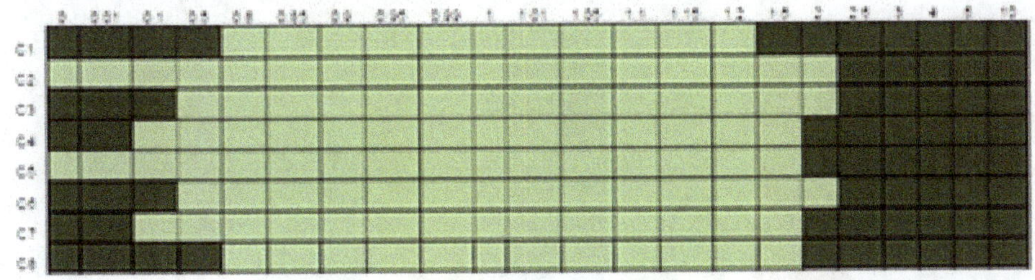

Figure 6. The sensitivity of COPRAS to the criteria weight.

Figure 7. The Sensitivity of EDAS to the Criteria Weight

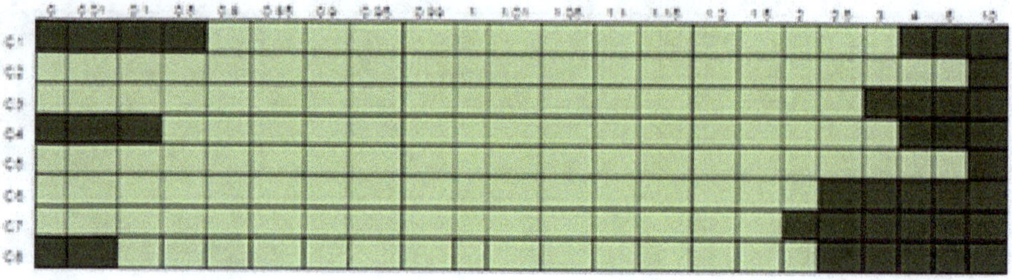

Figure 8. The comparison of sensitivity to the criteria weight.

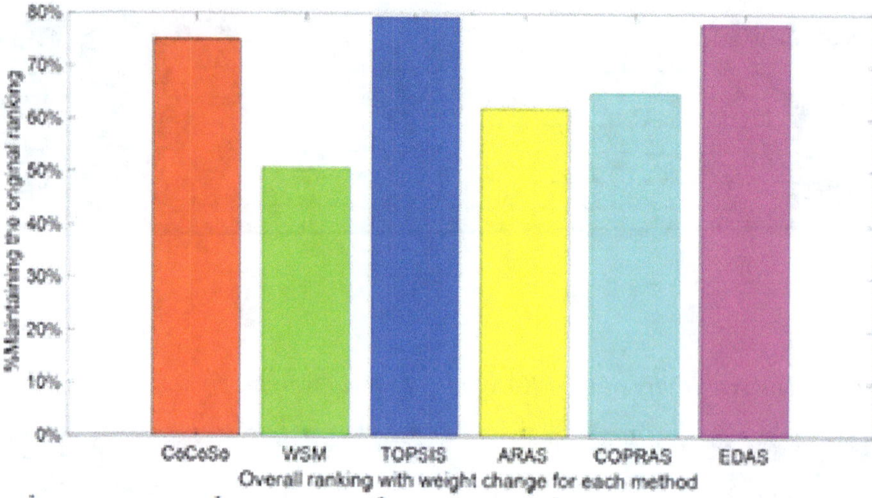

The first sensitivity analysis measured the sensitivity to measurement errors as well as the compatibility and joint sensitivity of the methods with MEREC, while the second sensitivity analysis measured the extent of their resistances to maintaining the criteria weight change. In the first, the EDAS, CoCoSo, and TOPSIS methods came to the fore, while in the second, it was again these three methods that maintained the highest level of performance. Accordingly, the MEREC and CoCoSo methods demonstrated superior performance in terms of sensitivity compared to the other methods in ranking the MCDM problem in the study.

Sensitivity analysis findings can be translated into two critical policy lessons for the fund's design. First, the substantial preservation of the best and worst alternatives under different methods and measurement errors demonstrates that the fund's core prioritization logic is robust and provides a reliable signal in terms of general direction in policy discussions. However, since method-dependent shifts in rankings can be observed in the middle tiers, classifying countries in the fund design using gradual tiers (e.g., high/medium/low priority groups) rather than definitive rankings will reduce the risk of disagreement.

Secondly, the fact that certain criteria are more sensitive to weight changes indicates which indicators require stricter verification in terms of the fund's data governance. Therefore, (i) a standard definition and measurement protocol linked to the source for sensitive criteria, (ii) a smoothing rule such as multi-year average in the annual update, and (iii) an independent audit/appeal mechanism should be incorporated into the design. Thus, methodological sensitivity can be transformed into "institutional assurance" tools in practice, thereby enhancing the fund mechanism's applicability.

CONCLUSION

The ranking results indicate that the proposed fund should be designed not only as a post-crisis compensatory tool, but also as a financial resilience mechanism that strengthens the preparation and rapid recovery phases of the tourism crisis management cycle. Therefore, pre-defining the fund's trigger criteria (e.g., a certain percentage drop in tourist numbers, an increase in the proportion of burned forest area, a surge in pandemic costs) and basing them on transparent data sources will reduce delays during a crisis.

According to the findings of the study, based on the data from 2019 and 2020, if the Mediterranean Basin Tourism Association emergency fund is established, France would have benefited the most. Then, Italy, Turkey, Spain, and Greece would receive the benefits following France. France's participation in such a fund would increase cooperation in both France and among the countries of the region and launch an innovative and pioneering movement in the field of sustainable tourism. In fact, since the establishment of such a fund would serve as an example in a regional sense, economic damage may be minimized in any extraordinary event that may occur. When the economic damage is minimized, the financing of the tourism sector, which is damaged by unforeseen events, will be easily realized. This is because, in this way, a kind of state budget would be created and asset transfers could be provided, like a wealth transfer from the state budget to the tourism sector. Therefore, positive results would be achieved in terms of various economic elements, such as employment, service production, and profit maximization. Positive results in the aforementioned economic elements would undoubtedly provide positive effects for the countries located in the Mediterranean basin, which forms the focus of this study.

In particular, if it is desired to minimize the cost of COVID-19 and burnt forest area, the most important criterion is GDP, followed by total employment, number of incoming tourists, burnt forest area, total forest area, total tourism employment, cost of COVID-19, and tourism revenues in terms of compensation for the damage. Considering the importance of GDP for the countries, the establishment of this fund would provide both economic sustainability and emergency response assurance for tourists against possible bad scenarios. If GDP increases, the level of forestation also increases. This is because it is possible for the level of forestation to increase with the fund being established. Therefore, it would also be possible to compensate for the areas reduced by the forest fires. With the appropriate provision of tourism services to such areas, it would be possible to increase the national incomes of the countries again. Considering especially the points of view of France, Italy, Turkey, Spain, and Greece, which are in the Mediterranean basin, an increase in national income would contribute to the revival of the tourism sector in these countries. This circumstance may even reach the capacity of elevating health tourism. In particular, in extraordinary circumstances that bring the economy to a standstill, such as COVID-19, the expansion of health tourism would ensure the survival of the tourism sector. This circumstance is also important from the point of view of the states. In this way, the emergence of a stagflation phenomenon, such as unemployment or inflation, which has the capacity to create a cost for the states, could also be prevented. In such cases, where most sectors come to a standstill, the survival of the tourism sector would also reflect positively on other sectors. Maintaining tourism revenues through such a fund in times of crises would help in solving any employment problem that may arise in other sectors. The main reason for this is that a fund created for tourism has the potential to increase production opportunities beyond the current capacity. Therefore, in order to reduce a negative supply shock to a minimum level in the Mediterranean basin and even carry it beyond the possibilities, some kind of state budget should be established. It is possible that this state budget could also be provided by creating a fund for forest fires in the Mediterranean basin.

According to the results of the sensitivity analysis, in which six different methods are considered using the MEREC weights, France would benefit the most from the fund to be established. The minimum benefit would be gained by Greece. According to four different methods, Italy would be the second country benefiting the most, while Turkey and Spain would be the second countries benefiting once. Turkey is found to be the country with the third highest benefit among five countries, with Spain being the fourth. Since it cannot be known in advance which country would face a bad scenario or even when, the establishment of the relevant fund would provide each member country with additional time until it implements its own policies and would mitigate the effects of possible bad consequences. The main problem here is the ratio at which all the countries would contribute to the financing of the fund.

If all the countries had to contribute to the financing at an equal rate, crises might arise during the creation of such a fund. In fact, the countries in the Mediterranean basin would be reluctant to create a fund against forest fires considering various possible economic crises. Therefore, it is necessary to resort to methods that would increase the willingness of countries to make payments for contributions to the fund. From this point of view, a proper financing method, proportional to the level of benefits that the countries would receive as per the MEREC-CoCoSo method, may create an increase in the willingness of countries to make payment for the fund. According to the sensitivity analysis, which covered six different methods, France would pay the highest in the financing of the forest fire fund since it would receive the greatest benefit. After that, Italy would contribute to the financing more than the remaining four countries, as it would also benefit the most. Following France and Italy, the financing would be provided by Turkey, Spain, and Greece, respectively, in descending order. France would be the one most affected by any costs that would arise following the creation of the fund for forest fires, with Greece being the one least affected. Such a circumstance could lead to France financing Greece for the forest fires. However, any fund to be established profitably is an insurance against forest fires just like any insurance. Therefore, the only way for the financing to be fair should be in direct proportion to the benefit to be obtained according to the MEREC-CoCoSo method.

Focusing only on the favorableness of the financing of a fund may cause unintended consequences. Making cooperation obligatory against forest fires among the countries in the Mediterranean basin may help to bring about a more equitable approach to financing. For instance, if a forest fire occurs in France, which would receive the greatest benefit according to the MEREC-CoCoSo method, cooperation with other countries, especially Greece, which would receive the lowest benefit, would ensure fairness in financing. In this way, although other countries would contribute less to the fund compared to France, they would use their resources for forest fires in France in case of a possible risk. Therefore, justice could be achieved in financing, and it would be possible to benefit from this fund in diverse ways in sustainable tourism projects and in the protection of natural resources. For instance, a joint fire response unit to be established with this fund would provide significant support to countries, especially during the summer months. Considering the geographical proximity of these countries, the speed of search and rescue and interventions would increase in the face of disasters, such as fire or flood, and possible losses would be reduced. Emergency response points, which would be established especially in locations close to areas with a substantial risk of fire or flood, are of vital importance in times of crisis. Considering the losses suffered by both people and nature in such disasters, it will draw attention to how vital emergency intervention is. Therefore, the cooperation of countries that have common consequences both in terms of harm and intervention is important. As a result, according to the MEREC-CoCoSo method, the finances should be weighted from the country that contributes the most to the financing of the fund to the country that contributes the least, and a balance could be achieved in contributing to the financing through cooperation against the risk of possible forest fires. In this way, on the one hand, an effective measure would be taken against forest fires that would occur in the Mediterranean basin, and on the other hand, a construct that would set a precedent for other regions would emerge.

In this respect, the study directly contributes to discussions on regional risk sharing and policy coordination in conditions where multiple crises have become the new normal in Mediterranean tourism. Thus, the methodological findings are interpreted in the context of the transformation of global tourism governance after the pandemic, in line with a focus on crisis governance and sustainability.

MANAGERIAL IMPLICATIONS

The results of the study have important managerial implications for tourism managers and policy makers in the Mediterranean basin. The distribution of the Fund among countries in a fair and transparent manner according to the principles of proportional financing balances the relationship between contributions and benefits and increases the willingness to pay. The MEREC-CoCoSo analysis's benefit scores can be used as a neutral standard to calculate national contributions. At the managerial level, it is recommended to create separate departments for the fund within state budgets, to establish rapid appropriation transfer mechanisms and to develop budget codes to protect tourism employment. In this context, it is recommended that a certain portion of the fund resources be set aside as a 'social resilience window' and directed towards wage/employment support for tourism workers and rapid micro-support mechanisms for small-scale tourism businesses (SMEs) during times of crisis. National implementation units can report simple output indicators such as protected employment, number of supported businesses, and local supply chain continuity in a standard format to make inclusive recovery visible. In addition, the existence of the fund will contribute to the preservation of the brand value of destinations, the compensation of income losses of tourism enterprises in times of crisis and the continuity of sustainable tourism investments. In this context, the fund is a crisis management tool that aims to increase destination resilience not only at the moment of intervention but also during the recovery and reconstruction phases.

The impact of the fund depends not only on the size of the financing but also on its ability to strengthen crisis management capacity. In this context, establishing minimum common standards for joint crisis communication between countries, pre-season multi-scenario exercises (such as simultaneous pandemic and fire scenarios), and rapid budget transfer/spending monitoring increases the speed of recovery. Thus, the fund combines short-term liquidity support with long-term destination resilience and regional policy coordination.

At the implementation level, the fund's governance structure should consist of a multi-stakeholder model that includes not only financial authorities but also tourism ministries, destination management organizations, and relevant disaster management agencies. This will enhance the coordination and learning capacity emphasized in the resilience literature. Furthermore, pre-established protocols on how the fund's operational support, employment protection measures, and destination improvement investments will be distributed will accelerate resource allocation along the intervention–recovery continuum.

The MEREC–CoCoSo outputs of this study can be translated into two distinct decision problems in the fund's policy design: (i) calculating countries' annual contribution shares and (ii) allocating fund usage/payment shares during a crisis. Explaining this transformation with a concrete rule would enhance the policy value of the fund proposal. Therefore, the following set of rules is proposed:

Step 1. Definition of fund size: The fund's annual target size F can be indexed to regional tourism volume or economic capacity. For example, F=τ×∑Tourism Revenue (or F=τ×∑GDP); τ can be set at a low rate within a certain range (e.g., at the per mille (‰) level) as a policy choice.

Step 2. Conversion of CoCoSo points to shares: The normalized benefit share for the integrated CoCoSo score Si of each country is defined as follows:

$$b_i = \frac{S_i}{\sum_j S_j}$$

This can be interpreted as the relative benefit/priority share of countries from the fund mechanism.

Step 3. Contribution share (financing) rule: A contribution based solely on b_i provides the pay-as-you-benefit principle; however, the payment capacity dimension is also necessary for the sustainability of the fund. Therefore, a hybrid rule is recommended.

$$k_i = \lambda \cdot g_i + (1 - \lambda) \cdot b_i$$

Here, $g_i = \frac{GDP_i}{\sum_i GDP_i}$ is the payment capacity share. $\lambda \in [0,1]$ is a policy parameter. The annual contribution of countries is calculated as follows:

$$C_i = k_i \cdot F$$

Step 4. Reducing the risk of floor-ceiling disagreements: The following band is defined for the political feasibility of contribution shares:

$$k_i \in [k_{min}, k_{max}]$$

Values that exceed the band are pulled back into the band; shares for the remaining countries are rescaled to total 1. This step increases the fund's feasibility and the likelihood of consensus.

Step 5. Payment/benefit rule during a crisis (activation + damage index): Measurable triggers are defined for the activation of the fund. Post-activation payment should be linked not only to the "benefit share" but also to crisis damage. For this purpose, a damage index (d_i) can be calculated (using the 2019–2020 exchange rates) and normalized, and payments can be made based on $p_i = \frac{d_i}{\sum d_i}$. Thus, the fund will have both a predictable contribution rule and a distribution rule that prioritizes needs during a crisis.

Damage index example: $DI_i = \text{mean}\ \{\text{norm}(C3_i), \text{norm}(C8_i/C7_i), \text{norm}(-\Delta C2_i), \text{norm}(-\Delta C4_i), \text{norm}(-\Delta C5_i)\}$

Crisis payment/allocation: $Share_i = (1-\lambda).S_i + \lambda.DI_i\ (0 \leq \lambda \leq 1)$

PRACTICAL IMPLICATIONS

This study highlights the common challenges faced by Mediterranean tourism when simultaneously exposed to two distinct types of shocks (health crisis and environmental disaster) during the same period: a collapse in demand, the need for crisis communication/trust management, liquidity pressure on businesses, and job losses. The findings suggest that, due to the cross-border nature of these problems, individual country policies may sometimes be insufficient; without regional coordination, common triggers, and an accountable institutional structure, resource allocation may be delayed. The governance architecture proposed below aims to reduce this coordination gap in line with the book's theme of multiple crisis pressures. The feasibility of the proposed Mediterranean Region Tourism Emergency Fund depends not only on the size of the fund but also on its institutional architecture and decision-making mechanism designed in accordance with multi-level governance principles. Therefore, the fund should be operated under a structure that incorporates (i) strategic direction and political legitimacy, (ii) technical verification and data governance, and (iii) rapid operation and accountability components. In this context, the fund's corporate structure is proposed to be four-tiered:

- **Ministerial/Council Level (Strategic Decision):** A fund council composed of representatives from member countries at the tourism and/or finance level determines the fund's annual target size, contribution rules, activation (trigger) thresholds, and oversight standards. A predefined voting rule, such as a qualified majority (e.g., a certain percentage of member countries + a certain percentage of total contributions), is used for large consistent payments and rule changes.
- **Technical Secretariat (Operations + Data):** It is a permanent secretariat function responsible for data collection, standardization, calculation, and reporting processes. The secretariat is responsible for standardizing data definitions, versioning calculation protocols, publishing annual updates, and operating rapid payment processes during crises.
- **Scientific/Technical Committee (Verification + Methodology):** It monitors the representativeness of the indicators used (tourism revenue, number of tourists, employment, burned area ratio, etc.) in terms of crisis impact and data quality; develops recommendations for revisions to the indicators. This committee ensures methodological consistency and comparability over time with the fund's technical infrastructure.
- **Independent Audit and Compliance Mechanism (Accountability):** Fund contributions, payments, and expenditure items are monitored in accordance with independent audit principles; results are published in publicly available reports. A "complaint–correction–arbitration" process with time limits is defined for disputes (e.g., data objections, payment objections).

The division of authority and responsibility must be clarified. Member countries report on the use of fund resources (employment protection, business support, destination improvement, etc.) through national implementation units; the Secretariat verifies the data; the Fund Council approves the final decisions; and the audit mechanism also checks the appropriateness of spending. Thus, unlike ad hoc post-crisis packages, the fund becomes a policy instrument that operates according to predefined rules, is transparent, and institutionalizes regional cooperation. Therefore, the findings contribute directly to the discussion on multi-level governance, inter-agency coordination, and long-term recovery capacity, rather than merely representing a technical ranking. In this sense, the proposed fund should be interpreted not only as an emergency liquidity tool but also as a governance instrument aligned with post-pandemic recovery and climate-adaptation priorities (e.g., UNDRR/UNDP/EU policy frameworks).

Within the scope of this study, the proposed regional fund can be positioned as an institutional tool that goes beyond the post-crisis compensation only approach, aligning with UNDRR's risk reduction–preparedness–recovery cycle and the orientation of the European Green Deal (green transition and climate adaptation). In this context, the fund could function as a complementary risk-sharing and adaptation financing mechanism at the regional level, coordinated with multilateral frameworks such as UNEP/UNDP, without replacing national policies.

To enhance the fund's feasibility, an intergovernmental agreement signed among member countries during the establishment phase could define a framework (role–authority–voting–trigger–oversight); in the medium term, institutional continuity could be strengthened through a regulatory framework that makes common reporting and oversight standards binding. It is recommended that the technical secretariat receive technical support from existing international expert institutions such as UN Tourism in terms of data standards and comparability capacity; in terms of accountability, independent audit reports should be published publicly. Solidarity mechanisms such as the EU Solidarity Fund, eligibility and expenditure monitoring; and UNDP's climate resilience financing practices can provide a reference framework for fund design in terms of rapid payment + recovery/adaptation investment windows.

Practical contributions of the study are directly related to the implementation dimension. Through the regional emergency fund, countries will have the capacity to respond quickly to crises; joint response teams can be established in disasters such as forest fires and emergency deployment points can be established. Designing the fund in such a way that it can be activated with certain trigger indicators (such as proportional decrease in the number of tourists, percentage of burnt forest area, pandemic-related costs) within the state budget architecture will accelerate resource allocation. This system will make sure that countries share the burden fairly and reduce both economic and social losses. The fund can help support alternative tourism models like reforestation projects, investments in infrastructure, and health tourism. This will make the sector more resilient in the long run.

THEORETICAL IMPLICATIONS

This study makes an important theoretical contribution to the sustainable tourism literature by addressing the concept of regional contingency fund against unexpected crises such as pandemics and forest fires from a public finance perspective. Furthermore, the analysis using the MEREC-CoCoSo method objectively demonstrates the impact of the fragile structure of the tourism sector on economic sustainability and provides a new perspective on how proportional financing models can work within a state budget-like mechanism. The findings show that GDP is the most critical criterion and emphasize the importance of budget carrying capacity and counter-cyclical fiscal capacity in times of crisis. Thus, the study fills the gap in the literature by proposing a regional fund that operates in the logic of a state budget against tourism shocks and builds a theoretical bridge between sustainable tourism and public finance. Additionally, the emergence of macro-level transportation/financial capacity indicators as decisive factors demonstrates that tourism vulnerability must be addressed not only through sectoral indicators but also in conjunction with public finance and governance capacity. This emphasis is consistent with the adaptive capacity approach in the destination resilience literature and strengthens the theoretical basis for regional risk-sharing mechanisms in an era of multiple crises.

REFERENCES

Aslan, A. (2016). Does tourism cause growth? Evidence from Turkey. *Current Issues in Tourism, 19*(12), 1176–1184. DOI: 10.1080/13683500.2015.1015970

Boughton, J. M. (2002). Why white, not Keynes? Inventing the postwar international monetary system. *IMF working Paper.* https://www.imf.org/external/pubs/ft/wp/2002/wp0252.pdf

Boustras, G., & Boukas, N. (2013). Forest fires' impact on tourism development: A comparative study of Greece and Cyprus. *Management of Environmental Quality, 24*(4), 498–511. DOI: 10.1108/MEQ-09-2012-0058

Brezina, C. (2012). *Understanding the gross domestic product and the gross national product.* Rosen Publishing.

Canadell, J. G., Meyer, C. P. M., Cook, G. D., Dowdy, A., Briggs, P. R., Knauer, J., Pepler, A., & Haverd, V. (2021). Multi-decadal increase of forest burned area in Australia is linked to climate change. *Nature Communications, 12*(1), 1–11. DOI: 10.1038/s41467-021-27225-4 PMID: 34836974

Çelik, A. K., Özcan, S., Topcuoğlu, A., & Yıldırım, K. E. (2013). Effects of the tourism industry on the balance of payments deficit. *Anatolia, 24*(1), 86–90. DOI: 10.1080/13032917.2013.772529

Cortes-Jimenez, I., & Pulina, M. (2010). Inbound tourism and long run economic growth. *Current Issues in Tourism, 13*(1), 61–74. DOI: 10.1080/13683500802684411

Dann, G. M. S. (1977). Anomie, ego-enhancement and tourism. *Annals of Tourism Research, 4*(4), 184–194. DOI: 10.1016/0160-7383(77)90037-8

Davis, D., Allen, J., & Cosenza, R. M. (1988). Segmenting local residents by their attitudes, interests, and opinions toward tourism. *Journal of Travel Research, 27*(2), 2–8. DOI: 10.1177/004728758802700201

Demir, E., & Bahar, O. (2021). The impact of tourism revenues on economic growth: Empirical analysis on Turkey. *International Journal of Social Sciences and Education Research, 7*(2), 162–167. DOI: 10.24289/ijsser.699497

Duro, J. A., Perez-Laborda, A., Turrion-Prats, J., & Fernández-Fernández, M. (2021). Covid-19 and tourism vulnerability. *Tourism Management Perspectives, 38*, 1–12. DOI: 10.1016/j.tmp.2021.100819 PMID: 34873568

Faulkner, B. (2001). Toward a framework for tourism disaster management. *Tourism Management, 22*(2), 135–147. DOI: 10.1016/S0261-5177(00)00048-0

Fields, G. S. (1988). Employment and economic growth in Costa Rica. *World Development, 16*(12), 1493–1509. DOI: 10.1016/0305-750X(88)90222-7

Gelashvili, I. A. (2022). The Negative impact of pandemics on the tourism sector and solution ways. *Globalization and Business*, *7*(13), 138–142. DOI: 10.35945/gb.2022.13.021

Gössling, S., Scott, D., & Hall, C. M. (2021). Pandemics, tourism and global change: A rapid assessment of Covid-19. *Journal of Sustainable Tourism*, *29*(1), 1–20. DOI: 10.1080/09669582.2020.1758708

Holling, C. S. (1973). Resilience and stability of ecological systems. *Annual Review of Ecology and Systematics*, *4*(1), 1–23. DOI: 10.1146/annurev.es.04.110173.000245

IMF (2019). Building resilience in developing countries vulnerable to large natural disasters. *Policy Papers* 2019, No:020. https://doi.org/DOI: 10.5089/9781498321020.007

Jafari, J. (2001). *Encyclopedia of Tourism*. Routledge.

Keshavarz-Ghorabaee, M., Amiri, M., Zavadskas, E. K., Turskis, Z., & Antucheviciene, J. (2021). Determination of objective weights using a new method based on the removal effects of criteria (MEREC). *Symmetry*, *13*(4), 1–20. DOI: 10.3390/sym13040525

Khan, A., Bibi, S., Lorenzo, A., Lyu, J., & Babar, Z. U. (2020). Tourism and development in developing economies: A policy implication perspective. *Sustainability (Basel)*, *12*(4), 1–19. DOI: 10.3390/su12041618

Kumar, J., & Hussain, K. (2014). Evaluating tourism's economic effects: Comparison of different approaches. *Procedia: Social and Behavioral Sciences*, *144*, 360–365. DOI: 10.1016/j.sbspro.2014.07.305

Lee, H. C., & Chang, C. T. (2018). Comparative analysis of MCDM methods for ranking renewable energy sources in Taiwan. *Renewable & Sustainable Energy Reviews*, *92*, 883–896. DOI: 10.1016/j.rser.2018.05.007

Li, C. C., Mahmood, R., Abdullah, H., & Chuan, O. S. (2013). Economic growth, tourism and selected macroeconomic variables: A triangular causal relationship in Malaysia. *Margin*, *7*(2), 185–206. DOI: 10.1177/0973801013483503

Marin, D. (1992). Is the export-led growth hypothesis valid for industrialized countries? *The Review of Economics and Statistics*, *74*(4), 678–688. DOI: 10.2307/2109382

McKibbin, W., & Fernando, R. (2020). The economic impact of COVID-19. In Baldwin, R., & Fernando, R. (Eds.), *Economics in the time of COVID-19* (1st ed., pp. 45–51). CEPR Press.

Miller, G. A., & Ritchie, B. W. (2003). A Farming Crisis or a Tourism Disaster? An Analysis of the Foot and Mouth Disease in the UK. *Current Issues in Tourism*, *6*(2), 150–171. DOI: 10.1080/13683500308667949

Mou, N., Yuan, R., Yang, T., Zhang, H., Tang, J., & Makkonen, T. (2020). Exploring spatio-temporal changes of city inbound tourism flow: The case of Shangai, China. *Tourism Management*, *76*, 1–14. DOI: 10.1016/j.tourman.2019.103955

Mukhametzyanov, I., & Pamucar, D. (2018). A sensitivity analysis in MCDM problems: A statistical approach. *Decision Making: Applications in Management and Engineering*, *1*(2), 51–80. DOI: 10.31181/dmame1802050m

Nene, G., & Taivan, A. (2017). Causality between tourism and economic growth: Evidence from Sub Saharan African (SSA). *Journal of Developing Areas*, *51*(2), 155–169. DOI: 10.1353/jda.2017.0037

OECD. (2020) Tourism trends and policies 2020. https://www.oecd.org/cfe/tourism/OECD-Tourism-Trends-Policies%202020-Highlights-ENG.pdf

OECD. (2024) Employment by activity. https://data.oecd.org/emp/employment-by-activity.htm#indicator-chart

Qin, Y., Luo, Y., Zhao, Y., & Zhang, J. (2018). Research on relationship between tourism income and economic growth based on meta-analysis. *Applied Mathematics and Nonlinear Sciences*, *3*(1), 105–114. DOI: 10.21042/AMNS.2018.1.00008

Ritchie, B. W., & Jiang, Y. (2019). A review of research on tourism risk, crisis and disaster management: Launching the annals of tourism research curated collection on tourism risk, crisis and disaster management. *Annals of Tourism Research*, *79*, 102812. DOI: 10.1016/j.annals.2019.102812

Ritchie, J. R. B., & Goeldner, C. R. (1994). *Travel, tourism, and hospitality research: A handbook for managers and researchers* (2nd ed.). Wiley.

Sánchez-Cañizares, S. M., Cabeza-Ramírez, L. J., Muñoz-Fernández, G., & Fuentes-García, F. J. (2021). Impact of the perceived risk from Covid-19 on intention to travel. *Current Issues in Tourism*, *24*(7), 970–984. DOI: 10.1080/13683500.2020.1829571

Scott, D., & Lemieux, C. (2010). Weather and climate information for tourism. *Procedia Environmental Sciences*, *1*, 146–183. DOI: 10.1016/j.proenv.2010.09.011

Stevens, R. (2023, 27 July). Wildfires raging across southern Europe and North Africa kill more than 40 people. World Socialist Web Site. https://www.wsws.org/en/articles/2023/07/27/rkzb-j27.html

Tourism, U. N. (2020) UNWTO launches a call for action for tourism's COVID-19 mitigation and recovery. https://www.unwto.org/news/unwto-launches-a-call-for-action-for-tourisms-covid-19-mitigation-and-recovery

UNDP. (2025) Social protection, risk finance and insurance. https://irff.undp.org/sites/default/files/2025/Sep/Social-Protection-Risk-Finance-and-Insurance.pdf

Vanhove, N. (1981). Tourism and employment. *International Journal of Tourism Management*, *2*(3), 162–175. DOI: 10.1016/0143-2516(81)90003-7

World Bank. (2024) World Bank national accounts data, and OECD national accounts data files. https://databank.worldbank.org/metadataglossary/world-development-indicators/series/NY.GDP.MKTP.KD.ZG

World Bank & GFDRR. (2014) Financial protection against natural disasters. https://www.gfdrr.org/sites/default/files/documents/Financial%20Protection.pdf

World Travel and Tourism Council. (2019) Benchmarking research trends 2019: How does travel & tourism compare to other sectors? https://wttc.org/Portals/0/Documents/Reports/2019/Benchmarking%20Trends%20Report%202019.pdf?ver=2021-02-25-182807-540

Yamin, M. (2020). Counting the cost of Covid-19. *International Journal of Information Technology : an Official Journal of Bharati Vidyapeeth's Institute of Computer Applications and Management*, *12*(2), 311–317. DOI: 10.1007/s41870-020-00466-0 PMID: 32412538

Yazdani, M., Zarate, P., Kazimieras Zavadskas, E., & Turskis, Z. (2019). A combined compromise solution (CoCoSo) method for multi-criteria decision-making problems. *Management Decision*, *57*(9), 2501–2519. DOI: 10.1108/MD-05-2017-0458

Zhang, Y., Tachibana, S., & Nagata, S. (2006). Impact of socio-economic factors on the changes in forest areas in China. *Forest Policy and Economics*, *9*(1), 63–76. DOI: 10.1016/j.forpol.2005.02.005

KEY TERMS AND DEFINITIONS

CoCoSo (Combined Compromise Solution): It is a method that compares alternatives according to different criteria and finds the most appropriate solution.

Forest Fires: It is the destruction of the forest ecosystem by fires that go out of control.

MEREC (Method Based on the Removal Effects of Criteria): It is a method that determines the weights of the criteria in the decision process by subtracting the effect of each criterion on the result.

Pandemic: It is the global spread of an infectious disease such as COVID-19.

Proportional Financing: It means financing a project, program or post-disaster expenditure in a fair and balanced manner according to resources and needs.

State Budget: It is the process of organizing and approving public revenues and public expenditures within the framework of the annual plan.

Sustainable Tourism: It is an understanding that balances the economic, social and environmental dimensions of tourism activities and secures the transfer of natural resources to future generations.

Chapter 12
Extreme Weather Linked to Climate Change:
Risk, Vulnerability, and Exposure

Mohd Faiz Afzal
Afzal Multispeciality Clinic, India

Pradosh Kumar Sharma
Chinmaya Degree College, India

S. V. Subrahmanyam
Koneru Lakshmaiah Education Foundation, India

Asha Bala
Shri Guru Ram Rai University, India

Baburao Gaddala
http://orcid.org/0000-0003-4190-8325
School of Studies of Engineering and Technology, Bilaspur, India

Manasi Vyankatesh Ghamande
Vishwakarma Institute of Technology, India

V. Bhoopathy
http://orcid.org/0000-0003-2175-6328
Sree Rama Engineering College, India

ABSTRACT

Climate change has made extreme weather events happen more often, more severely, and with less warning all around the planet. Heatwaves, floods, cyclones, droughts, and wildfires are all getting worse and are bad for ecosystems, economies, and people's health. This chapter looks at extreme weather events from the points of view of risk, susceptibility, and exposure. It looks at the scientific reasons why climate change is linked to extreme occurrences, the social, economic, and environmental aspects that make people more vulnerable, and the patterns of exposure across different areas and groups of people. The chapter goes on to talk about evaluation frameworks, real-world case studies, adaptation and mitigation techniques, and policy responses that are meant to lessen the effects of disasters. To make societies that can survive climate change and move forward with sustainable development in a climate that is changing quickly, it is important to understand these characteristics.

DOI: 10.4018/979-8-3373-6274-8.ch012

1. INTRODUCTION

An important problem that humans face is the anthropogenic climate crisis. Changes in many extreme weather and climate events have been observed since about 1950 and human influence on the climate system is clear were the main conclusions drawn from the AR5. Adaptation to a changing climate is becoming an increasingly pressing concern for societies worldwide. Combining adaptation to certain inevitable hazards with climate change mitigation through reduction of greenhouse gas emissions might help reduce such risks, which necessitates considering susceptibility and exposure to climate-related hazards. The process of evaluating the relative contributions of multiple causal factors to a change or event with an assignment of statistical confidence is one definition of attribution, which is used to describe climate change. Since many climate change hazards are linked to the incidence of severe weather or climate events, attrition is an essential part of comprehending these risks. Separate episodes of extreme weather or unusual climate conditions, often associated with detrimental impacts on society or natural systems, defined using some metric to characterise either the meteorological characteristics of the event or the consequent impacts. is one way to describe these occurrences (Stott et al., 2023). Timescales for events can range from minutes to seasons or even longer, and spatial ranges from a few kilometres to the size of continents are equally flexible. To that end, attribution can be used for extremes that could be categorised as either weather-related (such as an exceptionally high daily rainfall total) or climate-related (such as an exceptionally high yearly mean temperature). A fundamental assumption is that certain extremes will see different occurrences as a result of climate change. In regions where average temperatures rise, scorching heat waves are likely to become more often. Changing the distribution can either amplify or counteract this trend, but a change in the mean alone is enough. As an example, when soils dry out, they no longer offer evaporative cooling to moderate temperatures, which can lead to an increase in temperature extremes and a wider dispersion of summer maximum daily temperatures in the interiors of continental regions (Çelik, 2026). Extreme rainfall events may become more common as a result of atmospheric warming, which enhances the atmosphere's moisture-holding capacity. The observed record shows these trends in temperature and rainfall extremes predicted by thermodynamic considerations, and climate models indicate that human-caused forcings are responsible for the increased occurrence of such occurrences globally (D. G. Huber & Gulledge, 2021).

2. UNDERSTANDING CLIMATE RISK

In light of the patterns shown by the aforementioned severe events, we can predict the kinds of dangers that society will face in the years to come. You can evaluate the level of risk and be ready for it if you have a record of patterns and reliable predictions for the future. You might think of risk as a continuous range of possibilities, each with its own probability distribution; the low-probability tails of the range or distribution include the most extreme events. One effect of climate change is an increase in the frequency and severity of heat waves, as well as a broadening of the probability distribution for temperature extremes. Playing with loaded dice, which are weighted to roll high numbers more frequently, is quite similar to the increasing danger of extreme events. Thus, a (i.e., a greater maximum achievable temperature) may be rolled (due to the loaded dice), and it would be more likely (compared to rolling a with two regular dice) than before (Bolan et al., 2024). The translated risk of changing climate from the loaded dice's probability distribution compared to that of normal dice. The odds of rolling snake eyes (the cold extreme) and double sixes (the hot extreme) on a standard dice roll are roughly identical. But the odds are stacked against cold extremes (as defined in the previous climate) and in favour of hot extremes (as defined in the current climate) due to climate change. There is a far higher chance of record heat under this revised risk profile. The catastrophic heat wave that hit Europe in 2003 is a real-life example that fits this new standard.

A new, hotter, and more variable climatic profile greatly increases the likelihood of such a massive occurrence, compared to the old, unaltered climate regime. This event's actual occurrence proves that such a powerful event is not out of the question, and our best estimates indicate that the likelihood of such an occurrence will increase substantially in the decades to come as a result of climate change (Ummenhofer & Meehl, 2017). It is worth noting that the 2003 heat wave could have occurred 60 years ago instead of 500 years ago under the previous system if the variance were to increase by 50% without changing the average temperature. According to some studies, the likelihood of a heat wave similar to the one in 2003 occurring in Europe is already two times higher due to the increase in global temperatures over the past few decades. This kind of disaster may become more common by the middle of this century if global warming continues. Unpredictable increases have occurred in a variety of weather events, not only hot extremes. While variations in daily precipitation are within predictions, the observed increases in extreme hourly precipitation are beyond projections (D. G. Huber & Gulledge, 2021). This suggests that there is a discrepancy between the scaling of precipitation with increases in atmospheric moisture for sudden downpours and for cumulative amounts over longer time periods. Observed increases in one-hour precipitation extremes have been twice as rapid as temperature increases in the Netherlands, according to a study. This is in line with expectations when temperatures surpass 12°C. When the risk distribution is not only moved, but also shows greater variance, as this example shows, the number of extreme events might increase rapidly.

3. CLIMATE CHANGE AND ECE

The frequency and intensity of extreme weather events have increased due to climate change and global warming (Amsalu et al., 2026). According to these occurrences are deemed extreme when they deviate considerably from 90% to 95% of previously documented meteorological events in the same region. The pollution of different ecosystems can be affected, either directly or indirectly, by these severe weather events caused by climate change, including droughts, floods, and wildfires (Hulme, 2020; Çelik et al., 2025).

3.1 Flooding

Climate change is making extreme weather events, such flooding, increasingly worse. Warming temperatures and higher atmospheric moisture content lead to heavier rainfall, which in turn increases the likelihood of floods. Tropical cyclones and hurricanes also become more common as a result of warmer skies. According to the Intergovernmental Panel on Climate Change (IPCC) special statement on extreme weather events, current climate change has affected several water-related variables, such as rainfall and snowmelt, which contribute to floods (McBean, 2020). However, there is a lack of data on historical flooding, which makes it difficult to compare it to trends in flooding caused by humans and climate change. On the other hand, the frequency of extraordinary precipitation occurrences has increased significantly over the past 30 years, as pointed out by (Howe, 2021). Between 1981 and 2010, rainfall occurrences increased by 12% due to this tendency, suggesting a 26% chance that the newly-established unique rainfall peak was linked to human-caused climate change. Climate change is a major contributor to floods because it raises the relative humidity of the air as a function of temperature. Atmospheric moisture capacity increases by about 7% for each degree of warming (NASA, 2022). According to NASA (2023), the average global temperature has risen around 1 o Celsius due to human activity, and according to the amount of water vapour in the atmosphere has climbed by 17%. In their 2018 study, examined the association between climate change and the frequency of flood episodes on both a global and latitudinal scale.

Additionally, a number of studies have shown that extreme weather events are more likely to be severe, last longer, and be more intense when there is more water in the atmosphere. The planetary water cycle has also been altered by climate change, which has intensified the present-day climatic patterns. More concentrated precipitation with strong downpours is a consequence of changing climate patterns; this has raised the risk of flooding in both flood-prone and previously flood-free places. There will be up to 50% more precipitation in the future due to studies predicting that future severe rainfall events will be up to three times the average of the past century (Jones & Boer, 2023). According to likelihood of rain-on-snow events increases in areas where seasonal snowmelt is accompanied by heavier and warmer precipitation in the form of rain. This phenomenon has been observed in the western United States, where rivers that are fed by snowmelt have experienced an earlier peak flow in the spring since at least 1950. Levels of precipitation can cause significant damage, especially in areas that are more likely to experience urban flooding. Anthropogenic climate change has increased the likelihood of extreme weather events like the devastating 2016 Louisiana floods by 40% and made them 10% more likely, according to an analytical study by the National Oceanic and Atmospheric Administration (NOAA) (NOAA, 2016). Climate change is a big worry because it increases the likelihood of flooding. Floods not only cause physical damage, but they also transport pollutants and toxins such as sewage, chemicals, and garbage, which can poison waterways and potentially injure both humans and wildlife (Zscheischler et al., 2018).

3.2 Wildfires

The frequency of forest fires and wildfires around the world has been affected by the increase in temperatures caused by human activity and climate change. America and Canada are not immune to wildfires and forest fires; in fact, they happen all over the globe. The global number of wildfires declined little from 5.09 million in 2004 to 4.51 million in 2019. U.S. and Canadian wildfire occurrences fell at around the same rate. Possible explanations for the decline in wildfire occurrence include better education of the public, more legislation for fire prevention, and innovative technological solutions (Magnan et al., 2023). The environmental and public health impacts of wildfires are a major issue in many parts of the world, including Asia, Australia, Europe, North America, and South America, it has been shown in studies by (Jurgilevich et al., 2017). They are just a few of the numerous controlled experiments, field studies, and modelling efforts that have gone into understanding wildfires and their causes, effects, and ignition sources. Forest fuel, terrain, and weather are three of the many variables that influence the frequency, severity, and long-term effects of wildfires on the ecology and climate. The frequency, intensity, and impact of wildfires have been magnified due to climate change and the continuous extension of fire-prone landscapes. Severe wildfires can have far-reaching social and economic effects. Significant human casualties and property damage were caused by the California wildfires. Globally, fires are becoming more frequent, extensive, and devastating due to a combination of reasons such as changing land use patterns, longer fire seasons, more severe droughts, insufficient fire suppression techniques, and higher average temperatures. The four main factors that determine the occurrence of fires are the following: the nature of the fuel, the surrounding environment and weather, the sources of ignition, and human activities. Fire occurrence and progression are affected by factors such as fuel quantity, kind, availability, and structural and moisture conditions. Above a hot source, such a wildfire, pyrocumulonimbus clouds form (Manatsa et al., 2023).

Surface fire weather conditions and extreme atmospheric circumstances work together to cause extreme fire events. The time it takes for a fire to start depends on where it started. According to lightning-induced fires tend to happen in the summer, while human-caused fires tend to be more common in the afternoon. Intentional and unintentional fires are associated with socioeconomic factors, including unemployment rates and agricultural activity. It found that variables like as elevation, slope, and closeness to both habitations and transportation were associated with the frequency of wildfires in eastern Kentucky, USA. Human activity accounted for 49% of the variation in wildfire patterns in Canada between 1990 and 2016, with lightning accounting for 47% and other unknown causes accounting for 3%. Extreme weather, including high temperatures, low humidity, and strong winds, was associated with devastating wildfires in eastern Australia and the United States (Ekwurzel et al., 2023). While human-caused fires were either flat or declined across various regions of Canada from 1959 to 2015, lightning-caused fires increased. According to modelling research conducted for New South Wales, the likelihood of extreme fires is expected to increase by over 25% by 2050 under two scenarios with high emissions of greenhouse gases and aerosols, and by an additional 20% under a scenario with low emissions by 2100 found that variables such as elevation, land cover, population density, and human accessibility impacted the spatial distributions of fire ignitions in Portugal. Prevention of wildfires or decrease in the frequency and severity of related risks, such as loss of life, property, and resources, could be achieved by the development and promotion of suitable risk management strategies and mitigation methods. Effective fire management and control techniques can be developed by fire managers by understanding the behaviour and hazard of fires at multiple spatial and temporal dimensions, such as regional, global, daily, and seasonal (Keller et al., 2021).

3.3 Drought

According to predictions, global temperatures are projected to surpass pre-industrial levels by more than 2 degrees Celsius by the year 2100, even though the

Paris Agreement aims to keep the increase in global temperature below 2 or 1.5 degrees Celsius. There is some evidence that droughts are becoming more often, intense, and prolonged as a result of climate change, even though this phenomenon is not always caused by it. This finding is supported by a trend analysis of drought frequency in locations like the Haihe River Basin in China (Taylor et al., 2024). The timing, frequency, and quantity of water availability are affected by the rise in world average temperatures caused by anthropogenic greenhouse gas emissions, which in turn influence climate change. Meteorological droughts, which are brought on by climate change, are preceded by hydrologic droughts, which affect streams, lakes, and groundwater; agricultural droughts, which affect the moisture of agricultural topsoil; and socio-economic droughts, which affect water resources for human use estimates that the United States has incurred damages above $250 billion due to drought, making it the costliest natural disaster in the country. Droughts will become more common as a result of climate change. It found that soil temperatures can be raised by elevated CO2 levels, which in turn increases water loss due to enhanced plant photosynthesis. But there are academics who think that plants could require less water if atmospheric CO2 concentrations were to rise. Warmer weather causes soil moisture trends to decline, even though a CO 2 fertilisation effect could partially compensate for predicted drought. It found that warm and dry conditions are more likely to occur at the same time found that soil moisture and temperature can trigger droughts and wildfires when they interact. Soil surface moisture and soil water retention are both expected to decrease at an accelerated rate due to climate change (Li et al., 2024).

The water resources that the 1.9 billion people living in the Northern Hemisphere rely on—winter snowfall and snowpack—are also being affected by climate change. To ensure a sufficient water supply, it is necessary to collect snowfall throughout the winter and let it melt during the summer. The availability of snow resources is decreasing, and their storage durations are getting shorter, due to increasingly severe climate conditions (LI et al., 2021). For instance, the snowpack in the Sierra Nevada in the United States was 5% of the historical average in April 2015, marking the lowest level in 500 years. In that year, California had a severe drought due in part to the fact that the snowpack from the Sierra Nevada did not reach the level required to fill 30% of the state's reservoirs. It observed that the intense wildfires of 2020 caused the snowpack to melt around 18 days earlier than typical in the western United States, demonstrating that snowmelt can be accelerated by climate change-induced droughts and wildfires. Because snow is so reflective, a lot of the sun's heat doesn't end up warming the Earth; instead, it's reflected back into space.

As a result, a domino effect that worsens dryness can be created when snowpack areas are rapidly reduced, which in turn raises surface temperatures. Changing storm paths is another manner in which atmospheric warming impacts precipitation. Storm tracks, which include both temperate and subtropical cyclones, generally develop between 30 and 60 degrees north and south of the equator. Due to rising CO 2 levels and ozone degradation, the southern storm track experiences noticeable shifts during the summer (Ginglinger & Moreau, 2023). The two-stage shift in storm tracks, which includes nonlinear horizontal advection and diabatic heating linked with latent heat release, could result in less rainfall in regions that need it, since most mid-latitude rainfall extremes occur in cyclones and their accompanying fronts and warm conveyor belts. On the other hand, severe rainfall from extra-tropical cyclones might cause floods in certain areas of the same region cite a striking example of this phenomenon: southern Australia has seen a dramatic drop in cool-season rainfall, whereas northern Australia has seen an increase in warm-season precipitation and thunderstorms. Moreover, climate change has the potential to amplify Oscillation occurrences, which in turn affect hydroclimates on a more frequent and unpredictable basis across large geographic scales. There are two sides to every coin, and climate change is no exception. In regions that are already prone to floods, it might cause even more rain to fall (Eckstein et al., 2019).

4. CONCEPTUAL FRAMEWORK-RISK, VULNERABILITY AND EXPOSURE

Individual extreme weather events can and do serve as learning tools for policymakers and planners to evaluate economic and social vulnerabilities and devise strategies to mitigate future costs and suffering caused by increasing risk. For instance, during the 1995 Chicago heat wave, which killed hundreds of people, the city made effective preparations for future heat waves. Nobody in Europe had ever considered the possibility that a heat wave of this magnitude could hit until the 2003 event. Preparations have begun now that the European public is aware of the likelihood have been implemented to mitigate potential economic harm and alleviate future pain. As demonstrated, a large American metropolis can remain inaccessible for weeks due to a lack of proper communication, security, restrooms, or medical treatment (Aven, 2020). As a means of adaptation, New Orleans was spared Katrina's devastating storm surge by sealing up the Mississippi River-Gulf Outlet Canal. Lessons on the most pressing areas for flood control and emergency response systems, as well as the worth of the necessary investments, should be gleaned from other recent instances of severe rainfall and flooding. Since it is of the utmost importance to update estimates of both existing and future danger, severe occurrences also provide data points for updating trends and improving risk predictions. Managing the dangers of extreme weather in a warming environment requires both adapting to the inevitable climate change and reducing future emissions of greenhouse gases. Cutting down on carbon dioxide emissions helps keep nonlinear increases in extremes to a minimum because doing so reduces the overall impact of climate change (Arbuckle et al., 2021).

This quality makes mitigation activity a benefit multiplier, since it can cause emissions to decline in a nonlinear fashion even when emissions are reduced linearly. The flip side is that civilisation must adapt to some repercussions of climate change since it is already happening. For adaptation measures to work, they need to be proportional to the severity of the threat. Adaptation efforts must go above and beyond what is anticipated in response to changes in ordinary climate conditions in order to address the nonlinear increases in risk posed by weather extremes. Additionally, if the climate changes too much, many adaption techniques will become infeasible (Rising et al., 2022). As a result, the amount of adaptation that can be implemented should be proportional to the mitigating efforts. Although science cannot predict the future, it does provide useful resources for assessing the dangers posed by climate change. By looking at current trends and future predictions from climate models, scientists can determine if the risk of particular catastrophes is on the rise or not. When these two signs come together, it's a good idea to take precautions against these kinds of disasters. Climate models are being used by an increasing amount of research to predict potential future increases in risk (Churi et al., 2022).

While models can't foretell every possible outcome, they can show how risk profiles can shift in the future in response to certain extremes. This method shines when compared to real-life threats that everyone believes should be avoided. Devastating autumn floods, for instance, struck the UK in 2000, and climate models do a good job of simulating the corresponding climatic circumstances. Severe fall flooding is 20–90% more likely under current climatic conditions than it was before industrialisation, according to a climate model (Nieto, 2019). While models did not show that climate change has raised the likelihood of the catastrophic 2010 Russian heat wave, they did predict that comparable waves will be more common later this century due to sustained warming. So, as the world's temperature continues to warm, the likelihood of future heat waves like that one will increase, regardless of what caused it. The event's high death toll and economic impact highlight the need to give considerable thought to how to mitigate and adapt to future disasters of a similar kind. Considering the dangers and unknowns, it is pointless to wonder if certain occurrences are accelerated by global warming. Unabated emissions of greenhouse gases are increasing the likelihood of extreme weather occurrences; thus, it is prudent to learn from past mistakes and strengthen our defences against them. In order to prioritise and evaluate investments in adaptation and prevention, decision makers can utilise risk-based information that climate science provides to understand how the risk is evolving (Stroebel & Wurgler, 2021).

5. CHANGES IN ECE WITH ECOLOGICAL RELEVANCE

Changes in many ECEs were quite strong in the second part of the 20th century. Various modifications can cause shifts in the frequency of ECEs; one such change for a temperature distribution: Changes in the distribution's variability, symmetry, or skewness, along with changes in the mean, might increase the frequency of ECEs (B. Howard et al., 2022).

5.1 Ocean

5.1.1 Additional Extreme Climatic Phenomena Impacting Marine Biological Systems

When reviewing the ways in which coastal marine species are impacted by climate change, differentiated physical changes associated with rising sea levels, altered circulation, pH, carbon dioxide, and ultraviolet light. Alterations to distributional patterns (such as biogeographic ranges and zonation patterns), species composition, biodiversity, and community structure; changes to primary and secondary production; and population dynamics are all examples of emergent ecological responses. When it comes to the impact that physical environment changes have on marine biological systems, ECEs are probably a regulating factor (Quandt et al., 2023). Severe storms extreme wave activity extreme sea level salinity fluctuations and maritime heatwaves and cold snaps are all examples of ECEs that are not tied to weather. It conducted a synthesis study that looked at changes in extratropical storm, wind, and wave ECEs. They discovered that storms have gotten stronger and more frequent during the cold season in the Northern Hemisphere since the 1950s, and that extreme winds over the oceans have been on the rise since the 1980s. While there is conflicting evidence for other US coastlines, the Pacific coast has shown a moderate increase in extreme waves during the 1950s (Kreft, 2023).

Several parts of the globe have seen an uptick in extremely high waves, including the North Atlantic over the twentieth century the South American Pacific coast during the 1980s and the US Pacific Northwest. According to who used a multi-model ensemble, the world's annual mean significant wave height will fall by 25% between 2070 and 2100. However, just 7% of ocean areas, largely in the Southern Ocean, showed an increase during the same timeframe. On the other hand, predict that by the end of the twenty-first century, significant wave height is expected to rise by 15% worldwide, which is more than the predicted changes in surface pressure and wind speed (Mirza, 2023). Similarly, towards the end of the twenty-first century, discovered that the eastern Pacific and the extratropic of the Southern Hemisphere had large increases in wave height. Coastal areas, including the Gulf of Mexico,

South and East Asian coasts, the Gulf of Bengal, and Chile, are also projected to see extreme wave heights double or even triple as a result of higher surface winds and sea-level pressure gradients. Because coastal and near-shore biological communities undergo physical disturbances as a result of intense wave action, changing storm features can significantly affect natural systems. As an example, alterations in intense storms and the resulting wave action can alter the structure of kelp forests (Konisky et al., 2020).

While low levels of severe storms (i.e., one storm per 3-4 years) assist sustain complexity in kelp forest food webs, demonstrated that gigantic kelp forests experience a reduction in diversity when severe storms occur more frequently (i.e., at an annual frequency). According to subtidal algal forests in the Adriatic have been decimated due to a combination of factors, including a number of severe storms and the persistent instability of habitats caused by humans. It devastating coastal storms can impact ecosystems far beyond the shoreline towns (Meehl et al., 2023). The benthic communities in the Western Mediterranean were abrasion and buried as a result of the movement and redistribution of shelf sediments induced by an abnormally violent storm that hit the Spanish coast in December 2008. Coralliferous outcrops in the Northwest Mediterranean were among the areas hit by the storm, which caused a dramatic change in species composition at exposed and affected sites shortly after the storm passed and a loss of cover of benthic species ranging from 22-58%, with vulnerable species being hit the hardest. Also, marine ecosystems are susceptible to disruptions caused by biogeochemical property extremes. The amplitude of the annual CO_2 cycle is growing in relation to increasing emissions of greenhouse gases, according to research by who looked into instances of high CO_2 concentrations in seawater. Periodic exposure to CO2 concentrations that have negative physiological and neurological consequences on marine species may become a reality in the Southern Ocean, Pacific, and Atlantic fisheries by the middle of the 21st century (Zhang, 2023).

5.1.2 Marine Heatwaves and Cols Events

Marine heatwaves are periods of exceptionally high temperatures that occur in the ocean, just as heatwaves in the atmosphere. Its spatial range, intensity, length, and pace of evolution are hallmarks of this long-term discrete event characterised by abnormally warm water. When marine heatwaves disrupt aquaculture or significant fishing species, it can have economic ramifications cause local extinctions and alter the range of certain species. A few of the most recent maritime heatwaves that had significant biological effects happened in 2011 in Western Australia in the northwest Atlantic in 2012, and in the northern Mediterranean in 2003. Examining the North Atlantic and Pacific maritime heatwave frequency since 1950. The extent, duration, and intensity of maritime heatwaves are influenced by a variety of factors, including the particular variability of each region, different types of climate variability, and human-caused warming (D. Huber & Gulledge, 2023). The frequency of marine heatwaves has increased over the past several decades and is expected to rise further as a result of climate change however decadal variability is likely to have an impact as well. Extreme marine heatwaves and ongoing ocean warming have been associated with regime transitions in temperate reef ecosystems in Western Australia, according to research by (Cheng et al., 2023).

There was a dramatic shift away from kelp forests and towards communities more typical of tropical and subtropical waters as a result of these factors. In order to better understand both sudden reactions and more gradual changes (such as long-term warming) in biological systems, demonstrated that the cumulative responses to atmospheric forcing can shed light on marine ecosystems' large-amplitude state transitions. The North Pacific heat wave of 2013–2015 is an example of a record-setting ocean heat wave. For two years in a row, sea surface temperature anomalies were more than three standard deviations, making it one of the worst maritime heatwaves in recorded history (Romm, 2018). The most ecologically significant event in recorded history may have been this record-breaking ocean heat wave. Among the effects were the deaths of sea lions, whales, and sea birds, extremely low primary productivity in the ocean and the worst algal bloom ever recorded, which had a detrimental effect on shellfish along the western coast of North America. This event was marked by a general warming trend in the North Pacific Ocean overlaid with unusually warm interannual SST in the Gulf of Alaska and along the western coast of North America (Wehner, 2023).

These SSTs persisted and increased as a result of alternating interactions between the mid-latitudes and the tropical regions, and there's a chance that they could happen more often in the future as a result of more winter-time climate variability in the North Pacific. In which caused exceptionally high tropical sea surface temperatures in areas where coral reefs went through their third major bleaching event on record, is another instance of an ocean heatwave's influence (Clarke et al., 2022). The first mass bleaching happened in 1998, and the second was in 2010; These mass bleaching have happened every time an has happened since 1997, and they have happened when the gradually increasing sea surface temperatures due to human-caused warming have been accompanied by warm sea surface temperature anomalies, which have crossed the tolerance threshold and caused the reefs to bleach. The possibility of widespread coral reef death due to this ECE associated with the event of 2015–2016 is increased because a certain proportion of bleached reefs have perished in past mass bleaching events. As an example, by early 2016, more than 90% of Australia's Great Barrier Reef has reportedly bleached. Future episodes are likely to bring about even more severe mass bleaching due to the continued warming of the environment, which will have devastating effects on coral reef ecosystems (Ornes, 2018).

5.2 Atmosphere and Land

5.2.1 Extreme Precipitation

An increase in intense precipitation is anticipated in a world that is warming, as the water-holding capacity of the atmosphere scales with temperature, according to theoretical thermodynamic understanding. As a result of increased availability of moisture and stronger rainfall, the saturation-specific humidity rises by around 7% every 8C of warming, as stated in the famous Clausius-Clapeyron relationship. The yearly maximum daily rainfall rate also increased by 5.9–7.7 percent for the average surface temperature over the world. It reviewed changes in sub daily extreme rainfall and found that, at temperatures below 208C, the actual and simulated rates of increases in extreme precipitation were twice as high as what Clausius-Clapeyron had projected (Curtis et al., 2017). The likelihood of storms with short durations (less than one day) increased, which could lead to more frequent and severe flash floods. Although confidence in heavy precipitation over other land areas was only medium, it is likely that its frequency has increased over land globally since the 1950s, particularly over North America and Europe. Reduced rain use efficiency has been observed throughout biomes as a result of a change to more extreme precipitation patterns characterised by heavier rainfall and longer dry periods. This is most noticeable in dry grasslands and Mediterranean woodlands, where the reduced vegetation output and higher water stress conditions account for 16-20%. Both static (including changes in circulation, modes of variability, and teleconnections) and dynamic (including thermodynamic) factors can contribute to changes in regional precipitation. The significance of studying the impacts of local scale thermodynamics and large-scale atmospheric circulation on sub daily severe rainfall intensity was emphasised by (Sánchez-Almodóvar et al., 2022). Extreme precipitation occurrences have been observed to be intensified by regional SST warming, according to a series of recent studies. As an example, the event caused severe flooding in the northeast of Australia due to the country's extremely heavy rainfall. Atmospheric circulation model tests using ocean conditions in 2010 and 2011 with and without long-term warming showed that both dynamic and thermodynamic factors intensified the circumstances that produced rain. In 2010 and 2011, the world's semi-arid vegetation had record-breaking growth due to the widespread rainy conditions that occurred in the interior of the continent of Australia. This included the rare filling (Perkins-Kirkpatrick et al., 2022).

5.2.2 Wildfires

Wildfires are more likely to break out during the final stages of a drought, when circumstances are particularly favourable because to the high temperatures and low humidity that sometimes accompany thunderstorms. A study conducted by examined changes in fire severity in the western US. They discovered that mean and maximum fire size, as well as the annual area burned, had significantly increased since the 1980s. This suggests that forest fuels are no longer a limiting factor for fire occurrence. Midway through the 1980s, there was a dramatic shift in wildfire activity in the forests of the northern Rocky Mountains. Large-scale fires became more common, wildfires lasted longer, and the wildfire season lasted longer. This shift was accompanied by earlier spring snowmelt and higher temperatures in the spring and summer. Given the present trends of growing wildfire intensity, shifting precipitation patterns, and warming regional temperatures, it is crucial to discuss the consequences of continuing to suppress fires (Demski et al., 2017). For example, forest fragmentation, erosion rates, carbon sequestration, wildlife habitat availability, and post-fire seedling recruitment are just a few of the many ecological consequences that can result from intense wildfires. When trying to understand the dynamics between climate change and wildfires, multiple studies have stressed the significance of the interplay between the physical climate system and biological processes on both temporal and spatial scales: According to data on climate change and human activity at different time scales are essential for a better comprehension of the reasons and effects of forest wildfires in the western United States. In order to better understand the direct causal factors that drive wildfires across different spatial scales, called for additional research. In a study conducted by it was found that dense forests in the Amazon basin were more prone to fires during severe ENSO-related droughts. The researchers observed that when the end of the dry season was delayed by a month, the forests burned more extensively and left larger burn scars. On the other hand, forests that were open or transitional, with higher rates of deforestation, burned more frequently, suggesting that landscape fragmentation worsened the climate-mediated forest flammability (Hay & Mimura, 2023). Rather than linked the devastating bushfires that occurred in February 1983 and 2009 in Southeast Australia to the typical conditions of the Indian Ocean, specifically a positive Indian Ocean Dipole, which had an effect on soil moisture, as demonstrated by the long droughts that occurred in the region.

5.2.3 Temperature Extremes

An increase in either the maximum or minimum temperature causes a noticeable shift in the distribution of temperatures. The daily temperature range may fall as a result of strong evidence from many datasets showing that global minimum temperatures are increasing at a faster rate than previously thought. It is widely anticipated that the number of cold days and nights would decrease, while the number of warm days and nights will rise, on a global scale. Although there are regional differences, the coldest nights tend to be shorter in Asia and Europe. By 2070, the overall area experiencing freezing days is predicted to reduce by around 6% in North America, with a decline of 10-90 days in the number of freezing days depending on the region. Results for the United States, Australia, and Europe demonstrate that the ratio of average daily high temperature has been rising while average daily low temperature has been falling throughout the first ten years of in these areas, the ratio of daily record highs to record lows has been roughly 2:1 in the twenty-first century (Mann et al., 2017). The rise in average temperatures throughout this time period in these regions is the reason why the chances of more record highs than record lows have shifted. It is anticipated that this percentage will rise in the coming years due to the ongoing warming of the environment. For regions like North and Central America, Europe, Asia, Southeast Asia, and Oceania, there is a lot of faith in the trends of severe temperature. The frequency of extremely hot and cold days and nights have been trending in climate models since the 1950s, which is in line with observations, and is expected to keep changing throughout the twenty-first century. The majority of the changes in the United States climate, including less frost days, longer growing seasons, more warm nights, and more intense heatwaves, have been linked to higher levels of greenhouse gases produced by humans. Some extremes in precipitation and heat have also been linked to human-caused climate change (Kumar, 2023). There has been a correlation between changes in regional extreme temperatures and shifts in global and regional patterns of atmospheric circulation and SST. Ecosystems can experience reductions or changes in the distribution range of species' habitats due to the exceeding of critical temperature thresholds, but individual species can be significantly impacted by such extremes, as demonstrated by the negative effects on clutch size for one bird species and the overwintering success of butterfly species in the UK.

6. POLITICAL DIMENSIONS OF WEATHER ATTRIBUTION

In the midst of heated political discussions on the origins, effects, and solutions to climate change, a new field of study known as extreme weather attribution is taking shape. There is no escaping the political implications of the framing, funding, and motivation of weather event attribution studies. Different potential research funders and the subsequent interpretation and use of scientific results are affected by the justifications given for investing in this field of study, which can be categorised as scientific curiosity, adaptation guidance, liability for damages, or visualising climate change (Ummenhofer & Meehl, 2017). According to what Nature's Designers of [such weather attribution] services must think very clearly, the editorial stated in 2012 regarding the potential applications of the knowledge that climate scientists generate. Think about a scenario where the field of weather attribution is presented as a way to help with guiding adaptation decisions. There are many who believe this field of study must be advanced immediately so that new international adaptation monies can be more wisely distributed. Quantifying the impacts of anthropogenic climate change in this way is important in guiding the allocation of resources available for adaptation as stated by others, has similar sentiment. Nonetheless, the reasons why such ambitious promises are improbable are demonstrated by (Sisco et al., 2017). According to their argument, funding for climate adaptation should be prioritised in areas where vulnerability to meteorological hazard is high, rather than in areas where such hazards are primarily caused by humans.

There are complex political, social, and economic systems that mediate physical dangers; yet, extreme weather attribution provides little information regarding the extent to which damages are caused by meteorological risks as opposed to risk exposure. Additionally, there are practical and ethical questions regarding the future of investment allocation standards that would preclude the victims of tough-luck weather from receiving adaptation funds, when weather is divided into human-caused and tough-luck categories. An opposing viewpoint held by several weather attrition experts is that the growing loss and damage agenda of the UNFCCC does not differentiate between tough-luck and human-caused weather (Solow, 2017). From events linked to variability around present-day climate norms (such as weather-related natural disasters) to [slow-onset] processes linked to future projected changes in climate norms, the effects of loss and damage lie along a continuum. Even if standards and definitions have not been officially approved, it appears that the forensic expertise provided by extreme weather attribution science will not be useful. Think about the $38.8 billion in damage that the 2011 Thai floods produced. This amount of rain was not exceptional, according to the weather attribution study by hence it couldn't have been caused by humans. Considering the heavy precursor rains that caused the high river flows through it is understandable to ask and answer the extreme-weather blame question (van de Pol et al., 2017).

However, it is doubtful that conclusion would change the practical or rhetorical use of these events to justify funding flood defence improvements or enhancing adaptive capacity. Whether or whether a disaster's meteorological component can be attributed to human agency is less important than the likelihood of lowering susceptibility and enhancing coping capacities when it comes to policies to address loss and damage. If sulphate aerosols are ever injected into the stratosphere on a global scale, weather attrition capabilities could be used to settle political disputes over who is responsible for dangerous weather events that follow (Maxwell et al., 2019). Concerned parties may wonder, was this catastrophic typhoon caused by naturally occurring tropical storms, by elevated concentrations of greenhouse gases, or by artificially injected aerosols Nevertheless, applying extreme weather attrition science in such situations could potentially intensify the already-present political and ethical conflicts caused by aerosol injection technology, as stated more broadly by regarding the role of science in environmental controversies. Asking for proof that precursor cloud-seeding actions caused local rainfall is a poor historical parallel (Palmer et al., 2017).

7. FUTURE CLIMATE CHANGE

Climate change and a militarised world were recently named as the top two threats to global security by 110 Nobel laureates. Assumptions about future population and economic growth, technology advancements, energy consumption, etc., are necessary for any climate change estimate that takes the future into account. These assumptions in turn depend on the future emissions of aerosols and greenhouse gases. For various societal, economic, and technical futures, the IPCC proposed 35 emission scenarios (Jentsch & Beierkuhnlein, 2023). The Intergovernmental Panel on Climate Change (IPCC) determined that the range of expected 2100 temperature would be determined by combining all emissions scenarios with global and other climate models. For a summary of the current state of climate change research. These predictions allow us to derive several important conclusions. Strong albedo feedbacks caused by melting snow and sea ice cause higher latitudes to warm faster than lower ones. Because the ocean is now directly exposed to the atmosphere, it can warm the atmosphere from below, which is an extra positive feedback effect of melting sea ice. The impact of the seas will cause land areas to warm faster than ocean areas, and the interiors of the continents to warm faster than their coasts. It is also predicted that winters will be warmer than summers (Bailey & van de Pol, 2023). Though there is room for debate when it comes to climate change projections, the majority of scientists agree with the following statement from Dr. R. Watson, Chair of the IPCC, delivered at the Sixth Conference of the Parties to the UN Framework Convention on Climate Change: While acknowledging the existence of scientific uncertainty, the vast majority of specialists in the field continue to hold the belief that climate change caused by humans will occur. Extreme weather phenomena such as heat waves, floods, droughts, fires, and the like have been devastating to many parts of the globe in recent years, causing enormous economic damage and human casualties. Although it is not possible to attribute specific weather occurrences to global warming, scientists do anticipate that such phenomena will become more common and more severe as the planet warms. From a risk-based decision-making perspective, have addressed the subject of whether Canada should ratify the Kyoto Protocol. In the section that follows, we will examine a comparable strategy in relation to the issue of how to prepare for more frequent and severe extreme weather events (Cinto Mejía & Wetzel, 2023).

8. CONCLUSION

Some broad themes and recommendations for future research may be inferred from the usually mixed results regarding the association between extreme weather experience and climate-related opinion or behaviour. Above all else, there is mounting evidence that increasingly extreme weather events have not, thus far, prompted those who were previously indifferent to climate change to change their minds, rally behind policies to mitigate and adapt, and alter their own behaviour. However, for many who were already worried about climate change, severe weather has only served to heighten their sentiments and actions. But this data might not be applicable to future extreme weather events, which, according to most climate models, will be more prevalent and severe. In addition to the effect asymmetry caused by the current climate change polarisation, there is some evidence that people who are already inclined to be less concerned about climate change may change their minds after experiencing severe weather events firsthand (Smith, 2023). The kind and severity of the extreme weather event also appear to have an impact on people's opinions on climate change. Consequences for a shared research approach among related studies using large representative survey datasets may result from this discovery. It is highly improbable that these studies will employ a cross-sectional representative sample of people, considering the large number of respondents who have likely suffered severe direct negative experiences due to extreme weather events, such as health effects or property damage. When compared to the total national or state-level population, the percentage of individuals who have recently experienced severe weather—including heavy property damage or personal or family injuries from floods, wildfires, drought, storms, or heat—is low in the countries where the majority of research has been carried out. In addition, conventional methods like web panels and mail surveys may not adequately represent individuals who have suffered severe harm from a disaster, like those who have recently lost their homes in a wildfire, due to the high rate of nonresponse and the ethical challenges that come with studying disaster survivors (Moreno & Møller, 2023).

REFERENCES

Amsalu, A., Garedew, W., Melka, G. A., Çelik, M. A., & Gemeda, D. O. (2026). Trends in climate extreme indices and implications for coffee farming in The Kafa Biosphere Reserve, southwest Ethiopia. *Climate Services*, *42*, 100637. DOI: 10.1016/j.cliser.2026.100637

Arbuckle, J. G.Jr, Morton, L. W., & Hobbs, J. (2021). Understanding Farmer Perspectives on Climate Change Adaptation and Mitigation. *Environment and Behavior*, *47*(2), 205–234. DOI: 10.1177/0013916513503832 PMID: 25983336

Aven, T. (2020). Climate change risk – what is it and how should it be expressed? *Journal of Risk Research*, *23*(11), 1387–1404. DOI: 10.1080/13669877.2019.1687578

Bailey, L. D., & van de Pol, M. (2023). Tackling extremes: Challenges for ecological and evolutionary research on extreme climatic events. *Journal of Animal Ecology*, *85*(1), 85–96. DOI: 10.1111/1365-2656.12451 PMID: 26433114

Bolan, S., Padhye, L. P., Jasemizad, T., Govarthanan, M., Karmegam, N., Wijesekara, H., Amarasiri, D., Hou, D., Zhou, P., Biswal, B. K., Balasubramanian, R., Wang, H., Siddique, K. H. M., Rinklebe, J., Kirkham, M. B., & Bolan, N. (2024). Impacts of climate change on the fate of contaminants through extreme weather events. *Science of The Total Environment*, *909*(August 2023), 168388. DOI: 10.1016/j.scitotenv.2023.168388

Çelik, M. A. (2026). Agro-climatic classification of Türkiye's Agricultural Basins using long-term ERA5-Land temperature data. *Physics and Chemistry of the Earth Parts A/B/C*, *142*, 104307. DOI: 10.1016/j.pce.2026.104307

Çelik, M. A., Elbeltagi, A., Mishra, R., Singh, S., Vishwakarma, D. K., Pandey, K., & Akiner, M. E. (2025). Flood Risk and Inundation Mapping in a Changing Climate. Advancing Environmental Research Through Applied GIS and Remote Sensing, 1–34. DOI: 10.4018/979-8-3373-6608-1.ch001

Cheng, C. S., Auld, H., Li, Q., & Li, G. (2023). Possible impacts of climate change on extreme weather events at local scale in south–central Canada. *Climatic Change*, *112*(3–4), 963–979. DOI: 10.1007/s10584-011-0252-0

Churi, A. J., Mlozi, M. R. S., Tumbo, S. D., & Casmir, R. (2022). Understanding Farmers Information Communication Strategies for Managing Climate Risks in Rural Semi-Arid Areas, Tanzania. *International Journal of Information and Communication Technology Research*, *2*(11), 838–845. http://www.esjournals.org

Cinto Mejía, E., & Wetzel, W. C. (2023). The ecological consequences of the timing of extreme climate events. *Ecology and Evolution, 13*(1), 1–10. DOI: 10.1002/ece3.9661 PMID: 36713483

Clarke, B., Otto, F., Stuart-Smith, R., & Harrington, L. (2022). Extreme weather impacts of climate change: An attribution perspective. *Environmental Research: Climate, 1*(1), 012001. DOI: 10.1088/2752-5295/ac6e7d

Curtis, S., Fair, A., Wistow, J., Val, D. V., & Oven, K. (2017). Impact of extreme weather events and climate change for health and social care systems. *Environmental Health : A Global Access Science Source, 16*(S1), 128. DOI: 10.1186/s12940-017-0324-3 PMID: 29219105

Demski, C., Capstick, S., Pidgeon, N., Sposato, R. G., & Spence, A. (2017). Experience of extreme weather affects climate change mitigation and adaptation responses. *Climatic Change, 140*(2), 149–164. DOI: 10.1007/s10584-016-1837-4 PMID: 32355377

Eckstein, D., Hutfils, M.-L., & Winges, M. (2019). Global climate risk index 2019. *Who Suffers Most from Extreme Weather Events*, 1–36. https://germanwatch.org/sites/germanwatch.org/files/Global Climate Risk Index 2019_2.pdf%0Ahttps://germanwatch.org/en/7677

Ekwurzel, B., Frumhoff, P. C., & McCarthy, J. J. (2023). Climate uncertainties and their discontents: Increasing the impact of assessments on public understanding of climate risks and choices. *Climatic Change, 108*(4), 791–802. DOI: 10.1007/s10584-011-0194-6

Ginglinger, E., & Moreau, Q. (2023). Climate Risk and Capital Structure. *Management Science, 69*(12), 7492–7516. DOI: 10.1287/mnsc.2023.4952

Hay, J., & Mimura, N. (2023). The changing nature of extreme weather and climate events: Risks to sustainable development. *Geomatics, Natural Hazards & Risk, 1*(1), 3–18. DOI: 10.1080/19475701003643433

Howard, B., Dietz, S., Ebi, K., Gollier, C., Gregory, R., Horton, B., Kriegler, E., Mach, K., Mastrandrea, M., Millner, A., Oppenheimer, M., Träger, C., Kunreuther, H., Heal, G., Allen, M., Edenhofer, O., Field, C. B., & Yohe, G. (2022). *Risk Management and Climate Change*. https://www.nber.org/papers/w18607

Howe, P. D. (2021). Extreme weather experience and climate change opinion. *Current Opinion in Behavioral Sciences, 42*, 127–131. DOI: 10.1016/j.cobeha.2021.05.005

Huber, D., & Gulledge, J. (2023). *Extreme weather and climate change: Understanding the link and managing the risk. Science and Impacts Program*. Center for Climate and Energy Solutions.

Huber, D. G., & Gulledge, J. (2021). Extreme Weather and Climate Change Understanding the Link, Managing the Risk. *Science and Impacts Program. Center for Climate and Energy Solutions, December*, 1–13. http://www.pewclimate.org/publications/extreme-weather-and-climate-change%5Cnhttp://www.c2es.org/publications/extreme-weather-and-climate-change

Hulme, M. (2020). Attributing weather extremes to 'climate change.'. *Progress in Physical Geography*, *38*(4), 499–511. DOI: 10.1177/0309133314538644

Jentsch, A., & Beierkuhnlein, C. (2023). Research frontiers in climate change: Effects of extreme meteorological events on ecosystems. *Comptes Rendus Geoscience*, *340*(9–10), 621–628. DOI: 10.1016/j.crte.2008.07.002

Jones, R., & Boer, R. (2023). *Assessing current climate risks* (Issue 4).

Jurgilevich, A., Räsänen, A., Groundstroem, F., & Juhola, S. (2017). A systematic review of dynamics in climate risk and vulnerability assessments. *Environmental Research Letters*, *12*(1), 013002. DOI: 10.1088/1748-9326/aa5508

Keller, K., Helgeson, C., & Srikrishnan, V. (2021). Climate Risk Management. *Annual Review of Earth and Planetary Sciences*, *49*(1), 95–116. DOI: 10.1146/annurev-earth-080320-055847

Konisky, D. M., Hughes, L., & Kaylor, C. H. (2020). Extreme weather events and climate change concern. *Climatic Change*, *134*(4), 533–547. DOI: 10.1007/s10584-015-1555-3

Kreft. (2023). GLOBAL CLIMATE RISK INDEX. In *Eco Watch*. http://ecowatch.com/2015/12/04/day-4-cop21%5Cnpapers3://publication/uuid/148F5B59-F94A-49D2-9FE2-F72730750610

Kumar, A. (2023). Climate Change and Damage from Extreme Weather Events. *Key Concepts in Modern Indian Studies*, 207–211.

Li, H.-M., Wang, X.-C., Zhao, X.-F., & Qi, Y.LI. (2021). Understanding systemic risk induced by climate change. *Advances in Climate Change Research*, *12*(3), 384–394. DOI: 10.1016/j.accre.2021.05.006

Li, Q., Shan, H., Tang, Y., & Yao, V. (2024). Corporate Climate Risk: Measurements and Responses. *Review of Financial Studies*, *37*(6), 1778–1830. DOI: 10.1093/rfs/hhad094

Magnan, A. K., O'Neill, B. C., & Garschagen, M. (2023). Further understanding "severe" climate risk. *Climate Risk Management*, *42*(July), 100538. DOI: 10.1016/j.crm.2023.100538

Manatsa, D., Chingombe, W., & Matarira, C. H. (2023). The impact of the positive Indian Ocean dipole on Zimbabwe droughts Tropical climate is understood to be dominated by. *International Journal of Climatology*, *2029*(March 2008), 2011–2029. https://doi.org/DOI: 10.1002/joc

Mann, M. E., Lloyd, E. A., & Oreskes, N. (2017). Assessing climate change impacts on extreme weather events: The case for an alternative (Bayesian) approach. *Climatic Change*, *144*(2), 131–142. DOI: 10.1007/s10584-017-2048-3

Maxwell, S. L., Butt, N., Maron, M., McAlpine, C. A., Chapman, S., Ullmann, A., Segan, D. B., & Watson, J. E. M. (2019). Conservation implications of ecological responses to extreme weather and climate events. *Diversity & Distributions*, *25*(4), 613–625. DOI: 10.1111/ddi.12878

McBean, G. (2020). Climate Change and Extreme Weather: A Basis for Action. *Natural Hazards*, *31*(1), 177–190. DOI: 10.1023/B:NHAZ.0000020259.58716.0d

Meehl, G. A., Zwiers, F., Evans, J., Knutson, T., Mearns, L., & Whetton, P. (2023). Trends in Extreme Weather and Climate Events: Issues Related to Modeling Extremes in Projections of Future Climate Change. *Bulletin of the American Meteorological Society*, *81*(3), 427–436. DOI: 10.1175/1520-0477(2000)081<0427:TIEWAC>2.3.CO;2

Mirza, M. (2023). Climate change and extreme weather events: Can developing countries adapt? *Climate Policy*, *3*(3), 233–248. DOI: 10.3763/cpol.2003.0330

Moreno, J., & Møller, A. P. (2023). Extreme climatic events in relation to global change and their impact on life histories. *Current Zoology*, *57*(3), 375–389. DOI: 10.1093/czoolo/57.3.375

Nieto, M. J. (2019). Banks, climate risk and financial stability. *Journal of Financial Regulation and Compliance*, *27*(2), 243–262. DOI: 10.1108/JFRC-03-2018-0043

Ornes, S. (2018). How does climate change influence extreme weather? Impact attribution research seeks answers. *Proceedings of the National Academy of Sciences of the United States of America*, *115*(33), 8232–8235. DOI: 10.1073/pnas.1811393115 PMID: 30108186

Palmer, G., Platts, P. J., Brereton, T., Chapman, J. W., Dytham, C., Fox, R., Pearce-Higgins, J. W., Roy, D. B., Hill, J. K., & Thomas, C. D. (2017). Climate change, climatic variation and extreme biological responses. *Philosophical Transactions of the Royal Society B: Biological Sciences, 372*(1723), 20160144. DOI: 10.1098/rstb.2016.0144

Perkins-Kirkpatrick, S. E., Stone, D. A., Mitchell, D. M., Rosier, S., King, A. D., Lo, Y. T. E., Pastor-Paz, J., Frame, D., & Wehner, M. (2022). On the attribution of the impacts of extreme weather events to anthropogenic climate change. *Environmental Research Letters, 17*(2), 024009. DOI: 10.1088/1748-9326/ac44c8

Quandt, A., Neufeldt, H., & Gorman, K. (2023). Climate change adaptation through agroforestry: Opportunities and gaps. *Current Opinion in Environmental Sustainability, 60*, 101244. DOI: 10.1016/j.cosust.2022.101244

Rising, J., Tedesco, M., Piontek, F., & Stainforth, D. A. (2022). The missing risks of climate change. *Nature, 610*(7933), 643–651. DOI: 10.1038/s41586-022-05243-6 PMID: 36289386

Romm, J. (2018). Extreme Weather and Climate Change. In *Climate Change*. Oxford University Press., DOI: 10.1093/wentk/9780190866112.003.0002

Sánchez-Almodóvar, E., Gómez-Trigueros, I. M., & Olcina-Cantos, J. (2022). Climate Change and Extreme Weather Events in the Education of the Citizens of the Twenty-First Century: The Perception of Secondary Education Students. *Social Sciences (Basel, Switzerland), 12*(1), 27. DOI: 10.3390/socsci12010027

Sisco, M. R., Bosetti, V., & Weber, E. U. (2017). When do extreme weather events generate attention to climate change? *Climatic Change, 143*(1–2), 227–241. DOI: 10.1007/s10584-017-1984-2

Smith, M. D. (2023). An ecological perspective on extreme climatic events: A synthetic definition and framework to guide future research. *Journal of Ecology, 99*(3), 656–663. DOI: 10.1111/j.1365-2745.2011.01798.x

Solow, A. R. (2017). On detecting ecological impacts of extreme climate events and why it matters. *Philosophical Transactions of the Royal Society B: Biological Sciences, 372*(1723), 20160136. DOI: 10.1098/rstb.2016.0136

Stott, P. A., Christidis, N., Otto, F. E. L., Sun, Y., Vanderlinden, J., van Oldenborgh, G. J., Vautard, R., von Storch, H., Walton, P., Yiou, P., & Zwiers, F. W. (2023). Attribution of extreme weather and climate-related events. *Wiley Interdisciplinary Reviews: Climate Change, 7*(1), 23–41. DOI: 10.1002/wcc.380 PMID: 26877771

Stroebel, J., & Wurgler, J. (2021). *Nber Working Paper Series What Do You Think About Climate Finance?* https://www.nber.org/papers/w29136

Taylor, A. L., Dessai, S., & Bruine de Bruin, W. (2024). Public perception of climate risk and adaptation in the UK: A review of the literature. *Climate Risk Management, 4–5*, 1–16. DOI: 10.1016/j.crm.2014.09.001

Ummenhofer, C. C., & Meehl, G. A. (2017). Extreme weather and climate events with ecological relevance: a review. *Philosophical Transactions of the Royal Society B: Biological Sciences, 372*(1723), 20160135. DOI: 10.1098/rstb.2016.0135

van de Pol, M., Jenouvrier, S., Cornelissen, J. H. C., & Visser, M. E. (2017). Behavioural, ecological and evolutionary responses to extreme climatic events: challenges and directions. *Philosophical Transactions of the Royal Society B: Biological Sciences, 372*(1723), 20160134. DOI: 10.1098/rstb.2016.0134

Wehner, M. (2023). Connecting extreme weather events to climate change. *Physics Today, 76*(9), 40–46. DOI: 10.1063/PT.3.5309

Zhang, Z. (2023). Weather and Climate Extreme Events in a Changing Climate. In *Climate Change 2021 – The Physical Science Basis* (pp. 1513–1766). Cambridge University Press., DOI: 10.1017/9781009157896.013

Zscheischler, J., Westra, S., van den Hurk, B. J. J. M., Seneviratne, S. I., Ward, P. J., Pitman, A., AghaKouchak, A., Bresch, D. N., Leonard, M., Wahl, T., & Zhang, X. (2018). Future climate risk from compound events. *Nature Climate Change, 8*(6), 469–477. DOI: 10.1038/s41558-018-0156-3

Chapter 13
Early Warning Systems and Real-Time Decision Support: Leveraging GIS and Remote Sensing for Global Disaster Preparedness

Sitesh Kumar Singh
http://orcid.org/0000-0002-7108-0808
College of Engineering, National University of Science and Technology, Muscat, Oman

Jaishri Gothania
http://orcid.org/0000-0002-2656-642X
Lingaya's Vidyapeeth, Faridabad, India

Sunil K. Sansaniwal
http://orcid.org/0000-0002-0997-5096
NITI Aayog, Government of India, India

ABSTRACT

The chapter analyzes how Geographic Information Systems (GIS) and remote sensing can help fortify the early warning system (EWS) and prompt response decision support to international disaster readiness. It gives a review on the technological fundamentals, e.g., remote sensing platforms, GIS-based hazard modeling, and/or AI-enhanced geospatial analytics. Its use in applications involving floods, earthquakes, tsunamis, wildfires, and cyclones indicates how real-time information can be used to reduce risks proactively, direct evacuation, and design resilient infra-

DOI: 10.4018/979-8-3373-6274-8.ch013

structure. Successes, challenges, and limitations in multi-hazard EWS are clear evidence of case studies at a global and regional level. The concluding chapter provides recommendations and future directions that are inclusive, reliable, and adaptive of systems and focuses on technology, policy, and community outlook on building resilient societies.

1. INTRODUCTION TO EARLY WARNING SYSTEMS

Natural disasters, including floods, cyclones, earthquakes and wildfires, as well as anthropogenic disasters, including industrial accidents, continue to test the security, infrastructure, and eco-systems of humans across the world. Increased frequency and severity of these hazards has further been aggravated by climate change, making it important to have in place effective preparedness mechanisms. Central to disaster preparedness is the Early Warning System (EWS) that aims to enhance the supply of timely, accurate, and actionable information to the stakeholders to initiate proactive measures to lessen loss of life and damages.

Modern EWS have four interconnected parts of which are risk knowledge, hazard monitoring, and warning dissemination, as well as response capability (UNDRR, 2020). When these factors work in coordination, it permits the authorities to transition away with reactive emergency responses to proactive ones. This has changed radically with the inclusion of Geographic Information Systems (GIS) and remote sensing tools, which provide a strong arm in the processing of hazard involvement, spatial analysis and real time surveillance. As an example, GIS enables the use of hazard maps in conjunction with socio-economic/demographic data to enhance evacuation design and the distribution of resources. Risk and hazard detection and tracking technologies, including satellites, drones, and LiDAR, improve monitoring and tracking at space and time domains. Collectively, these are the technologies that comprise the core of the current EWS and promote disaster preparedness on the global level (Coppola, 2020).

This section also examines how EWS can contribute to the disaster risk reduction and the global policy instruments that influence the development and application of EWS.

1.1 Role of EWS in Disaster Risk Reduction

Early Warning Systems are important components of the disaster risk reduction (DRR). Successful EWS fills this gap that exists between identifying a hazard and getting the community to act to address it, by converting scientific evidence into usable data. Through lead time, EWS allows governments, organizations, and people to prepare mitigation measures comprised of evacuations, strengthening of infrastructure as well as supply chain modification before the occurrence of disasters.

Evidence at the international range indicates reliably that the well-established EWS have experienced huge losses and fatalities anywhere lower as opposed to the systems that are not well-established (Basher, 2006). As an example, Bangladesh developed a significant improvement of cyclone preparedness through community-based warning prevalence that instead of hundreds of dying people as years ago, have experienced only a few deaths, after cyclone impacts since 1970s. Likewise, the earthquake and Tsunami warning systems in Japan incorporate seismic sensors and mass communication strategies so that millions of Japanese citizens receive the alerts within a few seconds after the moment of seismic detection. The above examples demonstrate that EWS performs the role of force multipliers of resilience.

GIS and remote sensing integration are effective methods to improve efficiency and effectiveness of the EWS in DRR. GIS platforms offer spatial modeling of hazard, the mapping of vulnerable people, and analysis of exposure of critical infrastructures. Remote sensing by contrast facilitates constant monitoring of the environment including rain patterns in the event of a flood or a drought of vegetation in the event of wildfire (Vishvakarma, 2019). When these technologies are matched with predictive analytics and AI, decision-makers will have the capability of predicting complex risk situations and the opportunity of acting in a targeted way.

Figure 1. Conceptual Framework of Early Warning Systems (EWS) Using GIS and Remote Sensing

Moreover, EWS have long-term effects of risk mitigation as they are used in land-use planning, infrastructure and environmental control. As shown in Figure 1, early warning systems integrate hazard detection, data processing, and communication to facilitate proactive disaster risk management. Through the integration of early warning data into policy and practice, societies enter a new era of disaster governance; a proactive approach to disaster governance that places a focus on prevention and preparedness rather than based on response (UNDRR, 2020). Therefore, it is not only technological systems that perform the role of EWS but also a part of resilient societies.

1.2 Global Frameworks and Policy Mandates

The creation and adoption of EWS are highly advised by international frameworks and policy instruments. The most relevant international agreement is the Sendai Framework for Disaster Risk Reduction (2015 - 2030), in which the setting up of multi-hazard and people-centered early warning systems was highlighted as a priority (UNDRR, 2015). The framework emphasizes inclusion, accessibility and community, and focuses on ensuring warnings are not only technically excellent but culturally applicable and socially relevant.

The disaster resilience is also highlighted under the Sustainable Development Goals (SDGs), especially under Goal 11 (Sustainable Cities and Communities) and Goal 13 (Climate Action). Early warning systems play a direct role towards the realization of these since they contribute to the protection of lives and livelihoods against any disasters related to climatic changes.

Such global commitments are also strengthened by regional efforts as an example, the Emergency Management Service of the European Union Copernicus is delivering rapidly delivered standardized mapping after a disaster, helping member states to reduce the risk and assist in recovery processes. In line with this, the African Union's Continental Early Warning System targets surveillance of conflicts and natural hazards with an aim of incorporating the same in larger resilience initiatives. In Asia, there is the Indian Ocean Tsunami Warning and Mitigation System which is a shared system between the coastal states to share seismic data and disseminate warning.

There are also framing policies into national policy frameworks with domestic stances emulating global solutions regarding national legislation in terms of disaster management, climate adaptation policies, and strategies. Nevertheless, the implementation of global commitments in local contexts, especially in low-income countries, is still a problem as technical, financial, and institutional capacity might be weak. The ways to deal with such gaps would demand long-term investments, international partnership, and innovation of geospatial technologies.

The international community aims to change disaster management into a proactive, resilience-oriented pursuit by integrating EWS into global and regional policy requirements. The mutual action of international structures, advancements in technology, and local socioeconomic demands will define how EWS will help protect communities regarding future risks.

2. TECHNOLOGICAL FOUNDATIONS OF GIS AND REMOTE SENSING

The herald of disasters preparedness has been technology making change and altering the shape of preparedness as Geographic Information Systems (GIS) and remote sensing are the pillars of current Early Warning Systems (EWS). Collectively the tools allow systematic surveillance, spatial surveillance and visualization of local, regional and global hazards. Remote sensing delivers observations in real time/near real time terms, and GIS aggregates these observations with socio-economic and environmental data to support decision making. It is crucial to understand the premises of these technologies to be able to place them in a positive light when it comes to improved resilience and facilitating disaster risk reduction (Goodchild & Li, 2021).

Here, the most relevant types of remote sensing platforms will be presented, and the potential of these platforms will be discussed with respect to GIS in terms of hazard modeling and visualization as the foundation of the insights on how they complement each other in full-scope EWS.

2.1 Overview of Remote Sensing Platforms (Satellites, UAVs, LiDAR)

Remote sensing: The term remote sensing is used to describe surface information retrieved on the ground without touch, most commonly by aerial or satellite-fed sensors. The platforms differ in their scale, resolution and application; thus, their applications are highly versatile in the context of disasters.

Satellite-based platforms: The commonly used platforms are satellite-based in monitoring disasters globally. The weather satellites (GOES, Geostationary Operational Environmental Satellites) and Himawari provide near real time images of cyclones and weather prediction. Earth observation satellites such as Landsat, Sentinel and MODIS (Moderate Resolution Imaging Spectroradiometer) also provide multispectral and hyperspectral imagery that is used to identify flood extent, monitor wildfires and assess droughts (Joyce et al., 2009). Advancements in CubeSats and commercial constellations like Planet Labs have enhanced the temporal coverage and provide daily or sub-daily global coverage.

Unmanned Aerial Vehicles (UAVs): Drones, or Unmanned Aerial Vehicles, are used to supplement satellites to offer locally focused-high resolution monitoring that is flexible and comparatively cheap. UAVs have proven efficiency of post-disaster damage assessment as well as mapping the inaccessible or hazardous environments. As an example, drones were heavily used in response to the 2015 Nepal earthquake to survey collapsed buildings and direct the rescue effort. Their capability to carry thermal sensors would also maintain surveillance of wildfire hotspots and capabilities to conduct search and rescue missions (Yucesoy et al., 2024).

Light Detection and Ranging (LiDAR): It is one more arc-critical remote sensing platform, able to create accurate 3D models of surface relief and infrastructure. TIDAR is particularly helpful in assessing flood risks where high quality elevation models are required to model hydrology, and map inundation. Storm surge modeling has also played a vital role on the coast where LiDAR has been used to assess the vulnerability of shores. Airborne LiDAR can also benefit from urban planning with resilience in mind since it can point to potential infrastructure weaknesses (Slatton et al., 2007).

Table 1. Overview of GIS and Remote Sensing Platforms for Disaster Monitoring

Platform Type	Key Features	Typical Applications in Disaster Preparedness	Spatial Resolution	Temporal Resolution
Satellite Imagery	Large-area coverage, multispectral sensors	Flood mapping, wildfire detection, cyclone tracking	1–30 m	Daily to weekly
UAV / Drone	High-resolution imagery, flexible deployment	Localized hazard assessment, landslide monitoring	<0.5 m	Hourly to daily
LiDAR	3D terrain mapping, canopy penetration	Earthquake fault line analysis, floodplain modeling	1–5 m	Event-based
Aerial Photography	Visual analysis, historical archives	Urban hazard planning, post-disaster damage assessment	0.3–2 m	Event-based / periodic

All these platforms are quite distinct in their benefits. Satellites provide immense scope of continuous monitoring whereas UAVs can offer agility and local details, and LiDAR will provide the high-accurate elevation and structural survey. Put together, they make a multi-level observation network that supports hazard detection, monitoring, and forecasting globally. Table 1 provides a comparative overview of GIS and remote sensing platforms used in disaster preparedness, highlighting their features, typical applications, spatial and temporal resolutions, which aid decision-makers in selecting appropriate tools for hazard monitoring.

2.2 GIS Capabilities for Hazard Modeling and Visualization

The most fitting technology to take raw spatial information obtained by remote sensing and convert it to useful intelligence is Geographic Information Systems (GIS). GIS combined hazard data with demographic, infrastructural, environmental data to allow decision-makers to simulate the method of disaster and develop effective preventive measures.

Hazard modeling is one of the major GIS uses in disaster planning. An example of this is flood hazard maps created by overlapping rainfall information, topography and land use that enable authorities to determine high-risk areas in the flood plains. Likewise, models of seismic hazard, which are based on GIS, make use of tectonic information, soil types, and population density to estimate the possible effects of earthquakes. They are useful models in that they assist policymakers in allocating their resources based on their priority and come up with a specific risk reduction strategy (Coppola, 2020).

The visualization and communication of risk information is also improved through GIS. Hazard maps and dashboards are interactive and enable up to date disaster information to be published to both authorities and the mass audience. As an example, the on-line platform such as the Global Disaster Alert and Coordination System (GDACS) would employ visualizations on a GIS system basis to deliver global alerts on earthquakes, tsunamis and tropical cyclones in near real-time. Through such tools, complicated geospatial information is displayed in an easy-to-read form in a manner that fulfills the situational awareness and decision-making needs under pressure.

Vulnerability and exposure analysis is another very crucial role of GIS. Superimposing hazard zones over population density, socio-economic information, and key infrastructure (e.g., hospitals, power plants, schools) with GIS, the at-risk communities can be identified and the priorities during the evacuation planning should receive attention. This integration in space is of particular importance in an urban setting where vulnerabilities are spatially imbalanced and risks could overlap.

GIS is also becoming a smart decision-aiding tool with the improvement of machine learning and cloud-based platforms. Web-based GIS services including Google Earth Engine and ArcGIS Online enable merging of large sets of data and cross-institute/cross-border collaborative analysis. Real-time modeling of the hazards and fast delivery of the results to various stakeholders become empowered with these platforms (Goodchild & Li, 2021).

Ultimately, the power of GIS consists in the fact that it helped to translate different geospatial information into comprehensive insights, which can be used to prepare against disasters. Used with remote sensing, it allows assessing risks multi-dimensionally, which can facilitate proactive and inclusive disaster governance (Vishvakarma, 2019).

3. INTEGRATION OF REAL-TIME DATA IN EARLY WARNING SYSTEMS

Early Warning Systems (EWS) effectiveness is dependent on capacity to process and share timely, reliable and context pertinent information. Conventional methods tended to use infrequent observations and slow distribution and thus were not well placed to respond to risks that evolved more quickly. With the incorporation of real-time data, this paradigm has been completely changed to allow continuous monitoring, quicker response and adaptive decision making (Kalogiannidis et al., 2025).

Increased access to sensors, geospatial technologies, and high-performing data structures has, in turn, made real-time data combination not only a possibility but a necessity of contemporary EWS. The key to this change lies in the use of geographic Information Systems (GIS) and remote sensing technology, and new technologies like Internet of Things (IoT), cloud-computing, and artificial intelligence. This section describes the role of real-time data integration in the enhancement of hazard detection, better predictive modeling and rapid decision support.

3.1 Data Sources: Sensors, IoT, and Satellite Feeds

In the modern EWS, the various real-time data used to shape it consider the socio-economic and environmental indication of risk or hazards. These will be ground sensors, IoT-sensed networks, and satellites. Figure 2 illustrates the workflow of integrating remote sensing and GIS data for spatial modeling and hazard visualization in disaster preparedness.

Figure 2. Remote Sensing Data Acquisition and GIS Integration Workflow

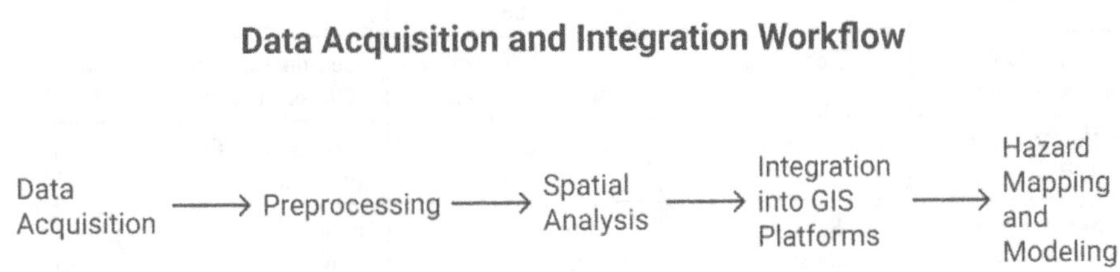

Ground-based sensors: Many hazard monitoring systems are built on ground-based sensors. As an example, seismographs measure motion on the ground in earthquake prone regions, and stream gauges together with rainfall sensors are important in prediction of floods. These sensors constantly send data to central tracking systems so that it can be detected quickly in case of anomalies. Improvements in sensor technologies lowering the cost of sensor technology have enabled the deployment of sensors especially in resource-deprived areas (Tao et al., 2024).

IoT-enabled networks: Networks based on IoT technologies can increase the reach of real-time monitoring by interlinking conspicuously distributed devices into smart telecommunication networks. IoT tools can facilitate an intelligent weather station, landslide observation equipment, and pre-flood systems that independently send information to a centralized network. As an example, IoT sensors have been integrated into systems that monitor smart river basins in South Asia to allow water level, rainfall and soil saturation levels to be measured, which facilitates proactive flood management.

Satellite feeds: It has never stopped to be needed to monitor large and distant areas. The Earth Observing System (EOS) operated by NASA and Sentinel constellation operated by the European Space Agency also offer near real-time data on vegetation heath, land surface temperatures, and atmospheric conditions. Storm trajectories, winds and the amount of precipitation within cyclones can be tracked by using satellite feeds at a high rate of periodicity. Combination of satellite information with ground observations allows ensuring comprehensive situational awareness (NASA's Earth Observing System, n.d.).

Table 2. Integration of Real-Time Data Sources in Early Warning Systems

Data Source	Type	Contribution to EWS	Integration with GIS/Remote Sensing	Frequency / Update Rate
Weather Stations	Ground-based sensors	Real-time precipitation, temperature, wind data	Overlaid on hazard maps	Hourly
Social media / Crowdsourcing	Text, images	Rapid hazard reporting, validation	Geo-tagging into GIS dashboards	Minutes to hourly
Satellite Remote Sensing	Multispectral / radar	Large-scale hazard detection	Layered into GIS for spatial analysis	Daily to weekly
IoT Sensors	Sensors on infrastructure (bridges, dams)	Structural stress monitoring, river levels	Integrated via real-time dashboards	Seconds to minutes
Historical & Socioeconomic Data	Static datasets	Vulnerability mapping, risk assessment	Baseline GIS layers	As available

Table 2 summarizes the major real-time data sources integrated into early warning systems, illustrating their contributions, interoperability with GIS and remote sensing platforms, and update frequency, emphasizing their role in enabling timely disaster decision support. Collectively, these sources deliver a multi-layered data ecosystem that has the potential to improve EWS, both due to localizing precision and extending coverage to a global scale.

3.2 Data Processing and Integration Techniques

Sensor and satellite data in their raw form should undergo processing, standardization and bending before they can be useful in providing decision support. This includes data cleaning, synchronicity of time, spatial integration and real-time analytics.

Data fusion: In particular, the important method in this regard is data fusion, which is used to synthesize measurements of different sensors and thus minimize uncertainty and enhance accuracies. As an example, combining radar rainfall information together with ground-based rainfall radar improves the accuracy of flood simulation models. Likewise, a combination of thermal satellite data and drone-based data gives the user detailed information about the dynamics of wildfire spread.

Cloud computing platforms: Google Earth Engine, Amazon Web Services (AWS) and Microsoft Azure are cloud computing platforms that have transformed the ability of EWS vis-a-vis its capacity to process large volumes of real-time datasets. These services have the capability to scale storage and processing, remote accessibility, and enable analysis across the stakeholders. As an illustration, DisasterAWARE by the Pacific Disaster Center is a multi-hazard integration platform used to embed real-time multi-hazard data to support governing bodies and humanitarian agencies in decision-making across the globe (Sharma et al., 2020).

Artificial Intelligence (AI) and Machine Learning (ML): The use of Artificial Intelligence (AI) and Machine Learning (ML) algorithms is actively being used to process streaming geospatial information. Figure 4 demonstrates a real-time decision support dashboard integrating GIS, AI-based analytics, and remote sensing data for rapid disaster response. ML models can recognize patterns and anomalies that indicate near-term dangers, including the precursors to landslides, warming/cooling of sea-surface temperatures that might lead to cyclones. In contrast to time-tested statistical models, predictive analytics based on AI provides an opportunity to increase the lead time and increase the accuracy of advancement (Diehr et al., 2025).

Figure 3. Real-Time Decision Support Dashboard for Disaster Management

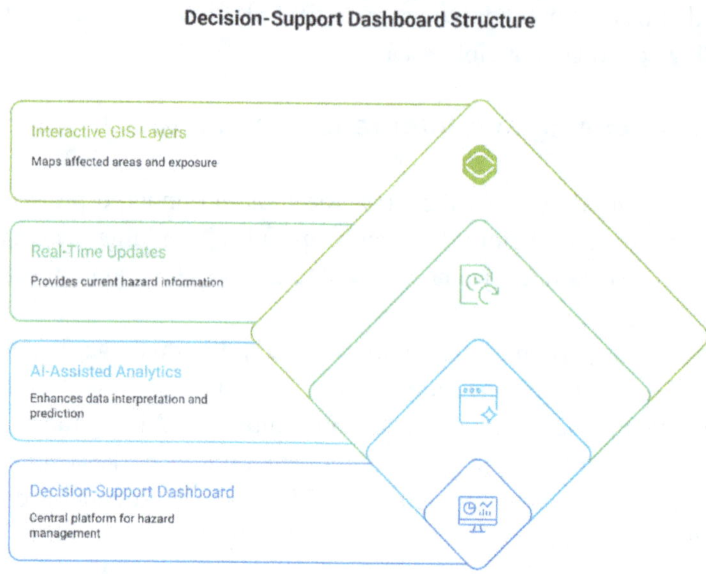

Another crucial technique is interoperability and standards that can be applied to ensure the use of heterogeneous datasets. A common standards cross-border effort like the Open Geospatial Consortium (OGC) facilitates data exchange across platforms and institutions with ease. This is to see to it that essential data moves freely across the borders of states and institutions in times of emergency.

The combination of such techniques converts a set of raw data into well-structured actionable information flows that improve the reliability and responsiveness of EWS.

3.3 Case Studies of Real-Time Disaster Monitoring

The real-life usefulness of data integration in real-time can be seen through one successful example in the example of disaster monitoring all over the world.

Cyclone tracking in the Indian Ocean region: The INCOIS mainly combines the satellite feeds with satellite-based buoy data and numerical weather models to provide early warnings of cyclones. This system gave more than 72 hours of lead times during Cyclone Fani (2019), which has resulted in mass evacuations and their resulting drastic decreases in the number of fatalities observed in previous events (Mohapatra & Sharma, 2021).

Flood monitoring in Europe: The European Flood Awareness System (EFAS) provides an amalgamation of radar-based rainfall, hydrological measurements and imaging of satellites in the context of a GIS environment. EFAS has become invaluable in cross-border flood management in Europe through its probabilistic forecasts and the visualization in real time (Smith et al., 2016).

Wildfire detection in California: It is the perfect example of the usefulness of real-time systems powered by IoT. The ALERTWildfire network installs more than 1,000 high-definition cameras and sensor solutions to fire-prone parts of the USA, delivering real-time data to emergency administrators. This combined with the satellite thermal data enables real-time fire outbreaks where their detection is done at a faster rate compared to 911 calls powered by this system (Rjoub et al., 2023).

Earthquake early warning in Japan: In Japan, earthquake early warning also identifies the incorporation of real-time information. Japan Meteorological Agency (JMA) has one of the most developed seismic monitoring networks in the world. The system has the ability of process seismic waves in real time, which can provide possible warning a few seconds before the urban centers receive a destructive shaking with the aim of allowing residents and strategic infrastructure to reduce themselves to protective measures (Kamigaichi, 2004).

Such examples affirm the fact that real-time data in EWS makes them far more timely, accurate, and effective in a reduction of disaster risks.

4. DECISION SUPPORT SYSTEMS FOR DISASTER PREPAREDNESS

Decision Support Systems (DSS) is a very important piece of equipment where complicated multidimensional data can be condensed into actionable information to be used by the decision-maker in the field of disaster preparedness. They fill the gap between unprocessed data sent by real-time monitoring systems and the decisions that should be taken regarding conditions of uncertainty and urgency (Liu et al., 2024).

A DSS is usually a combination of data management, an analysis modeling process, and intuitive interfaces used in facilitating knowledge-based decision-making at various stages of disaster management. DSS can enable governments, humanitarian organizations, and the local people on the ground to access information in a timely manner due to integration with GIS, remote sensing, and deployment of advanced computational procedures. The changing nature and development of DSS relates to the increased priority placed on resilience, inclusivity, and technological innovation on disaster preparation.

4.1 Overview of DSS in Disaster Management

There are many disaster management's Decision Support Systems, including hazard monitoring, post-disaster recovery and evacuation planning systems. Initial DSS were mainly based on static risk models of hazard, with historical data. But due to the advent of GIS, remote sensing and big data analytics in DSS it is dynamic and adaptive now (French & Niculae, 2005).

A classic DSS in disaster management combines three basic aspects:

- **Data management layer:** It receives and stores different types of data, such as geospatial data, demographic, and real-time sensor data.
- **Analytical layer:** It is the applicable models of flood inundation mapping, earthquake impact simulation, or cyclones trajectory forecast.
- **Interface layer:** It makes decision-ready dashboards and visualization tools, as well as communication modules, interactive and easier to understand.

The coronary power of contemporary DSSs is their capacity to consider the multi-hazard scenarios. An example is provided by the Pacific Disaster Center which has a platform called DisasterAWARE, which incorporates the data on earthquakes, cyclones, floods and pandemics and can provide in real-time a risk assessment to policymakers around the globe (DisasterAWARE: Multi-Hazard Early Warning, Risk Analytics, and Situational Awareness I Pacific Disaster Center, 2025). The Indian Ocean Tsunami Warning and Mitigation System is also based on DSS to model the advancement of the tsunami and evaluate the vulnerabilities in the region. As shown in Figure 3, it represents the multi-hazard early warning systems utilize hazard-specific sensors and GIS layers to optimize preparedness and response across disaster types such as earthquakes, floods, cyclones, wildfires, etc.

Figure 4. Multi-Hazard Early Warning System Process Across Different Disaster Types

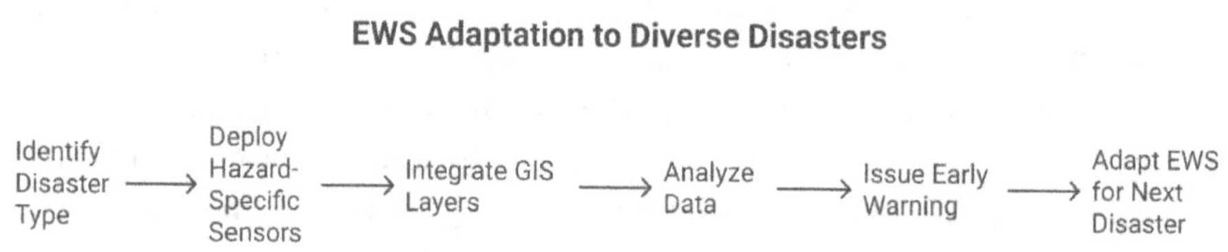

The addition of remote sensing also improves the functionality of DSS. There is situational awareness before, during and after the disaster using high-resolution imagery. As an example, during floods, synthetic aperture radar (SAR) data can be used to determine flood extents even through cloud cover to aid near real-time decision making in flood-stricken areas. The innovations demonstrate the centrality of DSS in translating early warning information to specific action measures.

4.2 Role of DSS in Risk Assessment and Response Planning

The main value of DSS is the evaluation of the risk and the creation of the preparedness strategies. DSS facilitates higher authorities in modeling and visualization of risks to help authorities gain insight into the chances of debris, vulnerability levels and possible outcomes.

- **Risk Assessment:** DSS incorporates spatial data to identify the landscape of the hazard-prone areas, as well as assess the threats of exposures to the population and infrastructure and ecosystem. As an example, the Copernicus Emergency Management Service of the European Union combines satellite evidence with geospatial models to create flood and wildfire risk maps down the countries of the European Union. The maps enable the authorities to focus investments on the mitigation of disaster and the increasing community preparedness.
- **Response Planning:** DSS also play an important role in the planning aspect of evacuation, resources deployment and emergency logistics. The use of scenario modeling helps the authorities to predict various outcomes of the hazard and develop contingencies plans. As an example, cities in Japan have earthquake DSS which simulates the effects of an earthquake on transport infrastructure, which is used to formulate evacuation routes in advance, including defects in the transport infrastructure.
- **Community-Centered DSS:** There is a rising trend that sees the need to incorporate local knowledge and participatory contributions into DSS systems. The method will guarantee that decision-making instruments depict the conditions of vulnerable populations and meet their needs. This is illustrated in Community-based Flood Early Warning System of Nepal. They merge local observation, hydrological models, and mobile warning into a DSS platform that enables the local actors to co-design and implement preparedness strategies (Trogrlić et al., 2022).
- **Uncertainty Management:** The other essential role of DSS is as an Uncertainty Manager in terms of hazard forecasting. Probabilistic model DSS offers confidence intervals, scenario comparisons and sensitivity analy-

sis, enabling decision-makers to consider various alternatives in circumstances where there is an uncertainty (Liu et al., 2024).

DSS with such capabilities can; through the aid of these functionalities, not only support technical decision-making but also support engendering trust/coordination/ timely response among institutions and communities.

4.3 Case Studies of DSS Applications in Disaster Preparedness

Some applicable examples in the real world show the role DSS play in increasing disaster preparedness across levels.

United States & Canada - HAZUS-MH: HAZUS-MH is a DSS created by FEMA, which is a comprehensive model to assess the potential losses in the event of earthquakes, floods, and hurricanes comparing its initial success in these types of events. HAZUS incorporates demographic and infrastructural data with the hazard models in a GIS platform to give policymakers an understanding of the anticipated harm, damage, and economic impacts. The planning of disaster preparedness, land-use, and investment aligns with its products (Nastev & Todorov, 2013).

Bangladesh - Flood Forecasting and Warning Centre (FFWC): Bangladesh has operationalized a DSS Sent to real time flood forecasting that combines upstream rainfall data river basin hydrology and satellite imagery. The FFWC provides probability-based prediction and flood-maps to the labor authorities and the population so that pre-mature evacuations can be done in the flood-affected areas. This DSS has also been credited with ensuring flood-related deaths have dropped considerably in the past 20 years (Rahman et al., 2012).

Japan - Earthquake Disaster DSS: Using DSS to integrate seismic monitoring data is one of the most developed in the world in Japan. Earthquake Early Warning is already available to residents with the assistance of the JMA, which alerts within nearly real-time. Such DSS can provide warnings to the citizens but also automatically initiate actions, like raising high-speed trains and closing gas lines would accomplish these goals, reducing the occurrence of the cascade effect (Cremen et al., 2021).

Africa -Locust forecasting DSS: DSS has not only been used in conventional hazards but also in biological disasters. The Food and Agriculture Organization (FAO) maintains the Desert Locust Information Service which combines ground based, remote sensing, and climatic modeling to forecast locust swarms and is a DSS. During the preliminary predictions, solutions related to pesticide deployment and prevention measures are taken and agricultural population losses alongside the lack of food become smaller (Klein et al., 2023).

Global – INFORM Risk Index: The INFORM DSS presents a multi-hazard risk assessment framework in the context of 190 countries that integrates hazard exposure, vulnerability, and coping capacity data. This is a device that helps international agencies and donors to allocate humanitarian investments and reinforce resilience in weak scenarios (Liu et al., 2024).

Such examples depict versatility and practicality of DSS in containing a variety of hazards, geographical area and aspect of governance.

5. CHALLENGES AND LIMITATIONS OF CURRENT SYSTEMS

Although phenomenal advances have taken place in both ruined readiness technologies and DSS, several problems occupy the ideal improvement and effectiveness of disaster readiness technologies. The limitations mentioned can be divided into technical, operational, and socio-political realms. The identification of such obstacles is critical on the way to develop disaster resiliency, increase the use of DSS, and achievements at the community level (Smith et al., 2016).

5.1 Technical Challenges

- **Data Quality and Availability:** This is one of the greatest technical issues in DSS as it requires the availability of good, timely data. High-resolution spatial and temporal data are required to develop a disaster model, but in numerous areas, especially low and middle-income economies, such data is incomplete or out-of-date. Lack of consistency in sensor networks, the absence of a history of hazards, and insufficient levels of data interoperability weaken DSS outputs.
- **Heterogeneous Data Integration:** Current DSS combine several sources, such as satellite images, social platform streams, IoT sensors and standard census information. These differing data streams can be very tricky to harmonize, particularly in terms of their standardization into a common format, alignment in terms of time resolution and consistency with respect to semantics. Without addressing these issues, it is possible to end up with inaccurate risk evaluation and ineffective decision support.
- **Model Uncertainty:** Uncertainties will always exist in predictive models built into DSS because of the unpredictability of stochastic threats, a lack of understanding of physical mechanisms, and a lack of fidelity in the computational methods. As an example, flood warning models can mis-predict the peak water level by correcting assumptions of the rainfall-off associations. DSS should, hence, give probabilistic outputs and sensitivity tests required to communicate the uncertainties inherent to a decision-maker (Liu et al., 2024).

- **Computational Complexity:** Sophisticated DSS, especially the AI and machine learning-based one's demand compute-intensive resources. Intense training of multi-hazardous events or those time-sensitive decisions resulted in high-resolution analytics on social media and IoT, which ended up overwhelming the infrastructure preventing the rapid decision-making in times of emergency.

5.2 Operational Challenges

- **Interagency Coordination:** Disaster preparedness can be efficiently done through multiple players, comprising of government, non-governmental organizations and local communities. The coordination of operations is often poor because of duplication of mandates, disorganized lines of communication, and divergent priorities. The implementation of DSS can be bogged down by the unclear position of the institutional roles or the lack of training in the interpretation of the outputs (Trogrlić et al., 2022).
- **DSS Training and Adoption:** DSS will not be that effective unless people utilizing them are trained and adopted. Disaster management professionals and local authorities lack sufficient training with the skills to interpret probabilistic outputs, and/or skills to incorporate DSS recommendations into planning operations. Lack of adoption to new technologies, particularly in low-digital literacy regions can drastically reduce DSS impact.
- **Real-Time Decision Making:** Disasters are dynamic, and this may leave very little time to make decisions. DSS generate forecasts, and the scenario simulation, but it is hard to interpret such results into real-time decision-making. There can be dilemma whereby automated alerts only lead to conflict with local perception and logistical factors (French & Niculae, 2005).
- **Maintenance and Sustainability:** DSS platforms need to maintain through updates, adjustment, and maintenance. Old software, break in sensor network, shortage of funds or technical expertise can also make these systems ineffective in the long term. The long run viability of DSS in resource limited context is also a serious issue.

5.3 Socio-Political and Ethical Limitations

- **Community Engagement:** A considerable number of DSS have been built on top-down methods that only consider the government or institutional requirements and not the view of the local communities. This may cause discrepancy between DSS suggestions, and what is happening in communities and hence they might not follow the prepared efforts or orders given to evacuate. Most regions have yet to adopt the participatory design methods (Trogrlić et al., 2022).

- **Equity and Accessibility**: Populations that are vulnerable to stressors, such as those living in informal settlements, the elderly and individuals with disability agencies may not have equal access to DSS outputs. GIS-based evacuation messages can be passed on as mobile alerts, which cannot reach those out of the loop to further entrench disparities in disaster risk (UNDRR, 2019).
- **Ethical Issues in AI and Predictive Modeling:** ethical issues concerning AI and AI-powered DSS involve the use of predictive algorithms that open the door to unintentional forms of bias. As an example, training models using past historical hazard data without taking account of socio-economic processes can underestimate informal settlement risks. Ethics is also needed to promote fair risk communication and distribution of available resources.
- **Policy and Governance Constraints:** Policy weaknesses or the absence of political will constrain the successful implementation of preparedness despite correct output of DSS. Policies should be consistent with the capabilities of technologies, community demands, and future objectives of resilience. DSS could offer valuable information and advice that fail to make any use or fall on deaf ears unless there are strong governance processes in place.

6. CONCLUSION, FUTURE DIRECTIONS, AND RECOMMENDATIONS

6.1 Conclusion

Risk management is an area that has been revolutionized using disaster preparedness systems and decision support systems as risk responders are able to quickly and more reliably react to nature and manmade hazards. As identified in this chapter, technology, data-driven modeling and community involvement were identified as some of the core pillars of successful disaster preparedness. Smart tools, such as AI-based predictive models, geospatial analytics, and real-time monitoring have also improved situational awareness by allowing the decision-makers to predict threats, efficiently target resources, and reduce the loss of life and property.

Notwithstanding these innovations, there are still some problems in a way that the potential of running systems is restrained. The technical challenges are majorly faced by data quality and interoperability problems, computational constraints, and model uncertainty. Other factors that limit preparedness strategy effectiveness are operational inefficiencies such as the lack of coordination between agencies and inefficient user training. Ethical and socio-political systems, including fair distributions of early warning opportunities and the active inclusion of destitute populations, also inform that a bespoke approach incorporating the social, institutional and technological aspects is required.

The synthesis of the lessons learnt in the context of case studies proves that research and technological solutions are not sufficient to contribute towards resilience. Adaptive, location-dependent methods demand disaster preparation that unite additional technologies with regional understanding, policy guidelines, and competencies of stakeholders. The rising layering of threats, such as urbanization, climate change, population pressures, and the complexity of hazards themselves all calls into question the need not only to have sophisticated systems but also flexible, inclusive, and sustainable systems.

To summarize, modern systems of disaster preparedness have advanced risk management greatly; nevertheless, to continue being successful, investments in infrastructure, human capacity, and community relations are a necessity. Integrated methods integrating predictive analytics and monitoring, in real-time, and participatory planning to achieve operationally effective and socially equitable early warning systems and decision support platforms are the main themes of the chapter. It is only with this kind of integration that various disaster preparedness efforts that are currently reactive can and will be transformed into proactive efforts to build resiliency, which will in the long run protect society and its infrastructure.

6.2 Future Directions

The future of disaster preparedness is also the further utilization of the emerging technologies accompanied by solving the limitations. Predictive modeling, taking into consideration the complex hazard interactions and the assumptions of socio-economic dynamics is becoming possible using artificial intelligence, big data analytics and machine learning. Situational awareness will always be boosted with real-time data provided by IoT gadgets or drones and satellite networks that will allow anticipatory interventions and greater accuracy in assigning assets.

The multi-hazard early warning systems integration will also be prioritized since disasters rarely happen individually. Future platforms will allow dynamic risk assessment to be provided with cascading impact, interdependencies and uncertainties to help decision-makers to have a more comprehensive view of what effects can happen. Improved interoperability in terms of systems and jurisdiction will be able to coordinate the local, regional, and national agencies more easily.

In addition, participatory practices will come into the limelight. Community knowledge and citizen science projects, as well as feedback infrastructures via mobile devices will enhance relevance of the system and promote trust. Ethical artificial intelligence systems and the availability of early warning systems fairly will make sure that vulnerable representatives will not be left behind.

Lastly, realistic training environments can be offered to emergency responders and planners through scenario-based simulations, digital twins, and augmented reality technologies. Technology will enable the stakeholders to run strategies in virtual spaces before activating them in the real world, majorly increasing their preparedness and minimizing operational risk. The nexus between technological innovation, policy reform, and community engagement will be the hallmarks of the new era of disaster preparedness, which will be more resilient, adaptive, and inclusive.

6.3 Recommendations

In a bid to ensure disaster preparedness and the response of the decision support systems, some of the main imperative's recommendations have been recommended.

- First, it is necessary to invest in data infrastructure, such as growth of sensor networks, high resolution spatial data, and data sharing protocols. It is a necessary base of proper modeling and real-time surveillance.
- Second, it would be necessary to focus on capacity building both on the level of government machines and emergency responders as well as on the local level. Workshops, simulation training, and training programs will be used to increase understanding of DSS output, and operational decision-making. Digital literacy and probabilistic reasoning should be tried to focus on so that the full potential of technology could be unlocked.
- Third, development of the system should be participatory and include planning. Ensuring relevance, compliance with early warnings and building trust, the local or vulnerable communities should be engaged in designing and implementing preparedness platforms. Mobile-based communication channels, localized knowledge integration, as well as feedback loop, represent the necessary elements of this strategy.
- Fourth, the ethical approaches should regulate the use of AI and predictive models. Transparency in the process of algorithmic decision-making, bias audits, and fair access to early warnings required to avoid inequalities and enhance social resilience.

Lastly, interdisciplinary teamwork involving technologists, policy makers, social scientists and emergency managers is essential. This kind of collaboration will facilitate innovation, better system integration and coordination of technological capacity with the needs of societies. The recommendations, when implemented, will enhance disaster preparedness to support the shift of reactive response to proactive, adaptive and inclusive resilience strategies to protect communities and infrastructure in both the immediate and future risks being faced.

REFERENCES

Basher, R. (2006). Global early warning systems for natural hazards: Systematic and people-centred. *Philosophical Transactions of the Royal Society A: Mathematical, Physical and Engineering Sciences, 364*(1845), 2167–2182. https://doi.org/DOI: 10.1098/rsta.2006.1819

Coppola, D. P. (2020). *Introduction to international disaster management* (4th ed.). Elsevier., https://shop.elsevier.com/books/introduction-to-international-disaster-management/coppola/978-0-12-817368-8

Cremen, G., Bozzoni, F., Pistorio, S., & Galasso, C. (2021). Developing a risk-informed decision-support system for earthquake early warning at a critical seaport. *Reliability Engineering & System Safety*, *218*, 108035. DOI: 10.1016/j.ress.2021.108035

Diehr, J., Ogunyiola, A., & Dada, O. (2025). Artificial intelligence and machine learning-powered GIS for proactive disaster resilience in a changing climate. *Annals of GIS*, *31*(2), 1–14. DOI: 10.1080/19475683.2025.2473596

Disaster, A. W. A. R. E. Multi-Hazard early warning, risk analytics, and situational awareness | Pacific Disaster Center. (2025, July 16). Pacific Disaster Center. https://www.pdc.org/disasteraware/

French, S., & Niculae, C. (2005). Believe in the model: Mishandle the emergency. *Journal of Homeland Security and Emergency Management*, *2*(1). Advance online publication. DOI: 10.2202/1547-7355.1108

Goodchild, M. F., & Li, W. (2021). Replication across space and time must be weak in the social and environmental sciences. *Proceedings of the National Academy of Sciences of the United States of America*, *118*(35), e2015759118. Advance online publication. DOI: 10.1073/pnas.2015759118 PMID: 34417345

Joyce, K. E., Belliss, S. E., Samsonov, S. V., McNeill, S. J., & Glassey, P. J. (2009). A review of the status of satellite remote sensing and image processing techniques for mapping natural hazards and disasters. *Progress in Physical Geography*, *33*(2), 183–207. DOI: 10.1177/0309133309339563

Kalogiannidis, S., Kalfas, D., Kontsas, S., Papaevangelou, O., & Chatzitheodoridis, F. (2025). Evaluating the Effectiveness of Early Warning Systems in Reducing Loss of Life in Natural Disasters: A case study of Greece. *Journal of Risk Analysis and Crisis Response*, *15*(1), 33. DOI: 10.54560/jracr.v15i1.547

Kamigaichi, O. (2004). JMA EARTHQUAKE EARLY WARNING. *Journal of Japan Association for Earthquake Engineering*, *4*(3), 134–137. DOI: 10.5610/jaee.4.3_134

Klein, I., Uereyen, S., Eisfelder, C., Pankov, V., Oppelt, N., & Kuenzer, C. (2023). Application of geospatial and remote sensing data to support locust management. *International Journal of Applied Earth Observation and Geoinformation, 117*, 103212. DOI: 10.1016/j.jag.2023.103212

Liu, Y., Gu, Y., & Zhang, H. (2024). Seismic risk assessment and damage analysis: Emerging trends and new developments. *Journal of Safety Science and Resilience = An Quan Ke Xue Yu Ren Xing (Ying Wen), 5*(3), 365–381. DOI: 10.1016/j.jnlssr.2024.04.005

Mohapatra, M., & Sharma, M. (2021). Cyclone warning services in India during recent years: A review. *Mausam (New Delhi), 70*(4), 635–666. DOI: 10.54302/mausam.v70i4.204

NASA's Earth observing system. (n.d.). https://eospso.nasa.gov/content/nasas-earth-observing-system-project-science-office

Nastev, M., & Todorov, N. (2013). Hazus: A standardized methodology for flood risk assessment in Canada. Canadian Water Resources Journal / Revue Canadienne Des Ressources Hydriques, 38(3), 223–231. https://doi.org/DOI: 10.1080/07011784.2013.801599

Rahman, M., Goel, N., & Arya, D. (2012). Study of early flood warning dissemination system in Bangladesh. *Journal of Flood Risk Management*, *6*(4), 290–301. DOI: 10.1111/jfr3.12012

Rjoub, D., Alsharoa, A., & Masadeh, A. (2023). Unmanned-Aircraft-System-Assisted Early Wildfire Detection with Air Quality Sensors. *Electronics (Basel), 12*(5), 1239. DOI: 10.3390/electronics12051239

Sharma, P., Wang, J., Zhang, M., Woods, C., Kar, B., Bausch, D., Chen, Z., Tiampo, K., Glasscoe, M., Schumann, G., Pierce, M., & Eguchi, R. (2020). DisasterAWARE – a GLOBAL ALERTING PLATFORM FOR FLOOD EVENTS. ISPRS Annals of the Photogrammetry, Remote Sensing and Spatial Information Sciences, VI-3/W1-2020, 107–113. https://doi.org/DOI: 10.5194/isprs-annals-VI-3-W1-2020-107-2020

Slatton, K. C., Carter, W. E., Shrestha, R. L., & Dietrich, W. (2007). Airborne Laser Swath Mapping: Achieving the resolution and accuracy required for geosurficial research. *Geophysical Research Letters*, *34*(23), 2007GL031939. Advance online publication. DOI: 10.1029/2007GL031939

Smith, P., Pappenberger, F., Wetterhall, F., Del Pozo, J. T., Krzeminski, B., Salamon, P., Muraro, D., Kalas, M., & Baugh, C. (2016). On the Operational Implementation of the European Flood Awareness System (EFAS). In Elsevier eBooks (pp. 313–348). https://doi.org/DOI: 10.1016/B978-0-12-801884-2.00011-6

Tao, Y., Tian, B., Adhikari, B. R., Zuo, Q., Luo, X., & Di, B. (2024). A review of Cutting-Edge Sensor Technologies for improved flood monitoring and damage assessment. *Sensors (Basel)*, *24*(21), 7090. DOI: 10.3390/s24217090 PMID: 39517986

Trogrlić, R. Š., Van Den Homberg, M., Budimir, M., McQuistan, C., Sneddon, A., & Golding, B. (2022). Early warning systems and their role in disaster risk reduction. In Springer eBooks (pp. 11–46). https://doi.org/DOI: 10.1007/978-3-030-98989-7_2

UNDRR. (2015). *Sendai framework for disaster risk reduction 2015–2030*. United Nations Office for Disaster Risk Reduction. https://www.undrr.org/publication/sendai-framework-disaster-risk-reduction-2015-2030

UNDRR. (2020). *Words into action: Multi-hazard early warning systems*. United Nations Office for Disaster Risk Reduction. https://www.undrr.org/words-into-action/guide-multi-hazard-early-warning

United Nations Office for Disaster Risk Reduction (UNDRR). (2019). *Global assessment report on disaster risk reduction 2019*. Geneva: UNDRR. https://www.undrr.org/publication/global-assessment-report-disaster-risk-reduction-2019

Vishvakarma, M. D. P. (2019). THE ROLE OF GEOSPATIAL TECHNOLOGIES IN DISASTER MANAGEMENT: a REVIEW. https://psychologyandeducation.net/pae/index.php/pae/article/view/8666

Yucesoy, E., Balcik, B., & Coban, E. (2024). The role of drones in disaster response: A literature review of operations research applications. *International Transactions in Operational Research*. Advance online publication. DOI: 10.1111/itor.13484

Chapter 14
Climate Change, Disasters, Adaptation, Immobility, and Trapped Populations in Türkiye

Merve Altundal Öncü
http://orcid.org/0000-0003-2402-9134
Ankara University, Turkey

ABSTRACT

This chapter examines the complex effects of climate change on human mobility in Türkiye, a Mediterranean climate hotspot highly vulnerable to floods, droughts, storms, and sea-level rise. Climate impacts act as threat multipliers, exacerbating existing socio-economic and spatial inequalities. Sudden-onset events trigger short-term, local mobility, while slow-onset hazards drive long-term relocation. Migration can serve as adaptation, yet many "trapped" populations remain immobile due to vulnerability and lack of resources. Climate-related counter-urbanization reflects privileged adaptation, deepening inequalities. Policy-induced displacement, arising from conservation-development tensions, further threatens livelihoods. Using social inequality and environmental justice frameworks, the study concludes that climate-induced mobility in Türkiye is a deeply social issue, requiring holistic, equitable, and resilient policy responses.

1. INTRODUCTION

Climate change is unequivocally recognized today as one of the most paramount and challenging global issues of our time, a phenomenon that fundamentally tran-

DOI: 10.4018/979-8-3373-6274-8.ch014

scends national borders and has firmly positioned itself at the epicenter of both the scientific community's research agenda and high-level political deliberations worldwide. The overwhelming and undisputed body of evidence concerning human-induced global warming, meticulously documented by authoritative bodies such as the Intergovernmental Panel on Climate Change (IPCC, 2014), robustly confirms that our planet is now irreversibly embarked upon a complex and highly dynamic trajectory of profound environmental transformation. This escalating crisis extends far beyond a simple or singular metric, such as merely an average increase in global temperatures. Instead, it ushers in and accelerates a destructive series of unprecedented forms of environmental degradation that fundamentally destabilize the inherent equilibrium of natural ecosystems and, consequently, vital human support systems. The most immediate and tangible manifestations of this new environmental disaster era include a marked increase in the frequency and intensity of extreme weather and climate events. These encompass phenomena such as severe and prolonged heatwaves, sudden and destructive cold snaps, torrential high precipitation events leading to large-scale flooding, catastrophic wind storms, multi-year droughts that decimate food security, and the persistent, critical threat of accelerating sea-level rise along global coastlines (IPCC, 2021). The cascading effects of these climatic shocks specifically the onset of chronic water scarcity, significant reduction in arable land productivity, and pervasive ecological destruction are progressively rendering critical regions of the world economically unviable for traditional livelihoods or, more critically, physically uninhabitable for large populations. Therefore, the resultant environmental degradation caused by anthropogenic climate change emerges as one of the most pivotal and urgent societal consequences of our generation, directly generating the risk of widespread human mobility and forced displacement, thereby solidifying its urgent place at the very focal point of intensive scientific inquiry and strategic policy formulation (Piguet, 2022; IDMC, 2024).

The linkages between climate change and human mobility constitute a multidimensional and highly complex area of debate within the existing literature (Rigaud et al., 2018). Moving away from the early-period predictions of mass, linear migratory flows, the contemporary consensus acknowledges this relationship as a multicausal, context-specific process deeply intertwined with socio-economic, political, and demographic factors (Black et al., 2011; McLeman, 2013). The current body of literature often struggles to fully articulate the complex dynamics between forced displacement, migration viewed as an adaptation mechanism, and the existence of trapped populations individuals unable to relocate due to severe socio-economic incapacities. Specifically, in a country like Türkiye, which is recognized as a climate change hotspot within the Mediterranean basin and is already exposed to escalating climate risks alongside long-standing rural-to-urban migration trends, there is a critical need for a holistic analysis of how climate impacts transform local mobility flows, exacerbate inequalities, and reshape governance practices. This study aims to fill this analytical gap by elucidating how climate risks operate as a threat multiplier within the Turkish context and by demonstrating the necessity of addressing mobility as a profound social problem. The primary objective of this book chapter is to systematically investigate the complex and multi-layered effects of climate change trends on human mobility flows within Türkiye. The scope of the study encompasses the climate crisis itself, the multi-causal nature of mobility, the perception of migration as an adaptive strategy, and the ongoing terminological controversies surrounding the subject. For Türkiye specifically, the analysis includes the region's climate change vulnerability, existing rural to urban migration patterns, policy-induced pressures particularly around protected areas and the influence of social inequalities on mobility outcomes. Furthermore, the scope includes both forms of forced displacement and voluntary mobility types, such as climate-related counter-urbanization, which often represents a more privileged adaptation strategy.

This book chapter is structured as a conceptual and theoretical synthesis, meticulously compiling and analyzing the existing literature. The study systematically encompasses current academic publications, international institutional reports, and relevant case studies in the interdisciplinary fields of climate change, disasters, human mobility, and Türkiye focused social sciences. The methodological approach relies on integrating and critically synthesizing the main currents within the literature to illuminate key analytical junctures, specifically debates surrounding causality, terminology (e.g., the preference for 'environmental migrants' over 'climate refugees'), adaptation strategies, and the critical issue of trapped populations. The interpretation of the findings utilizes the social inequality and environmental justice framework as a normative basis, thereby scrutinizing the underlying political and societal processes that shape climate migration outcomes.

Following this introductory section, the book chapter is divided into four main sections. Section 2, titled "Climate Change, Mobility and Disasters" will address the challenging scientific and political dimensions of climate change, specifically examining the differential impacts of sudden-onset and slow-onset events on migration and focusing on the ongoing debates regarding terms that risk environmental determinism (such as climigration and climate refugee). Subsequently, Section 3, which is titled "Findings" will concentrate on Türkiye's particular vulnerability as a climate change hotspot, discuss anticipated climate projections, and detail Türkiye-specific mobility dynamics like climate-related counter-urbanization. Section 4, "Components of Climate-Induced Mobility," establishes the analytical foundation of the study by providing detailed explanations of the conceptual and analytical framework used, elucidating the relationships between environmental triggers, types of mobility, causality, and the overarching normative frameworks (environmental justice/equity and the conservation-development dilemma). Finally,

Section 5, "Conclusions and Recommendations," will summarize the study's core findings and conclude the chapter by presenting holistic, equitable, and resilient policy and research recommendations tailored for addressing climate-induced mobility in Türkiye.

2. CLIMATE CHANGE, MOBILITY AND DISASTERS

Climate change is widely acknowledged as one of the most challenging scientific and political issues of our era. Scientifically, the evidence for human-induced warming is unequivocal (IPCC, 2014). This situation demonstrates that a complex and dynamic process of unprecedented environmental change has reached severe levels, signaling the emergence of a new epoch of environmental disaster (Scott, et al., 2024). The anticipated changes brought about by climate change do not merely signify shifts in global average temperatures; they also lead to alterations in the frequency and intensity of extreme weather and climate events. These events encompass threats such as severe flooding, high precipitation incidents, major storms, prolonged droughts, intense heat and cold waves, and accelerated sea-level rise (Ayaz, 2022). The potential global consequences of climate change, resulting in certain parts of the world becoming less habitable due to extreme heat, persistent drought, or inundation stemming from sea-level rise, are leading to widespread human displacement. The linkages between climate change and human mobility remain a significant subject of debate within the scholarly literature. Researchers working in this field are generally categorized into two principal schools of thought. Firstly, some experts who focus on the impacts of climate change predict that vast numbers of people will be compelled to relocate as a result of rising sea levels, increases in the severity and/or frequency of climate-related extreme events, or other forms of environmental degradation caused by global warming. These observers tend to view migration in a negative light. Climate change further exacerbates the urgency of traditional migration drivers such as economic hardship, political turmoil, oppression, and inequitable social conditions. Data highlighting the severity of disaster impacts reveal that 9.8 million people were displaced by disasters in the first half of 2020 (Adger et al., 2020). Moreover, certain aid organizations, such as Christian Aid (2007) have suggested that the number of people displaced due to climate change could exceed one billion. Additionally, it is estimated that every 1°C increase in global warming elevates the risk of global displacement due to flooding by approximately 50 percent. Conversely, social scientists, particularly migration experts, are often reluctant to attribute a predominant role to environmental factors in migration decisions and tend to view migration as a positive outcome in all circumstances. One of the core arguments of this critical perspective is the scarcity of empirical evidence substantiating projected massive migratory flows directly attributable to climate change.

Environmental events that trigger migration within the context of climate change are generally categorized into two main types based on their speed of onset (Foresight, 2011). The first category comprises sudden onset events, which include rapid disasters such as severe floods, major storms, hurricanes/cyclones, tsunamis, and extensive wildfires. These rapid incidents typically tend to result in short-distance, temporary, and internal mobility. Sudden disasters exert distinct and immediate effects on both the type and the duration of the ensuing migratory movement. The second category consists of slow onset events, which are protracted processes that include drought, sea-level rise, cumulative global warming effects, desertification, water scarcity, and the sustained degradation of land productivity. These slow moving changes, in contrast, tend to increase migration over the long term, often leading to more permanent relocations. For instance, the drastic reduction in agricultural output caused by prolonged drought can frequently serve as a more effective and compelling driver of migration than a sudden flood event. The displacement figures resulting from the impacts of sudden-onset events are notably significant. For example, during the six month period between the second half of 2020 and the first half of 2021, an overwhelming 10.3 million people were displaced due to extreme climate events a figure nearly four times greater than the number of people displaced by war and conflict during that same timeframe (Kam et al., 2021).

The relationship between climate change and migration has generated numerous debates within scientific and political circles (Obokata et al., 2014). One of the most significant controversies in the context of climate-forced mobility is the terminology utilized to define the people who are moving. The concepts employed in this context include:

- The term "environmental refugee" emerged in the 1970s and gained popularity with a 1985 United Nations Environment Programme (UNEP) report (Aghazarm and Laczko, 2009). The estimation that up to 200 million people would migrate due to environmental changes by the year 2050 garnered widespread attention in scientific and political environments and heavily influenced the following decades (Myers, 2002).
- Terms such as "climate migrants," "environmental migrants," and "ecological refugees" are used interchangeably in the literature, despite the absence of a clear demarcation between these concepts. The International Organization for Migration (IOM), however, recommends the use of the term "environmental migrants" (Black and Collyer, 2014).

- Although the concept of "climate refugee" does not reflect a legal framework, it is often employed to draw attention to a Human Rights-Based Approach (Piguet, 2010; Gemenne, 2011).
- Academic interest in the term "climate migrant" has significantly increased since 2010 (İçduygu & Gören, 2023).
- The term "climigration," coined by Bronen (2010), defines a situation where a community becomes unsustainable solely due to climate-related events, necessitating permanent relocation (Ferris, 2020; Hamilton et al., 2016).

The terminology employed to describe the movement induced by climate change inherently reflects the complexity of this relationship. Terms like climate migration risk invoking environmental determinism by implying a simplistic and linear causality where global warming directly triggers migratory outcomes. However, the dynamics governing human population movements have historically been recognized as invariably intricate and multi-factorial. Empirically establishing a direct and definitive link between climate change and migration is fraught with numerous challenges. Given that climate change is often considered to be in its nascent stages, or alternatively, because isolating individual climate events that prompt current migration and definitively attributing them to long-term climate change is scientifically difficult, there is currently a relative scarcity of robust empirical evidence that can conclusively support or decisively refute the existence of such singular, climate driven migration. Much of the existing research relies heavily on historical analogies, which are often predicated on responses to extreme events that occurred under conditions where long-term environmental contexts were relatively stable and permitted eventual return. In contrast, climate change involves a sustained shift in the average conditions over time, coupled with an increase in the magnitude or frequency of extreme events. Under these novel conditions, the possibility of a return to a stable normal state, which would enable spontaneous return migration, may simply not materialize. Consequently, the use of definitive terms such as climigration, climate change refugees, or climate migrants carries the inherent risk of emphasizing a linear and singular causal link leading from climate change directly to forced migration, thereby potentially overlooking the complex interplay of socio-economic and political drivers.

The nature of the causal relationship between climate change and migration remains a significant subject of debate. Early studies in the 1990s often presupposed a simple and linear relationship where climate/environmental change directly led to migration, frequently predicting massive displacements primarily from the Global South towards the North (Lilleør and Van den Broeck, 2011). However, it is now widely accepted that this relationship is far more complex, context-specific, and multi-causal (Warren, 2016). Climate and environmental factors are deeply intertwined with socioeconomic, political, and demographic determinants. Whether climate shocks actually lead to migration is largely contingent upon the capacity and socioeconomic circumstances of the individuals or households affected (Nawrotzki et al. 2017). The question of whether migration should be viewed as an adaptation mechanism to environmental change, alongside the precarious situation of those who cannot migrate, occupies a crucial place in the literature. Migration can be a key strategy for coping with environmental change and is increasingly recognized as a potential adaptation mechanism against climate change (Cantor and Apollo 2020). The focus on migration as an adaptation strategy has experienced an unprecedented surge in the literature since 2010. Nonetheless, the literature indicates that migration is frequently emphasized as an unplanned and individual form of adaptation, with notably less focus placed on planned adaptation strategies. Conversely, high levels of vulnerability and socio-economic deficiencies can heighten the need to migrate while simultaneously minimizing the capacity to do so (McLeman and Smit, 2006). This critical situation leads to the existence of trapped populations people unable to move from their location particularly among the poorest and most vulnerable segments of society.

The framing of climate migration as a security issue by states also constitutes a critical area of debate. Maximalist studies during the 1990s, which forecasted exceptionally high migration figures, catalyzed the tendency to treat climate-induced migration as a national security threat (Selby et al. 2017). Traditional theories were largely inadequate in recognizing climate migration as a security problem unless it was directly linked to pre-existing national security interests. However, growing empirical evidence in recent years has increasingly refuted this securitization narrative, demonstrating only a weak connection between climate change and violent conflict (Soest 2020). Correspondingly, the number of academic studies focusing primarily on securitization implications has decreased as of 2020. The literature highlights a persistent lack of adequate focus on the broader social and political processes, underlying inequalities, and vulnerabilities that underpin climate migration. It is now widely acknowledged that climate migration is not merely an ecological problem but fundamentally a social issue deeply rooted in processes of social inequality.

3. ADAPTATION, IMMOBILITY, AND TRAPPED POPULATIONS IN THE CONTEXT OF TÜRKIYE

Climate change and the projected future climate disasters have emerged as one of the most challenging scientific and political issues of our time, fundamentally reshaping demographic and spatial processes globally. Migration, as a response and a coping strategy to environmental changes, can assume a significant role. Türkiye, in particular, which is situated within the Middle East and North Africa region and is recognized as a climate change hotspot, is highly susceptible to these complex interactions (Toros et al., 2019). According to the Revision Plan published by the T.C. Ministry of Environment, Urbanisation and Climate Change (MoEUCC, 2021), climate projections for the wider region, including Türkiye, indicate that in addition to changes in global average temperatures, the frequency and intensity of extreme weather and climate events such as severe floods, storms, droughts, and sea-level rise will increase. Maximum temperatures are expected to rise between 1°C and 4 °C for future periods, accompanied by a significant warming trend in minimum temperatures (Çeliktopuz, 2024; Erlat et al., 2021). Along with this increase, a rise in the number of tropical nights is also anticipated. In Northern Mediterranean countries, particularly Türkiye and Southern European nations, the rise in tropical nights and hot days may result in adverse health consequences for the population and could trigger the need for individuals without access to cooling systems to seek new settlement areas due to heat-induced unproductivity.

When examining the effects of environmental degradation on migration, statistically significant relationships have been identified between key climate variables and net migration rates. Analyses conducted across 65 countries in the Middle East and North Africa (MENA) region reveal a negative relationship, showing that a 1% increase in minimum temperature leads to a 51% increase in the net migration rate, while a 1% increase in precipitation results in a 56% decrease in the net migration rate (Khanchel and Lassoued,2025). In countries like Türkiye, where agriculture is heavily reliant on rainfall, the anticipated dangerous levels of precipitation decline and the increase in the number of dry days may potentially trigger migration due to the deterioration of agricultural conditions and the reduction of livelihoods (Kadirbeyoglu, 2010). Historically, Türkiye has already been experiencing rural-to-urban migration stemming from worsening agricultural conditions and rural poverty (Akıner et al., 2025).

Climate-related human mobility manifests not only in forced mass flows but also in the form of more individual adaptation strategies. For instance, climate related counter-urbanization is defined as a largely voluntary type of movement that occurs downwards in the settlement hierarchy from urban to rural areas as a direct or indirect response to climate breakdown (Korkmaz and Meşhur, 2022). This movement is often centered on the perception and positive representation of the rural area as a refuge or sanctuary (Sözen and Kılıç, 2025). However, such relocations typically represent a privileged form of adaptation that can only be executed by households possessing greater financial resources, thus reflecting and deepening socio-spatial inequalities (Toros et al., 2019). Although such relocation may be a rational decision for individual households, these movements can ultimately be considered a form of societal maladaptation because they may generate adverse effects on rural host communities or potentially undermine community-wide adaptation efforts (Adger et al., 2020).

Disasters induced by climate change are not the sole or simple cause of migration; these decisions are embedded in a complex interplay of economic, social, political, and demographic factors. However, climatic impacts can serve as a threat multiplier, exacerbating existing inequalities and insecurities (Witmer et al., 2017). High vulnerability can increase the imperative for communities most affected by climate change to migrate, while simultaneously minimizing their capability to do so due to socio-economic insufficiencies, a situation that gives rise to trapped populations (Zickgraf, 2019). Presently, displacement caused by environmental factors has surpassed displacement due to conflict and violence: 56% of global internal displacement in 2023 resulted from disasters. In the Middle East and North Africa region, nearly a quarter of the total internal displacement is attributed to disasters (IDMC, 2024). At the societal level, a study involving social work students in Türkiye demonstrated that those who expressed a high level of anxiety regarding climate change believed that migration due to environmental factors would increase (Doğan and Buz, 2024). Furthermore, the majority of these students reported a negative perception, stating that the measures taken in Türkiye to combat climate change and its impacts are inadequate. This perception aligns with the low overall performance rating documented in the annual Climate Change Performance Index (CCPI, 2023) report, where Türkiye was ranked 47th among the 59 highest-emitting countries. This evidence underscores that the risks associated with climate change are deeply interwoven with the complex and multi-layered socio-political and economic context that determines migratory flows.

In recent years, the nexus between climate change, environmental degradation, and human mobility has increasingly become central to research agendas at both global and national levels. Studies focused on Türkiye underscore that the impacts of the climate crisis are not confined solely to environmental transformation; rather, they are becoming a structural driving force behind societal, spatial, and economic changes (Turhan et al., 2016). In this context, climate change is treated not merely as an environmental problem but as a source of multi-dimensional societal problems and a fundamental impetus for research (Table 1). In the literature, the climate and biodiversity crises are conceptualized as a dual dynamic that simultaneously mobilizes conservation efforts and triggers displacement processes (İçduygu & Gören, 2023).

Table 1. Conceptual Framework of Climate Change and Mobility Studies in Türkiye

Conceptual Framework	Approach	Topic
Climate Change/Climate Crisis	Source of the Problem	Conservation awareness, climate change and biodiversity, resource scarcity, social conflict (Aktaş, 2014; Okutan and Otay, 2025; Biehl Öztuzcu and Aslan, 2022).
Migration/Displacement	Outcome/Impact	The direct result of conservation policies or conflicts (Keleş, 2022; Yenal,2023; Kızıldere, 2025).
Protected Areas	Spatial Focus	The spatial locus of conflict and displacement (Khan and Somuncu, 2019; Taşeli, 2007; Arslan and Kaymaz; 2020).
Conservation-Development Dilemma/Conflict	Interaction	The tension between conservation policies and local communities (Sayan, 2019; Ward and Ruckstuhl, 2017; Yıldız and Atmiş, 2023).
Livelihoods	Social Impact	The socio-economic basis restricted by conservation policies, leading to migration (Doğan and Buz, 2024; Atmiş, 2018; 2018; Sardar et al., 2020).
Equity/Environmental Justice	Normative Framework	The ethical/theoretical perspective used in resolving the conservation and development dilemma (Altan, 2024; Ökten, 2017; Aygün, 2015).
Local People/Community	Actor	The primary experiencer of conflict and displacement (Kadirbeyoglu, 2017; Wissink et al., 2013; Uzelgun and Castro,2014).

A growing body of research indicates that displacement is the most visible social outcome of the climate crisis (Table1). Studies conducted across Türkiye reveal that disasters such as drought, floods, wildfires, and land degradation lead to forms of temporary or permanent migration/displacement in rural and semi-arid regions. In this literature, climate-induced displacement is debated not only as a result of environmental pressures but also as a direct outcome/ımpact of policy-related pressures. Specifically, compulsory or indirect displacement processes arise when conservation policies or resource management practices restrict local communities' access to land and water (Keleş, 2022; Yenal, 2023; Kızıldere, 2025). This forced removal from living spaces solidifies the crisis's devastating impact on social structures, making visible how governance, inequality, and vulnerability shape mobility outcomes.

At the spatial level, much research shows that these dynamics are concentrated around protected areas(Table1). Studies conducted in the Mediterranean, Black Sea, and Eastern Anatolia regions reveal that these areas often function as spatial focus points of conflict where conservation goals intersect with local livelihood needs (Khan and Somuncu, 2019; Taşeli, 2007; Arslan and Kaymaz; 2020). These areas serve as a critical spatial mechanism for understanding displacement processes because policy-based exclusion, livelihood restrictions, and environmental degradation converge there. Tensions between local communities and central authorities are inevitable where natural resources are limited and conservation measures are strictly enforced. Thus, protected areas become both the symbolic and material centers of socio-ecological conflicts.

In this literature, conflict is treated as an interaction and a mediating mechanism that explains the relationship between environmental governance and mobility (Table1). Research shows that tensions between centralized conservation policies and the local people/community who depend on these areas for livelihoods frequently generate conflict (Sayan, 2019; Ward and Ruckstuhl, 2017; Yıldız and Atmiş, 2023). Studies on forest management, fisheries regulations, or rural land use indicate that conservation measures sometimes lead to livelihood losses and indirect migration. In this context, the local population acts as the primary actor and direct experiencer of these conflicts, bearing the daily reflections of environmental injustice (Kadirbeyoglu, 2017; Wissink et al., 2013; Uzelgun and Castro,2014).

The societal impact of these complex processes is observed particularly through transformations in livelihoods (Table1). The literature reveals that the sustainability of traditional livelihood forms such as agriculture, animal husbandry, and small-scale forestry is reduced with the restriction of access to natural resources and the increase in climate shocks (Doğan and Buz, 2024; Atmiş, 2018; Sardar et al., 2020). Livelihood insecurity emerges as one of the essential socio-economic factors compelling mobility; migration, in turn, increasingly becomes an adaptive response to environmental and institutional constraints.

At the normative framework level, this literature relies on the frame of equity/environmental justice (Table1). Studies examining environmental governance in Türkiye (Altan, 2024; Ökten, 2017; Aygün, 2015) emphasize that environmental costs and benefits are unevenly distributed among social groups, reflecting a broader dynamic of power and exclusion. This perspective offers an ethical and theoretical foundation for interpreting environmental conflicts shaped by the conservation-development dilemma.

Consequently, the conservation-development dilemma forms the foundational theoretical framework for analyzing the relationships between climate change, displacement, and livelihoods (Table1). This dilemma is a structural tension that encompasses, on one hand, the imperative to protect ecological systems and, on the other, the necessity to support the socio-economic sustainability of the local population. As climate-related disasters such as drought, floods, and wildfires increasingly intersect with rural and protected areas in Türkiye, this dilemma is not merely a theoretical discussion but a tangible social reality shaping vulnerability, adaptation, and human mobility. The literature generally demonstrates that climate-induced mobility must be addressed holistically with strategies that are equitable, inclusive, and resilient.

4. COMPONENTS OF CLIMATE-INDUCED MOBILITY

The foundation of climate-induced mobility lies in the climate change/climate crisis, recognized as the most challenging scientific and political issue of our time. Defined by the unequivocal evidence of human-induced warming (IPCC, 2014), this situation triggers an unprecedented process of environmental degradation and constitutes the primary impetus for research. As shown in *Figure 1*, climate-induced mobility arises from diverse environmental triggers and interacts with complex socioeconomic and political factors. The environmental triggers initiating mobility are categorized into two main types based on their speed of onset: sudden-onset events, such as severe floods and wildfires, which typically lead to short-distance, temporary, and internal mobility; and slow-onset events, such as droughts, desertification, and the sustained degradation of land productivity, which tend to increase migration over the long term, leading to more permanent relocations. For instance, the drastic reduction in agricultural output caused by drought can serve as a significant driver of migration. The types and directions of mobility resulting from these triggers are diverse. While forced migration/displacement due to extreme events occurs, mobility can also be viewed as a key strategy and potential adaptation mechanism for coping with environmental change. However, trapped populations individuals unable to migrate due to high vulnerability and socio-economic deficiencies remain a critical part of this picture. An interesting type is climate-related counter-urbanization, a largely voluntary and privileged movement from urban to rural areas in response to climate disruption. While historical predictions foresaw mass flows from the Global South to the North, current mobility trends also include the prevalent rural-to-urban shift.

Figure 1. Schematic Overview of Climate-Induced Mobility Drivers and Outcomes

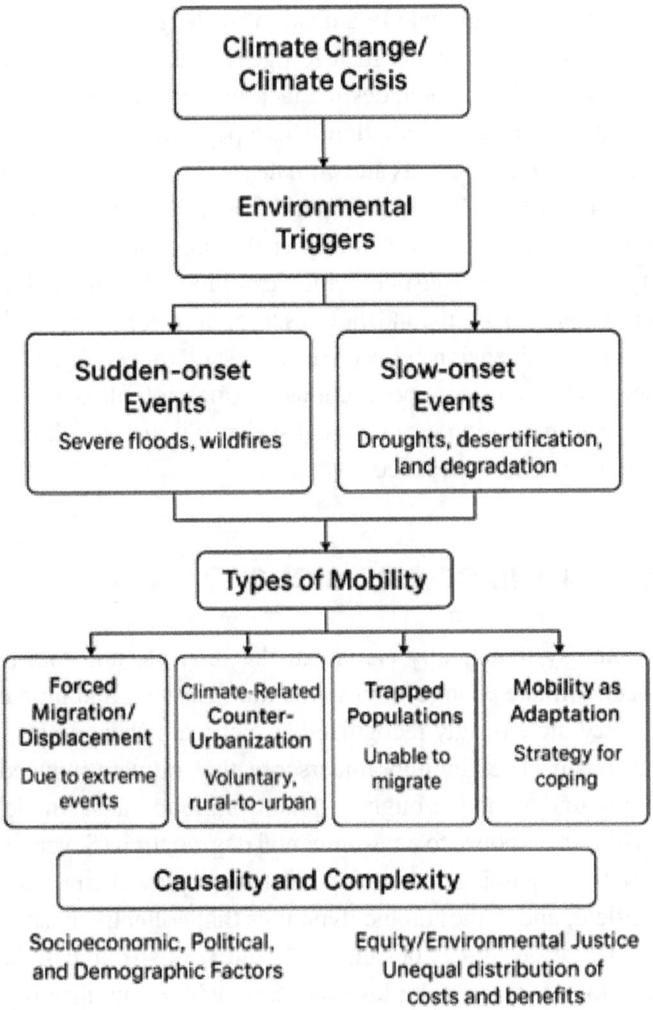

The causality and complexity behind mobility constitute a major area of debate. The prevailing consensus is that climate factors are not the sole or simple cause of migration; rather, the relationship is multi-causal and deeply intertwined with socioeconomic, political, and demographic factors. In this context, climatic impacts serve as a threat multiplier, exacerbating existing inequalities and insecurities. The outcomes of this relationship are context-specific, depending on the capabilities and socioeconomic circumstances of the affected individuals or households. Furthermore, policy-induced pressures, arising when conservation policies or resource management practices restrict local communities' access to land and water, also trigger indirect displacement processes. The terminology used to define movement induced by climate change is also contentious: while the earliest term was environmental refugee, the International Organization for Migration (IOM) recommends environmental migrants. The term climate refugee is often employed to draw attention to a human rights-based approach, despite lacking a legal framework. Other terms include climigration defining a situation where permanent relocation is necessary solely due to climate-related events and climate migrants, which has seen increased academic interest since 2010 (Matthews and Potts, 2018). Finally, the normative and theoretical frameworks used to interpret this phenomenon are crucial: these include equity/environmental justice, which examines the unequal distribution of environmental costs and benefits, and the conservation-development dilemma, which represents the structural tension between the necessity for ecological protection and the requirement to support local populations. A final notable debate is the tendency to frame climate migration as a national security issue, which has been largely refuted by recent empirical evidence.

5. DISCUSSION AND CONCLUSIONS

This study has systematically examined the intricate and multi-layered relationship between climate change, environmental disasters, and human mobility in Türkiye, a country increasingly recognized as a climate change hotspot within the Mediterranean basin. The findings underscore that climate-induced mobility in Türkiye cannot be understood through simplistic or mono-causal models that reduce movement to a direct response to environmental degradation (Sever, 2023). Instead, it emerges from the dynamic and interdependent interplay of environmental, socio-economic, political, and demographic dynamics that collectively shape patterns of vulnerability, adaptation, and displacement. In the Turkish context, the intersection of these dimensions forms a complex landscape where mobility becomes both a symptom and a strategy an outcome of risk exposure as well as an expression of resilience and adaptation to environmental stress.

The analysis reveals that both sudden-onset and slow-onset climate events ranging from floods, storms, and wildfires to droughts, land degradation, and desertification have tangible and increasingly visible impacts on human mobility patterns across the country. These impacts are not limited to direct or forced displacement following disasters but also encompass indirect, policy-induced, and adaptive forms of movement. The empirical evidence suggests that mobility related to climate stress in Türkiye is not uniform; it varies spatially and socially, influenced by local environmental conditions, livelihood dependencies, and the institutional capacity to cope with or adapt to change (Gümüş et al., 2023). Importantly, migration in this context can function as a coping mechanism for households seeking to diversify income sources or reduce exposure to environmental risks. In some cases, migration represents a proactive adaptation strategy that enables individuals and communities to anticipate future threats. However, migration as adaptation also exposes deep structural inequalities within society, reflecting differential access to financial resources, information, and social networks that enable mobility in the first place.

A particularly salient dimension of this inequality is the phenomenon of trapped populations those who are unable to migrate despite being exposed to environmental stressors. In rural regions of Türkiye, where livelihoods are heavily dependent on agriculture, water scarcity and land degradation exacerbate economic precarity. Yet, limited financial capacity, weak institutional support, and socio-cultural attachment to place often restrict mobility options. These populations remain physically and economically immobilized, illustrating how adaptive capacity is unevenly distributed across space and social strata (Tekin & Deniz, 2025). The presence of trapped populations underscores the structural injustices embedded in the geography of vulnerability. It reveals how socio-economic constraints intersect with environmental pressures to produce immobility as a form of exclusion, where remaining in place is not an expression of choice but a reflection of constraint.

The study therefore concludes that climate-induced mobility in Türkiye operates as a threat multiplier that intensifies existing social, economic, and regional inequalities. Vulnerable groups particularly rural, low-income, and marginalized communities bear disproportionate burdens of climate related disruptions, while wealthier populations often employ mobility as a privileged adaptation strategy. For instance, counter-urbanization trends, where affluent groups relocate from urban centers to less environmentally stressed or more desirable rural and coastal regions, exemplify the use of mobility as a form of selective adaptation. Such movements reflect not only differential exposure to climate risk but also differential access to the means of adaptation. This unequal capacity to move reveals how climate mobility cannot be adequately addressed as a purely environmental or demographic issue. Rather, it constitutes a profound question of social equity and environmental justice, demanding policy responses that explicitly confront inequality as a central dimension of adaptation.

Governance and policy-induced pressures further complicate this landscape. In Türkiye, conservation, land use, and resource management policies especially those implemented in and around protected areas have at times restricted local communities' access to vital natural resources. These policies, though motivated by ecological preservation, have inadvertently triggered displacement and livelihood insecurity among local populations (Pelek et al., 2025). The resulting conflicts illustrate the broader conservation development dilemma: the persistent tension between the imperative of environmental protection and the socio-economic sustainability of communities living within or adjacent to protected ecosystems. Without participatory governance mechanisms and equitable benefit-sharing frameworks, conservation efforts risk reproducing historical patterns of marginalization and displacement. The challenge, therefore, lies not only in balancing ecological and human priorities but also in transforming governance structures to be inclusive, transparent, and socially responsive.

From a policy perspective, these findings indicate that any effective framework addressing climate induced mobility in Türkiye must be holistic, inclusive, and justice oriented. Policymakers should move beyond reactive or securitized approaches that frame climate migration as a threat to national stability and instead adopt human-centered frameworks grounded in rights, equity, and participation (Yavuz & Çelik, 2023). Integrating climate adaptation, disaster risk reduction, and social protection mechanisms is crucial to ensure that the most vulnerable populations are not left behind in the transition toward climate resilience. This integration requires cross-sectoral coordination between environmental agencies, social welfare institutions, and local governments. Moreover, it necessitates the inclusion of community voices in policy design and implementation to ensure that interventions reflect local realities and priorities rather than top and down technocratic visions.

Equally important is the reframing of climate mobility from a national security concern to a developmental and human rights issue (Birpınar & Tuğaç, 2022). Early policy discourses in Türkiye often interpreted migration driven by environmental stress through a security lens, emphasizing border control, internal stability, and population management. While these concerns are not entirely unfounded, such framings tend to obscure the structural drivers of vulnerability and the agency of affected populations. A human centered approach would recognize mobility as both a potential form of adaptation and a site of structural inequality, calling for policies that enhance rather than restrict people's capacity to make voluntary and dignified mobility choices. This perspective aligns with the broader principles of climate justice, which emphasize responsibility, participation, and fairness in the distribution of adaptation burdens and benefits.

From a research standpoint, the study identifies an urgent need for more empirical, interdisciplinary, and regionally focused investigations into the dynamics of climate induced mobility in Türkiye. Current scholarship remains fragmented, often privileging quantitative projections over qualitative insights. While numerical estimates of potential climate migrants are valuable for policy planning, they must be complemented by ethnographic, sociological, and political analyses that capture the lived experiences, perceptions, and adaptive strategies of affected communities. Such interdisciplinary approaches can illuminate how governance structures, social norms, and historical inequalities shape adaptive outcomes. Building a robust and context-specific evidence base will not only enhance scholarly understanding but also guide the formulation of equitable and effective policy interventions.

Ultimately, Türkiye's experience mirrors the broader global reality that climate change is not only an environmental crisis but also a profound process of social transformation. It reshapes the geography of vulnerability, mobility, and justice, compelling societies to rethink the relationship between people, place, and the environment (Ortakaya, 2023). Addressing these challenges requires transcending disciplinary silos and policy compartmentalization. Collaboration between climate scientists, geographers, sociologists, economists, and policymakers is essential to co-produce knowledge and solutions that are scientifically sound, socially equitable, and ecologically sustainable. Only through such integrative approaches can Türkiye and the global community develop governance frameworks that not only respond to the immediate risks of climate-induced mobility but also foster long-term resilience, justice, and sustainability in an era of accelerating climate change.

REFERENCES

Adger, W., Crépin, A., Folke, C., Ospina, D., Chapin, F. S.III, Segerson, K., & Wilen, J. (2020). Urbanization, migration, and adaptation to climate change. *One Earth*, *3*(4), 396–399. DOI: 10.1016/j.oneear.2020.09.016

Aghazarm, C., & Laczko, F. (2009). *Migration, environment and climate change: Assessing the evidence Geneva.* International Organization for Migration (IOM), Institute for Environment and Human Security and United Nations University.

Akıner, M. E., Akıner, N., & Akıner, İ. (2025). Interplay between climate change and the migration-induced population growth: A case study in Konyaaltı, Antalya, Turkey. *Sürdürülebilir Çevre Dergisi*, *5*(1), 1–18. DOI: 10.62816/cevder.1581432

Aktaş, Ö. (2014). Impacts of climate change on water resources in Turkey. *Environmental Engineering and Management Journal*, *13*(4), 881–889. DOI: 10.30638/eemj.2014.092

Altan, C. (2024). *Critical evaluation of climate action plans in Türkiye from the perspective of climate justice* (Master's thesis). Middle East Technical University, Ankara, Turkey.

Arslan, E. S., & Kaymaz, I. (2020). Visitor perception of recreational ecosystem services and their role in landscape management of Gölcük Nature Park, Turkey. *International Journal of Sustainable Development and World Ecology*, *27*(3), 202–213. DOI: 10.1080/13504509.2019.1711247

Atmiş, E. (2018). A critical review of the (potentially) negative impacts of current protected area policies on the nature conservation of forests in Turkey. *Land Use Policy*, *70*, 675–684. DOI: 10.1016/j.landusepol.2017.10.054

Ayaz, A. (2022). Unsustainable systems. *Insight Turkey*, *24*(1), 37–54. DOI: 10.25253/99.2022241.3

Aygün, A. (2015). *Climate change and urban resilience: Vulnerability and risk assessment for Istanbul* (Unpublished master's thesis). Istanbul Technical University, Department of Urban and Regional Planning.

Biehl Öztuzcu, K., & Aslan, Ö. (2022). *Migration and environment from a gender perspective: Examples from civil society in Turkey.*

Birpınar, M. E., & Tuğaç, Ç. (2022). Climate security and migration: An evaluation for the world and Turkey. *Insight Turkey*, *24*(1), 105–133. DOI: 10.25253/99.2022241.7

Black, R., Adger, W. N., Arnell, N. W., Dercon, S., Geddes, A., & Thomas, D. (2011). The effect of environmental change on human migration. *Global Environmental Change, 21*(S1), S3–S11. DOI: 10.1016/j.gloenvcha.2011.10.001

Black, R., & Collyer, M. (2014). Populations 'trapped' at times of crisis. *Forced Migration Review, 45*, 52–56.

Bronen, R. (2010). Forced migration of Alaskan indigenous communities due to climate change. In Afifi, T., & Jäger, J. (Eds.), *Environment, forced migration and social vulnerability* (pp. 87–98). Springer., DOI: 10.1007/978-3-642-12416-7_7

Cantor, D. J., & Apollo, J. O. (2020). Internal displacement, internal migration, and refugee flows: Connecting the dots. *Refugee Survey Quarterly, 39*(4), 647–664. DOI: 10.1093/rsq/hdaa016

Çeliktopuz, E. (2024). Forecasting some climate parameters of Türkiye using the SSP3-7.0 scenario for the years 2040–2059. *International Journal of Agriculture Environment and Food Sciences, 8*(1), 62–71. DOI: 10.31015/jaefs.2024.1.7

Christian Aid. (2007). *Human tide: The real migration crisis.* https://apo.org.au/node/9086

Climate Change Performance Index (CCPI). (2023). *Climate Change Performance Index 2023.* Germanwatch/NewClimate Institute/Climate Action Network. https://ccpi.org/download/climate-change-performance-index-2023/

Doğan, H., & Buz, S. (2024). The effect of climate change anxiety on social work students in Turkey. *International Social Work, 67*(4), 1045–1058. DOI: 10.1177/00208728231208007

Erlat, E., Türkeş, M., & Aydın-Kandemir, F. (2021). Observed changes and trends in heatwave characteristics in Turkey since 1950. *Theoretical and Applied Climatology, 145*(3), 137–157. DOI: 10.1007/s00704-021-03620-1

Ferris, E. (2020). Research on climate change and migration: Where are we and where are we going? *Migration Studies, 8*(4), 612–625. DOI: 10.1093/migration/mnaa028

Foresight: Government Office for Science. (2011). *Migration and global environmental change* (Final Project Report). Government Office for Science.

Gemenne, F. (2011). Climate-induced population displacements in a 4°C+ world. *Philosophical Transactions of the Royal Society A: Mathematical, Physical and Engineering Sciences, 369*(1934), 182–195.

Gümüş, B., Örüç, S., Yücel, İ., & Yılmaz, M. T. (2023). Impacts of climate change on extreme climate indices in Türkiye driven by high-resolution downscaled CMIP6 climate models. *Sustainability (Basel), 15*(9), 7202. DOI: 10.3390/su15097202

Hamilton, L. C., Saito, K., Loring, P. A., Lammers, R. B., & Huntington, H. P. (2016). Climigration? Population and climate change in Arctic Alaska. *Population and Environment, 38*(2), 115–133. DOI: 10.1007/s11111-016-0259-6 PMID: 27829694

Intergovernmental Panel on Climate Change (IPCC). (2014). *Climate Change 2014: Synthesis Report.* Cambridge University Press.

Intergovernmental Panel on Climate Change (IPCC). (2021). *Climate Change 2021: The Physical Science Basis.* Cambridge University Press.

Internal Displacement Monitoring Centre (IDMC). (2024). *Global Report on Internal Displacement (GRID 2024).* Norwegian Refugee Council. https://api.internal-displacement.org/sites/default/files/publications/documents/IDMC-GRID-2024-Global-Report-on-Internal-Displacement.pdf

Internal Organization for Migration (IOM). (2007). *Discussion note: Migration and the environment.* IOM.

Kadirbeyoglu, Z. (2010). In the land of ostriches: Developmentalism, environmental degradation, and forced migration in Turkey. In Afifi, T., & Jäger, J. (Eds.), *Environment, forced migration and social vulnerability* (pp. 223–234). Springer. DOI: 10.1007/978-3-642-12416-7_17

Kadirbeyoglu, Z. (2017). The impact of power and civic engagement in the decentralized management of natural resources: The case of Turkey. *Public Administration and Development, 37*(4), 277–291. DOI: 10.1002/pad.1809

Kam, P. M., Aznar-Siguan, G., Schewe, J., Milano, L., Ginnetti, J., Willner, S., McCaughey, J. W., & Bresch, D. N. (2021). Global warming and population change both heighten future risk of human displacement due to river floods. *Environmental Research Letters, 16*(4), 044026. DOI: 10.1088/1748-9326/abd26c

Keleş, R. (2022). The effects of external migration on regional development in Turkey. In *Uneven development in Southern Europe* (pp. 54–75). Routledge. DOI: 10.4324/9781003290766-2

Khan, A. A., & Somuncu, M. (2019). Is rural out migration constructive for sustainable development? A case of mountain community of Kıbrıscık, Turkey. *İnsan ve Toplum Bilimleri Araştırmaları Dergisi, 8*(4), 3337–3352.

Khanchel, I., & Lassoued, N. (2025). Climate vulnerability in Middle East and North African countries: Do social dimensions shape resilience? *Journal of Economic Studies (Glasgow, Scotland)*. Advance online publication. DOI: 10.1108/JES-10-2024-0679

Kızıldere, D. (2025). Earthquake disaster-induced urban policies: Paradoxes and challenges in Turkey. *Urban Research & Practice*, *18*(1), 130–141. DOI: 10.1080/17535069.2024.2422627

Korkmaz, C., & Meşhur, H. F. A. (2022). Counter-urbanisation experience in developing countries: The case of Istanbul metropolitan area. In *CPUD'22 Conference Proceedings.*

Lilleør, H. B., & Van den Broeck, K. (2011). Economic drivers of migration and climate change in LDCs. *Global Environmental Change*, *21*(S1), S70–S81. DOI: 10.1016/j.gloenvcha.2011.09.002

Matthews, T., & Potts, R. (2018). Planning for climigration: A framework for effective action. *Climatic Change*, *148*(4), 607–621. DOI: 10.1007/s10584-018-2205-3

McLeman, R., & Smit, B. (2006). Migration as an adaptation to climate change. *Climatic Change*, *76*(1–2), 31–53. DOI: 10.1007/s10584-005-9000-7

McLeman, R. A. (2013). *Climate and human migration: Past experiences, future challenges*. Cambridge University Press. DOI: 10.1017/CBO9781139136938

Ministry of Environment, Urbanization and Climate Change (MoEUCC). (2021). *Turkey's national climate change strategy and action plan (2011–2023) revision.* Official Gazette of the Republic of Türkiye.

Myers, N. (2002). Environmental refugees: A growing phenomenon of the 21st century. *Philosophical Transactions of the Royal Society of London. Series B, Biological Sciences*, *357*(1420), 609–613. DOI: 10.1098/rstb.2001.0953 PMID: 12028796

Nawrotzki, R. J., DeWaard, J., Bakhtsiyarava, M., & Ha, J. T. (2017). Climate shocks and rural-urban migration in Mexico: Exploring nonlinearities and thresholds. *Climatic Change*, *140*(2), 243–258. DOI: 10.1007/s10584-016-1849-0 PMID: 28435176

Obokata, R., Veronis, L., & McLeman, R. (2014). Empirical research on international environmental migration: A systematic review. *Population and Environment*, *36*(1), 111–135. DOI: 10.1007/s11111-014-0210-7 PMID: 25132701

Ökten, Ş. (2017). Environmental justice, dams and displacement in Southeastern Anatolia region, Turkey. *Journal of International Social Research*, *10*(50), 414–419. DOI: 10.17719/jisr.2017.1674

Okutan, P., & Otay, E. N. (2025). Climate change risk perception, adaptive capacity and psychological distance in urban vulnerability: A district-level case study in Istanbul, Türkiye. *Sustainability (Basel), 17*(12), 5358. DOI: 10.3390/su17125358

Ortakaya, Ü. İ. (2023). Examining climate migration risks through the lens of social equality: An evaluation of Germany, France and Türkiye. *Hacettepe Üniversitesi Sosyal Bilimler Dergisi, 5*(2), 186–210.

Pelek, D., Yıldızcan, C., & Turhan, E. (2025). Permanently temporary: Unveiling the im/mobility and intersecting vulnerabilities of migrant seasonal agricultural workers in disaster-affected areas of Türkiye. *Agriculture and Human Values, 42*(4), 2541–2557. DOI: 10.1007/s10460-025-10761-w

Piguet, E. (2010). Climate and migration: A synthesis. In Afifi, T., & Jäger, J. (Eds.), *Environment, forced migration and social vulnerability* (pp. 73–85). Springer. DOI: 10.1007/978-3-642-12416-7_6

Piguet, E. (2022). Linking climate change, environmental degradation, and migration: An update after 10 years. *Wiley Interdisciplinary Reviews: Climate Change, 13*(1), e746. DOI: 10.1002/wcc.746

Rigaud, K. K., de Sherbinin, A., Jones, B., Bergmann, J., Clement, V., Ober, K., Schewe, J., Adamo, S., Kuriakose, A., Weber, U., & McClure, B. (2018). *Groundswell: Preparing for internal climate migration.* World Bank. DOI: 10.1596/29461

Sardar, A., Kıanı, A. K., Kuslu, Y., & Bılgıc, A. (2020). Examining the role of livelihood diversification as a part of climate-smart agriculture (CSA) strategy. *Atatürk Üniversitesi Ziraat Fakültesi Dergisi, 51*(1), 79–87. DOI: 10.17097/ataunizfd.604937

Sayan, R. C. (2019). Exploring place-based approaches and energy justice: Ecology, social movements, and hydropower in Turkey. *Energy Research & Social Science, 57*, 101234. DOI: 10.1016/j.erss.2019.101234

Scott, M., Gkartzios, M., & Halfacree, K. (2024). Introducing climate-related counterurbanisation: Individual adaptation or societal maladaptation? *Habitat International, 143*, 102970. DOI: 10.1016/j.habitatint.2023.102970

Selby, J., Dahi, O. S., Fröhlich, C., & Hulme, M. (2017). Climate change and the Syrian civil war revisited. *Political Geography, 60*, 232–244. DOI: 10.1016/j.polgeo.2017.05.007

Sever, B. C. (2023). An evaluation of the possibilities of human mobility and immobility in the context of the climate crisis in Türkiye. *İnsan Hareketliliği Uluslararası Dergisi (International Journal of Human Mobility),* Special Issue, 34–55.

Soest, C. V. (2020). *A heated debate: Climate change and conflict in Africa* (Working Paper). Hamburg: GIGA German Institute of Global and Area Studies – Leibniz-Institut für Globale und Regionale Studien, Institut für Afrika Studien.

Sözen, B., & Kılıç, S. E. (2025). Tourism-led rural gentrification in multi-conservation rural settlements: Yazıköy/Datça case. *Sustainability (Basel), 17*(18), 8439. DOI: 10.3390/su17188439

Taşeli, B. K. (2007). Nature conservation efforts set for specially protected areas of Turkey. *The Environmentalist, 27*(3), 315–327. DOI: 10.1007/s10669-007-9030-y

Tekin, M. K., & Deniz, O. (2025). Climate change-induced migration potential: A multi-dimensional field research on Türkiye. [Journal of Geographical Sciences]. *Coğrafi Bilimler Dergisi, 23*(1), 236–263. DOI: 10.33688/aucbd.1517859

Toros, H., Mokari, M., & Abbasnia, M. (2019). Regional variability of temperature extremes in the maritime climate of Turkey: A case study to develop agricultural adaptation strategies under climate change. *Modeling Earth Systems and Environment, 5*(3), 857–865. DOI: 10.1007/s40808-019-00572-4

Turhan, E., Cerit Mazlum, S., Şahin, Ü., Şorman, A. H., & Cem Gündoğan, A. (2016). Beyond special circumstances: Climate change policy in Turkey 1992–2015. *Wiley Interdisciplinary Reviews: Climate Change, 7*(3), 448–460. DOI: 10.1002/wcc.390

Uzelgun, M. A., & Castro, P. (2014). The voice of science on climate change in the mainstream Turkish press. *Environmental Communication, 8*(3), 326–344. DOI: 10.1080/17524032.2014.898674

Ward, C., & Ruckstuhl, S. (2017). *Water scarcity, climate change and conflict in the Middle East: Securing livelihoods, building peace.* Bloomsbury Publishing. DOI: 10.5040/9781350989719

Warren, P. D. (2016). Forced migration after Paris COP21: Evaluating the "Climate Change Displacement Coordination Facility.". *Columbia Law Review, 116*, 2103–2139.

Wissink, M., Düvell, F., & Van Eerdewijk, A. (2013). Dynamic migration intentions and the impact of socio-institutional environments: A transit migration hub in Turkey. *Journal of Ethnic and Migration Studies, 39*(7), 1087–1105. DOI: 10.1080/1369183X.2013.778026

Witmer, F. D., Linke, A. M., O'Loughlin, J., Gettelman, A., & Laing, A. (2017). Subnational violent conflict forecasts for sub-Saharan Africa, 2015–65, using climate-sensitive models. *Journal of Peace Research, 54*(2), 175–192. DOI: 10.1177/0022343316682064

Yavuz, K., & Çelik, A. (2023). Exploring the intersection of climate change and social policy in Türkiye: Challenges and opportunities for a sustainable future. *Journal of Management and Economics Research*, *21*(1), 244–265.

Yenal, Z. (2023). Buckling under ageing and climate crisis: Migration flows in Turkish agriculture. *International Migration (Geneva, Switzerland)*, *61*(6), 349–353. DOI: 10.1111/imig.13205

Yıldız, D., & Atmiş, E. (2023). Conflict management in protected areas in Turkey: Küre Mountains National Park example. *Forestist, 73*(2).

Zickgraf, C. (2019). Keeping people in place: Political factors of (im)mobility and climate change. *Social Sciences (Basel, Switzerland)*, *8*(8), 228. DOI: 10.3390/socsci8080228

Chapter 15
Chad's Global Disaster Risk Map

Fatma Şencan
http://orcid.org/0009-0007-3223-6463
Medipol Üniversitesi, Turkey

ABSTRACT

This study examines the transformation of global disasters into social catastrophes in Chad through Beck's "Risk Society" and Giddens's "Reflexive Modernity" theories. Despite being non-industrialized, Chad bears "manufactured risks" from climate change, greenhouse gas emissions, and global inequalities. Geographic vulnerability (Sahara-Sahel location, 90% Lake Chad shrinkage, 4% arable land) creates cycles of drought, hunger, water scarcity, mass migration, and epidemics. Using qualitative document analysis from World Bank, UN, and academic sources, a disaster risk map was constructed. Findings show tribal structures provide both resilience and conflict, while weak state capacity increases international aid dependency. The study recommends institutional reform, sustainable resource management, climate-resilient livelihoods, and locally-focused international cooperation. Chad exemplifies how unequally distributed global risks create social disaster laboratories.

INTRODUCTION

The hierarchy of beings refers to the entirety of systems that functionally contribute to the preservation of ecological integrity, thereby supporting the continuity of diverse species. Theories regarding the origin of life on Earth are scientifically grounded through the analysis of ecological traces and their surviving remnants. These data support the argument that life on Earth has been interrupted at different points in time and subsequently reemerged. While humans were historically passive

DOI: 10.4018/979-8-3373-6274-8.ch015

actors in the face of natural disasters, technological development and changing living conditions have transformed them into active agents shaping both the scale and frequency of disasters. Unconscious human intervention in natural systems has disrupted ecological balances, pushing both humanity and other species toward catastrophic consequences. This disruption has been instrumental in transforming recent disasters into a global phenomenon that threatens all forms of life.

As these threats intensify, global disasters continue to generate destructive impacts across physical and human geography, becoming increasingly difficult to manage over time. With the growing influence of human and technological factors, disasters now occur more frequently and over wider areas, posing threats not only to human populations but also to ecosystems and other species. The social dimension of this crisis is particularly visible in countries where access to basic necessities is already limited. In such contexts, environmental disasters such as drought, water scarcity, floods, wildfires, hurricanes, and soil degradation accelerate social transformation processes, particularly by triggering migration and deepening social vulnerability. This vicious cycle gradually depletes essential resources for sustainable life—namely clean water, sufficient food, and breathable air. Consequently, the concept of catastrophe as a social phenomenon has acquired new and increasingly destructive dimensions.

Recognizing the urgency of these challenges, the escalation of global disasters has placed environmental issues at the center of international agendas. Emergency action plans, environmental platforms, and both individual and institutional initiatives with high levels of awareness play a crucial role in enhancing the effectiveness of disaster policies. Similarly, recycling projects and large-scale global campaigns require coordinated action plans aimed at expanding environmentally conscious social groups. For these efforts to succeed, designing initiatives in accordance with region-specific disaster impacts while embedding them within a global framework is essential for developing holistic and sustainable approaches.

African countries exemplify regions where the physical and social destruction caused by global disasters is most visible (). Within this continental context, this study focuses on Chad, where multiple intersecting factors such as geographical location, climate-sensitive livelihoods, widespread poverty, and unequal access to basic needs interact with global disasters to heighten the risk of social catastrophe. The country faces a persistent cycle of deprivation marked by drought, water scarcity, hunger, and epidemics that intensifies social vulnerability. As one of the world's poorest countries, Chad provides a critical case for examining how global disaster risks translate into localized social disasters. To illuminate these dynamics, this study employs qualitative research methods to analyze visual, written, and electronic documents from the literature in order to explore the relationship between global disaster risks and social crises in Chad. The findings are thematically categorized and presented in sub-sections, culminating in the development of a disaster risk map that highlights the adverse effects of these processes on social life and underscores the importance of sustainable action plans.

CONCEPTUAL AND THEORETICAL FRAMEWORK

The Evolution and Global Dimension of the Disaster Concept

In general terms, a disaster can be defined as a complex phenomenon that emerges as a result of the interaction between natural and human factors, causing severe destruction on social structures and ecosystems (Çelik et al., 2018; Peduzzi, 2019; Bricker & Pelupessy, 2025). Until recently, the traditional notion of disaster focused on the physical destructions caused by suddenly occurring events and emphasized response and recovery activities (Alexander, 2002). In the modern approach, however, a broader understanding of disaster has come to the fore. Accordingly, there is a process-oriented method that prioritizes managing the destructive/harmful effects occurring before, during, and after disasters in a gradual and coordinated manner (UNDRR, 2022). The World Health Organization (WHO), which approaches disasters in the context of health systems, also defines disasters as complex events and situations that exceed the health infrastructure capacity of the affected region, disrupt the vital flow of life, and require national or international assistance. In addition, the integrated effects of disasters on public health, social stability, and governance systems are emphasized (WHO, 2023).

With the widespread adoption of the modern understanding of disasters, concepts such as early warning systems and emergency preparedness have gained significant importance in disaster management. According to this paradigm shift, four fundamental stages, preparedness, response, risk communication, and recovery and reconstruction, must be planned collaboratively (UNDRR, 2022). The preparedness stage encompasses the establishment of early warning systems, the development of risk communication strategies, and the formulation of emergency plans. The response stage focuses on life-saving activities and meeting basic needs immediately following a disaster. Risk communication, on the other hand, plays a decisive role in managing not only the physical but also the social impacts of disasters by directly influencing how individuals perceive hazards and the behaviors they develop in response to them.

The recovery and reconstruction phases of the disaster cycle involve long-term support processes that enable survivors to return to their normal lives and the restoration of the socio-economic structure (Alexander, 2002). However, the effective functioning of all stages of this cycle depends largely on the institutional capacity of the state and the ability of local communities to cope with disasters. Especially in regions where state capacity is weak, the inadequacy of early warning systems, emergency preparedness, and institutional coordination mechanisms negatively affects all stages of disaster management. In this context, local capacity building emerges as a critical element that enhances community resilience when central administration remains limited. It plays a determining role in preventing natural events from escalating into social catastrophes (Wisner et al., 2004; Cutter et al., 2014)

The multidimensional impacts of disasters on human life and the environment have become more visible along with their increasing frequency. With the addition of human-induced factors such as climate change, greenhouse gas emissions, rapid population growth, and unplanned urbanization, the destructions caused by disasters continue to produce consequences that take a long time to reverse. For instance, climate-based changes stand out as serious problems that show their effects in many areas, from agriculture to tourism, from biodiversity to glacial systems (Lama and Devkota, 2009; Wang et al., 2023). These global-scale effects, which have become one of the focal points of scientific research and international academic discussions, can be given as striking examples of the multifaceted and profound impacts of climate change: mountaineering tourism in the Himalayas (Nyaupane and Chhetri, 2009), the transformations in the life practices of nomadic communities in Turkey (Sevgi and Yulu, 2024), the increasing disease cases among livestock in Mongolia (DiPietro et al., 2023), food security problems in China (Lee et al., 2024), and the distribution and population of bird species in tropical regions (Şekercioğlu et al., 2012).

The destructive effects of disasters are not limited to the physical environment but also create multidimensional impacts on social structure and functioning. Mass migration movements and economic crises caused by climate change trigger global-scale social disasters, leading to the deepening of social inequalities and the weakening of social cohesion (Watts et al., 2018; Cutter et al., 2014). The gradual deterioration of cohesion in society causes the social bonds that hold individuals and communities together to be damaged. As Wisner and his colleagues (2004) also point out, global disasters cause problems such as the rapid depletion of existing natural resources and access to safe food and drinking water, which further deepen existing social inequalities (Wisner et al., 2004). The fact that global problems have become so influential can trigger multidimensional crises that transform social, economic, and political structures. Due to these crises that affect the society as a whole, vulnerable and fragile communities—such as the elderly, the disabled, children, the poor, women, and migrants—are disproportionately affected by these problems (Flanagan, Kim & Bostwick, 2020; Cutter et al., 2003; Wisner et al., 2004).

Disasters create profound effects on the political and economic structures of societies, reshaping resource distribution, governance processes, and economic activities. The measures taken by governments before disasters in order to reduce their negative impacts lead to the examination of governments' capacities in terms of the confirmation of management capacity during and after disasters. The disruption of economic activities or the increase of grievances during this period makes it necessary to re-evaluate social trust and governance mechanisms. The changes shaped by disasters in traditional societies make both short- and long-term adaptation compulsory and require the strengthening of social bonds and community resilience. Therefore, disasters should be considered not only as natural events but also as multidimensional phenomena that shape the transformation of economic and political structures. As the effects of global disaster events become more widespread, the continuity of economic and political activities will become increasingly important in terms of national and regional development goals (Peduzzi, 2019). In this context, theoretical debates in modern disaster literature, particularly Beck's 'risk society' and Giddens' 'reflexive modernization' approaches, provide an explanatory framework for how disaster risks are produced and how they interact with social vulnerabilities.

Disasters in the Light of Risk Society and Reflexive Modernity Theories

Throughout history, humanity has sought to develop its ability to cope with environmental disasters, while the new risks that emerged alongside the modernization process have continued to deepen their effects on a global scale. Particularly since the Industrial Revolution, these risks have become inevitable with the expansion of the modern social structure. This situation renders Ulrich Beck's "risk society" approach—emphasizing that modernity has transformed into a structure that reproduces risks through its own internal dynamics—meaningful, and provides an important theoretical framework for analyzing the vulnerabilities of societies. Similarly, Anthony Giddens explains this inevitable condition faced by the modern world through the concept of "reflexive modernity."

According to Beck's "risk society" approach, global disaster risks in modern societies are no longer the result of external and unpredictable natural processes but are instead the direct product of human activities. Therefore, the risks that have emerged along with modernization are defined as "manufactured risks," which generally appear in technological, environmental, and social dimensions (Beck, 2011). In this context, the types of global risks that modern humans are compelled to confront in various forms manifest themselves as "manufactured risks," arising from technological advancements, scientific innovations, and processes of global economic integration (Beck, 2011). According to Beck, these risks are not driven by natural hazards but are triggered by the internal dynamics of modernization itself. From this perspective, Beck's conceptualization also provides an explanatory basis for countries such as Chad. Although Chad is not, in itself, an industrialized or modernized society, it is in a position of bearing the risks generated by global modernization. Problems such as drought, migration, hunger, and water scarcity are not merely the results of natural disasters; they are also direct reflections of the manufactured risks produced by the modern world's climate policies, carbon-based production systems, and global economic inequalities. Therefore, in the case of Chad, Beck's "risk society" concept is significant in explaining how risks produced at the core of modernity are transferred to peripheral societies and how "natural disasters" are transformed into "social disasters" in this process.

Giddens's reflexive modernity, on the other hand, is based on the awareness within modern societies of recognizing and attempting to control the damages caused by their own actions on the environment—such as the disruption of ecological balance or the risks resulting from technological developments (Beck, Giddens, & Lash, 1994; Giddens, 1994). This theory, which explains the reflections of such awareness in modern societies, underlines, as in the case of Chad, the necessity of confronting the more destructive experiences and resulting grievances occurring in societies left behind by modernity. These countries, which have not yet acquired the technological equipment brought by modernity, become directly exposed to the adverse effects of global crises (Beck, Giddens, & Lash, 1994; Giddens, 1994). Furthermore, according to this theory, the risks faced by modern societies are not limited solely to the physical destructions caused by disasters. The dissolution of solidarity, cooperation, and other social values that hold societies together also constitutes a part of this process (Giddens, 1994). Modernization is not only a process of progress and rationalization but also a process of transformation in which the dangers within itself are recognized and critical perspectives emerge (Giddens, 1994).

The joint application of Beck and Giddens's theories to the case of Chad reveals why disaster risks produce more devastating outcomes in countries where modernization processes are experienced in unequal forms. In this context, Beck's "risk society" approach and Giddens's "reflexive modernity" theory provide a complementary theoretical ground for understanding Chad's deep-rooted problems such as the cycle of poverty, environmental issues, and political instability. The two theories, which are addressed to reveal how the risks produced by modernization materialize in Chad and how they shape the nature of the disaster phenomenon, not only provide a theoretical framework but also form a basis for developing practically applicable solutions. Therefore, Beck's "risk society" approach demonstrates that although Chad does not possess the characteristics of a modern society, it is still affected by the risks produced by modernity. Giddens's conceptualization of reflexive modernity, on the other hand, enables a clearer understanding of the opportunities and risks brought about by modernization in the case of Chad, where global risks intersect across environmental, economic, and social dimensions. This theoretical framework is employed to explain why, in the case of **Chad**, environmental risks do not merely remain natural disasters but instead evolve into **social catastrophes** through weak institutional capacity and global inequalities.

RESEARCH DESIGN: METHOD AND SCOPE

This study is a qualitative research project that examines the transformation process of global disasters into social catastrophes specifically within the context of **Chad**. The research is framed within **Beck's "Risk Society"** and **Giddens' "Reflexive Modernity"** theories and is based on a **descriptive document analysis** method.

Data Collection Process and Source Selection

The data collection process was conducted through a systematic literature review and document analysis to present Chad's structural vulnerabilities and disaster risks. Data sources were gathered into three main categories:

- Primary Sources: Scientific reports, statistical data, and country profiles related to Chad from international organizations such as the World Bank, UNDP, UNICEF, WHO, IOM, AfDB, the Fragile States Index (FSI), and Transparency International were utilized. These organizations represent the most reliable and systematic information sources in areas where the Chadian government's own institutional data generation capacity is insufficient.
- Secondary Sources: Searches were conducted via Web of Science, Scopus, Google Scholar, and JSTOR databases using keywords such as *"Chad"*, *"climate change"*, *"food security"*, *"migration"*, *"Lake Chad"*, *"drought"*, *"conflict"*, *"vulnerability"*, and *"disaster risk"*. Out of 156 academic studies screened, 68 peer-reviewed articles and book chapters published between 2000 and 2025 that directly address Chad's environmental, social, and economic vulnerabilities were included in the analysis.
- Observational Data: To complement the quantitative and documentary data, qualitative insights derived from the author's field observationsin Chad were utilized. These observations provide a contextual grounding for the structural vulnerabilities identified in the literature, offering first-hand evidence of the local population's living conditions and environmental challenges.

Source Selection Criteria

The following criteria were considered in the selection of the sources:

- Studies directly referencing Chad's geographical, climatic, socio-economic, or institutional vulnerabilities.
- Data produced by international organizations with transparent and reproducible methodologies.
- Recent (last 25 years), empirically-based research published in peer-reviewed academic journals.
- Studies offering regional comparisons within the context of the Lake Chad Basin, the Sahel region, and Central Africa.

Data Analysis

The collected data were analyzed using content and thematic analysis methods. In the first stage, findings regarding Chad's disaster risks were divided into two main categories: "environmental and climatic crises" and "social, institutional, and political crises." Subsequently, sub-themes such as "disasters created by geographical location and physical conditions," "migration," "hunger and malnutrition," "access to clean water," "the dual role of tribal structures," and "the state's lack of institutional capacity" were established.

Limitations and Scope of the Research

The most significant limitation of the research is the extreme inadequacy of the Republic of Chad's official institutions (Ministry of Interior, Statistical Office, Ministry of Health, etc.) in terms of systematic data production and sharing. The near-total absence of up-to-date, accessible, and reliable state-sourced datasets has made the research dependent on international reports and academic literature. This situation limits access to primary data produced from the country, thereby leading to a predominance of an external viewpoint in the interpretation of findings. However, since institutions like the World Bank and UN agencies work in cooperation with the Chadian government and produce data based on fieldwork, these sources were deemed the most reliable foundation of information under current conditions.

Another limitation is that primary field data were not collected. However, this was a deliberate methodological decision. In a country like Chad, which faces large-scale, multi-layered, and structural issues, data obtained from a limited field study might lack generalizability. Instead, systematic analysis of secondary data was preferred to perform a holistic analysis of socio-ecological processes spread over a wide geographical area and to develop a consistent argument within a theoretical framework. This method was found more suitable for modeling complex relationships, mapping multidimensional risks, and discussing the local manifestations of global disasters on a theoretical ground.

The scope of the research is limited to the social manifestations observed in Chad in the modern era, particularly in recent years (2000–2025), which are directly linked to global disasters. The study treats the country's historical colonial experience and political past as context, while concentrating its focus on contemporary disaster dynamics and their social consequences.

AN OVERVIEW OF AFRICA, THE SAHEL REGION, AND CHAD

The African continent stands as a geography disproportionately affected by global climate-oriented disasters and the socioeconomic problems triggered by these events (WMO, 2024). The fact that 27 of the 33 countries expected to be most affected by global climate change are located in Africa highlights the magnitude of the disaster risks the continent faces (Independent, 2021). Environmental threats such as drought, rising temperatures, irregular precipitation regimes, floods, and desertification, which dominate the continent, have already been long-standing issues (Çelik et al., 2025; Demir, 2025; Çelik et al., 2026). The food insecurity, epidemics, forced migration, displacement, and internal instabilities caused by these problems are directly linked to social inequalities. Many countries on the continent are unable to effectively manage disaster risks due to limited economic resources and weak institutional capacity (Okunola, 2025). This situation leads to the deepening of humanitarian crises and the geographical expansion of disaster zones. Consequently, climate-induced disasters in Africa are shifting from being merely environmental issues to becoming structural and multidimensional security concerns.

The African continent is divided into different climatic and geographical zones. Specifically, the Sahel belt is among the regions where climate change and environmental vulnerabilities are felt most intensely. The semi-arid ecological structure of the Sahel limits its natural resilience against climatic fluctuations. Countries located in the Sahel basin, such as Cameroon, Niger, Nigeria, and Sudan, share similar global risk profiles with Chad in terms of environmental and structural characteristics. While environmental vulnerabilities in Chad and Niger primarily manifest through drought, population pressure, and food insecurity, these risks are intertwined with security issues, armed groups, and political instability in Nigeria and Cameroon (Nagarajan et al., 2024; Ruppel & Funteh, 2019). Although Cameroon has relatively lower risk indicators, it is indirectly affected by these threats due to **transboundary instabilities** and regional inequalities in the Sahel. As a result, similar climate-induced risks in the Sahel combine with different structural internal problems in each country to produce unique areas of vulnerability.

SOCIAL MANIFESTATIONS OF GLOBAL DISASTERS IN CHAD

In the research, using data obtained from the World Bank, UN agencies, and relevant academic publications, the social disasters directly affecting the lives of the population have been categorized thematically as follows. Within this framework, the focus has been placed on the primary social issues shaping the social life of the Chadian people, and the manifestations of this classification in daily life have been revealed.

The fundamental problems such as hunger, thirst, drought, and poverty, to which the Chadian people have been exposed for many years, constitute a primary area of debate in understanding the country's structural vulnerabilities and governance capacity. In order to evaluate this multidimensional problem area from a holistic perspective, a distinction has been made between primary (triggering/structural) and secondary (consequential/manifested) crises.

Figure 1. Risk Map of Disaster-Centered Crises in Chad

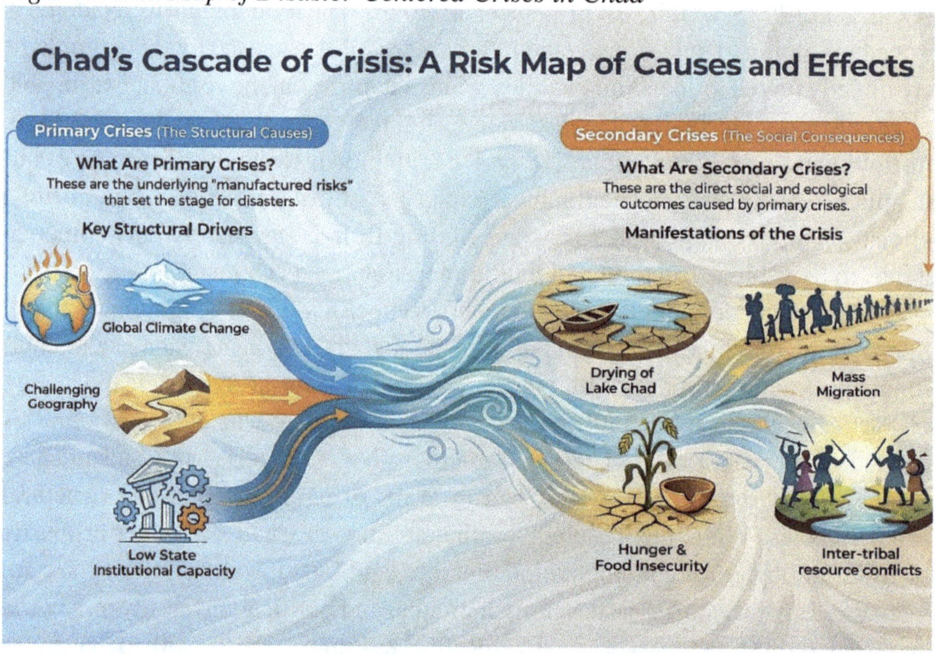

In this context;

- **Primary Crises:** These include structural elements and "manufactured risks" that set the stage for and trigger the disaster process, such as global climate change, challenging geographical location, and the low institutional capacity of the state.

- **Secondary Crises:** These represent the social and ecological consequences resulting from the interaction of primary crises, such as the shrinking of Lake Chad, mass migration movements, hunger, food insecurity, and resource-based inter-tribal conflicts.

Figure 2. Visual Schema of the Types, Degrees, and Causal Relationships of Crises in Chad

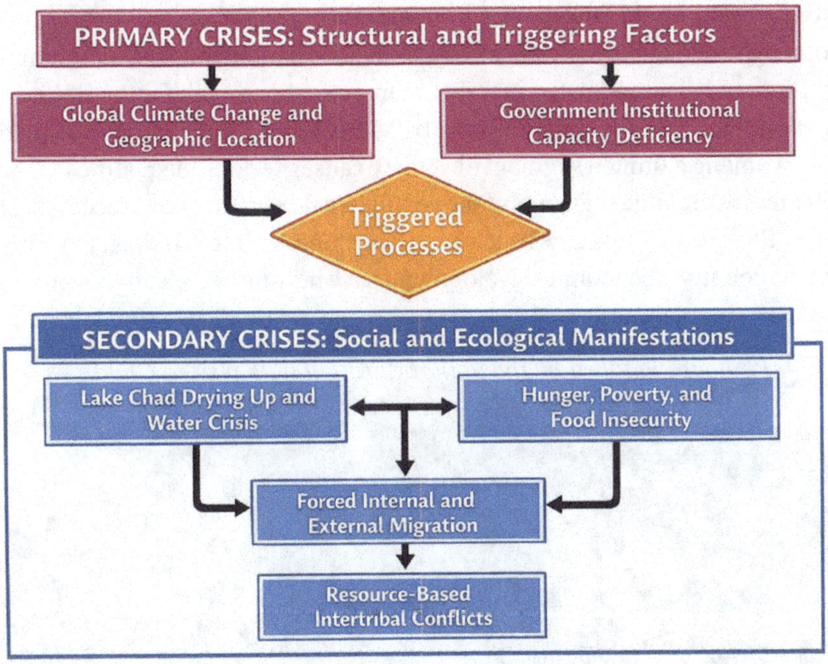

The following subheadings detail the thematic reflections in Chad's disaster risk map, based on this hierarchical relationship.

Environmental Crises and Ecological Vulnerabilities

Disasters Induced by Geographical Location and Climatic Conditions

The impacts of disasters vary regionally depending on geographical, social, and economic conditions. However, addressing disasters in conjunction with the entire African continent, while facilitating the identification of obstacles to regional development goals, makes it more difficult to highlight the specific problems faced by Chad. As a Central African country, Chad represents a striking example for understanding the social consequences of problems that are both local and interconnected with those of the continent.

Although the country is rich in natural resources (such as oil, uranium, gold, natural gas, and diamonds), the inability to manage this wealth for the welfare of its population reinforces the cycle of poverty. Moreover, the concentration of national resources among a limited segment of society causes serious inequalities in income distribution, while large segments of the population face difficulties in accessing basic services. Structural imbalances such as these, along with low living standards that hinder the country's economic development, have become increasingly widespread.

Figure 3. Photograph taken by the author during a field visit to Chad.

The geographical location and physical conditions of the country exert a decisive influence on the background of its economic and social vulnerabilities. The fact that Chad is largely situated between the Sahara Desert and the Sahel belt transforms the country into a geography dominated by arid and semi-arid climatic characteristics. Chad stands out as one of the most vulnerable countries to climate change, already feeling its impacts profoundly (World Bank Group, 2022). Ranked among the hottest countries in the world, the average annual temperature increase in Chad has exceeded 0.5°C since the 1990s, reaching approximately twice the global average (Republic of Chad, 2021). Unless effective and decisive measures are taken, global temperatures are expected to rise by an additional 2.2°C by the year 2100 (World Bank, 2021). Alongside the heat, the inadequate and irregular precipitation regimes, combined with limited water resources and high evaporation rates, directly constrain agricultural production and livestock activities.

Figure 4. Mean Temperature and Precipitation Trends in Chad, 1901–2021

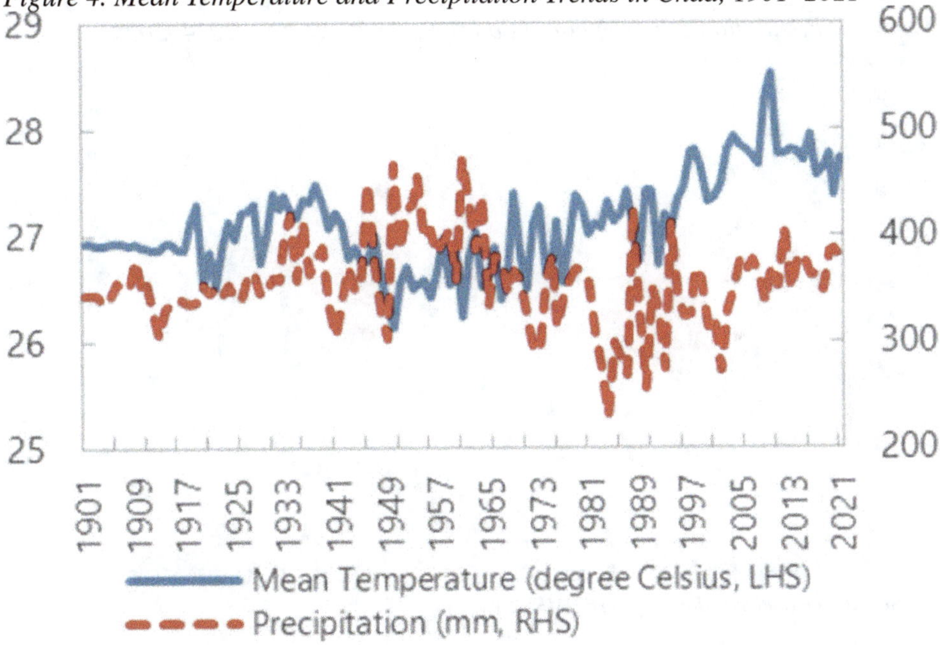

Source: *International Monetary Fund [IMF], 2024*

These climatic issues not only hinder the country's economic activities but also trigger social problems such as food insecurity, domestic migration, and resource-based conflicts. This situation can be directly linked to the limitations imposed by geographical conditions and climatic risks. In particular, increasing desertification and soil degradation in the northern regions threaten livelihoods, further weakening the population's resilience to disasters. Especially in regions inhabited by the impoverished, the situation can reach the proportions of a humanitarian crisis. Data for the 2014–2024 period indicate that the population experiencing severe food insecurity during the lean season in Chad has entered a distinct and sustained upward trend, both in absolute numbers and as a share of the total population. The sharp rise observed particularly after 2020 suggests that food insecurity has evolved from a seasonal vulnerability into a structural crisis (IMF, 2024). This trend reveals that agricultural production and household livelihoods are increasingly being eroded under the combined impact of climate change-related droughts, precipitation irregularities, global and regional economic shocks, and security issues (Çelik, 2026). The fact that approximately one-fifth of the population has been exposed to food crisis, emergency, or famine conditions in recent years demonstrates that food security in Chad can no longer be addressed through short-term humanitarian interventions alone; instead, long-term adaptation and resilience policies have become mandatory.

Figure 5. Population Experiencing Food Insecurity in Chad, 2014–2024

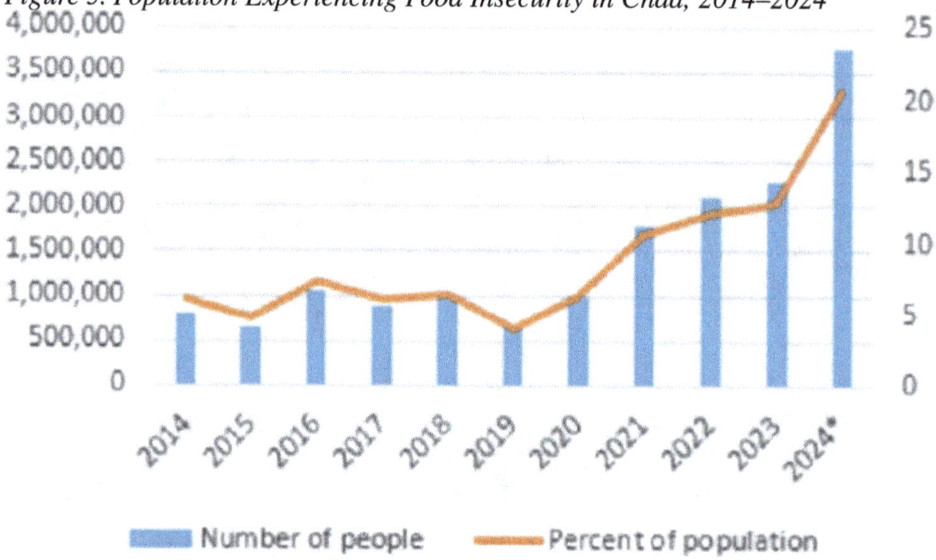

Source: *International Monetary Fund [IMF], 2024*

Chad's geographical location is one of the primary factors determining the social life, economic activities, and societal structure of its population. As a landlocked country, its economy relies heavily on agriculture and livestock. However, World Bank (2023) data indicate that these activities, conducted using traditional methods and without modernization, are insufficient to meet the basic needs of the population and contribute only minimally to national income. The sharp north-south differences in climate conditions, which are critical for agricultural production, also reduce the areas suitable for farming (Mfewou, Tata, & Nadji, 2020). Additionally, the almost negligible irrigation infrastructure and the shrinking of Lake Chad due to climate change seriously threaten the sustainability of production (Arfasa et al., 2024). These conditions, which hinder food security and accessibility, frequently expose the population to hunger crises. Therefore, the limitations arising from the country's natural conditions directly affect both production capacity and the population's livelihood security. This situation is even more evident when considering Chad's land area and land-use imbalances.

Despite being Africa's fifth-largest country with an area of 1,284,000 km^2, only approximately 40% of Chad's land is suitable for agricultural activities (United Nations Environment Programme, n.d.). When narrowed down to arable land, this proportion drops to as low as 4% (The Global Economy, 2025). This indicates that usable land for both settlement and production is extremely limited. Since much of the country's territory is under the influence of the Sahara Desert, habitable areas are naturally constrained. The diversity of climatic zones, difficulties in accessing water resources, increasing desertification, and natural disaster risks negatively affect the productivity of activities conducted in these areas (Kazeem, 2024; World Bank, 2023). Consequently, a significant portion of the population is concentrated in the southern and central regions, where living conditions are relatively more favorable (Jean-Richard et al., 2015). However, the vital risks and the persistent disaster cycle in this narrow area continuously subject the population to hunger, poverty, and migration pressures (Musa et al., 2022). Thus, Chad's geographical constraints shape not only production areas but also population distribution and social mobility. In this context, the differences between the north–south climatic zones play a decisive role.

There are marked differences among the north–south climatic zones of the country. For instance, the relative humidity in the Saharan region is approximately 10%, while it rises to 60–80% toward the south (Worlddata.info, 2023). This climatic diversity affects not only natural conditions but also all socio-economic dynamics, including agriculture, livestock, production, settlement density, and migration. Another study indicates that annual humidity rates decrease by an average of 2 percentage points per decade, with seasonal declines reaching 5–10 points (MDPI, 2023). These variations increase the vulnerability of a country whose economy relies heavily on agriculture and livestock. Consequently, climatic diversity and change are among the most significant triggers of Chad's economic and social vulnerabilities. When combined with the country's low level of development, these vulnerabilities result in more severe consequences.

Chad ranks among the lowest countries in terms of development level, with approximately half of the population living below the national poverty line according to

World Bank (2022) data. According to a report by the Institute for Security Studies (2023), one-third of the population lives in extreme poverty (ISS, 2023). These figures clearly demonstrate the magnitude of the humanitarian crisis in the country.

When high poverty rates are combined with climate-related disasters, the impact of social problems intensifies. Ongoing political conflicts, ethnic tensions, and the activities of armed groups such as Boko Haram exacerbate the humanitarian crisis, undermining arguments and mechanisms that could help the population cope with socially induced stress. These challenges also indicate that state capacity is insufficient to ensure security and stability, leaving the population dependent on humanitarian organizations and external interventions. Consequently, poverty, political instability, and security issues in Chad amplify the effects of natural disasters, rendering the country more vulnerable. These conditions also magnify the impacts of global disaster risks. In recent years, Chad has become one of the countries most acutely affected by the consequences of global disaster risks. The government, insufficient in coping with existing problems, lacks the resources necessary to mitigate the effects of global-scale disasters. Natural hazards arising from geographical location, combined with socio-economic and political problems, reach a level that threatens human life, increasing social vulnerability and pushing the population to the brink of social disasters.

CRISES IN ACCESS TO WATER RESOURCES: THE IMPACTS OF LAKE CHAD'S SHRINKAGE

Water, as a fundamental requirement for life, exhibits proportional disparities in terms of availability and utilization across the globe. Although more than half of the Earth's surface is covered by water, freshwater resources account for only about 2.5% of this total, representing a relatively limited supply (Srivastava & Yadav, 2023). Access to these scarce resources can result in severe humanitarian crises, particularly in areas affected by drought or where infrastructure for water transportation is inadequate.

Individuals and animals often must traverse hazardous routes over long distances to reach water sources, placing their lives at risk. As environmental conditions deteriorate, this challenge evolves beyond a mere issue of physical access and exposes structural problems related to the sustainable management and use of water resources.

Figure 6. Photograph taken by the author during a field visit to Chad.

People and animals, who must face numerous hazards to access water resources, risk their lives while traveling long distances. In Chad, one of the countries where these challenges are particularly severe, water crises stand out on the nation's disaster risk map due to irregular rainfall patterns, difficulties in accessing existing resources, resource management issues, droughts, and desertification. The country's location within the Sahel belt threatens the sustainability of both surface and groundwater resources (NourEldeen et al., 2020). Climate change–related increases in drought and temperature particularly disrupt the water cycle in semi-arid and arid regions, resulting in reductions in surface and underground water reserves. In addition, rapid population growth, increased irrigation demand in agriculture, and water pollution from industrial and domestic waste place further pressure on existing water resources. Therefore, the efficient use of water in Chad is not only an ecological issue but also a strategic factor that triggers social challenges such as food security, migration dynamics, and societal stability.

Lake Chad, which holds critical importance for the people of Chad and neighboring communities, has experienced multifaceted problems over the past half-century with economic, social, and political dimensions. The lake has shrunk by more than 90% since the 1960s, resulting in a decline in biodiversity and threatening the habitats of fish species and waterfowl (European Space Agency [ESA], 2019). Changes in the lake's water regime have limited irrigation opportunities for regional agricultural lands, leaving local populations unable to meet their own food needs (Bantin, Jun, & Si, 2017). This process has accelerated desertification, causing surface waters to recede further underground (Olowoyeye & Kanwar, 2023). Pastoralists struggling to find pasture and water for their livestock have experienced increased conflicts with settled communities over resource sharing, thereby exacerbating pre-existing social problems (Olowoyeye & Kanwar, 2023).

Figure 7. Trends of the declining volume of Lake Chad

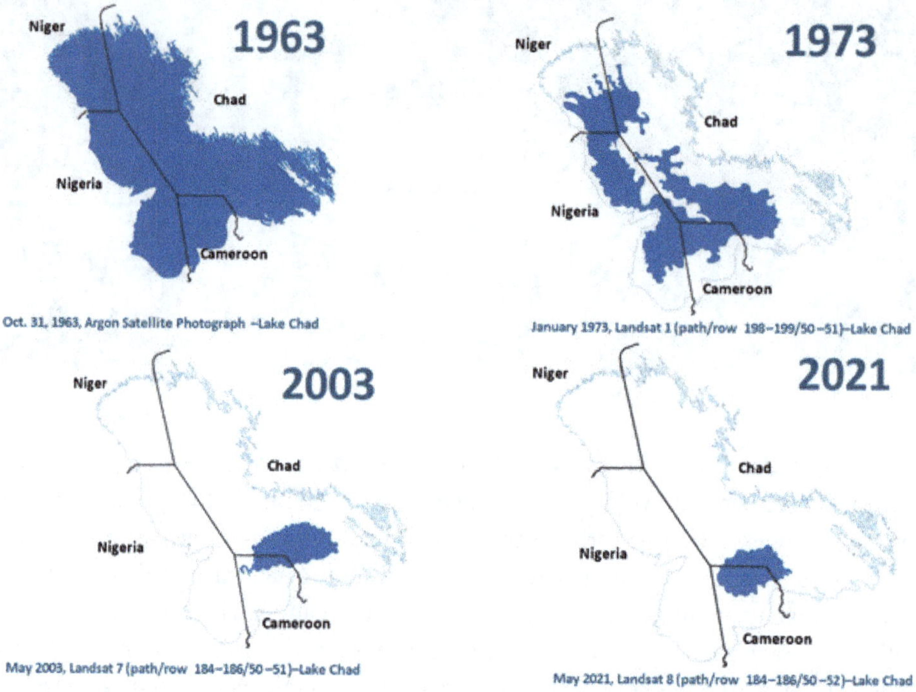

Source: *(Olowoyeye & Kanwar, 2023)*

In areas where the lake's waters have receded or in marshes formed around ponds, the proliferation of mosquitoes and other harmful species increases the risk of malaria, cholera, dysentery, typhoid, and other vector-borne diseases (Semenza, Rocklöv, & Ebi, 2022). In countries with limited access to clean water, consumption of contaminated water and insufficient sanitation lead to the spread of Giardia, Schistosoma, and other intestinal parasites (Siwila, Mwaba, Chidumayo, & Mubanga, 2020). Furthermore, during the rainy season, ponds—some artificially created by local populations and some naturally formed—can become contaminated over time, leading to similar health issues. These problems, which are more frequently observed in rural areas, can have particularly severe consequences in settlements where health services are inadequate or entirely absent. The local population and economy, directly affected by these adverse developments, face serious challenges due to shrinking livelihoods.

Water resources, which are vital for both the people of Chad and neighboring nations, require urgent action plans that are compatible with changing physical and human conditions and that maintain ecological balance. Global developments, which also undermine the political and economic stability of the region, have reached levels that exceed the capacity of the local population to cope. These challenges, which disrupt national development plans, lead to social disintegration and make vulnerable groups even more susceptible to adverse effects.

Social Crises and Societal Transformations

Migrations Triggered by Global and Local Disasters

Every individual establishes a living space by adapting the opportunities offered by their geographical environment in a manner compatible with the ecosystem. While geography is not an absolute determinant, it functions as a criterion that shapes the opportunities available to individuals and the constraints they may face. In particular, populations residing in areas with limited natural resources or inhospitable conditions are motivated by the desire to improve their living conditions or relocate to safer areas, which constitutes one of the primary drivers of migration (Lee, 1966; Castles, de Haas & Miller, 2014). In the context of migration motivations, factors such as natural disasters, political instability, ethnic and religious pressures, and resource scarcity act as push factors, whereas employment opportunities and the prospect of a secure life serve as pull factors (Adıgüzel, 2024). In this sense, migration evolves into a dynamic social phenomenon that necessitates both individual and societal transformation.

Migration patterns affecting Chad's social structure and way of life are shaped by both internal and external dynamics. The factors driving local populations to migrate are multidimensional, yet environmental conditions, climatic pressures, production losses in agriculture and livestock, rural unemployment, security issues, lack of access to drinking water, and insufficient agricultural outputs primarily direct internal migration from rural areas to urban centers (Niedenführ & Steinbrink, 2019; Kamta et al., 2020). Limited access to education and healthcare services constitutes another strong motivation for internal migration (Mercandalli et al., 2019). For some communities, internal migration serves as a livelihood strategy. Groups primarily dependent on livestock migrate seasonally to secure water and pasture resources suitable for their animals. These nomadic groups constitute a significant component of Chad's social and economic structure. However, in recent years, climate change has increasingly impacted these communities, threatening sustainable livelihood practices.

Nomadic communities, compelled to adjust their traditional migration routes in response to changing environmental conditions, have recently faced severe economic and subsistence challenges due to increasing livestock disease outbreaks. Rising animal mortality rates, lack of resources to combat diseases, and insufficient state support have further exacerbated these issues. These pressures affect the relationships between nomadic and sedentary populations, placing the sustainability of the nomadic lifestyle at risk. This situation indirectly affects the functionality of social relations, increasing tensions between nomads and settled communities over shared water and grazing resources. This factor, which plays a significant role in internal migration dynamics (Owonikoko & Momodu, 2020), drives populations to move toward urban centers, including the capital, N'Djamena, despite the associated risks (Niedenführ & Steinbrink, 2019). This trend places considerable pressure on national resources, deepens social inequalities, and strains the capacity of local governments to deliver social services (Ojiego-Okoro et al., 2024).

The drivers of mass migration in Chad are not limited to internal dynamics. External migration flows resulting from internal security issues, economic insufficiencies, and environmental pressures create new social, economic, and security challenges not only for the migrants themselves but also for the host countries (Ojiego-Okoro et al., 2024; Allarané et al., 2023). Consequently, migration profoundly affects the social structures of Chad and its neighboring countries, providing the foundation for ongoing social disasters. For instance, in recent years, hundreds of thousands of Sudanese have sought refuge in neighboring countries, particularly Chad, due to humanitarian crises resulting from civil war and political instability in Libya or coups and civil conflicts in Sudan (Cohen, 2007). Chad is similarly exposed to recurring challenges due to migratory routes originating from populations in nearby countries.

From a holistic perspective, both internal and external migration in Chad constitutes not only demographic mobility but also a structural issue directly influencing regional stability (Owonikoko & Momodu, 2020; Singh, 2022). The chaotic environment created by these migrations plays a decisive role in the transformation of global climatic and environmental hazards into social disasters within Chad. Populations displaced suddenly or caught in the midst of conflicts face severe difficulties not only in terms of shelter and livelihoods but also in accessing essential health, education, and security services. In this context, increasing social tensions between regions of origin and destination, along with intensifying conflicts among ethnic groups and clans, undermine societal and political cohesion. Therefore, migration dynamics in Chad do not represent merely a temporary crisis but sustain a multidimensional social disaster process that produces enduring structural problems through the interaction of environmental and social factors.

Hunger, Food Security, and the Inadequacy of the Healthcare System

Hunger and the Problem of Food Security

Among the consequences of the ongoing natural disaster cycles in Chad, the hunger crisis stands out as a significant factor adversely affecting both individual and societal life. The impact of global disasters on existing environmental challenges directly reduces soil fertility and crop quality, thereby hindering the population's access to sufficient and safe food (Xia & Bantin, 2017). A substantial portion of the local population, unable to meet its nutritional requirements adequately, faces severe health problems associated with malnutrition. These issues include chronic and epidemic diseases, as well as demographic-specific conditions such as developmental disorders in children and frailty among the elderly (Sharma et al., 2024). Beyond physical health challenges, psychosocial problems such as stress, hopelessness, loss of productivity, and the compulsion to migrate are also widely observed throughout society (Iriti & Vitalini, 2025). While hunger constitutes a disaster across the country, it generates more profound and enduring problems among vulnerable groups, particularly children, the elderly, and women.

Figure 8. Photograph taken by the author during a field visit to Chad.

Infrastructure deficiencies, limited transportation networks, and inadequate cold storage facilities jeopardize the long-term preservation of available food (Mfewou, Tata, & Nadji, 2020). These challenges become more apparent in regions experiencing droughts or irregular rainfall (Mfewou, Tata, & Nadji, 2020). Excessive summer droughts and erratic rainfall patterns adversely affect agriculture, which serves as the country's primary source of income (Bantin, Jun, & Si, 2017). Depending on climatic conditions, Chad cultivates grains such as millet, rice, and maize; industrial crops such as peanuts, sesame, and cotton; and fruits like mangoes, bananas, dates, as well as various garden vegetables (Bantin, Jun, & Si, 2017; Adhikari, Nejadhashemi, & Woznicki, 2015). While most agricultural and livestock production is consumed domestically, limited exports exist for crops such as sesame, cotton, and peanuts (Ambadi, 2022). Cattle, camels, and small ruminants constitute major livelihood sources alongside agriculture. However, as a landlocked country with insufficient infrastructure, Chad faces serious risks to sustainable exportation.

Hunger and food security-related issues are a widespread public health concern in Chad. Inadequate and unhealthy nutrition weakens the immune system, particularly in children, making them vulnerable to infectious diseases, causing illnesses to follow a more severe course, or increasing the risk of mortality (WHO, 2024). In a country with a high risk of epidemic diseases, illnesses such as cholera, malaria, and diarrhea are especially prevalent among children and can result in fatal outcomes (UNICEF, 2023). Furthermore, physical developmental issues associated with malnutrition—such as stunting, excessive weight loss, and general weakness—adversely affect children's healthy social development and their ability to adapt to educational processes (Kirolos et al., 2022; Ssentongo et al., 2021; Sánchez et al., 2024). Pregnant women who do not receive adequate nutrition face increased risks of giving birth to low-birth-weight infants, higher rates of complications during childbirth, and maternal mortality, which remains a significant concern (WHO, 2024). Due to the difficulty in obtaining direct data on child and adult mortality caused by hunger, thematic reports indicate that approximately 3.5 million people in Chad were in need of emergency intervention as of June 2025, according to the African Health Cluster (WHO, 2025)

Inadequacy of the Healthcare System

In a country where healthcare infrastructure is underdeveloped, opportunities for disease prevention, early diagnosis, and treatment remain limited. This leads to high morbidity and mortality rates among both children and adults. The inadequacy of healthcare system capacity functions as a structural factor that intensifies the severity and spread of crises at every stage of the disaster cycle in Chad. Deficiencies in medical infrastructure, personnel, equipment, and pharmaceutical stocks make access to healthcare almost impossible, especially in rural areas (World Bank, 2023). The concentration of 55% of healthcare personnel in the capital, N'Djamena, deprives the rural population and nomadic communities of even basic health services; the resulting decline in quality of life significantly increases susceptibility to diseases. This situation delays the early detection and response to infectious diseases, paving the way for minor health issues to escalate into community-wide epidemics in a short period. For instance, preventable diseases such as cholera and malaria transform into seasonal disasters due to inadequate health surveillance and weak immunization programs, leading to thousands of fatalities (UNICEF, 2023; WHO, 2024).

The most critical impact of weak health infrastructure becomes more evident during natural disasters and climate-induced crises. During periods of drought, reduced access to clean water and deteriorating hygiene conditions bring the already fragile healthcare system to a breaking point. Children and the elderly, whose immune systems are weakened by malnutrition, are left defenseless against epidemics, while the lack of beds, personnel, and medicine in hospitals dramatically reduces treatment capacity. This creates a vicious cycle: disasters trigger health crises, and the inadequate health system deepens the social impacts of the disaster, prolonging the recovery process and further weakening societal resilience against future disasters (Wisner et al., 2004). Furthermore, the lack of emergency medical response capacity leads to the inability to manage trauma, injuries, and infectious disease cases in the post-disaster period, thereby increasing preventable deaths.

This structural inadequacy reinforces Chad's dependence on international humanitarian organizations and weakens the state's capacity to provide basic services to its citizens. This gap in the healthcare system leads not only to individual loss of life but also to an **erosion of trust** at the societal level, social disintegration, and the questioning of state legitimacy. Indeed, the failure to ensure equitable and effective access to healthcare services deepens social inequalities and causes vulnerable groups (children, women, the elderly, migrants) to be disproportionately affected by crises (Flanagan et al., 2020; Cutter et al., 2003). Therefore, the case of Chad demonstrates that healthcare system capacity is not merely a technical infrastructure problem but a **strategic governance issue** that directly impacts disaster resilience, social trust, and sustainable development. Under normal circumstances, these health-related problems carry the risk of producing much more severe consequences when combined with crises caused by natural or global disasters. Agricultural and climatic conditions that limit the country's food production capacity play a significant role in the deepening of these health and nutrition crises (Bantin, Jun, & Si, 2017). Hunger in Chad cannot be explained solely by income inequality or inadequate health services. At the same time, the direct dependence of the production chain on climatic conditions makes the food supply extremely fragile. Consequently, the problems of hunger and malnutrition affecting public health are closely linked to the country's agriculture-based economic structure.

These infrastructural and production limitations indicate that the hunger crisis deepens not only due to economic or agricultural factors but also through the intersection of social, political, and environmental elements. Persistent political instability, income inequality, widespread drought, the shrinking of Lake Chad as a primary water source, inter-ethnic conflicts, and insufficient agricultural production to meet national needs have rendered hunger one of the country's most pressing crises (Bantin, Jun, & Si, 2017). Consequently, approximately five million people live with hunger and its associated problems (Musa et al., 2022). Furthermore, migration and refugee pressures exacerbate the food crisis (Musa et al., 2022). This dependence on external aid represents a significant obstacle for an otherwise sovereign Chad. Environmental and natural conditions, along with production losses and quality issues, directly impact the hunger problem. This chronic challenge not only affects food security and access to essential nutrients but also represents a social issue with broader societal implications, undermining community-based mutual assistance and solidarity.

THE DUAL ROLE OF TRADITIONAL TRIBAL STRUCTURE IN RESILIENCE AND INTERNAL CONFLICTS

Each individual shapes their sense of belonging and identity through the network of relationships they maintain with the group or community to which they belong. In Chad and other societies with similar ethnic structures, tribalism serves an important social function in this regard. Tribes play a central role in socializing young individuals, enabling them to learn community norms, internalize value hierarchies, and align with traditional rules. Moreover, in rural areas where the institutional authority of the Chadian government is weak, tribal bonds provide a complementary function by ensuring social security and fostering a culture of solidarity within the community (Thissen, 2019). Tribalism, which shapes Chad's social and cultural fabric, thus functions not only as a form of belonging but also as a form of authority that supports social order and resilience. Likewise, during humanitarian crises induced by disasters, the solidarity exhibited within tribal communities acts as a traditional social protection mechanism, compensating for institutional deficiencies. In this sense, tribalism in Chad operates both as a carrier of cultural identity and as a crucial pillar of social resilience (Thissen, 2019).

Tribalism has dual effects in the context of global and local disasters. Strong bonds among tribal members facilitate mutual support in terms of shelter, nutrition, and material aid to quickly mitigate the impacts of such crises (Lin & Lin, 2020). Traumas caused by post-disaster losses are often alleviated through collective rituals and psychosocial practices. Tribal leaders may also play key roles in disseminating emergency notifications and implementing action plans (Quinn, Williamson, & Gibbs, 2022). However, alongside these positive functions, limited disaster aid directed toward tribes may be distributed preferentially to members or allies, creating unequal allocations that can generate tension both within and between tribes (Monteil, Simmons, & Hicks, 2020). Such competition and dissatisfaction can hinder effective aid distribution and, in some cases, escalate into armed conflicts. Particularly, tribes holding political power or aligned with ruling authorities often have easier access to resources and services, generating tensions in other communities (Joffé, 1997). In rural areas, legal gaps regarding resource-sharing between tribes whose primary livelihoods are agriculture and pastoralism also lead to disputes (Henrico & Doboš, 2024). Collective memories of past conflicts can be transmitted across generations, potentially triggering larger violent incidents (Seltzer, 2019).

Tribal-based social structures typically consist of networks grounded in kinship and blood relations, organized hierarchically according to attributes such as age, gender, and leadership. These structures, which define an individual's position within both the tribe and society, play a significant role in Chad's social construction. Such relationships are crucial for coordinating disaster response and enhancing community recovery capacities. However, mistrust and divisions among tribes can impede co-operation during disaster mitigation, recovery, and reconstruction processes, thereby exacerbating the adverse effects of disasters (Seltzer, 2019). This represents one of the unintended negative outcomes of tribalism with regard to social solidarity. Recent conflicts in Chad (2021–2024) between herders and farmers illustrate this dynamic (Joffé, 1997). Disputes over the use and sharing of limited resources have escalated into violent incidents resulting in injuries and fatalities. Similar conflicts have been reported across different regions of the country (see: Eberle, Rohner, & Thoenig, 2020; Joffé, 1997; Adams, Thill, Kuusaana, & Mittag, 2023). Enhancing the positive effects of tribal structures could play a key role in leveraging local capacities to respond to global disasters.

Political Crises and Institutional Fragility

The State's Lack of Institutional Capacity and Governance Issues

A state's political stability and institutional capacity directly influence its economic and social performance. Conversely, the absence of these qualities is equally consequential. From this perspective, the multifaceted economic, social, and disaster-related crises experienced by Chad are profoundly affected by its existing political instability and fragility. The country's low level of development and widespread poverty among its population are significant factors that impede the ability to respond effectively to and mitigate the impacts of disasters. Consequently, the Chadian government's capacity to manage disasters arising from both its geographic conditions and global challenges has become increasingly constrained.

The primary reason the Republic of Chad cannot adequately address these issues or implement the necessary political, economic, and social measures lies in its political weakness and structural fragility. Although the country is no longer officially a colony, traces of the colonial era persist within its institutional and political structures.

This legacy limits the effectiveness of state institutions and increases dependency on external actors in decision-making processes. Asymmetric relationships with former colonial powers, particularly France, hinder the country's ability to develop an internally driven development strategy. A substantial portion of the political elite relies on external support to maintain power, which undermines democratization, institutionalization, and social cohesion processes. Therefore, Chad's economic backwardness, poverty, and social inequalities are not only the product of internal dynamics but also closely linked to forms of dependency reproduced by historical colonial relations. This fragile structure exacerbates humanitarian crises triggered by disasters and reduces the long-term effectiveness of sustainable development policies.

Figure 9. Photograph taken by the author during a field visit to Chad.

Administrative challenges, such as the lack of a modern state structure and the inability to maintain political stability, structurally limit Chad's capacity to address ongoing crises and pursue development initiatives. According to the World Bank, public administration in Chad exhibits "very low institutional capacity." Approximately 55% of civil servants are concentrated solely in the capital, N'Djamena, creating significant disparities in the delivery of public services to rural areas (World Bank,

2023). Moreover, a substantial portion of the state budget is allocated to military expenditures and the management of oil revenues, limiting the allocation of resources to meet the population's basic needs. Data from Transparency International indicate that public oversight in these areas is extremely limited and anti-corruption mechanisms have not reached institutionalized levels over the past decade (Transparency International, 2025). The Fragile States Index (FSI) further confirms the weakness of state authority, the persistent lack of coordination between security forces and the central government, and the inability to manage social tensions—particularly ethnically based north–south divisions (The Fund for Peace, 2024). The transitional government's failure to establish institutional stability after 2021 corroborates this assessment.

INTERNATIONAL DEPENDENCY AND POST-COLONIAL STRUCTURAL PROBLEMS

The historical legacy of colonial administrative structures continues to shape the socio-economic landscape of many African nations, often manifesting as a persistent fragility in state institutions. In the case of Chad, this structural weakness creates a vacuum where national development is frequently dictated by external pressures rather than domestic priorities. Understanding the current humanitarian landscape requires an analysis of how these post-colonial vulnerabilities intertwine with modern systemic challenges.

These indicators reveal that the ongoing humanitarian crises affect nearly every aspect of life in Chad, from education and healthcare to food security and safety. In the absence of structural transformation, partial progress achieved through support from international institutions has not resulted in lasting improvements in the population's well-being. The state's limited capacity to manage and mitigate disaster risks has rendered Chad dependent on international aid organizations. In areas where the government falls short, local communities rely on clan-based solidarity networks to meet social security, food provision, and other essential needs. However, these local mechanisms are insufficient to mitigate the impacts of large-scale disasters, increasingly making the country reliant on external assistance. This reliance not only exacerbates the physical effects of disasters but also delays social and economic recovery processes.

Recent humanitarian crises have refocused international attention on Chad. Organizations such as the World Health Organization (WHO), the International Organization for Migration (IOM), the African Development Bank (AfDB), the World Bank Group, and the United Nations (UN) are implementing various aid and development projects in the country. While these interventions may yield positive short-term humanitarian outcomes, they carry the risk of deepening the country's long-term dependence on external support. In an environment where institutional capacity remains underdeveloped, foreign aid can weaken the state's ability to function autonomously rather than encouraging the development of local governance mechanisms. Consequently, prolonged international assistance may increasingly expose economic and political decision-making processes to external actors, undermining Chad's capacity to determine its own trajectory. In this sense, continuous aid may inadvertently reproduce a cycle of dependency akin to "informal colonialism."

Chad's Disaster Risk Map in the Light of Beck and Giddens: A Society at the Threshold of Manufactured Risks and Reflexive Capacity

The findings of this study demonstrate that the transformation of global hazards into societal disasters in Chad does not follow a linear trajectory. Rather, it exhibits a complex and intertwined continuity where each element triggers another. This process reveals a striking picture within the context of Beck's "Risk Society" and Giddens' "Reflexive Modernization" theories. Chad stands in a position where it pays the price for "manufactured risks," embodied by greenhouse gas emissions and climate change produced by industrialized societies, yet is deprived of the "reflexive capacity" required to manage these risks. In other words, while the country is exposed to the side effects and catastrophic consequences of modernization in the severest form, it lacks access to the modern institutional mechanisms and resources necessary to cope with them.

The concrete reflections of this lack of capacity and the risks involved are clearly visible in the risk map and infographic data presented in this study. Extreme droughts triggered by global climate change, combined with geographical vulnerability, systematically destroy the livelihoods of a population dependent on agriculture and livestock. This results directly in mass hunger and forced migration. The traditional knowledge of local communities proves inadequate in the face of these massive, modern-era "manufactured" climatic risks, which intensifies the severity of the destruction. Migration movements do not only strain urban infrastructure, they also transform traditional clan structures into support systems. On one other hand, they deepen social disintegration by turning them into potential sources of conflict over vital resources such as water and pasture.

In conclusion, the case of Chad proves that the destructive impact of disasters is determined by the existing social, economic, and political vulnerabilities within that society. The situation in Chad demonstrates that these events should not be read merely as "natural disasters," but as socio-political disasters emerging at the intersection of global inequalities, manufactured risks, and local capacity issues.

Therefore, an approach limited solely to emergency aid policies is insufficient. To mitigate these crises, it is imperative to strengthen local institutional capacity, ensure sustainable resource management, and establish a more equitable and reflexive global order that targets the injustice in global risk production.

Figure 10. Chad's Disaster Risk Map: From Global Risk to Social Catatrophe

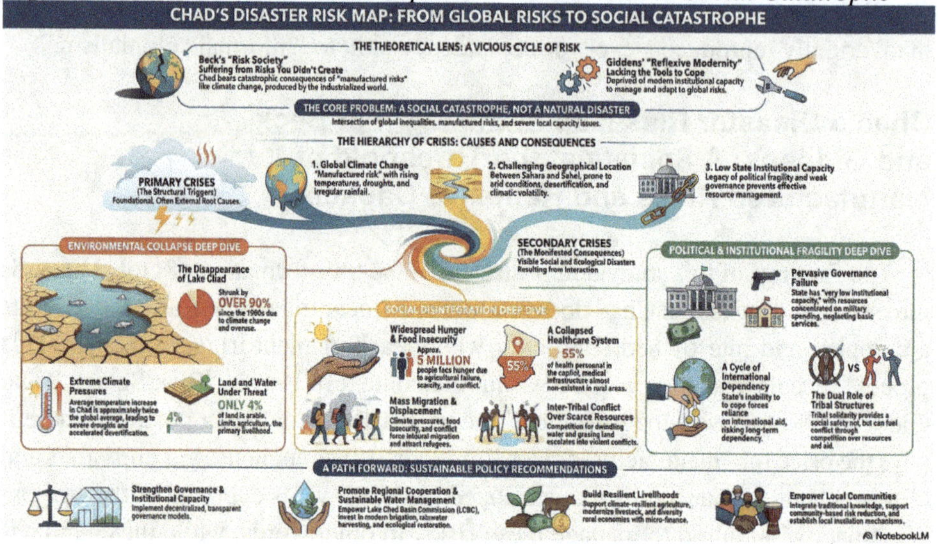

Sustainable Policy Recommendations and Potential Solutions

Chad, as one of the countries where the consequences of modernization emerge unevenly, faces a dual disadvantage in this process. Local challenges stemming from its geographic location and demographic structure—such as drought, desertification, irregular rainfall patterns, internal and external migration issues, infrastructure deficits, the shrinking of Lake Chad, and inefficient water resource utilization—represent major obstacles to the country's sustainable development plans. These structural problems not only exacerbate environmental and economic vulnerabilities but also, when combined with weak governance capacity, further reduce the country's resilience to crises.

The second disadvantage of Chad lies precisely in this context. The state's capacity to implement multidimensional solutions to crises is markedly low. The government's reliance on traditional governance approaches rather than enacting policies that address internal challenges contributes to the deepening of humanitarian crises. The exacerbation of the country's disaster risk profile when combined with global challenges underscores the urgent need for well-planned action strategies that are socially and culturally aligned.

Based on the analysis of the findings of this study, the development of integrated policies in the following areas could contribute to addressing both structural and governance-related challenges:

INSTITUTIONAL REFORM AND STRENGTHENING GOVERNANCE CAPACITY

Addressing the structural weaknesses inherited from the colonial era and establishing a accountable governance model, comprehensive political and administrative reforms must be implemented. These reforms should include the following elements:

- **Democratic Participation and Representation:** To develop a more democratic management approach, political mechanisms where local people and tribes are represented in a balanced manner must be established and strengthened. Replacing centralized structures with **decentralized models** that provide more authority and resources to local governments will increase social participation and reduce tribe-based divisions.
- **Fiscal Reform and Transparency:** Ensuring the country's income-expenditure balance by making the tax system fair and effective is critical for the efficient use of public resources. Ensuring transparency in the management of oil and natural resource revenues, reducing corruption, and directing these revenues toward the basic needs of the population (education, health, infrastructure) should be priority targets. The establishment of independent audit mechanisms in accordance with international standards is a cornerstone of this process.
- **Institutional Capacity Building:** It is necessary to increase the human resource capacity of public institutions, employ a sufficient number of qualified public officials capable of delivering services to rural areas, and organize training programs to improve service quality. Institutional expertise should be developed particularly in disaster management, early warning systems, and emergency response coordination.

Sustainable Management of Water Resources and Regional Cooperation

The sustainable management of the Lake Chad Basin is vital not only for Chad but also for riparian countries such as Nigeria, Niger, Cameroon, and the Central African Republic. Therefore, a regional approach is mandatory:

- **Strengthening Regional Cooperation Mechanisms:** The capacity of existing regional platforms, such as the Lake Chad Basin Commission (LCBC), must be increased, and effective coordination regarding water sharing, basin management, and joint action plans among member states must be ensured. Technical and financial support should be provided to these platforms by international financial institutions and environmental organizations.
- **Water Conservation and Modern Irrigation Techniques:** The adoption of modern irrigation technologies, such as drip and sprinkler systems, which reduce water consumption in agricultural production should be encouraged. Subsidy programs and educational campaigns should be organized to facilitate farmers' access to these technologies.
- **Alternative Water Sources and Infrastructure Development:** Hydrogeological research should be conducted for the sustainable use of groundwater resources, and infrastructure for water wells and drinking water supply should be strengthened. Rainwater harvesting systems should be popularized, especially in rural areas, to ensure more effective use of existing water resources.
- **Ecological Restoration:** Projects for afforestation, wetland restoration, and combating desertification should be implemented around Lake Chad. The participation of local communities in these projects should be encouraged to support both environmental recovery and the diversification of local livelihoods.

Creating Resilient and Sustainable Livelihoods

To break the cycle of hunger, poverty, and drought for the population, economic activities must be diversified and resilience must be enhanced:

- **Climate-Resilient Agricultural Practices:** The development and dissemination of drought-resistant seed varieties, the use of organic fertilizers, and the adoption of sustainable farming methods such as soil conservation techniques and agroforestry should be encouraged. Agricultural consultancy services should be strengthened to provide farmers with support on climate change adaptation.

- **Modernization of the Livestock Sector:** Alongside traditional livestock practices, the development of modern veterinary services, vaccination programs to prevent epidemics, animal health monitoring systems, and fodder production infrastructure should be supported. The integration of the traditional knowledge of nomadic communities with modern practices should be ensured.
- **Income Diversification and Micro-Finance:** Micro-credit and entrepreneurship programs should be established to develop small-scale trade, handicrafts, and non-agricultural economic activities in rural areas. Social resilience should be increased by encouraging the participation of women and youth in economic activities.
- **Education and Employment:** Priority should be given to minimizing social inequalities by emphasizing education to train qualified personnel, expanding vocational training programs, and making recruitment based on **merit-based criteria**. The participation of the young population in scholarship programs in Turkey and other countries should be increased, and incentives should be created for those who complete their education to return to their country.

Strengthening Local Resilience and Community-Based Solutions

In cases where state capacity is insufficient to cope with disasters, the solidarity networks within local communities play a critical role:

- **Supporting Positive Aspects of Tribal Solidarity:** The mutual aid and social protection mechanisms exhibited by traditional tribal structures during disaster periods can compensate for the lack of official systems. These solidarity networks should be strengthened, cooperation with tribal leaders should be established, and local knowledge should be integrated into disaster management and early warning systems.
- **Community-Based Disaster Risk Reduction (CBDRR):** To ensure local preparedness, community-based training programs, disaster drills, and the creation of emergency plans at the local level are required, alongside increasing the capacity of community leaders. Special protection mechanisms should be developed for vulnerable groups (women, children, the elderly, the disabled).
- **Evaluating Local Knowledge:** The traditional knowledge of nomadic communities regarding climate and the environment should be combined with

modern scientific approaches and included in adaptation strategies. Projects created with the participation of local people increase both acceptability and sustainability.

- **Social Integration and Conflict Resolution:** To reduce inter-tribal conflicts, mediation mechanisms, peacebuilding programs, and local-level consensus platforms for resource sharing should be established. Legal regulations should be made for the fair sharing of limited resources such as water, pasture, and land, bridging the gap between traditional and modern legal systems.

Redefining International Cooperation and Reducing Dependency

International aid must be integrated into long-term technical and financial cooperation projects that support the structural transformations mentioned above, moving beyond just improving food and health conditions.

- **Development-Oriented Structural Aid:** In addition to emergency humanitarian aid, long-term investments should be made in education, health, infrastructure, and institutional capacity building. International actors must focus on projects that support sustainable development and encourage local ownership rather than short-term interventions.
- **Technology Transfer and Knowledge Sharing:** Developed countries and international organizations should provide technical support for the transfer of appropriate technologies to Chad, the sharing of methods to increase agricultural productivity, water management technologies, and the establishment of renewable energy systems.
- **Climate Justice and Climate Finance:** As one of the countries that contributes the least to global climate change but suffers the most, Chad must receive its fair share of climate finance. It is vital that the climate funds pledged by wealthy countries under the Paris Agreement reach Chad to provide resources for adaptation and mitigation projects.
- **Exiting the Dependency Cycle:** To free the country from a structure dependent on foreign aid, strategies must be developed to transition to a development model that encourages self-sufficiency, environmental sustainability, and social stability. International cooperation should be restructured to support Chad in determining and implementing its own development priorities.

CONCLUSION

The impacts of global and local disasters in Chad, though varying according to the region's geographic and societal conditions, are not confined to the Chadian population alone. These effects extend beyond humans, posing significant threats to biodiversity that is unable to adapt to such changes. At this point, the destructive consequences of disasters have become visibly evident across environmental, economic, and social domains. Therefore, to mitigate the adverse effects of global disasters and implement timely preventive measures, it is necessary to develop and replicate localized disaster risk maps for Chad, as exemplified in this study. In this regard, the formulation of resilience policies that are compatible with communities' specific geographic and demographic conditions is crucial for achieving sustainability. Strengthening national capacities to respond to both natural and global disasters, raising public awareness, and supporting these initiatives through international funding are essential.

Failure to implement timely measures against global disasters that directly affect the country inevitably disrupts ecological balance. Consequently, climate change-induced challenges, which negatively impact social life, have pushed the population toward the threshold of social disasters. In Chad, where more than half of the population lives below the poverty line, urgent action plans must be activated. Ensuring access to basic needs requires the development of climate-resilient agriculture and livestock practices, the proper management of shrinking water resources especially Lake Chad and comprehensive initiatives to secure access to clean water. Prompt intervention, including the involvement of volunteer organizations, will also help prevent the spread of epidemic diseases.

Poverty, epidemic diseases, desertification, drought, water scarcity, climate variability, inadequate infrastructure, internal and external migration, and education deficits highlighted in Chad's disaster risk map represent deep rooted challenges that profoundly affect individuals and societal structures. Addressing these issues through integrated action plans, rational utilization of local resources, balanced allocation of international support, initiatives to stimulate the national economy, and the protection of social equilibrium in accessing basic needs is vital to minimizing social disasters.

In this context, the development and effective implementation of sustainable policies is a fundamental factor in strengthening the country's long-term resilience. A promising step forward is the increasing participation of the young population in educational programs in Turkey and other countries, which has the potential to enhance the capacity of future generations to generate scientific and innovative solutions to the country's challenges. Supporting these students in continuing their education through local resources and international funding is crucial for realizing this potential. Moreover, Chad's development and crisis management strategies should be constructed based on its own resources and institutions, while the international community must contribute to this process with a sense of responsibility and fairness.

REFERENCES

Adams, E., Thill, A., Kuusaana, E., & Mittag, A. (2023). Farmer–herder conflicts in sub-Saharan Africa: Drivers, impacts, and resolution and peacebuilding strategies. *Environmental Research Letters*, *18*(12), 123001. Advance online publication. DOI: 10.1088/1748-9326/ad0702

Adhikari, U., Nejadhashemi, A. P., & Woznicki, S. A. (2015). Climate Change and Eastern Africa: A Review of Impact on Major Crops. *Food and Energy Security*, *4*(2), 110–132. DOI: 10.1002/fes3.61

Adıgüzel, Y. (2024). *Göç sosyolojisi*. Nobel Akademik Yayıncılık. https://www.kitapyurdu.com/kitap/goc-sosyolojisi/407575.html

Alexander, D. (2002). *Confronting catastrophe: New perspectives on natural disasters*. Oxford University Press.

Allarané, N., Azagoun, V. V. A., Atchadé, A. J., Hetcheli, F., & Atela, J. (2023). Urban vulnerability and adaptation strategies against recurrent climate risks in Central Africa: Evidence from N'Djaména City (Chad). *Urban Science (Basel, Switzerland)*, *7*(3), 97. DOI: 10.3390/urbansci7030097

Ambadi, D. (2022). The roughage problem and solutions in animal feeding in Chad. *6th International Students Science Congress Proceedings Book*. https://doi.org/DOI: 10.52460/issc.2022.003

Arfasa, G. F., Owusu-Sekyere, E., Doke, D. A., & Ampofo, J. A. (2024). Impacts of climate and land use/cover changes on the sustainability of irrigation water in West Africa: A systematic review. *All Earth*. https://doi.org/DOI: 10.1080/27669645.2024.2308371

Bantin, A., Jun, X., & Si, H. (2017). The impact of climate change on water resource of agricultural landscape and its adaptation strategies: A case study of Chari Basin, Chad. *Journal of Earth Science & Climatic Change*, *8*(12), 1–7. DOI: 10.4172/2157-7617.1000437

Beck, U., Giddens, A., & Lash, S. (1994). *Reflexive modernization: Politics, tradition and aesthetics in the modern social order*. https://doi.org/DOI: 10.2307/3341775

Bricker, J., & Pelupessy, D. (2025). The nature of disasters. *Journal of Delta Urbanism*. https://doi.org/DOI: 10.59490/jdu.5.2024.8070

Castles, S., de Haas, H., & Miller, M. J. (2014). *The age of migration: International population movements in the modern world* (5th ed.). Guilford Press. DOI: 10.1007/978-0-230-36639-8

Çelik, M. A. (2026). Agro-climatic classification of Türkiye's Agricultural Basins using long-term ERA5-Land temperature data. *Physics and Chemistry of the Earth Parts A/B/C, 142*, 104307. DOI: 10.1016/j.pce.2026.104307

Çelik, M. A., Bayram, H., & Özüpekçe, S. (2018). An assessment on climatological, meteorological and hydrological disasters that occurred in Turkey in the last 30 years (1987-2017). *International Journal of Geography and Geography Education*, (38), 295–310. DOI: 10.32003/iggei.424675

Çelik, M. A., Bilik, A., & Türkeş, M. (2025). Spatio-Temporal Analysis of Observed Drought Events in the Tigris–Euphrates Basin during the 1960–2023 Period Via SPI and SPEI Drought Indices. *Pure and Applied Geophysics*. Advance online publication. DOI: 10.1007/s00024-025-03863-3

Çelik, M. A., Kıraç, A., & Gouveia, C. M. (2026). North Atlantic Oscillation modulation of drought and vegetation activity in Türkiye. *Theoretical and Applied Climatology, 157*(2), 74. Advance online publication. DOI: 10.1007/s00704-025-05971-5

Cohen, R. (2007). Response to Hathaway. *Journal of Refugee Studies, 20*(3), 370–376. Advance online publication. DOI: 10.1093/jrs/fem020

Cutter, S. L., Boruff, B. J., & Shirley, W. L. (2003). Social vulnerability to environmental hazards. *Social Science Quarterly, 84*(2), 242–261. DOI: 10.1111/1540-6237.8402002

Cutter, S. L., Burton, C. G., & Emrich, C. T. (2014). Disaster resilience indicators for benchmarking baseline conditions. *Journal of Homeland Security and Emergency Management, 7*(1), 1–22. DOI: 10.2202/1547-7355.1732

Demir, E. (2025). Climate change in Afrıca. *Uluslararası Afro-Avrasya Araştırmaları Dergisi, 10*(20), 117–120.

Eberle, U., ... (2020). *Heat and Hate, Climate Security and Farmer-Herder Conflicts in Africa* (ESOC Working Paper No. 22). Empirical Studies of Conflict Project. Retrieved [22.10.25], from https://esoc.princeton.edu/wp22

European Space Agency. (2019). *Lake Chad's shrinking waters*. https://www.esa.int/ESA_Multimedia/Images/2019/03/Lake_Chad_s_shrinking_waters

Giddens, A. (1994). *Runaway world: How globalization is reshaping our lives*. Profile Books.

Henrico, I., & Doboš, B. (2024). Shifting sands: The geopolitical impact of climate change on Africa's resource conflicts. *The South African Geographical Journal*. Advance online publication. DOI: 10.1080/03736245.2024.2441116

Institute for Security Studies. (2023). *Chad: Country profile.* ISS African Futures & Innovation. https://futures.issafrica.org/geographic/countries/chad/

International Monetary Fund. (2024). *Chad: Selected issues* (IMF Staff Country Report No. 24/336). https://www.elibrary.imf.org/view/journals/002/2024/336/002.2024.issue-336-en.xml

Iriti, M., & Vitalini, S. (2025). *Climate change, natural disasters, armed conflicts and migrations at the crossroads between food and nutrition insecurity and undernourishment.* Functional Food Science., DOI: 10.31989/ffs.v5i1.1547

Jean-Richard, V., Crump, L., Abicho, A., Abakar, A., Mahamat, A., Béchir, M., Eckert, S., Engesser, M., Schelling, E., & Zinsstag, J. (2015). Estimating population and livestock density of mobile pastoralists and sedentary settlements in the south-eastern Lake Chad area.. *Geospatial health,* 10 1, 307 . .DOI: 10.4081/gh.2015.307

Joffé, G. (1997). Chad: Power vacuum or geopolitical focus. *Geopolitics, 2*(1), 25–39.

Kamta, F. N., Schilling, J., & Scheffran, J. (2020). Insecurity, resource scarcity, and migration to camps of internally displaced persons in Northeast Nigeria. *Sustainability (Basel), 12*(17), 6830. DOI: 10.3390/su12176830

Kazeem, O. S. (2024). Climate change and violent conflict in the Sahel and Lake Chad Basin. *Journal of Contemporary International Relations and Diplomacy, 5*(1), 74–85. DOI: 10.53982/jcird.2024.0501.05-j

Kirolos, A., Goyheneix, M., Eliasz, M., Chisala, M., Lissauer, S., Gladstone, M., & Kerac, M. (2022). Neurodevelopmental, cognitive, behavioural and mental health impairments following childhood malnutrition: A systematic review. *BMJ Global Health,* 7(7), e009330. Advance online publication. DOI: 10.1136/bmjgh-2022-009330 PMID: 35793839

Lama, S., & Devkota, B. (2009). Vulnerability of mountain communities to climate change and adaptation strategies. *Journal of Agriculture and Environment, 10,* 76–83. DOI: 10.3126/aej.v10i0.2133

Lee, C. C., Zeng, M., & Luo, K. (2024). How does climate change affect food security? Evidence from China. *Environmental Impact Assessment Review, 104,* 107324. DOI: 10.1016/j.eiar.2023.107324

Lee, E. S. (1966). A theory of migration. *Demography, 3*(1), 47–57. DOI: 10.2307/2060063

Lin, P., & Lin, W. (2020). Rebuilding relocated tribal communities better via culture: Livelihood and social resilience for disaster risk reduction. *Sustainability (Basel)*, *12*(11), 4538. Advance online publication. DOI: 10.3390/su12114538

MDPI. (2023). Changes in temperature and relative humidity in Equatorial Africa: Analysis of ERA5 reanalysis data (1991–2020). *Atmosphere*, *16*(7), 828.

Mercandalli, S., Losch, B., Belebema, M. N., Bélières, J.-F., Bourgeois, R., Dinbabo, M. F., Freguin-Gresh, S., Mensah, C., & Nshimbi, C. C. (2019). Rural migration in sub-Saharan Africa: Patterns, drivers and relation to structural transformation. *FAO Working Paper*. https://doi.org/DOI: 10.4060/ca7404en

Mfewou, A., Tata, J., & Nadji, P. (2020). Climate variability and peasant agriculture in Kélo in the south of Chad. *Advances in Social Sciences Research Journal*, *7*(11), 16–30. DOI: 10.14738/assrj.711.9294

Monteil, C., Simmons, P., & Hicks, A. (2020). Post-disaster recovery and socio-cultural change: Rethinking social capital development for the new social fabric. *International Journal of Disaster Risk Reduction*, *42*, 101356. Advance online publication. DOI: 10.1016/j.ijdrr.2019.101356

Musa, S. S., Bouri, T., & Manirambona, E. (2022). How climate change and insecurity pushed 5 million people to hunger in Chad, Africa. *Public Health Challenges*, *1*(2), 1–8. DOI: 10.1002/puh2.47 PMID: 40496685

Nagarajan, C., Vivekananda, J., Pham-Duc, B., Sylvestre, F., Pohl, B., & Munoz, H. (2024). Peace in an extreme climate: How climate-related security risks affect prospects for stability in Lake Chad. *PLOS Climate*, *3*(10), e0000314. Advance online publication. DOI: 10.1371/journal.pclm.0000314

Niedenführ, H., & Steinbrink, M. (2019). *Africa on the move: Migration, translocal livelihoods and rural development in Sub-Saharan Africa*. https://www.amazon.com/Africa-Move-Livelihoods-Development-Sub-Saharan/dp/3030228436

Noureldeen, N., Mao, K., Mohmmed, A., Yuan, Z., & Yang, Y.NourEldeen. (2020). Spatiotemporal drought assessment over Sahelian countries from 1985 to 2015. *Journal of Meteorological Research*, *34*(4), 760–774. DOI: 10.1007/s13351-020-9178-7

Nyaupane, G. P., & Chhetri, N. (2009). Vulnerability to climate change of nature-based tourism in the Nepalese Himalaya. *The Journal of Climate Change and Health*, *10*, 100208.

Ojiego-Okoro, C., Umeh, L., & Nwodom, D. (2024). Boko Haram-induced forced migration: Implications for social and economic development in the Lake Chad Basin. *International Journal of Social Science and Human Research*, *7*(5). Advance online publication. DOI: 10.47191/ijsshr/v7-i05-94

Olowoyeye, O., & Kanwar, R. (2023). Water and food sustainability in the riparian countries of Lake Chad in Africa. *Sustainability (Basel)*, *15*(13), 10009. Advance online publication. DOI: 10.3390/su151310009

Owonikoko, S. B., & Momodu, J. A. (2020). Environmental degradation, livelihood, and the stability of Chad Basin Region. *Small Wars & Insurgencies*, *31*(6), 1295–1322. DOI: 10.1080/09592318.2020.1776092

Peduzzi, P. (2019). The disaster risk, global change, and sustainability nexus. *Sustainability (Basel)*, *11*(4), 957. Advance online publication. DOI: 10.3390/su11040957

Quinn, P., Williamson, B., & Gibbs, L. (2022). Indigenous-informed disaster recovery: Addressing collective trauma using a healing framework. *Progress in Disaster Science*, *16*, 100257. Advance online publication. DOI: 10.1016/j.pdisas.2022.100257

Republic of Chad. (2021). *First national climate change adaptation plan of Chad.* United Nations Framework Convention on Climate Change (UNFCCC). https://unfccc.int/sites/default/files/resource/CHAD-NAP_EN-web.pdf

Ruppel, O., & Funteh, M. (2019). Chapter 5: Climate change, human security and the humanitarian crisis in the Lake Chad Basin region: selected legal and developmental aspects with a special focus on water governance. *Law | Environment | Africa*. https://doi.org/DOI: 10.5771/9783845294605-105

Sánchez, A., Favara, M., Sheridan, M., & Behrman, J. (2024). Does early nutrition predict cognitive skills during later childhood? Evidence from two developing countries. *World Development*, *176*, 106480. Advance online publication. DOI: 10.1016/j.worlddev.2023.106480 PMID: 38249341

Şekercioğlu, Ç. H., Primack, R. B., & Wormworth, J. (2012). The effects of climate change on tropical birds. *Biological Conservation*, *148*(1), 1–18. DOI: 10.1016/j.biocon.2011.10.019

Seltzer, N. (2019). The evolution of tribalism: A social-ecological model of cooperation and inter-group conflict under pastoralism. *Journal of Artificial Societies and Social Simulation*, *22*(2), 6. Advance online publication. DOI: 10.18564/jasss.3963

Semenza, J., Rocklöv, J., & Ebi, K. (2022). Climate change and cascading risks from infectious disease. *Infectious Diseases and Therapy*, *11*(4), 1371–1390. DOI: 10.1007/s40121-022-00647-3 PMID: 35585385

Sevgi, M. A., & Yulu, A. (2024). "The old rains are gone, so are we": Kurdish nomads of Mount Ağrı (Ararat). *British Journal of Middle Eastern Studies*, *52*(2), 503–529. DOI: 10.1080/13530194.2024.2313752

Sharma, P., Kumari, A., Shivali, S., & Sharma, V. (2024). Impact of food and health in 21st century era: A global economic crises. *International Journal of Technology*. https://doi.org/DOI: 10.52711/2231-3915.2024.00016

Singh, A. (2022). Environment–conflict nexus in the Sahel region: Lake Chad as a case in point. *Journal of African Conflicts and Peace Studies*, *5*(1). Advance online publication. DOI: 10.54103/2612-6672/19469

Siwila, J., Mwaba, F., Chidumayo, N., & Mubanga, C. (2020). Food and waterborne protozoan parasites: The African perspective. *Food and Waterborne Parasitology*, *20*, e00088. Advance online publication. DOI: 10.1016/j.fawpar.2020.e00088 PMID: 32995582

Srivastava, P., & Yadav, U. (2023). Mode of the mechanism of biogenic nanomaterials involved in the adsorption of pollutants. In Adetunji, C. O., Singh, R. P., Singh, J., & Singh, K. R. B. (Eds.), *Nanobiotechnology for bioremediation* (pp. 273–296). Elsevier., DOI: 10.1016/B978-0-323-91767-4.00009-4

Ssentongo, P., Ssentongo, A., Ba, D., Ericson, J., Na, M., Gao, X., Fronterrè, C., Chinchilli, V., & Schiff, S. (2021). Global, regional and national epidemiology and prevalence of child stunting, wasting and underweight in low- and middle-income countries, 2006–2018. *Scientific Reports*, *11*(1), 5204. Advance online publication. DOI: 10.1038/s41598-021-84302-w PMID: 33664313

The Fund for Peace. (2024). *Country dashboard | Fragile States Index*. Retrieved October 20, 2025, from https://fragilestatesindex.org/country-data/

The Global Economy. (2025). *Chad: Percent agricultural land*. Retrieved September 17, 2025, from https://www.theglobaleconomy.com/Chad/Percent_agricultural_land

Thissen, P. (2019). Legacy institutions and political order in weak states: Evidence from Chad. SSRN *Electronic Journal*. https://doi.org/DOI: 10.2139/ssrn.3630086

Transparency International. (2025). *Chad*. https://www.transparency.org/en/countries/chad

UNICEF. (2023). *Chad: Humanitarian situation report*. https://www.unicef.org/chad

United Nations Environment Programme. (n.d.). *Chad | Interactive Country Fiches*. https://dicf-unepgrid-ch.translate.goog/chad?_x_tr_sl=en&_x_tr_tl=tr&_x_tr_hl=tr&_x_tr_pto=tc

Wang, L., Zhang, F., Shi, X., Zeng, C., Ahmad, I., Wang, G., Thapa, S., & Xu, X. (2023). Water resources system vulnerability in high mountain areas under climate change. *Journal of Cleaner Production, 403*, 136789. DOI: 10.1016/j.jclepro.2023.136789

Watts, N., Amann, M., Arnell, N., Ayeb-Karlsson, S., Belesova, K., Berry, H., Bouley, T., Boykoff, M., Byass, P., Cai, W., Campbell-Lendrum, D., Chambers, J., Daly, M., Dasandi, N., Davies, M., Depoux, A., Dominguez-Salas, P., Drummond, P., Ebi, K. L., & Costello, A. (2018). The 2018 report of the Lancet Countdown on health and climate change: Shaping the health of nations for centuries to come. *Lancet, 392*(10163), 2479–2514. DOI: 10.1016/S0140-6736(18)32594-7 PMID: 30503045

Wisner, B., Blaikie, P., Cannon, T., & Davis, I. (2004). *At risk: Natural hazards, people's vulnerability and disasters* (2nd ed.). Routledge.

World Bank. (2021). *Chad poverty assessment: Investing in rural income growth, human capital, and resilience to support sustainable poverty reduction.* https://documents1.worldbank.org/curated/en/642041633397593711/pdf/Chad-Poverty-Assessment-Investing-in-Rural-Income-Growth-Human-Capital-and-Resilience-to-Support-Sustainable-Poverty-Reduction.pdf

World Bank. (2023a). *Africa Country Policy and Institutional Assessment (CPIA): Assessing Africa's policies and institutions – Country tables (English).* Washington, D.C.: World Bank Group. https://documents.worldbank.org/curated/en/099634410132333843

World Bank. (2023b). *The World Bank in Chad: Overview.* https://www.worldbank.org/en/country/chad/overview

World Bank Group. (2022). *G5 Sahel Region: Country Climate and Development Report.* Washington, D.C.: World Bank Group. https://documents.worldbank.org/curated/en/099150006302219454

World Health Organization. (2023). *Health emergency and disaster risk management framework.* https://apps.who.int/iris/handle/10665/326106

World Health Organization. (2024). *Chad health profile.* https://www.who.int/countries/tcd

World Health Organization. Health Cluster. (2025). *Chad [Data dashboard].* WHO Health Cluster. https://healthcluster.who.int/countries-and-regions/chad

World Meteorogiçal Organization. (2024). *Africa faces disproportionate burden from climate change and adaptation costs.* https://wmo.int/news/media-centre/africa-faces-disproportionate-burden-from-climate-change-and-adaptation-costs https://www.indyturk.com/node/438166/t%C3%BCrki%CC%87yeden-sesler/i%CC%87klim-de%C4%9Fi%C5%9Fikli%C4%9Fi-ve-afrika

Worlddata.info. (2023). *Climate: Chad.* https://www.worlddata.info/africa/chad/climate.php

Xia, J., & Bantin, A. B. (2017). The impact of hydro climatic variability on water resources of Lake Chad hydrographic basin. *Hydrology: Current Research*, *8*(3), 1–10. DOI: 10.4172/2157-7587.1000282

Compilation of References

Abdelkrim, Z., & Eslamian, S. (2026). An evaluation of spatial and temporal variations of land surface temperature in Biskra, Algeria using remote sensing and GIS. *Journal of African Earth Sciences*, *235*, 105942. DOI: 10.1016/j.jafrearsci.2025.105942

Acharya, A. (2009). *Constructing a security community in Southeast Asia: ASEAN and the problem of regional order*. Routledge., DOI: 10.4324/9780203939239

Adams, E., Thill, A., Kuusaana, E., & Mittag, A. (2023). Farmer–herder conflicts in sub-Saharan Africa: Drivers, impacts, and resolution and peacebuilding strategies. *Environmental Research Letters*, *18*(12), 123001. Advance online publication. DOI: 10.1088/1748-9326/ad0702

Adger, W., Crépin, A., Folke, C., Ospina, D., Chapin, F. S.III, Segerson, K., & Wilen, J. (2020). Urbanization, migration, and adaptation to climate change. *One Earth*, *3*(4), 396–399. DOI: 10.1016/j.oneear.2020.09.016

Adhikari, U., Nejadhashemi, A. P., & Woznicki, S. A. (2015). Climate Change and Eastern Africa: A Review of Impact on Major Crops. *Food and Energy Security*, *4*(2), 110–132. DOI: 10.1002/fes3.61

Adıgüzel, Y. (2024). *Göç sosyolojisi*. Nobel Akademik Yayıncılık. https://www.kitapyurdu.com/kitap/goc-sosyolojisi/407575.html

ADPC. (2019). *Disaster risk reduction in Cambodia*. Bangkok: Asian Disaster Preparedness Center. https://www.preventionweb.net/files/68249_682301cambodiaupdaed16oct2019.pdf

Aghazarm, C., & Laczko, F. (2009). *Migration, environment and climate change: Assessing the evidence Geneva*. International Organization for Migration (IOM), Institute for Environment and Human Security and United Nations University.

Ajaero, C. K. (2017). A gender perspective on the impact of flood on the food security of households in rural communities of Anambra state, Nigeria. *Food Security*, *9*(4), 685–695. DOI: 10.1007/s12571-017-0695-x

Ajibade, I., McBean, G., & Bezner-Kerr, R. (2013). Urban flooding in Lagos, Nigeria: Patterns of vulnerability and resilience among women. *Global Environmental Change*, *23*(6), 1714–1725. DOI: 10.1016/j.gloenvcha.2013.08.009

Akiner, M. E., Akiner, N., & Akiner, İ. (2025a). From Historical Management to Modern Sustainability: A Deep Dive Into Antalya's Environmental Evolution and Climate Resilience. In Community Climate Justice and Sustainable Development (pp. 181-230). IGI Global Scientific Publishing. DOI: 10.4018/979-8-3373-0619-3.ch009

Akiner, M. E. (2025). Integrated Modeling Approach to Predict the Morphodynamic Impacts of the Boğaçay Project on Konyaaltı Beach. International Journal of Environment. *Engineering and Education*, *7*(2), 97–108. DOI: 10.55151/ijeedu.v7i2.215

Akıner, M. E., Akıner, N., & Akıner, İ. (2025b). Interplay Between Climate Change and The Migration-Induced Population Growth: A Case Study in Konyaaltı, Antalya, Turkey. *Sürdürülebilir Çevre Dergisi*, *5*(1), 1–18. DOI: 10.62816/cevder.1581432

Akiner, M. E., & Lačnjevac, Č. (2022). Managing Saltwater Intrusion and Agricultural Practices along the Boğaçay River, Turkey: Effects from Excavation and Land Source Pollution. *Journal of Coastal Research*, *38*(3), 567–577. DOI: 10.2112/JCOASTRES-D-21-00101.1

Aktaş, Ö. (2014). Impacts of climate change on water resources in Turkey. *Environmental Engineering and Management Journal*, *13*(4), 881–889. DOI: 10.30638/eemj.2014.092

Aldrich, D. P. (2012). *Building resilience: Social capital in post-disaster recovery*. University of Chicago Press. DOI: 10.7208/chicago/9780226012896.001.0001

Alexander, R. (2023). A Critical Introduction to gender and disaster: Learning from women survivors in northeast Japan. (19-40). Editor: Ronni Alexander & Siriporn Wajjwalku, *Making Disaster: SaferA Gender and Vulnerability Approach,* Singapore: Springer Pte Ltd.

Alexander, D. (2002). *Confronting catastrophe: New perspectives on natural disasters*. Oxford University Press.

Alexander, D. (2020). Disaster risk reduction. In Devuyst, Y. (Ed.), *The European Union's community method: Foundations and evolutionlocked*. Oxford University Press., DOI: 10.1093/acrefore/9780190228637.013.1538

Alexander, D. E. (2014). Social media in disaster risk reduction and crisis management. *Science and Engineering Ethics*, *20*(3), 717–733. DOI: 10.1007/s11948-013-9502-z PMID: 24306994

Alkhammash, R. (2023). Bibliometric, network, and thematic mapping analyses of metaphor and discourse in COVID-19 publications from 2020 to 2022. *Frontiers in Psychology*, *13*, 1062943. DOI: 10.3389/fpsyg.2022.1062943

Allarané, N., Azagoun, V. V. A., Atchadé, A. J., Hetcheli, F., & Atela, J. (2023). Urban vulnerability and adaptation strategies against recurrent climate risks in Central Africa: Evidence from N'Djaména City (Chad). *Urban Science (Basel, Switzerland)*, *7*(3), 97. DOI: 10.3390/urbansci7030097

Almagro, A., Thomé, T. C., Colman, C. B., Pereira, R. B., Junior, J. M., Rodrigues, D. B. B., & Oliveira, P. T. S. (2019). Improving Cover and Management Factor (C-factor) Estimation Using remote Sensing Approaches for Tropical Regions. *International Soil and Water Conservation Research*, *7*(4), 325–334. Advance online publication. DOI: 10.1016/j.iswcr.2019.08.005

Almutairi, A., Mourshed, M., & Ameen, R. F. M. (2020). Coastal community resilience frameworks for disaster risk management. *Natural Hazards*, *101*(2), 595–630. DOI: 10.1007/s11069-020-03875-3

Alongi, D. M. (2008). Mangrove forests: Resilience, protection from tsunamis, and responses to global climate change. *Estuarine, Coastal and Shelf Science*, *76*(1), 1–13. DOI: 10.1016/j.ecss.2007.08.024

Alonso, E. E. (2021). Triggering and motion of landslides. *Geotechnique*, *71*(1), 3–59. DOI: 10.1680/jgeot.20.RL.001

Altan, C. (2024). *Critical evaluation of climate action plans in Türkiye from the perspective of climate justice* (Master's thesis). Middle East Technical University, Ankara, Turkey.

Altıok, B., Arıcı, A., & Özüçelik, D. N. (2025). 6 Şubat 2023 Kahramanmaraş depremlerinin kadınlar ve çocuklar üzerindeki etkileri ve sosyal hizmet müdahaleleri. *Socrates Journal of Interdisciplinary Social Researches*, *11*(52). Advance online publication. DOI: 10.5281/zenodo.15244885

Altıparmak, İ. B., & Birel, E. (2024). 6 Şubat 2023 depremi sonrasında sosyal yaşamın toplumsal cinsiyet bağlamında değerlendirilmesi: Nurdağı örneği, *Hacettepe Üniversitesi. Edebiyat Fakültesi Dergisi*, *41*(1), 113–134. DOI: 10.32600/huefd.1304991

Alvarez, P. J., Chan, C. K., Elimelech, M., Halas, N. J., & Villagrán, D. (2018). Emerging opportunities for nanotechnology to enhance water security. *Nature Nanotechnology*, *13*(8), 634–641. DOI: 10.1038/s41565-018-0203-2

Ambadi, D. (2022). The roughage problem and solutions in animal feeding in Chad. *6th International Students Science Congress Proceedings Book*. https://doi.org/DOI: 10.52460/issc.2022.003

Amir Ahmadian, B. and Naseri, M. (2013). "Water crisis in central asia with emphasis on regional conflict over water issue." Journal of central eurasia studies, 6(12), 1-20.

Amnesty International. (2011). *Aftershocks women speak out against sexual violence in Haiti's camps*. Amnesty International Ltd.

Amsalu, A., Garedew, W., Melka, G. A., Çelik, M. A., & Gemeda, D. O. (2026). Trends in climate extreme indices and implications for coffee farming in The Kafa Biosphere Reserve, southwest Ethiopia. *Climate Services*, *42*, 100637. DOI: 10.1016/j.cliser.2026.100637

Andrew, M. (2018). Revisiting community-based disaster risk management. In I. Davis, S. Bender, F. Krimgold, & F. McDonald (Eds.), *Reducing disaster risks: Progress and challenges in the caribbean region* (pp. 42–52). Routledge. DOI: 10.4324/9781315067261-5

Anggraini, E., Yuliastuti, N., & Budihardjo, M. A. (2024, April). Analysis of Water Conservation Priority Locations in Upper Sub Watershed. []. IOP Publishing.]. *IOP Conference Series. Earth and Environmental Science*, *1323*(1), 012001. DOI: 10.1088/1755-1315/1323/1/012001

Anwar, N., Rahaman, R., Widodo, A. M., Sekti, B. A., Erzed, N., Santiago Tangkudung, R. R., Muchlis, M., Yuhefizar, Y., & Budhisantosa, N. (2025). Sustainable cybersecurity in the AI era: Innovations in green computing and security. In *Sustainable Information Security in the Age of AI and Green Computing* (pp. 603–620). DOI: 10.4018/979-8-3693-8034-5.ch030

Arabi, M., Frankenberger, J. R., Engel, B. A., & Arnold, J. G. (2008). Representation of agricultural conservation practices with SWAT. *Hydrological Processes*, *22*(16), 3042–3055. DOI: 10.1002/hyp.6890

Arbuckle, J. G.Jr, Morton, L. W., & Hobbs, J. (2021). Understanding Farmer Perspectives on Climate Change Adaptation and Mitigation. *Environment and Behavior*, *47*(2), 205–234. DOI: 10.1177/0013916513503832 PMID: 25983336

Arcaya, M., Raker, E. J., & Waters, M. C. (2020). The socialconsequences of disasters: Individual and communitychange. *Annual Review of Sociology, 46*(1), 671–691. DOI: 10.1146/annurev-soc-121919-054827

Arfasa, G. F., Owusu-Sekyere, E., Doke, D. A., & Ampofo, J. A. (2024). Impacts of climate and land use/cover changes on the sustainability of irrigation water in West Africa: A systematic review. *All Earth.* https://doi.org/DOI: 10.1080/27669645.2024.2308371

Arı Kovancı, Y. (2024). Afet ve kırılgan gruplar. *Siirt Üniversitesi Sosyal Bilimler Enstitüsü Dergisi, 12*(1), 13–30. DOI: 10.53586/susbid.1474653

Aria, M., & Cuccurullo, C. (2017). bibliometrix: An R-tool for comprehensive science mapping analysis. *Journal of Informetrics, 11*(4), 959–975. DOI: 10.1016/j.joi.2017.08.007

Arimoro, A. E., & Arinze-Umobi, C. (2026). Sustainable public-private partnerships in Sub-Saharan Africa: A conceptual framework for low carbon development and domestic financing. *Public Works Management & Policy, 31*(1), 40–76. DOI: 10.1177/1087724X251356462

Arora, S. (2022). Intersectional vulnerability in postdisaster contexts: Lived experiences of Dalit women after the Nepal earthquake, 2015. *Disasters, 46*(2), 329–347. DOI: 10.1111/disa.12471 PMID: 33314308

Arpacı, M. (2013). Modernitenin eşiğinde toplumsal cinsiyet rejimi: Pastoral iktidar, beden politikaları ve evlilik, *Doğu-Batı Düşünce Dergisi*, Yıl:16. *Sayı, 63*, 131–146.

Arslan, E. S., & Kaymaz, I. (2020). Visitor perception of recreational ecosystem services and their role in landscape management of Gölcük Nature Park, Turkey. *International Journal of Sustainable Development and World Ecology, 27*(3), 202–213. DOI: 10.1080/13504509.2019.1711247

Asio, J. (2021). *Disaster awareness and level of compliance to disaster programs in a highly urbanized city*. Aquademia., DOI: 10.21601/aquademia/9581

Aslan, A. (2016). Does tourism cause growth? Evidence from Turkey. *Current Issues in Tourism, 19*(12), 1176–1184. DOI: 10.1080/13683500.2015.1015970

Aslan, R., & Şahinöz, S. (2023). The experiences of people with disabilities in the 2020 Izmir earthquake: A phenomenological research. *International Journal of Disaster Risk Reduction, 95*, 103868. DOI: 10.1016/j.ijdrr.2023.103868

Ataol, M., Kale, M. M., & Tekkanat, İ. S. (2019). Assessment of the changes in shoreline using digital shoreline analysis system: A case study of Kızılırmak Delta in northern Turkey from 1951 to 2017. *Environmental Earth Sciences*, *78*(19), 579. DOI: 10.1007/s12665-019-8591-7

Atmiş, E. (2018). A critical review of the (potentially) negative impacts of current protected area policies on the nature conservation of forests in Turkey. *Land Use Policy*, *70*, 675–684. DOI: 10.1016/j.landusepol.2017.10.054

Aven, T. (2020). Climate change risk – what is it and how should it be expressed? *Journal of Risk Research*, *23*(11), 1387–1404. DOI: 10.1080/13669877.2019.1687578

Aversa, S., Cascini, L., Picarelli, L., & Scavia, C. (Eds.). (2018). *Landslides and Engineered Slopes. Experience, Theory and Practice* (1st ed.). CRC Press., DOI: 10.1201/9781315375007

Avrupa Konseyi (2023). Deprem bölgelerinde toplumsal cinsiyet analizi: Kadınların adalete ve adli yardıma erişimi, (Haz: Prof. Dr. Gülriz Uygur), Aralık, 2023

Ayaz, A. (2022). Unsustainable systems. *Insight Turkey*, *24*(1), 37–54. DOI: 10.25253/99.2022241.3

Aygün, A. (2015). *Climate change and urban resilience: Vulnerability and risk assessment for Istanbul* (Unpublished master's thesis). Istanbul Technical University, Department of Urban and Regional Planning.

Baig, M. N., & Sharif, R. (2013). Gender perspective considerations in disasters like earthquakes and floods of Pakistan. *International Journal of Business, Human and Social Sciences, Vol*, *7*(6), 1487–1490.

Bailey, L. D., & van de Pol, M. (2023). Tackling extremes: Challenges for ecological and evolutionary research on extreme climatic events. *Journal of Animal Ecology*, *85*(1), 85–96. DOI: 10.1111/1365-2656.12451 PMID: 26433114

Bakic, H., & Ajdukovic, D. (2021). Resilienceafternaturaldisasters: The process of harnessingresources in communitiesdifferentiallyexposed to a flood. *European Journal of Psychotraumatology*, *12*(1), 1891733. DOI: 10.1080/20008198.2021.1891733 PMID: 34992751

Baldassarre, L., Natali, V., De Pascale, F., Vezzi, A., Banchi, E., Bazzaro, M., Relitti, F., Tagliapietra, D., & Cibic, T. (2023). The impact of MOSE (Experimental electromechanical module) flood barriers on microphytobenthic community of the Venice Lagoon. *Microorganisms*, *11*(4), 936. DOI: 10.3390/microorganisms11040936 PMID: 37110359

Balis, B., Bubak, M., Harezlak, D., Nowakowski, P., Pawlik, M., & Wilk, B. (2017). Towards an operational database for real-time environmental monitoring and early warning systems. *Procedia Computer Science, 108*, 2250–2259. DOI: 10.1016/j.procs.2017.05.193

Bantin, A., Jun, X., & Si, H. (2017). The impact of climate change on water resource of agricultural landscape and its adaptation strategies: A case study of Chari Basin, Chad. *Journal of Earth Science & Climatic Change, 8*(12), 1–7. DOI: 10.4172/2157-7617.1000437

Barbier, E. B., Hacker, S. D., Kennedy, C., Koch, E. W., Stier, A. C., & Silliman, B. R. (2011). The value of estuarine and coastal ecosystem services. *Ecological Monographs, 81*(2), 169–193. DOI: 10.1890/10-1510.1

Basher, R. (2006). Global early warning systems for natural hazards: Systematic and people-centred. *Philosophical Transactions of the Royal Society A: Mathematical, Physical and Engineering Sciences, 364*(1845), 2167–2182. https://doi.org/DOI: 10.1098/rsta.2006.1819

Başkan, B. (2024). Deprem, toplumsal cinsiyet ve depremzede kadınlar. İ. Erdoğan Tarakçı (Ed.), *Disiplinlerarası Yaklaşımlarla Kadın Çalışmaları 3* (ss. 51-70) içinde. İstanbul: Efe Akademi Yayınları.

Batiuk, R. A., Blankenship, K., Boesch, D. F., Dennison, W. C., Harrison, V., Hershner, C., Hoagland, R. A., Horton, T. W., Matuszeski, W., Miller, T. J., Swanson, A. P., Todd, A. H., & Wardrop, D. H. (2025). Restoration of the Chesapeake Bay: Lessons for other ecosystems. *Frontiers in Marine Science, 12*, 1581261. DOI: 10.3389/fmars.2025.1581261

Beck, U., Giddens, A., & Lash, S. (1994). *Reflexive modernization: Politics, tradition and aesthetics in the modern social order.* https://doi.org/DOI: 10.2307/3341775

Behrman, J. A., & Weitzman, A. (2016). Effects of the 2010 Haiti earthquake on women's reproductive health. *Studies in Family Planning, 47*(1), 3–17. https://www.jstor.org/stable/24720394. DOI: 10.1111/j.1728-4465.2016.00045.x PMID: 27027990

Benavidez, R., Jackson, B., Maxwell, D., & Norton, K. (2018). A Review of the (Revised) Universal Soil Loss Equation (R/USLE) in the Context of Land Use and Management Changes. *Hydrology and Earth System Sciences, 22*(11), 4001–4025. https://doi.org/https://doi.org/10.5194/hess-22-6059-2018

Benevolenza, M. A., & DeRigne, L. (2019). The impact of climate change and natural disasters on vulnerable populations: A systematic review of literature. *Journal of Human Behavior in the Social Environment, 29*(2), 266–281. DOI: 10.1080/10911359.2018.1527739

Bennett, E. M., Peterson, G. D., & Gordon, L. J. (2009). Understanding relationships among multiple ecosystem services. *Ecology Letters, 12*(12), 1394–1404. DOI: 10.1111/j.1461-0248.2009.01387.x

Bezak, N., Mikoš, M., Borrelli, P., Alewell, C., Alvarez, P., Anache, J. A. A., Baartman, J., Ballabio, C., Biddoccu, M., Cerdà, A., Chalise, D., Chen, S., Chen, W., De Girolamo, A. M., Gessesse, G. D., Deumlich, D., Diodato, N., Efthimiou, N., Erpul, G., & Panagos, P. (2021). Soil Erosion Modelling: A Bibliometric Analysis. *Environmental Research, 197*(March), 111087. Advance online publication. DOI: 10.1016/j.envres.2021.111087 PMID: 33798514

Bhuiya, M. M. R., Shao, W., & Jones, S. (2023). Measuring accessibility of movement challenged persons during earthquake evacuation of Dhaka City, Bangladesh: A participatory GIS-Based approach. *Applied Geography (Sevenoaks, England), 156*, 1–16.

Biehl Öztuzcu, K., & Aslan, Ö. (2022). *Migration and environment from a gender perspective: Examples from civil society in Turkey.*

Bilham, R. (2010). Lessons from the Haiti earthquake. *Nature, 463*(7283), 878–879. DOI: 10.1038/463878a PMID: 20164905

Biricik, A. S. (2001). YERYUVARI'NDA DOĞAL OLAYLAR VE AFETLER. *Marmara Coğrafya Dergisi*, (3), 7–26.

Birkland, T. A. (2006). *Lessons of disaster: Policy change after catastrophic events.* Georgetown University Press.

Birpınar, M. E., & Tuğaç, Ç. (2022). Climate security and migration: An evaluation for the world and Turkey. *Insight Turkey, 24*(1), 105–133. DOI: 10.25253/99.2022241.7

Black, R., Adger, W. N., Arnell, N. W., Dercon, S., Geddes, A., & Thomas, D. (2011). The effect of environmental change on human migration. *Global Environmental Change, 21*(S1), S3–S11. DOI: 10.1016/j.gloenvcha.2011.10.001

Black, R., & Collyer, M. (2014). Populations 'trapped' at times of crisis. *Forced Migration Review, 45*, 52–56.

Blaikie, P., Cannon, T., Davis, I., & Wisner, B. (2014). *At risk: natural hazards, people's vulnerability and disasters.* Routledge. DOI: 10.4324/9780203714775

Bogardi, J. J., Dudgeon, D., Lawford, R., Flinkerbusch, E., Meyn, A., Pahl-Wostl, C., Vielhauer, K., & Vörösmarty, C. (2012). Water security for a planet under pressure: Interconnected challenges of a changing world call for sustainable solutions. *Current Opinion in Environmental Sustainability, 4*(1), 35–43. DOI: 10.1016/j.cosust.2011.12.002

Bolan, S., Padhye, L. P., Jasemizad, T., Govarthanan, M., Karmegam, N., Wijesekara, H., Amarasiri, D., Hou, D., Zhou, P., Biswal, B. K., Balasubramanian, R., Wang, H., Siddique, K. H. M., Rinklebe, J., Kirkham, M. B., & Bolan, N. (2024). Impacts of climate change on the fate of contaminants through extreme weather events. *Science of The Total Environment, 909*(August 2023), 168388. DOI: 10.1016/j.scitotenv.2023.168388

Borrelli, P., Alewell, C., Alvarez, P., Anache, J. A. A., Baartman, J., Ballabio, C., Bezak, N., Biddoccu, M., Cerdà, A., Chalise, D., Chen, S., Chen, W., De Girolamo, A. M., Gessesse, G. D., Deumlich, D., Diodato, N., Efthimiou, N., Erpul, G., Fiener, P., & Panagos, P. (2021). Soil Erosion Modelling: A Global Review and Statistical Analysis. *The Science of the Total Environment, 780*, 146494. Advance online publication. DOI: 10.1016/j.scitotenv.2021.146494 PMID: 33773346

Borrelli, P., Robinson, D. A., Fleischer, L. R., Lugato, E., Ballabio, C., Alewell, C., Meusburger, K., Modugno, S., Schütt, B., Ferro, V., Bagarello, V., Van Oost, K., Montanarella, L., & Panagos, P. (2017). An assessment of the global impact of 21st century land use change on soil erosion. *Nature Communications, 8*(1), 2013. Advance online publication. DOI: 10.1038/s41467-017-02142-7 PMID: 29222506

Borrelli, P., Robinson, D. A., Panagos, P., Lugato, E., Yang, J. E., Alewell, C., Wuepper, D., Montanarella, L., & Ballabio, C. (2020). Land use and climate change impacts on global soil erosion by water (2015-2070). *Proceedings of the National Academy of Sciences of the United States of America, 117*(36), 21994–22001. DOI: 10.1073/pnas.2001403117 PMID: 32839306

Bottelin, P., Jongmans, D., Daudon, D., Mathy, A., Helmstetter, A., Bonilla-Sierra, V., Cadet, H., Amitrano, D., Richefeu, V., Lorier, L., Baillet, L., Villard, P., & Donzé, F. (2014). Seismic and mechanical studies of the artificially triggered rockfall at Mount Néron (French Alps, December 2011). *Natural Hazards and Earth System Sciences, 14*(12), 3175–3193. DOI: 10.5194/nhess-14-3175-2014

Boudouresque, C. F., Blanfuné, A., Pergent, G., & Thibaut, T. (2021). Restoration of seagrass meadows in the Mediterranean Sea: A critical review of effectiveness and ethical issues. *Water (Basel), 13*(8), 1034. DOI: 10.3390/w13081034

Boughton, J. M. (2002). Why white, not Keynes? Inventing the postwar international monetary system. *IMF working Paper.* https://www.imf.org/external/pubs/ft/wp/2002/wp0252.pdf

Bournez, P., Salzinger, J., Cella, M., Vultaggio, F., d'Apolito, F., & Fanta-Jende, P. (2024). Pushing the limit to near real-time indoor LiDAR-based semantic segmentation. *International Archives of the Photogrammetry, Remote Sensing and Spatial Information Sciences - ISPRS Archives, 48*(2/W8-2024), 45–52. DOI: 10.5194/isprs-archives-XLVIII-2-W8-2024-45-2024

Boustras, G., & Boukas, N. (2013). Forest fires' impact on tourism development: A comparative study of Greece and Cyprus. *Management of Environmental Quality, 24*(4), 498–511. DOI: 10.1108/MEQ-09-2012-0058

Bozzolan, E., Holcombe, E., Pianosi, F., & Wagener, T. (2020). Including informal housing in slope stability analysis – an application to a data-scarce location in the humid tropics. *Natural Hazards and Earth System Sciences, 20*(11), 3161–3177. DOI: 10.5194/nhess-20-3161-2020

Bradley, T., Martin, Z., Upreti, B. R., Subedu, B., & Shrestha, S. (2023). Gender and disaster: The impact of natural disasters on violence against women in Nepal. *Journal of Asian and African Studies, 58*(3), 354–371. DOI: 10.1177/00219096211062474

Brezina, C. (2012). *Understanding the gross domestic product and the gross national product*. Rosen Publishing.

Bricker, J., & Pelupessy, D. (2025). The nature of disasters. *Journal of Delta Urbanism.*https://doi.org/DOI: 10.59490/jdu.5.2024.8070

Bronen, R. (2010). Forced migration of Alaskan indigenous communities due to climate change. In Afifi, T., & Jäger, J. (Eds.), *Environment, forced migration and social vulnerability* (pp. 87–98). Springer., DOI: 10.1007/978-3-642-12416-7_7

Bruce, D. G., Cornwell, J. C., Harris, L., Ihde, T. F., Kellogg, M. L., Knoche, S., . . . Vogt, B. (2021). A synopsis of research on the ecosystem services provided by large-scale oyster restoration in the Chesapeake Bay. NOAA Tech. Memo. NMFS-OHC, 8, 52. https://spo.nmfs.noaa.gov/tech-memos/ (Accessed 19 August 2025)

Bruun, P. (1962). Sea-level rise as a cause of shore erosion. *Journal of the Waterways and Harbors Division, 88*(1), 117–132. DOI: 10.1061/JWHEAU.0000252

Bu, J., Wang, Q., Wang, Z., Fan, S., Liu, X., & Zuo, X. (2024). Land Remote Sensing Applications Using Spaceborne GNSS Reflectometry: A Comprehensive Overview. *IEEE Journal of Selected Topics in Applied Earth Observations and Remote Sensing, 17*, 12811–12841. DOI: 10.1109/JSTARS.2024.3415754

Bulut, C., & Yıldız, F. (2024). Bibliometric analysis of laser powder bed additive manufacturing of high-entropy alloys: Where are we and how far have we come? *Progress in Additive Manufacturing*. Advance online publication. DOI: 10.1007/s40964-024-00840-5

Burrough, P. A. (1986). Principles of Geographical. In Information systems for land resource assessment.... https://www.academia.edu/download/2438559/9fjg8q78n4wux4l.pdf

Butler, J. (2005). *Cinsiyet belası: Feminizm ve kimliğin altüst edilmesi*. Metis Yayınları.

Buzan, B. (2008). "People, states & fear: an agenda for international security studies in the post-cold war era." Lynne Rienner Publishers, No. 1, Boulder, Colorado. https://doi.org/DOI: 10.1080/17449050802539344

Caballero-Anthony, M. (2015). *An introduction to non-traditional security studies: A transnational approach*. Sage Publications.

Campbell, D. W., Campbell, J. C., Yarandi, H. N., O'Connor, A. L., Dollar, E., Killion, C., Sloand, E., Callwood, G. B., Cesar, N. M., Hassan, M., & Gary, F. (2016). Violence and abuse of internally displaced women survivors of the 2010 Haiti earthquake. *International Journal of Environmental Research and Public Health*, *61*(8), 981–992. DOI: 10.1007/s00038-016-0895-8 PMID: 27624625

Canadell, J. G., Meyer, C. P. M., Cook, G. D., Dowdy, A., Briggs, P. R., Knauer, J., Pepler, A., & Haverd, V. (2021). Multi-decadal increase of forest burned area in Australia is linked to climate change. *Nature Communications*, *12*(1), 1–11. DOI: 10.1038/s41467-021-27225-4 PMID: 34836974

Cantor, D. J., & Apollo, J. O. (2020). Internal displacement, internal migration, and refugee flows: Connecting the dots. *Refugee Survey Quarterly*, *39*(4), 647–664. DOI: 10.1093/rsq/hdaa016

Castles, S., de Haas, H., & Miller, M. J. (2014). *The age of migration: International population movements in the modern world* (5th ed.). Guilford Press. DOI: 10.1007/978-0-230-36639-8

Çelik, E., & Sarıboğa, L. (2023). Assessment of the role of birds in seed distribution: a bibliometric analysis of studies on this subject. Recent advances in natural and engineering Sciences. Livre de lyon, Lyon, 29-53.

Çelik, M. A. (2020). Kuraklık Araştırmalarında Yeni Eğilimler, Kullanılan Teknikler ve Kavramlar Üzerine Bibliyometrik Ağ Analizi. International Journal of Geography and Geography Education, (42), 602-630.

Çelik, M. A., Elbeltagi, A., Mishra, R., Singh, S., Vishwakarma, D. K., Pandey, K., & Akiner, M. E. (2025). Flood Risk and Inundation Mapping in a Changing Climate. Advancing Environmental Research Through Applied GIS and Remote Sensing, 1–34. DOI: 10.4018/979-8-3373-6608-1.ch001

Çelik, A. K., Özcan, S., Topcuoğlu, A., & Yıldırım, K. E. (2013). Effects of the tourism industry on the balance of payments deficit. *Anatolia*, *24*(1), 86–90. DOI: 10.1080/13032917.2013.772529

Celik, E., Durmus, A., Adizel, O., & Nergiz Uyar, H. (2021). A bibliometric analysis: What do we know about metals (loids) accumulation in wild birds? *Environmental Science and Pollution Research International*, *28*(8), 10302–10334. DOI: 10.1007/s11356-021-12344-8

Çelik, M. A. (2026). Agro-climatic classification of Türkiye's Agricultural Basins using long-term ERA5-Land temperature data. *Physics and Chemistry of the Earth Parts A/B/C*, *142*, 104307. DOI: 10.1016/j.pce.2026.104307

Çelik, M. A., Bayram, H., & Özüpekçe, S. (2018). An assessment on climatological, meteorological and hydrological disasters that occurred in Turkey in the last 30 years (1987-2017). *International Journal of Geography and Geography Education*, (38), 295–310. DOI: 10.32003/iggei.424675

Çelik, M. A., Bilik, A., & Türkeş, M. (2025). Spatio-Temporal Analysis of Observed Drought Events in the Tigris–Euphrates Basin during the 1960–2023 Period Via SPI and SPEI Drought Indices. *Pure and Applied Geophysics*. Advance online publication. DOI: 10.1007/s00024-025-03863-3

Çelik, M. A., & Çelik, E. (2024). Are Urbanisation and Biodiversity Antithetical? A Bibliometric Analysis. *Coğrafya Dergisi*, (48), 121–135.

Çelik, M. A., Kıraç, A., & Gouveia, C. M. (2026). North Atlantic Oscillation modulation of drought and vegetation activity in Türkiye. *Theoretical and Applied Climatology*, *157*(2), 74. Advance online publication. DOI: 10.1007/s00704-025-05971-5

Çelik, M. A., Kizilelma, Y., Gülersoy, A. E., & Denizdurduran, M. (2013). Farkli Uzaktan Algilama Teknikleri Kullanilarak Aşaği Seyhan Ovasi Güneyindeki Sulak Alanlarda Meydana Gelen Değişimin Incelenmesi (1990-2010). *Electronic Turkish Studies*, *8*(12), 263. Advance online publication. DOI: 10.7827/TurkishStudies.6011

Çelik, M. A., Kopar, İ., & Çelik, E. (2021). Doğubayazıt sazlığının (Ağrı-Türkiye) arazi örtüsü deseninde meydana gelen değişimlerin ekolojik sonuçları üzerine bir analiz. *Doğu Coğrafya Dergisi*, *26*(46), 193–210.

Çeliktopuz, E. (2024). Forecasting some climate parameters of Türkiye using the SSP3-7.0 scenario for the years 2040–2059. *International Journal of Agriculture Environment and Food Sciences*, *8*(1), 62–71. DOI: 10.31015/jaefs.2024.1.7

Centre, A. H. A. (2023). *AHA centre annual report 2022*. Jakarta: ASEAN Coordinating Centre for Humanitarian Assistance on disaster management (AHA Centre). https://ahacentre.org/wp-content/uploads/2023/09/AHA-Centre-Annual-Report-2022-FA-Digital.pdf

Cerrada Ayllón, A. M., Cocero Matesanz, D., Cerrillo Cuenca, E., & Jordá Pardo, J. F. (2023). SPATIAL STUDY BY GEOGRAPHICAL INFORMATION SYSTEMS OF THE MEGALITHS IN THE REGION TIERRA DE LEDESMA (SALAMANCA, SPAIN). *GeoFocus*, *2023*(32), 5–26. DOI: 10.21138/GF.826

Chaudhary, R., Sharma, B. D., Shrestha, S., Gautam, S., & Dhital, R. (2024). Impact of improper shelters on women on the aftermath of Jajarkot earthquake in West Rukum, Nepal. *Journal of Environmental Sciences (China)*, *10*(1), 154–162. DOI: 10.3126/jes.v10i1.67033

Chen, C. (2017). Science mapping: A systematic review of the literature. *Journal of Data and Information Science*, *2*(2), 1–40. DOI: 10.1515/jdis-2017-0006

Cheng, C. S., Auld, H., Li, Q., & Li, G. (2023). Possible impacts of climate change on extreme weather events at local scale in south–central Canada. *Climatic Change*, *112*(3–4), 963–979. DOI: 10.1007/s10584-011-0252-0

Chiu, W. T., & Ho, Y. S. (2007). Bibliometric analysis of tsunami research. *Scientometrics*, *73*(1), 3–17. DOI: 10.1007/s11192-005-1523-1

Chmutina, K., Von Meding, J., Sandoval, V., Boyland, M., Forino, G., Cheek, W., Williams, D. A., Gonzalez-Muzzio, C., Tomassi, I., Páez, H., & Marchezini, V. (2021). What we measure matters: The case of the missing development data in Sendai framework for disaster risk reduction monitoring. *International Journal of Disaster Risk Science*, *12*(6), 779–789. DOI: 10.1007/s13753-021-00382-2

Cho, D. I., & Ogwang, T. (2014). "Water Poverty Index." Springer, No. 1, Dordrecht, Netherlands. DOI: 10.1007/978-94-007-0753-5_3703

Chopra, A. K., & Chakrabarti, P. (1972). The earthquake experience at Koyna dam and stresses in concrete gravity dams. *Earthquake Engineering & Structural Dynamics*, *1*(2), 151–164. DOI: 10.1002/eqe.4290010204

Chowdhury, J. R., Parida, Y., & Agarwal, P. (2022). How flood affects rural employment in India: A gender analysis. *International Journal of Disaster Risk Reduction*, *73*, 102881. Advance online publication. DOI: 10.1016/j.ijdrr.2022.102881

Christian Aid. (2007). *Human tide: The real migration crisis.* https://apo.org.au/node/9086

Churi, A. J., Mlozi, M. R. S., Tumbo, S. D., & Casmir, R. (2022). Understanding Farmers Information Communication Strategies for Managing Climate Risks in Rural Semi-Arid Areas, Tanzania. *International Journal of Information and Communication Technology Research, 2*(11), 838–845. http://www.esjournals.org

Cinto Mejía, E., & Wetzel, W. C. (2023). The ecological consequences of the timing of extreme climate events. *Ecology and Evolution, 13*(1), 1–10. DOI: 10.1002/ece3.9661 PMID: 36713483

Çıtak, G., & Toker, S. (2025). Exploring gender-based challenges after the earthquake: A phenomenological study of women in Turkey. *BMC Public Health, 25*(1), 3555. DOI: 10.1186/s12889-025-23907-6 PMID: 41121101

Clarke, B., Otto, F., Stuart-Smith, R., & Harrington, L. (2022). Extreme weather impacts of climate change: An attribution perspective. *Environmental Research: Climate, 1*(1), 012001. DOI: 10.1088/2752-5295/ac6e7d

Climate Change Performance Index (CCPI). (2023). *Climate Change Performance Index 2023.* Germanwatch/NewClimate Institute/Climate Action Network. https://ccpi.org/download/climate-change-performance-index-2023/

Coelho, C., Lima, M., Alves, F., Roebeling, P., Pais-Barbosa, J., & Marto, M. (2023). Assessing Coastal Erosion and Climate Change Adaptation Measures: A Novel Participatory Approach. *Environments (Basel, Switzerland), 10*(7), 110. DOI: 10.3390/environments10070110

Cohen, R. (2007). Response to Hathaway. *Journal of Refugee Studies, 20*(3), 370–376. Advance online publication. DOI: 10.1093/jrs/fem020

Congleton, R. D. (2006). The story of Katrina: New Orleans and the political economy of catastrophe. *Public Choice, 127*(1-2), 5–30. DOI: 10.1007/s11127-006-7729-9

Cook, C. (2016). Implementing drinking water security: The limits of source protection. *WIREs. Water, 3*(1), 5–12. DOI: 10.1002/wat2.1117

Cook, C., & Bakker, K. (2012). Water security: Debating an emerging paradigm. *Global Environmental Change, 22*(1), 94–102. DOI: 10.1016/j.gloenvcha.2011.10.011

Coppola, D. P. (2020). *Introduction to international disaster management* (4th ed.). Elsevier., https://shop.elsevier.com/books/introduction-to-international-disaster-management/coppola/978-0-12-817368-8

Cortes-Jimenez, I., & Pulina, M. (2010). Inbound tourism and long run economic growth. *Current Issues in Tourism, 13*(1), 61–74. DOI: 10.1080/13683500802684411

Costanza, R., d'Arge, R., de Groot, R., Farber, S., Grasso, M., Hannon, B., Limburg, K., Naeem, S., O'Neill, R. V., Paruelo, J., Raskin, R. G., Sutton, P., & van den Belt, M. (1997). The value of the world's ecosystem services and natural capital. *Nature, 387*(6630), 253–260. DOI: 10.1038/387253a0

Costanza, R., Perez-Maqueo, O., Martinez, M. L., Sutton, P., Anderson, S. J., & Mulder, K. (2008). The value of coastal wetlands for hurricane protection. *Ambio, 37*(4), 241–248. DOI: 10.1579/0044-7447(2008)37[241:TVOCWF]2.0.CO;2

Cowell, E. B. (1978). Ecological effects of dispersants in the United Kingdom. In *Chemical Dispersants for the Control of Oil Spills*. ASTM International. DOI: 10.1520/STP35727S

Cowell, P. J., Stive, M. J., Niedoroda, A. W., Swift, D. J., de Vriend, H. J., Buijsman, M. C., & de Boer, P. L. (2003). The coastal-tract (part 2): Applications of aggregated modeling of lower-order coastal change. *Journal of Coastal Research*, •••, 828–848. Retrieved August 19, 2025, from https://www.jstor.org/stable/4299223

Cremen, G., Bozzoni, F., Pistorio, S., & Galasso, C. (2021). Developing a risk-informed decision-support system for earthquake early warning at a critical seaport. *Reliability Engineering & System Safety, 218*, 108035. DOI: 10.1016/j.ress.2021.108035

Cruz, A. M. D., Maniquiz-Redillas, M. C., Tanhueco, R. M., & De Leon, M. P. (2025). Estimation of the Rainfall Erosivity Factor (R-Factor) for Application in Soil Loss Models. *Water (Basel), 17*(6), 837. https://doi.org/https://doi.org/10.3390/w17060837. DOI: 10.3390/w17060837

Cüce, H. (2023). Climate Change-related natural disasters: Environmental effects and sustainable monitoring. [IJENT]. *International Journal of Environmental Trends, 7*(2), 113–125.

Cunha, A. P. M., Buermann, W., & Marengo, J. A. (2024). Changes in compound drought-heat events over Brazil's Pantanal wetland: An assessment using remote sensing data and multiple drought indicators. *Climate Dynamics, 62*(1), 739–757. DOI: 10.1007/s00382-023-06937-x

Curtis, S., Fair, A., Wistow, J., Val, D. V., & Oven, K. (2017). Impact of extreme weather events and climate change for health and social care systems. *Environmental Health : A Global Access Science Source, 16*(S1), 128. DOI: 10.1186/s12940-017-0324-3 PMID: 29219105

Cutter, S. L. (1995). The forgotten casualties: Women, children, and environmental change. *Global Environmental Change*, *5*(3), 181–194. DOI: 10.1016/0959-3780(95)00046-Q PMID: 12320249

Cutter, S. L. (2017). The forgotten casualties redux: Women, children, and disaster risk. *Global Environmental Change*, *42*, 117–121. DOI: 10.1016/j.gloenvcha.2016.12.010

Cutter, S. L., Boruff, B. J., & Shirley, W. L. (2003). Social vulnerability to environmental hazards. *Social Science Quarterly*, *84*(2), 242–261. DOI: 10.1111/1540-6237.8402002

Cutter, S. L., Burton, C. G., & Emrich, C. T. (2014). Disaster resilience indicators for benchmarking baseline conditions. *Journal of Homeland Security and Emergency Management*, *7*(1), 1–22. DOI: 10.2202/1547-7355.1732

Damm, B., & Klose, M. (2015). The landslide database for Germany: Closing the gap at national level. *Geomorphology*, *249*, 82–93. DOI: 10.1016/j.geomorph.2015.03.021

Danielsson, E., & Eriksson, K. (2022). Women's invisible work in disaster contexts: Gender norms in speech on women's work after a forest fire in Sweden. *Disasters*, *46*(1), 141–161. DOI: 10.1111/disa.12464 PMID: 32895981

Danila, M. (2023). Mainstreaming gender into disaster recovery policy and practice: The Case Study of post-earthquake in Yogyakarta special province, Indonesia. (41-60). Editor: Ronni Alexander & Siriporn Wajjwalku, *Making Disaster: SaferA Gender and Vulnerability Approach,* Singapore: Springer Pte Ltd.

Dann, G. M. S. (1977). Anomie, ego-enhancement and tourism. *Annals of Tourism Research*, *4*(4), 184–194. DOI: 10.1016/0160-7383(77)90037-8

Das, A., & Swain, P. K. (2024). Navigating the sea level rise: Exploring the interplay of climate change, sea level rise, and coastal communities in India. *Environmental Monitoring and Assessment*, *196*(11), 1010. DOI: 10.1007/s10661-024-13191-z PMID: 39361047

Das, S., & Vincent, J. R. (2009). Mangroves protected villages and reduced death toll during Indian super cyclone. *Proceedings of the National Academy of Sciences of the United States of America*, *106*(18), 7357–7360. DOI: 10.1073/pnas.0810440106

Davidson, N. C., Van Dam, A. A., Finlayson, C. M., & McInnes, R. J. (2019). Worth of wetlands: Revised global monetary values of coastal and inland wetland ecosystem services. *Marine and Freshwater Research*, *70*(8), 1189–1194. DOI: 10.1071/MF18391

Davis, D., Allen, J., & Cosenza, R. M. (1988). Segmenting local residents by their attitudes, interests, and opinions toward tourism. *Journal of Travel Research, 27*(2), 2–8. DOI: 10.1177/004728758802700201

De Benedetti, L., & Marzano, M. (2024). Coastal Protection in the Mediterranean Sea: An Overview of International and European Coastal Defense Regulations. In Maccarrone, V., & Fadzil Akhir, M. (Eds.), *Coastal Sustainability. Coastal Research Library* (Vol. 39, pp. 273–296). Springer., DOI: 10.1007/978-3-031-75749-5_13

De Feudis, M., Trenti, W., & Manfredi, P.. (2024). Negative impact of alluvial sediments on physical properties of agricultural soils affected by flooding in May 2023 in Emilia Romagna Region (Northern Italy). *Journal of Soils and Sediments*. Advance online publication. DOI: 10.1007/s11368-024-03931-6

De Schipper, M. A., Ludka, B. C., Raubenheimer, B., Luijendijk, A. P., & Schlacher, T. A. (2021). Beach nourishment has complex implications for the future of sandy shores. *Nature Reviews. Earth & Environment, 2*(1), 70–84. DOI: 10.1038/s43017-020-00109-9

Dean, R. G. (2018). Principles of beach nourishment. Handbook of coastal processes and erosion, 217-232.

Deasy, A., Puji, H. K., Herry, P., & Nugroho, P. (2023). Disaster-resilient villages: Strengthening community capacity in flood disasters managing in wetland areas. *Disaster Advances, 16*(4), 1–7.

Demirci, K., & Avcu, T. (2021). Afet süreçlerinde kadın bireylerin yaşadığı sorunlar ve çözüm önerileri: İzmir ili örneği. *Batman University Journal of Life Sciences, 11*(1), 86–105.

Demir, E. (2025). Climate change in Afrıca. *Uluslararası Afro-Avrasya Araştırmaları Dergisi, 10*(20), 117–120.

Demir, E., & Bahar, O. (2021). The impact of tourism revenues on economic growth: Empirical analysis on Turkey. *International Journal of Social Sciences and Education Research, 7*(2), 162–167. DOI: 10.24289/ijsser.699497

Demski, C., Capstick, S., Pidgeon, N., Sposato, R. G., & Spence, A. (2017). Experience of extreme weather affects climate change mitigation and adaptation responses. *Climatic Change, 140*(2), 149–164. DOI: 10.1007/s10584-016-1837-4 PMID: 32355377

Desmet, P. J., & Govers, G. (1996). A GIS Procedure for Automatically Calculating the USLE LS Factor on Topographically Complex Landscape Units. *Journal of Soil and Water Conservation, 51*(5), 427–433. DOI: 10.1080/00224561.1996.12457102

Diehr, J., Ogunyiola, A., & Dada, O. (2025). Artificial intelligence and machine learning-powered GIS for proactive disaster resilience in a changing climate. *Annals of GIS*, *31*(2), 1–14. DOI: 10.1080/19475683.2025.2473596

Dikmenli, Y., Yakar, H., & Konca, A. S. (2018). Development of a disaster awareness scale. *Review of International Geographical Education Online*, *8*(2), 206–220.

Dinu, I., Gràcia, V., García-León, M., Lin-Ye, J., Stănică, A., & Sánchez-Arcilla, A. (2025). Feasibility of artificial reefs as coastal protection measures at the Danube Delta coast, Romania. *Frontiers in Environmental Science*, *13*, 1568364. DOI: 10.3389/fenvs.2025.1568364

Dionísio António, S., van der Werf, J., Horstman, E., Cáceres, I., Alsina, J., van der Zanden, J., & Hulscher, S. (2023). Influence of beach slope on morphological changes and sediment transport under irregular waves. *Journal of Marine Science and Engineering*, *11*(12), 2244. DOI: 10.3390/jmse11122244

Disaster, A. W. A. R. E. Multi-Hazard early warning, risk analytics, and situational awareness | Pacific Disaster Center. (2025, July 16). Pacific Disaster Center. https://www.pdc.org/disasteraware/

Doğan, H., & Buz, S. (2024). The effect of climate change anxiety on social work students in Turkey. *International Social Work*, *67*(4), 1045–1058. DOI: 10.1177/00208728231208007

Dominey-Howes, D., Gorman-Murray, A., & McKinnon, S. (2014). Queering disasters: On the need to account for LGBTI experiences in natural disaster contexts. *Gender, Place and Culture*, *21*(7), 905–918. DOI: 10.1080/0966369X.2013.802673

Donthu, N., Kumar, S., Mukherjee, D., Pandey, N., & Lim, W. M. (2021). How to conduct a bibliometric analysis: An overview and guidelines. *Journal of Business Research*, *133*, 285–296. DOI: 10.1016/j.jbusres.2021.04.070

Downs, P. W., Cui, Y., Wooster, J. K., Dusterhoff, S. R., Booth, D. B., Dietrich, W. E., & Sklar, L. S. (2009). Managing reservoir sediment release in dam removal projects: An approach informed by physical and numerical modelling of non-cohesive sediment. *International Journal of River Basin Management*, *7*(4), 433–452. DOI: 10.1080/15715124.2009.9635401

Duda, A. M. (2017). Co-managing land and water for sustainable development. In United Nations Convention to Combat Desertification: Bonn, Germany.

Dudgeon, D., Arthington, A. H., Gessner, M. O., Kawabata, Z. I., Knowler, D. J., Lévêque, C., Naiman, R. J., Prieur-Richard, A.-H., Soto, D., Stiassny, M. L. J., & Sullivan, C. A. (2006). Freshwater biodiversity: Importance, threats, status and conservation challenges. *Biological Reviews of the Cambridge Philosophical Society*, *81*(2), 163–182. DOI: 10.1017/S1464793105006950

Duran, V., & Çelik, M. A. (2022). Çeşitli Yöntemler Kullanılarak İklim Değişikliği ve Çevre Sorunlarına Yönelik Tutum Analizi. *International Journal of Geography and Geography Education*, (46), 79–103. DOI: 10.32003/igge.1069665

Durap, A., & Balas, C. E. (2024). Towards sustainable coastal management: A hybrid model for vulnerability and risk assessment. *Journal of Coastal Conservation*, *28*(4), 66. DOI: 10.1007/s11852-024-01065-y

Durmuş, E., Kinaci, M., & Kuruyer, D. (2024). *Examining the disaster awareness levels of secondary school students*. Uluslararası Türk Eğitim Bilimleri Dergisi., DOI: 10.46778/goputeb.1494675

Duro, J. A., Perez-Laborda, A., Turrion-Prats, J., & Fernández-Fernández, M. (2021). Covid-19 and tourism vulnerability. *Tourism Management Perspectives*, *38*, 1–12. DOI: 10.1016/j.tmp.2021.100819 PMID: 34873568

Du, Y., Li, Y., Yang, M., Wang, Y., Lian, Q., & Hong, R. (2025). Bibliometric analysis of advances in heritage resilience at the district scale. npj. *Heritage Science*, *13*(1), 606.

Eberle, U., ... (2020). *Heat and Hate, Climate Security and Farmer-Herder Conflicts in Africa (*ESOC Working Paper No. 22). Empirical Studies of Conflict Project. Retrieved [22.10.25], from https://esoc.princeton.edu/wp22

Echendu, A. J. (2021). Women, development, and flooding disaster research in Nigeria: A scoping review. *Environmental and Earth Sciences Research Journal*, *8*(4), 147–152. DOI: 10.18280/eesrj.080401

Eckstein, D., Hutfils, M.-L., & Winges, M. (2019). Global climate risk index 2019. *Who Suffers Most from Extreme Weather Events*, 1–36. https://germanwatch.org/sites/germanwatch.org/files/Global Climate Risk Index 2019_2.pdf%0Ahttps://germanwatch.org/en/7677

Eduljee, G. H. (2000). Trends in risk assessment and risk management. *The Science of the Total Environment*, *249*(1–3), 13–23. DOI: 10.1016/S0048-9697(99)00507-0 PMID: 10813443

EEA. (2018). Copernicus Land Monitoring Service. European Environmental Agency. https://land.copernicus.eu/en/products/corine-land-cover/clc2018#download

Ehrlich, M., Mozafari, S., Glatza, M., Starost, L., Velychko, S., Hallmann, A.-L., Cui, Q.-L., Schambach, A., Kim, K.-P., Bachelin, C., Marteyn, A., Hargus, G., Johnson, R. M., Antel, J., Sterneckert, J., Zaehres, H., Schöler, H. R., Baron-Van Evercooren, A., & Kuhlmann, T. (2017). Rapid and efficient generation of oligodendrocytes from human induced pluripotent stem cells using transcription factors. *Proceedings of the National Academy of Sciences of the United States of America, 114*(11). Advance online publication. DOI: 10.1073/pnas.1614412114 PMID: 28246330

Ejem, A. A., & Ben-Enukora, C. A. (2025). Gendered impacts of 2022 floods on livelihoods and health vulnerability of rural communities in select Southern states in Nigeria. *Discover Social Science and Health, 5*(67), 67. Advance online publication. DOI: 10.1007/s44155-025-00211-7

Ekwurzel, B., Frumhoff, P. C., & McCarthy, J. J. (2023). Climate uncertainties and their discontents: Increasing the impact of assessments on public understanding of climate risks and choices. *Climatic Change, 108*(4), 791–802. DOI: 10.1007/s10584-011-0194-6

Elgert, L., Austin, P., & Picchione, K. (2016). Improving water security through rainwater harvesting: A case from Guatemala and the potential for expanding coverage. *International Journal of Water Resources Development, 32*(5), 765–780. DOI: 10.1080/07900627.2015.1104499

Ellegaard, O. (2018). The application of bibliometric analysis: Disciplinary and user aspects. *Scientometrics, 116*(1), 181–202. DOI: 10.1007/s11192-018-2765-z

Elsaeidy, A., Saboor, A., Hassan, A., & Redwan, A. (2016). Disaster-resistant communities, public awareness program. *Journal of Emergency Medicine. Trauma & Acute Care, 137*(2). Advance online publication. DOI: 10.5339/jemtac.2016.icepq.137

Elsaid, K., Sayed, E. T., Abdelkareem, M. A., Baroutaji, A., & Olabi, A. G. (2020). Environmental impact of desalination processes: Mitigation and control strategies. *The Science of the Total Environment, 740*, 140125. DOI: 10.1016/j.scitotenv.2020.140125 PMID: 32927546

Emmers, R. (2017). *Non-traditional security in Asia: Dilemmas in securitization.* Routledge. DOI: 10.4324/9781315247878

Enarson, E. (2007). Toward a feminist perspective on women and disaster. *Women's Rights Law Reporter, 28*(1), 3–12.

Enarson, E., & Chakrabarti, P. D. (2009). *Women, gender and disaster: Global issues and initiatives.* SAGE Publications India Pvt Ltd.

Endter-Wada, J., Kettenring, K. M., & Sutton-Grier, A. E. (2018). Sustaining wetlands to mitigate disasters and protect people. *Frontiers in Ecology and the Environment, 16*(8), 431. DOI: 10.1002/fee.1959

Ergünay, O. (2007). Türkiye'nin afet profili. *TMMOB afet sempozyumu bildiriler kitabı, 5*(7), 1-14.

Erkal, S., Şafak, Ş., Yertutan, C., & Coşkuner, S. (2009). Bir Hastanede Fiziksel Çevre ve Yönetimi (Kurum Ev İdaresi) Hizmetlerini Yürüten Personelin İş Bölümüne İlişkin Görüşlerinin İncelenmesi. *Hacettepe Sağlık İdaresi Dergisi, 12*(2), 157–176.

Erlat, E., Türkeş, M., & Aydın-Kandemir, F. (2021). Observed changes and trends in heatwave characteristics in Turkey since 1950. *Theoretical and Applied Climatology, 145*(3), 137–157. DOI: 10.1007/s00704-021-03620-1

Erpul, G., İnce, K., Demirhan, A., Küçümen, A., Akdağ, M. A., Demirtaş, İ., Sarıhan, B., Çetin, E., & Şahin, S. (2020). *Su Erozyonu İl İstatistikleri Toprak Erozyonu Kontrol Stratejileri.* Çevre Şehircilik ve İklim Bakanlığı, Çölleşme ve Erozyonla Mücadele Müdürlüğü.

Erpul, G., Şahi, S., İnce, K., Küçümen, A., Akdağ, M. A., Demirtaş, İ., & Çetin, E. (2018). *Türkiye Su Erozyonu Atlası.* Çölleşme ve Erozyonla Mücadele Genel Müdürlüğü.

Erwin, K. L. (2009). Wetlands and global climate change: The role of wetland restoration in a changing world. *Wetlands Ecology and Management, 17*(1), 71–84. DOI: 10.1007/s11273-008-9119-1

European Space Agency. (2019). *Lake Chad's shrinking waters.* https://www.esa.int/ESA_Multimedia/Images/2019/03/Lake_Chad_s_shrinking_waters

Falkenmark, M., Berntell, A., Jägerskog, A., Lundqvist, J., Matz, M., & Tropp, H. (2007). "On the verge of a new water scarcity: a call for good governance and human ingenuity." Stockholm International Water Institute (SIWI), No. 1, Sweden.

Falkenmark, M., & Molden, D. (2008). Wake up to realities of river basin closure. *International Journal of Water Resources Development, 24*(2), 201–215. DOI: 10.1080/07900620701723570

FAO. (2013). *Climate-smart agriculture sourcebook: Agriculture, forestry and fisheries.* Rome: Food and Agriculture Organization of the United Nations. https://www.fao.org/3/i3325e/i3325e.pdf

FAO. (2016). Report of the Forty-fourth Session of the Codex Committee on Food Labelling. FAO.

Fatema, S. R., East, L., Islam, S., & Usher, K. (2023). Gender-based vulnerabilities for women during natural disasters in Bangladesh, *Sec. Frontiers in Communication*, *8*, 1180406. Advance online publication. DOI: 10.3389/fcomm.2023.1180406

Faulkner, B. (2001). Toward a framework for tourism disaster management. *Tourism Management*, *22*(2), 135–147. DOI: 10.1016/S0261-5177(00)00048-0

Fawzy, S., Osman, A. I., Doran, J., & Rooney, D. W. (2020). Strategies for mitigation of climate change: A review. *Environmental Chemistry Letters*, *18*(6), 2069–2094. DOI: 10.1007/s10311-020-01059-w

Fayaz, M., Meraj, G., Khader, S. A., & Farooq, M. (2022). ARIMA and SPSS statistics based assessment of landslide occurrence in western Himalayas. *Environmental Challenges*, *9*, 100624. DOI: 10.1016/j.envc.2022.100624

Feng, S., Zhao, W., Yan, J., Xia, F., Wang, H., Zhou, A., Pereira, P. (2025). Vegetation Restoration Contributes to a Reduction in Wind and Water Erosion in China's Drylands. Applied Geography, 176(May 2024). DOI: 10.1016/j.apgeog.2025.103517

Ferrarin, C., Bonaldo, D., Bergamasco, A., & Ghezzo, M. (2024). Sea level and temperature extremes in a regulated Lagoon of Venice. *Frontiers in Climate*, *5*, 1330388. DOI: 10.3389/fclim.2023.1330388

Ferreira, C. S., Kašanin-Grubin, M., Solomun, M. K., Sushkova, S., Minkina, T., Zhao, W., & Kalantari, Z. (2023). Wetlands as nature-based solutions for water management in different environments. *Current Opinion in Environmental Science & Health*, *33*, 100476. DOI: 10.1016/j.coesh.2023.100476

Ferreira, M., Coelho, C., & Silva, P. A. (2022). Littoral drift gradients on the Portuguese coastal sector Esmoriz-Nazare: Past and future trends. *Coastal Engineering Proceedings*, (37), 22–22. DOI: 10.9753/icce.v37.papers.22

Ferris, E. (2020). Research on climate change and migration: Where are we and where are we going? *Migration Studies*, *8*(4), 612–625. DOI: 10.1093/migration/mnaa028

Fibriantika, E., Mujahidin, A., Maryamah, S., & Rasyid, I. A. (2021, July). Analysis of water balance conditions as the impact of climate and land-use changes in Kapuas Watershed, Indonesia. [). IOP Publishing.]. *IOP Conference Series. Earth and Environmental Science*, *824*(1), 012018. DOI: 10.1088/1755-1315/824/1/012018

Fields, G. S. (1988). Employment and economic growth in Costa Rica. *World Development*, *16*(12), 1493–1509. DOI: 10.1016/0305-750X(88)90222-7

Filkins, D. (2016). "A Bigger Problem Than ISIS?" <www.newyorker.com> (July. 27, 2019).

Fischer, H. W. (1998). The role of the new information technologies in emergency mitigation, planning, response and recovery. *Disaster Prevention and Management, 7*(1), 28–37. DOI: 10.1108/09653569810206262

Foley, J. A., Ramankutty, N., Brauman, K. A., Cassidy, E. S., Gerber, J. S., Johnston, M., Mueller, N. D., O'Connell, C., Ray, D. K., West, P. C., Balzer, C., Bennett, E. M., Carpenter, S. R., Hill, J., Monfreda, C., Polasky, S., Rockström, J., Sheehan, J., Siebert, S., & Zaks, D. P. M. (2011). Solutions for a cultivated planet. *Nature, 478*(7369), 337–342. DOI: 10.1038/nature10452 PMID: 21993620

Folke, C., Hahn, T., Olsson, P., & Norberg, J. (2005). Adaptive governance of social-ecological systems. *Annual Review of Environment and Resources, 30*(1), 441–473. DOI: 10.1146/annurev.energy.30.050504.144511

Foresight: Government Office for Science. (2011). *Migration and global environmental change* (Final Project Report). Government Office for Science.

French, S., & Niculae, C. (2005). Believe in the model: Mishandle the emergency. *Journal of Homeland Security and Emergency Management, 2*(1). Advance online publication. DOI: 10.2202/1547-7355.1108

Gaillard, J.-C., & Mercer, J. (2013). From knowledge to action: Bridging gaps in disaster risk reduction. *Progress in Human Geography, 37*(1), 93–114. https://doi .org/OI. DOI: 10.1177/0309132512446717

Gain, A., Wada, Y., & Giupponi, C. (2016). Measuring global water security towards sustainable development goals. *Environmental Research Letters, 11*(12), 2–13. DOI: 10.1088/1748-9326/11/12/124015

Gallina, V., Torresan, S., Critto, A., Sperotto, A., Glade, T., & Marcomini, A. (2016). A review of multi-risk methodologies for natural hazards: Consequences and challenges for a climate change impact assessment. *Journal of Environmental Management, 168*, 123–132. DOI: 10.1016/j.jenvman.2015.11.011 PMID: 26704454

Ganasri, B. P., & Ramesh, H. (2015). Geoscience Frontiers Assessment of Soil Erosion by RUSLE Model Using Remote Sensing and GIS - A Case Study of Nethravathi Basin. *Geoscience Frontiers, 7*(6), 1–9. DOI: 10.1016/j.gsf.2015.10.007

Gardner, R. C., & Finlayson, C. (2018, October). Global wetland outlook: state of the world's wetlands and their services to people. In Ramsar convention secretariat (pp. 2020-5). DOI: 10.69556/GWO-2018-eng

Gavel, Y., & Iselid, L. (2008). Web of Science and Scopus: A journal title overlap study. *Online Information Review, 32*(1), 8–21. DOI: 10.1108/14684520810865958

Gelashvili, I. A. (2022). The Negative impact of pandemics on the tourism sector and solution ways. *Globalization and Business*, *7*(13), 138–142. DOI: 10.35945/gb.2022.13.021

Gellman, B. (2002). "Cyber-attacks by Al Qaeda feared." <www.washingtonpost.com> (July. 27, 2019).

Gemenne, F. (2011). Climate-induced population displacements in a 4°C+ world. *Philosophical Transactions of the Royal Society A: Mathematical, Physical and Engineering Sciences, 369*(1934), 182–195.

Gerdan, S. (2014). Determination of Disaster Awareness, Attitude Levels and Individual Priorities at Kocaeli University. *Dergipark. Eurasian Journal of Educational Research, 2014*(55), 159–176. DOI: 10.14689/ejer.2014.55.10

Ghani, M. W., & Hidayati, I. (2024, March). Adapting to sea level rise: Migration decisions and disaster mitigation strategies in small-scale fishing communities. [). IOP Publishing.]. *IOP Conference Series. Earth and Environmental Science, 1314*(1), 012013. DOI: 10.1088/1755-1315/1314/1/012013

Ghosh, P., Sudarsan, J. S., & Nithiyanantham, S. (2024). Nature-based disaster risk reduction of floods in Urban Areas. *Water Resources Management, 38*(6), 1847–1866. DOI: 10.1007/s11269-024-03757-4

Gibson, A. L., & Sokolov, D. (2026). A Modular Cognitive Architecture for Collective Intelligence Systems. *Lecture Notes in Computer Science, 16057 LNAI*, 181–191. DOI: 10.1007/978-3-032-00686-8_17

Giddens, A. (1994). *Runaway world: How globalization is reshaping our lives.* Profile Books.

Gilbreath, D. (2011). *The Indian Ocean tsunami: The global response to a natural disaster*. University Press of Kentucky.

Ginglinger, E., & Moreau, Q. (2023). Climate Risk and Capital Structure. *Management Science, 69*(12), 7492–7516. DOI: 10.1287/mnsc.2023.4952

Gleick, , P. H. (2019). Water as a weapon and casualty of conflict: Freshwater and international humanitarian law. *Water Resources Management, 33*, 1737–1751.

Gleick, P. H. (2006). Water and terrorism. *Water Policy*, *8*(6), 481–503. DOI: 10.2166/wp.2006.035

Goldsmith, L., Raditz, V., & Mendez, M. (2022). Queer and present danger: Understanding the disparate impacts of disasters on LGBTQ+ communities. *Disasters, 46*(4), 946–973. DOI: 10.1111/disa.12509 PMID: 34498778

Goldstein, E. B., Coco, G., & Plant, N. G. (2019). A review of machine learning applications to coastal sediment transport and morphodynamics. *Earth-Science Reviews*, *194*, 97–108. DOI: 10.1016/j.earscirev.2019.04.022

Goltz, J. D., Flores, P. J., Chang, S. E., & Atsumi, T. (2001). Emergency Response and Early Recovery. *Earthquake Spectra*, *17*(1, suppl), 173–183. DOI: 10.1193/1.1586198

Goodchild, M. F., & Li, W. (2021). Replication across space and time must be weak in the social and environmental sciences. *Proceedings of the National Academy of Sciences of the United States of America*, *118*(35), e2015759118. Advance online publication. DOI: 10.1073/pnas.2015759118 PMID: 34417345

Google. (n.d.). Boğaçay River and Konyaaltı Shoreline digital imaginary, satellite map. https://earth.google.com/web/@36.85211757,30.62297089,1512.2599538a,0d,35y,-0.6805h,2.0574t,-0r/data=CgRCAggBQgIIAEoNCP__________wEQAA?utm_source=earth7&utm_campaign=vine&hl=en (accessed 19 August 2025).

Gorman-Murray, A., McKinnon, S., & Dominey-Howes, D. (2014). Queer domicide. *Home Cultures*, *11*(2), 237–261. DOI: 10.2752/175174214X13891916944751

Gössling, S., Scott, D., & Hall, C. M. (2021). Pandemics, tourism and global change: A rapid assessment of Covid-19. *Journal of Sustainable Tourism*, *29*(1), 1–20. DOI: 10.1080/09669582.2020.1758708

Grant, A., & Cooker, M. J. (2023). Reductions in water level over coastal wetlands during storm surges and tsunamis: An analytical result and a critical review of the literature. *Coastal Engineering*, *183*, 104328. DOI: 10.1016/j.coastaleng.2023.104328

Graveline, M. H., & Germain, D. (2022). Disaster Risk Resilience: Conceptual Evolution, Key Issues, and Opportunities. *International Journal of Disaster Risk Science*, *13*(3), 330–341. DOI: 10.1007/s13753-022-00419-0

Green, H. K., Lysaght, O., Saulnier, D. D., Blanchard, K., Humphrey, A., Fakhruddin, B., & Murray, V. (2019). Challenges with disaster mortality data and measuring progress towards the implementation of the Sendai framework. *International Journal of Disaster Risk Science*, *10*(4), 449–461. DOI: 10.1007/s13753-019-00237-x

Grumbine, R. E., & Xu, J. (2011). Mekong hydropower development.

Güler, H. G., Sözdinler, C. Ö., Arikawa, T., & Yalçıner, A. C. (2018). Tsunami Afeti Sonrası Yapısal ve Yapısal Olmayan Önlemler ve Farkındalık Çalışmaları: Japonya Örneği. *Teknik Dergi*, *8605*(8629), 514.

Gümüş, B., Örüç, S., Yücel, İ., & Yılmaz, M. T. (2023). Impacts of climate change on extreme climate indices in Türkiye driven by high-resolution downscaled CMIP6 climate models. *Sustainability (Basel), 15*(9), 7202. DOI: 10.3390/su15097202

Gündüz, F. (2022). Afetlerde kadın ve toplumsal cinsiyet perspektifi ile çıkarılması gereken dersler (Haiti ve Japonya depremi örneği). *IBAD Sosyal Bilimler Dergisi*, (12), 440–461. DOI: 10.21733/ibad.1039215

Gündüz, F., & Atalay, E. (2025). Determining the effect of women's earthquake fear on earthquake risk perception. *Fırat Üniversitesi Sosyal Bilimler Dergisi, 35*(2), 801–813. DOI: 10.18069/firatsbed.1599856

Güngör, E., & Serttaş, S. (2024). EOG Region Detection using Fuzzy k-NN for Virtual Reality Applications. *Sakarya University Journal of Science, 28*(6), 1146–1157. DOI: 10.16984/saufenbilder.1386568

Guo, G., Pan, Y., Kuai, J., Dai, Q., Zhu, Y., Xu, Y., Jia, X., Zeng, G., & Lin, J. (2023). A More Accurate Approach to Estimate the C-factor of RUSLE by Coupling Stratified Vegetation Cover Index in Southern China. *Forest Ecology and Management, 541*(120979), 120979. https://doi.org/https://doi.org/10.1016/j.foreco.2023.120979. DOI: 10.1016/j.foreco.2023.120979

Gürkan, S. (2024). The feminine and masculine dimensions of earthquakes: A gender-focused sociological research on the victims of the 2023 Kahramanmaraş / Türkiye earthquakes. *Sted, 33*(5), 351–362. DOI: 10.17942/sted.1456836

Gurung, D. (2019). Through women's eyes: A study of vulnerability in 2015 earthquake. *Dhaulagiri, 13*, 76–83. DOI: 10.3126/dsaj.v13i0.23361

Haacke, J. (2013). *ASEAN's diplomatic and security culture: Origins, development and prospects*. Routledge. DOI: 10.4324/9780203037539

Haddaway, N. R., Woodcock, P., Macura, B., & Collins, A. (2015). Making literature reviews more reliable through application of lessons from systematic reviews. *Conservation Biology, 29*(6), 1596–1605. DOI: 10.1111/cobi.12541

Hamilton, L. C., Saito, K., Loring, P. A., Lammers, R. B., & Huntington, H. P. (2016). Climigration? Population and climate change in Arctic Alaska. *Population and Environment, 38*(2), 115–133. DOI: 10.1007/s11111-016-0259-6 PMID: 27829694

Hargono, A., Artanti, K. D., Astutik, E., Widodo, P. P., Trisnawati, A. N., Wardani, D. K., & Lioni, E. (2023). Relationship between disaster awareness and disaster preparedness: Online survey of the community in Indonesia. *Journal of Public Health in Africa, 14*(9), 2376. DOI: 10.4081/jphia.2023.2376 PMID: 37927360

Hay, J., & Mimura, N. (2023). The changing nature of extreme weather and climate events: Risks to sustainable development. *Geomatics, Natural Hazards & Risk, 1*(1), 3–18. DOI: 10.1080/19475701003643433

Henrico, I., & Doboš, B. (2024). Shifting sands: The geopolitical impact of climate change on Africa's resource conflicts. *The South African Geographical Journal.* Advance online publication. DOI: 10.1080/03736245.2024.2441116

Hernández, L. M., Gómara, B., Fernández, M., Jiménez, B., González, M. J., Baos, R., Hiraldo, F., Ferrer, M., Benito, V., Suñer, M. A., Devesa, V., Muñoz, O., & Montoro, R. (1999). Accumulation of heavy metals and As in wetland birds in the area around Donana National Park affected by the Aznalcollar toxic spill. *The Science of the Total Environment, 242*(1-3), 293–308. DOI: 10.1016/S0048-9697(99)00397-6

Hinrichsen, D. (2002). "Freshwater: lifeblood of the planet." <http://www.peopleandtheplanet.com> (Sep. 23, 2019).

Hoang, L. P., Biesbroek, R., Tri, V. P. D., Kummu, M., van Vliet, M. T., Leemans, R., Kabat, P., & Ludwig, F. (2018). Managing flood risks in the Mekong Delta: How to address emerging challenges under climate change and socioeconomic developments. *Ambio, 47*(6), 635–649. DOI: 10.1007/s13280-017-1009-4 PMID: 29478182

Hohenthal, J., & Minoia, P. (2017). Social aspects of water scarcity and drought. In *Handbook of Drought and Water Scarcity* (pp. 607-625). CRC Press.

Holling, C. S. (1973). Resilience and stability of ecological systems. *Annual Review of Ecology and Systematics, 4*(1), 1–23. DOI: 10.1146/annurev.es.04.110173.000245

Hooghe, L., & Marks, G. (2003). Unraveling the central state, but how? Types of multi-level governance. *The American Political Science Review, 97*(2), 233–243. https://garymarks.web.unc.edu/wp-content/uploads/sites/13018/2016/09/hooghe.marks_.unravelingcentralstate.apsr_.2003.pdf

Horton, J. (2012). After the earthquake: Gender inequality and transformation in post-disaster Haiti. *Gender and Development, 20*(2), 295–308. DOI: 10.1080/13552074.2012.693284

Howard, B., Dietz, S., Ebi, K., Gollier, C., Gregory, R., Horton, B., Kriegler, E., Mach, K., Mastrandrea, M., Millner, A., Oppenheimer, M., Träger, C., Kunreuther, H., Heal, G., Allen, M., Edenhofer, O., Field, C. B., & Yohe, G. (2022). *Risk Management and Climate Change.* https://www.nber.org/papers/w18607

Howe, P. D. (2021). Extreme weather experience and climate change opinion. *Current Opinion in Behavioral Sciences, 42*, 127–131. DOI: 10.1016/j.cobeha.2021.05.005

Hradecký, J., & Pánek, T. (2008). Deep-seated gravitational slope deformations and their influence on consequent mass movements (case studies from the highest part of the Czech Carpathians). *Natural Hazards*, *45*(2), 235–253. DOI: 10.1007/s11069-007-9157-7

Huber, D. G., & Gulledge, J. (2021). Extreme Weather and Climate Change Understanding the Link, Managing the Risk. *Science and Impacts Program. Center for Climate and Energy Solutions, December*, 1–13. http://www.pewclimate.org/publications/extreme-weather-and-climate-change%5Cnhttp://www.c2es.org/publications/extreme-weather-and-climate-change

Huber, D., & Gulledge, J. (2023). *Extreme weather and climate change: Understanding the link and managing the risk. Science and Impacts Program.* Center for Climate and Energy Solutions.

Hugo, G. (2011). Future demographic change and its interactions with migration and climate change. *Global Environmental Change*, *21*, S21–S33. DOI: 10.1016/j.gloenvcha.2011.09.008

Hu, H., Yan, K., Zhu, Z., & Lv, T. (2024). Building forwards: Bibliometric analysis and framework construction for community resilience. *Journal of Contingencies and Crisis Management*, *32*(3), e12604. DOI: 10.1111/1468-5973.12604

Hulme, M. (2020). Attributing weather extremes to 'climate change.'. *Progress in Physical Geography*, *38*(4), 499–511. DOI: 10.1177/0309133314538644

Huot, S. (2025a). Cultural diplomacy across the Mekong subregion: Cambodia's soft power diplomacy in Southeast Asia. *Literary Enigma: The International Journal of English Language. Literature and Culture*, *1*(6), 80–90. DOI: 10.5281/zenodo.15803843

Huot, S. (2025b). Digital tutors in the kingdom: Exploring ChatGPT's influence on Cambodian students' academic and cognitive growth. In Pandya, J. D., & Patel, K. C. (Eds.), *Interdisciplinary insights: Exploring multiple perspectives* (Vol. 2, pp. 36–54). Meadow Publication.

Huot, S. (2026a). Digital diplomacy in Cambodia: Embracing artificial intelligence for strategic engagement and national development. In Turan, A., & Ucuzal, H. (Eds.), *Diplomacy in the age of artificial intelligence* (pp. 307–330). IGI Global Scientific Publishing., DOI: 10.4018/979-8-3373-4937-4.ch013

Huot, S. (2026b). Emerging voices from Cambodia: International student journeys and educational aspirations in a globalized academic landscape. In Irham, I., Huda, M., & Yaqin, L. N. (Eds.), *International student narratives in globalized higher education* (pp. 365–400). IGI Global Scientific Publishing., DOI: 10.4018/979-8-3373-4212-2.ch014

Huot, S., & Chheang, S. (2026). Artificial intelligence and human rights in Cambodia: Pathways to social justice and positive developmental progress. In Pasupuleti, S., & Yadav, U. (Eds.), *Safeguarding social justice and human rights in the age of AI* (pp. 355–386). IGI Global Scientific Publishing., DOI: 10.4018/979-8-3373-6598-5.ch016

Ibanez, C., & Caiola, N. (2021). Sea-level rise, marine storms and the resilience of Mediterranean coastal wetlands: Lessons learned from the Ebro Delta. *Marine and Freshwater Research, 73*(10), 1246–1254. DOI: 10.1071/MF21140

İli, B., & Bakar, H. (2024). Veri Gazeteciliğine Yönelik Araştırmaların Bibliyometrik Analizi. *Erciyes İletişim Dergisi, 11*(2), 607–626. DOI: 10.17680/erciyesiletisim.1430017

Iloglu, S., & Albert, L. A. (2018). An integrated network design and scheduling problem for network recovery and emergency response. *Operations Research Perspectives, 5*, 218–231. DOI: 10.1016/j.orp.2018.08.001

Imamoglu, A., & Dengiz, O. (2017). Determination of Soil Rrosion Risk Using RUSLE Model and Soil Organic Carbon Loss in Alaca Catchment (Central Black Sea Region, Turkey). *Rendiconti Lincei. Scienze Fisiche e Naturali, 28*(1), 11–23. DOI: 10.1007/s12210-016-0556-0

IMF (2019). Building resilience in developing countries vulnerable to large natural disasters. *Policy Papers* 2019, No:020. https://doi.org/DOI: 10.5089/9781498321020.007

Institute for Security Studies. (2023). *Chad: Country profile.* ISS African Futures & Innovation. https://futures.issafrica.org/geographic/countries/chad/

Intergovernmental Panel on Climate Change (IPCC). (2014). *Climate Change 2014: Synthesis Report.* Cambridge University Press.

Intergovernmental Panel on Climate Change (IPCC). (2021). *Climate Change 2021: The Physical Science Basis*. Cambridge University Press.

Internal Displacement Monitoring Centre (IDMC). (2024). *Global Report on Internal Displacement (GRID 2024).* Norwegian Refugee Council. https://api.internal-displacement.org/sites/default/files/publications/documents/IDMC-GRID-2024-Global-Report-on-Internal-Displacement.pdf

Internal Organization for Migration (IOM). (2007). *Discussion note: Migration and the environment*. IOM.

International Monetary Fund. (2024). *Chad: Selected issues* (IMF Staff Country Report No. 24/336). https://www.elibrary.imf.org/view/journals/002/2024/336/002.2024.issue-336-en.xml

İnternet Kaynaklar Islam, R., & Walkerden, G. (2014). How bonding and bridging networks contribute to disaster resilience and recovery on the Bangladeshi coast. *International Journal of Disaster Risk Reduction, 10*, 281–291. DOI: 10.1016/j.ijdrr.2014.09.016

IPCC. (2022). *Climate change 2022: Impacts, adaptation and vulnerability.* Geneva: Intergovernmental Panel on Climate Change (IPCC). https://www.ipcc.ch/report/ar6/wg2/

Iriti, M., & Vitalini, S. (2025). *Climate change, natural disasters, armed conflicts and migrations at the crossroads between food and nutrition insecurity and undernourishment*. Functional Food Science., DOI: 10.31989/ffs.v5i1.1547

Irschik, H., & Belyaev, A. K. (Eds.). (2014). *Dynamics of Mechanical Systems with Variable Mass* (Vol. 557). Springer Vienna., DOI: 10.1007/978-3-7091-1809-2

Jackson, A., Corbett, B., & Tomlinson, R. (2007). Palm Beach Protection Strategy Update on recommendations as part of the Gold Coast Shoreline Management Plan. http://hdl.handle.net/10072/93914 (Accessed 19 August 2025).

Jafari, J. (2001). *Encyclopedia of Tourism*. Routledge.

Japan Aerospace Exploration Agency. (2021). ALOS Global Digital Surface Model. Distributed by OpenTopography. https://doi.org/Japan Aerospace Exploration Agency (2021). ALOS World 3D 30 meter DEM. V3.2, Jan 2021. Distributed by OpenTopography. DOI: 10.5069/G94M92HB

Jean-Richard, V., Crump, L., Abicho, A., Abakar, A., Mahamat, A., Béchir, M., Eckert, S., Engesser, M., Schelling, E., & Zinsstag, J. (2015). Estimating population and livestock density of mobile pastoralists and sedentary settlements in the south-eastern Lake Chad area.. *Geospatial health*, 10 1, 307 . .DOI: 10.4081/gh.2015.307

Jentsch, A., & Beierkuhnlein, C. (2023). Research frontiers in climate change: Effects of extreme meteorological events on ecosystems. *Comptes Rendus Geoscience, 340*(9–10), 621–628. DOI: 10.1016/j.crte.2008.07.002

Jeyaraj, R., Balasubramaniam, A., Kumara, M. A. A., Guizani, N., & Paul, A. (2023). Resource Management in Cloud and Cloud-influenced Technologies for Internet of Things Applications. *ACM Computing Surveys*, *55*(12). Advance online publication. DOI: 10.1145/3571729

Jia, X., Li, C., Cai, Y., Wang, X., & Sun, L. (2015). An improved method for integrated water security assessment in the Yellow River basin, China. *Stochastic Environmental Research and Risk Assessment*, *29*(8), 2213–2227. DOI: 10.1007/s00477-014-1012-2

JICA. (2021). *Building disaster and climate resilient cities in ASEAN*. Tokyo: Japan International Cooperation Agency. https://openjicareport.jica.go.jp/pdf/12321154_01.pdf

Joffé, G. (1997). Chad: Power vacuum or geopolitical focus. *Geopolitics*, *2*(1), 25–39.

Jones, R., & Boer, R. (2023). *Assessing current climate risks* (Issue 4).

Jones, R. N. (2001). An Environmental Risk Assessment/Management Framework for Climate Change Impact Assessments. *Natural Hazards*, *23*(2–3), 197–230. DOI: 10.1023/A:1011148019213

Joshi, A., Grover, J., Kanungo, D. P., & Panigrahi, R. K. (2019). Edge assisted Reliable Landslide Early Warning System. *2019 IEEE 16th India Council International Conference (INDICON)*, 1–4. DOI: 10.1109/INDICON47234.2019.9028898

Joyce, K. E., Belliss, S. E., Samsonov, S. V., McNeill, S. J., & Glassey, P. J. (2009). A review of the status of satellite remote sensing and image processing techniques for mapping natural hazards and disasters. *Progress in Physical Geography*, *33*(2), 183–207. DOI: 10.1177/0309133309339563

Jurgilevich, A., Räsänen, A., Groundstroem, F., & Juhola, S. (2017). A systematic review of dynamics in climate risk and vulnerability assessments. *Environmental Research Letters*, *12*(1), 013002. DOI: 10.1088/1748-9326/aa5508

K. C., L. & Hilhorst D. (2022). Gendered experience of disaster: Women's account of evacuation, relief and recovery in Nepal. *International Journal of Disaster Risk Reduction*, *72*, 102840. DOI: 10.1016/j.ijdrr.2022.102840

Kadirbeyoglu, Z. (2017). The impact of power and civic engagement in the decentralized management of natural resources: The case of Turkey. *Public Administration and Development*, *37*(4), 277–291. DOI: 10.1002/pad.1809

Kadir, S. B. (2021). Viewing disaster resilience through gender sensitive lens: A composite indicator based assessment. *International Journal of Disaster Risk Reduction, 62*, 102398. Advance online publication. DOI: 10.1016/j.ijdrr.2021.102398

Kajihara, C., Kashiwabara, M., Yamao, M., Aoki, Y., Yabe, K., Kaneko, M., Sano, M., & Munechika, M. (2023). Evaluation of the degree of crisis awareness and behavioral change through disaster exercise: A case study of a disaster response exercise in a pharmacy department. *Prehospital and Disaster Medicine, 38*(S1), s22–s22. DOI: 10.1017/S1049023X23000997

Kalogiannidis, S., Kalfas, D., Kontsas, S., Papaevangelou, O., & Chatzitheodoridis, F. (2025). Evaluating the Effectiveness of Early Warning Systems in Reducing Loss of Life in Natural Disasters: A case study of Greece. *Journal of Risk Analysis and Crisis Response, 15*(1), 33. DOI: 10.54560/jracr.v15i1.547

Kamigaichi, O. (2004). JMA EARTHQUAKE EARLY WARNING. *Journal of Japan Association for Earthquake Engineering, 4*(3), 134–137. DOI: 10.5610/jaee.4.3_134

Kam, P. M., Aznar-Siguan, G., Schewe, J., Milano, L., Ginnetti, J., Willner, S., McCaughey, J. W., & Bresch, D. N. (2021). Global warming and population change both heighten future risk of human displacement due to river floods. *Environmental Research Letters, 16*(4), 044026. DOI: 10.1088/1748-9326/abd26c

Kamta, F. N., Schilling, J., & Scheffran, J. (2020). Insecurity, resource scarcity, and migration to camps of internally displaced persons in Northeast Nigeria. *Sustainability (Basel), 12*(17), 6830. DOI: 10.3390/su12176830

Karanci, A. N., Aksit, B., & Dirik, G. (2005). Impact of a community disaster awareness training program in Turkey: Does it influence hazard-related cognitions and preparedness behaviors. *Social Behavior and Personality, 33*(3), 243–258. DOI: 10.2224/sbp.2005.33.3.243

Karger, D. N., Conrad, O., Böhner, J., Kawohl, T., Kreft, H., Soria-Auza, R. W., Zimmermann, N. E., Linder, H. P., & Kessler, M. (2017). CHELSA V2.1: Technical specification. *Scientific Data*, •••, 4.

Karger, D. N., Schmatz, D. R., Dettling, G., & Zimmermann, N. E. (2020). High-resolution monthly precipitation and temperature time series from 2006 to 2100. *Scientific Data, 7*(1), 1–10. DOI: 10.1038/s41597-020-00587-y PMID: 32703947

Karma, . (2024). *People vulnerability before, during and after a disaster: A dynamic taxonomic approach* (Di Pietro, A., Ed.). Climate Change and Risk Management - Strategies, Analysis, and Adaptation., DOI: 10.5772/intechopen.1005205

Katakidis, A., Kodonas, K., Fardi, A., & Gogos, C. (2023). A scientometric, bibliometric, and thematic map analysis of hydraulic calcium silicate root canal sealers. *Restorative Dentistry & Endodontics*, *48*(4), e41. DOI: 10.5395/rde.2023.48.e41

Kazeem, O. S. (2024). Climate change and violent conflict in the Sahel and Lake Chad Basin. *Journal of Contemporary International Relations and Diplomacy*, *5*(1), 74–85. DOI: 10.53982/jcird.2024.0501.05-j

Kehoe, L., Romero-Muñoz, A., Polaina, E., Estes, L., Kreft, H., & Kuemmerle, T. (2017). Biodiversity at risk under future cropland expansion and intensification. *Nature Ecology & Evolution*, *1*(8), 1129–1135. DOI: 10.1038/s41559-017-0234-3

Keleş, R. (2022). The effects of external migration on regional development in Turkey. In *Uneven development in Southern Europe* (pp. 54–75). Routledge. DOI: 10.4324/9781003290766-2

Keller, K., Helgeson, C., & Srikrishnan, V. (2021). Climate Risk Management. *Annual Review of Earth and Planetary Sciences*, *49*(1), 95–116. DOI: 10.1146/annurev-earth-080320-055847

Kelman, I. (2011). *Disaster diplomacy: How disasters affect peace and conflict*. Routledge., DOI: 10.4324/9780203806210

Kelman, I. (2017). Linking disaster risk reduction, climate change, and the sustainable development goals. *Disaster Prevention and Management*, *26*(3), 254–258. DOI: 10.1108/DPM-02-2017-0043

Kerr, R. A., & Stone, R. (2009). A human trigger for the great quake of Sichuan? *Science*, *323*(5912), 322–322. DOI: 10.1126/science.323.5912.322

Keshavarz-Ghorabaee, M., Amiri, M., Zavadskas, E. K., Turskis, Z., & Antucheviciene, J. (2021). Determination of objective weights using a new method based on the removal effects of criteria (MEREC). *Symmetry*, *13*(4), 1–20. DOI: 10.3390/sym13040525

Khan, A. A., & Somuncu, M. (2019). Is rural out migration constructive for sustainable development? A case of mountain community of Kıbrıscık, Turkey. *İnsan ve Toplum Bilimleri Araştırmaları Dergisi*, *8*(4), 3337–3352.

Khan, A., Bibi, S., Lorenzo, A., Lyu, J., & Babar, Z. U. (2020). Tourism and development in developing economies: A policy implication perspective. *Sustainability (Basel)*, *12*(4), 1–19. DOI: 10.3390/su12041618

Khanchel, I., & Lassoued, N. (2025). Climate vulnerability in Middle East and North African countries: Do social dimensions shape resilience? *Journal of Economic Studies (Glasgow, Scotland)*. Advance online publication. DOI: 10.1108/JES-10-2024-0679

Khanna, A., & Fujii, T. (2024). Do natural disasters cause domestic violence?: A study of the 2015 Nepal earthquake. Available at *SSRN*: https://ssrn.com/abstract=5084508 or http://dx.doi.org/DOI: 10.2139/ssrn.5084508

Kher, J., Aggarwal, S., & Punhani, G. (2015). Vulnerability of poor urban women to climate-linked water insecurities at the household level: A case study of slums in delhi. *Indian Journal of Gender Studies*, *22*(1), 15–40. DOI: 10.1177/0971521514556943

Kierzkowski, A., Kisiel, T., Milewski, M., Török, Á., Stosiak, M., & Wróbel, J. (2024). COMPUTER-AIDED SIMULATION OF UNMANNED AERIAL VEHICLE COMPOSITE STRUCTURE DYNAMICS. *Transport*, *39*(4), 302–312. DOI: 10.3846/transport.2024.23159

Kılıç, Z. (2020). The importance of water and conscious use of water. *International Journal of Hydrology*, *4*(5), 239–241. DOI: 10.15406/ijh.2020.04.00250

Kimberlin, L. W., & Moldenhauer, W. C. (1977). Predicting Soil Erosion. In ASAE Publ (Issues 4–77, pp. 31–42).

Kirolos, A., Goyheneix, M., Eliasz, M., Chisala, M., Lissauer, S., Gladstone, M., & Kerac, M. (2022). Neurodevelopmental, cognitive, behavioural and mental health impairments following childhood malnutrition: A systematic review. *BMJ Global Health*, *7*(7), e009330. Advance online publication. DOI: 10.1136/bmjgh-2022-009330 PMID: 35793839

Kisakye, V., & Van der Bruggen, B. (2018). Effects of climate change on water savings and water security from rainwater harvesting systems. *Resources, Conservation and Recycling*, *138*, 49–63. DOI: 10.1016/j.resconrec.2018.07.009

Kızıldere, D. (2025). Earthquake disaster-induced urban policies: Paradoxes and challenges in Turkey. *Urban Research & Practice*, *18*(1), 130–141. DOI: 10.1080/17535069.2024.2422627

Klein, I., Uereyen, S., Eisfelder, C., Pankov, V., Oppelt, N., & Kuenzer, C. (2023). Application of geospatial and remote sensing data to support locust management. *International Journal of Applied Earth Observation and Geoinformation*, *117*, 103212. DOI: 10.1016/j.jag.2023.103212

Klose, M. (2015). Landslide Databases—State of Research and the Case of Germany. In M. Klose, *Landslide Databases as Tools for Integrated Assessment of Landslide Risk* (pp. 25–43). Springer International Publishing. DOI: 10.1007/978-3-319-20403-1_2

Knabb, R. D., Rhome, J. R., & Brown, D. P. (2005). *Tropical cyclone report: Hurricane Katrina*. National Hurricane Center.

Koç, K., Koşun, E., & Karaman, M. E. (2015). Kıyı Sedimentlerinde Tane Boyunu Etkileyen Faktörler: Antalya Körfezi'nde Konyaaltı ve Lara Plajlarının (Antalya) Karşılaştırılması. *Türkiye Jeoloji Bülteni*, *58*(2), 85–102. DOI: 10.25288/tjb.298589

Kojm, C. (2012). "Global Water Security." Intelligence Community Assessment, Report No. Washington, DC, USA.

Köle, M. M., & Ataol, M. (2016). Coastal Erosion of Yeşilırmak Delta Between 2000–2016/Yeşilırmak Deltası'nda 2000-2016 Yılları Arasında Gözlenen Kıyı Erozyonu. In 4th International Geography Symposium-GEOMED (pp. 460-468). DOI: 10.17482/uumfd.1248184

Konisky, D. M., Hughes, L., & Kaylor, C. H. (2020). Extreme weather events and climate change concern. *Climatic Change*, *134*(4), 533–547. DOI: 10.1007/s10584-015-1555-3

Korkmaz, C., & Meşhur, H. F. A. (2022). Counter-urbanisation experience in developing countries: The case of Istanbul metropolitan area. In *CPUD'22 Conference Proceedings.*

Kreft. (2023). GLOBAL CLIMATE RISK INDEX. In *Eco Watch*. http://ecowatch.com/2015/12/04/day-4-cop21%5Cnpapers3://publication/uuid/148F5B59-F94A-49D2-9FE2-F72730750610

Kulak, M., Ozkan, A., & Bindak, R. (2019). A bibliometric analysis of the essential oil-bearing plants exposed to the water stress: How long way we have come and how much further? *Scientia Horticulturae*, *246*, 418–436. DOI: 10.1016/j.scienta.2018.11.031

Kumar, A. (2023). Climate Change and Damage from Extreme Weather Events. *Key Concepts in Modern Indian Studies*, 207–211.

Kumar, J., & Hussain, K. (2014). Evaluating tourism's economic effects: Comparison of different approaches. *Procedia: Social and Behavioral Sciences*, *144*, 360–365. DOI: 10.1016/j.sbspro.2014.07.305

Kyriacos, U., Jelsma, J., & Jordan, S. (2011). Monitoring vital signs using early warning scoring systems: A review of the literature: Monitoring vital signs - early warning scoring systems. *Journal of Nursing Management*, *19*(3), 311–330. DOI: 10.1111/j.1365-2834.2011.01246.x PMID: 21507102

Laflen, J. M., Lane, L. J., & Foster, G. R. (1991). WEPP: A New Generation of Erosion Prediction Technology. *Journal of Soil and Water Conservation*, *46*(1), 34–38. https://doi.org/https://doi.org/10.1080/00224561.1991.12456572. DOI: 10.1080/00224561.1991.12456572

Lama, S., & Devkota, B. (2009). Vulnerability of mountain communities to climate change and adaptation strategies. *Journal of Agriculture and Environment*, *10*, 76–83. DOI: 10.3126/aej.v10i0.2133

Lawanson, O. I., Proverbs, D., & Ibrahim, R. L. (2020). The impact of flooding on poor communities in Lagos State, Nigeria: The case of the Makoko urban settlement. *Journal of Flood Risk Management*. Advance online publication. DOI: 10.1111/jfr3.12838

Lee, C. C., Zeng, M., & Luo, K. (2024). How does climate change affect food security? Evidence from China. *Environmental Impact Assessment Review*, *104*, 107324. DOI: 10.1016/j.eiar.2023.107324

Lee, E. S. (1966). A theory of migration. *Demography*, *3*(1), 47–57. DOI: 10.2307/2060063

Lee, E., & Lee, H. (2019). Disaster awareness and coping: Impact on stress, anxiety, and depression. *Perspectives in Psychiatric Care*, *55*(2), 311–318. DOI: 10.1111/ppc.12351 PMID: 30648274

Lee, H. C., & Chang, C. T. (2018). Comparative analysis of MCDM methods for ranking renewable energy sources in Taiwan. *Renewable & Sustainable Energy Reviews*, *92*, 883–896. DOI: 10.1016/j.rser.2018.05.007

Liang, Y., & Fernandez, D. (2026). A bibliometric analysis of RPA and AI research: Current trends and future directions. *Multidisciplinary Science Journal*, *8*(3), 2026145. Advance online publication. DOI: 10.31893/multiscience.2026145

Liao, J., Xie, Y., Zhao, P., Xia, K., Xu, B., Wang, H., Li, C., Li, C., & Liu, H. (2024). Probabilistic assessment of the thermal performance of low-enthalpy geothermal system under impact of spatially correlated heterogeneity by using XGBoost algorithms. *Energy*, *313*, 133947. Advance online publication. DOI: 10.1016/j.energy.2024.133947

Li, C. C., Mahmood, R., Abdullah, H., & Chuan, O. S. (2013). Economic growth, tourism and selected macroeconomic variables: A triangular causal relationship in Malaysia. *Margin*, *7*(2), 185–206. DOI: 10.1177/0973801013483503

Lightfoot, E. S., Lesen, A. E., & Ferreira, R. J. (2020). Gender and resilience in Gulf Coast communities: Risk and protective factors following a technological disaster. *International Journal of Disaster Risk Reduction*, *50*, 101716. Advance online publication. DOI: 10.1016/j.ijdrr.2020.101716

Li, H.-M., Wang, X.-C., Zhao, X.-F., & Qi, Y.LI. (2021). Understanding systemic risk induced by climate change. *Advances in Climate Change Research*, *12*(3), 384–394. DOI: 10.1016/j.accre.2021.05.006

Lilleør, H. B., & Van den Broeck, K. (2011). Economic drivers of migration and climate change in LDCs. *Global Environmental Change*, *21*(S1), S70–S81. DOI: 10.1016/j.gloenvcha.2011.09.002

Lindell, M. K. (2013). Recovery and reconstructionafterdisaster. In *Encyclopedia of naturalhazards* (pp. 812–824). Springer.

Lin, P., & Lin, W. (2020). Rebuilding relocated tribal communities better via culture: Livelihood and social resilience for disaster risk reduction. *Sustainability (Basel)*, *12*(11), 4538. Advance online publication. DOI: 10.3390/su12114538

Li, Q., Shan, H., Tang, Y., & Yao, V. (2024). Corporate Climate Risk: Measurements and Responses. *Review of Financial Studies*, *37*(6), 1778–1830. DOI: 10.1093/rfs/hhad094

Liu, D., Zhao, Y., Ma, Z., & Ding, Q. (2025). Intelligent target detection and encrypted transmission system based on FPGA. *Computers & Electrical Engineering*, *128*, 110653. Advance online publication. DOI: 10.1016/j.compeleceng.2025.110653

Liu, G., Qiu, J., Huang, J., & Yuan, Y. (2026). GLGF-CR: A Gated Local-Global Fusion approach for cloud removal in real-world remote sensing. *Pattern Recognition*, *172*, 112319. Advance online publication. DOI: 10.1016/j.patcog.2025.112319

Liu, Y., Gu, Y., & Zhang, H. (2024). Seismic risk assessment and damage analysis: Emerging trends and new developments. *Journal of Safety Science and Resilience = An Quan Ke Xue Yu Ren Xing (Ying Wen)*, *5*(3), 365–381. DOI: 10.1016/j.jnlssr.2024.04.005

Li, Y., Zhong, X., Dai, P., Chen, B., Zhou, F., Fan, Y., Wu, J., & He, L. (2025). Construction and validation of a risk predictive model for fear of disease progression in patients after percutaneous coronary intervention. *Scientific Reports*, *16*(1), 1444. DOI: 10.1038/s41598-025-31099-7 PMID: 41353494

Llorente-Marrón, M., Díaz-Fernández, M., Méndez-Rodríguez, P., & Arias, R. G. (2020). Social vulnerability, gender and disasters: The case of Haiti in 2010. *Sustainability (Basel), 12*(9), 3574. DOI: 10.3390/su12093574

Llorente-Marrón, M., Fontanil-Gómez, Y., Díaz-Fernández, M., & García, P. S. (2021). Disasters, gender, and HIV infection: The impact of the 2010 Haiti earthquake. *International Journal of Environmental Research and Public Health, 2021*(18), 7198. DOI: 10.3390/ijerph18137198 PMID: 34281135

Lotka, A. J. (1926). The frequency distribution of scientific productivity. *Journal of the Washington Academy of Sciences, 16*(12), 317–323.

Lu, D., Li, G., Valladares, G. S., & Batistella, M. (2004). Mapping Soil Erosion Risk in Rondônia, Brazilian Amazonia: Using RUSLE, Remote Sensing and GIS. *Land Degradation & Development, 15*(5), 499–512. DOI: 10.1002/ldr.634

Lund, C. (2006). Twilight institutions: Public authority and local politics in Africa. *Development and Change, 37*(4), 685–705. DOI: 10.1111/j.1467-7660.2006.00497.x

Maganda, C. (2016). Water security debates in 'safe'water security frameworks: Moving beyond the limits of scarcity. *Globalizations, 13*(6), 683–701. DOI: 10.1080/14747731.2015.1133605

Magnan, A. K., O'Neill, B. C., & Garschagen, M. (2023). Further understanding "severe" climate risk. *Climate Risk Management, 42*(July), 100538. DOI: 10.1016/j.crm.2023.100538

Mahmood, A. N., El-Khazali, R., Çiçek, A., Assi, S. A., Al-Naimi, T. M., Majdi, H. S., & Hussain, A. S. T. (2025). Experimental Analysis of Fresnel Lens-Based Solar Desalination Systems with Copper Receivers for Enhanced Thermal and Electrical Performance. *International Journal of Robotics & Control Systems, 5*(2), 1191–1210. DOI: 10.31763/ijrcs.v5i2.1852

Manatsa, D., Chingombe, W., & Matarira, C. H. (2023). The impact of the positive Indian Ocean dipole on Zimbabwe droughts Tropical climate is understood to be dominated by. *International Journal of Climatology, 2029*(March 2008), 2011–2029. https://doi.org/DOI: 10.1002/joc

Manning, F. J., & Goldfrank, L. R. (2002). "Preparing for terrorism: tools for evaluating the metropolitan medical response system program." National Academy Press, No. 1, Washington, DC, USA.

Mann, M. E., Lloyd, E. A., & Oreskes, N. (2017). Assessing climate change impacts on extreme weather events: The case for an alternative (Bayesian) approach. *Climatic Change, 144*(2), 131–142. DOI: 10.1007/s10584-017-2048-3

Mariani, S., Zeighami, F., Marzani, A., & Palermo, A. (2025). PSAFE: A semi-analytical finite element formulation for spatially periodic structural systems. *Journal of Sound and Vibration*, *619*, 119391. Advance online publication. DOI: 10.1016/j.jsv.2025.119391

Mariki, S. B. (2016). Social impacts of protected areas on gender in West Kilimanjaro, Tanzania. *Open Journal of Social Sciences*, *4*(3), 220–235. DOI: 10.4236/jss.2016.43028

Marin, D. (1992). Is the export-led growth hypothesis valid for industrialized countries? *The Review of Economics and Statistics*, *74*(4), 678–688. DOI: 10.2307/2109382

Martin, A., Gunter, J. T., & Regens, J. L. (2003). Estimating Erosion in a Riverine Watershed: Bayou Liberty-Tchefuncta River in Louisiana. *Environmental Science and Pollution Research International*, *10*(4), 245–250. DOI: 10.1065/espr2003.05.153 PMID: 12943008

Masria, A., & Abouelsaad, O. (2025). Artificial intelligence applications in coastal engineering and its challenges – A review. *Continental Shelf Research*, *286*, 105425. DOI: 10.1016/j.csr.2025.105425

Massoud, M. A., Scrimshaw, M. D., & Lester, J. N. (2004). Integrated coastal zone and river basin management: A review of the literature, concepts and trends for decision makers. *Water Policy*, *6*(6), 519–548. DOI: 10.2166/wp.2004.0034

Matthews, T., & Potts, R. (2018). Planning for climigration: A framework for effective action. *Climatic Change*, *148*(4), 607–621. DOI: 10.1007/s10584-018-2205-3

Maxwell, S. L., Butt, N., Maron, M., McAlpine, C. A., Chapman, S., Ullmann, A., Segan, D. B., & Watson, J. E. M. (2019). Conservation implications of ecological responses to extreme weather and climate events. *Diversity & Distributions*, *25*(4), 613–625. DOI: 10.1111/ddi.12878

Mayer, B. (2019). A review of the literature on community resilience and disaster recovery. *Current Environmental Health Reports*, *6*(3), 167–173. DOI: 10.1007/s40572-019-00239-3 PMID: 31267349

McBean, G. (2020). Climate Change and Extreme Weather: A Basis for Action. *Natural Hazards*, *31*(1), 177–190. DOI: 10.1023/B:NHAZ.0000020259.58716.0d

McCall, R. T., van Thiel de Vries, J. S. M., Plant, N. G., van Dongeren, A. R., Roelvink, J. A., Thompson, D. M., & Reniers, A. J. H. M. (2010). Two-dimensional time dependent hurricane overwash and erosion modeling at Santa Rosa Island. *Coastal Engineering*, *57*(7), 668–683. DOI: 10.1016/j.coastaleng.2010.02.006

McColl, S. T. (2022). Landslide causes and triggers. In *Landslide Hazards, Risks, and Disasters* (pp. 13–41). Elsevier. DOI: 10.1016/B978-0-12-818464-6.00011-1

McDowell, L. (2003). *Gender, identity and place: Understanding feminist geographies* (2nd ed.). University of Minnesota Press.

McKibbin, W., & Fernando, R. (2020). The economic impact of COVID-19. In Baldwin, R., & Fernando, R. (Eds.), *Economics in the time of COVID-19* (1st ed., pp. 45–51). CEPR Press.

McLeman, R. A. (2013). *Climate and human migration: Past experiences, future challenges*. Cambridge University Press. DOI: 10.1017/CBO9781139136938

McLeman, R., & Smit, B. (2006). Migration as an adaptation to climate change. *Climatic Change*, *76*(1–2), 31–53. DOI: 10.1007/s10584-005-9000-7

Mcloughlin, C. (2015). When does service delivery improve the legitimacy of a fragile or conflict-affected state? *Governance: An International Journal of Policy, Administration and Institutions*, *28*(3), 341–356. DOI: 10.1111/gove.12091

MDH. (2000). "Ministerial declaration of the Hague on water security in the 21st century." Report No. 2, Hague, Netherlands.

MDPI. (2023). Changes in temperature and relative humidity in Equatorial Africa: Analysis of ERA5 reanalysis data (1991–2020). *Atmosphere*, *16*(7), 828.

Mechler, R., Bouwer, L. M., Schinko, T., Surminski, S., & Linnerooth-Bayer, J. (2019). *Loss and damage from climate change: Concepts, methods and policy options*. Springer Nature., DOI: 10.1007/978-3-319-72026-5

Meehl, G. A., Zwiers, F., Evans, J., Knutson, T., Mearns, L., & Whetton, P. (2023). Trends in Extreme Weather and Climate Events: Issues Related to Modeling Extremes in Projections of Future Climate Change. *Bulletin of the American Meteorological Society*, *81*(3), 427–436. DOI: 10.1175/1520-0477(2000)081<0427:TIEWAC>2.3.CO;2

Mendes, R. G., do Valle, R. F.Junior, Feitosa, T. H. S., de Melo, M. M. A. P., Fernandes, L. F. S., Pacheco, F. A. L., & Valera, C. A. (2024). Carbon footprints of tailings dams' disasters: A study in the Brumadinho region (Brazil). *The Science of the Total Environment*, *949*, 175026. DOI: 10.1016/j.scitotenv.2024.175026

Mercandalli, S., Losch, B., Belebema, M. N., Bélières, J.-F., Bourgeois, R., Dinbabo, M. F., Freguin-Gresh, S., Mensah, C., & Nshimbi, C. C. (2019). Rural migration in sub-Saharan Africa: Patterns, drivers and relation to structural transformation. *FAO Working Paper*. https://doi.org/DOI: 10.4060/ca7404en

Meulen, F., IJff, S., & van Zetten, R. (2022). Nature-based solutions for coastal adaptation management, concepts and scope, an overview. *Nordic Journal of Botany, 2023*(1), e03290. DOI: 10.1111/njb.03290

Mfewou, A., Tata, J., & Nadji, P. (2020). Climate variability and peasant agriculture in Kélo in the south of Chad. *Advances in Social Sciences Research Journal, 7*(11), 16–30. DOI: 10.14738/assrj.711.9294

Middleton, B. A. (1999). *Wetland Restoration, Flood Pulsing, and Disturbance Dynamics*. John Wiley & Sons.

Miller, G. A., & Ritchie, B. W. (2003). A Farming Crisis or a Tourism Disaster? An Analysis of the Foot and Mouth Disease in the UK. *Current Issues in Tourism, 6*(2), 150–171. DOI: 10.1080/13683500308667949

Millward, A. A., & Mersey, J. E. (1999). Adapting the RUSLE to Model Soil Erosion Potential in a Mountainous Tropical Watershed. *Catena, 38*(2), 109–129. DOI: 10.1016/S0341-8162(99)00067-3

Minamyer, S. (2008). Effective crisis communication during water security emergencies. *Journal - American Water Works Association, 100*(9), 180–184. DOI: 10.1002/j.1551-8833.2008.tb09719.x

Ministry of Environment, Urbanization and Climate Change (MoEUCC). (2021). *Turkey's national climate change strategy and action plan (2011–2023) revision.* Official Gazette of the Republic of Türkiye.

Ministry of Environment. (2014). *Cambodia climate change strategic plan 2014-2023.* Phnom Penh: Royal Government of Cambodia. https://www.undp.org/cambodia/publications/cambodia-climate-change-strategic-plan-2014-2023

Mirza, M. (2023). Climate change and extreme weather events: Can developing countries adapt? *Climate Policy, 3*(3), 233–248. DOI: 10.3763/cpol.2003.0330

Mitsch, W. J., & Gosselink, J. G. (2000). *Wetlands*. John Wiley & Sons.

Mohapatra, M., & Sharma, M. (2021). Cyclone warning services in India during recent years: A review. *Mausam (New Delhi), 70*(4), 635–666. DOI: 10.54302/mausam.v70i4.204

Mollamahmutoğlu, M., & Acar, Y. (2020). Human-Induced Landslide Risk Analysis at Durhasanlı Reservoir Site. In Shehata, H., Brandl, H., Bouassida, M., & Sorour, T. (Eds.), *Sustainable Thoughts in Ground Improvement and Soil Stability* (pp. 104–115). Springer International Publishing., DOI: 10.1007/978-3-030-34184-8_7

Molle, F., Foran, T., & Käkönen, M. (2009). Contested waterscapes in the Mekong Region.

Monreal-Jiménez, R., Carbajal, N., Contreras-Tereza, V. K., & Salas-Monreal, D. (2025). Sediment Dynamics and Erosion in a Complex Coastal Lagoon System in the Southern Gulf of Mexico. Water, 17(16), 2408. https://doi.org/10.3390/w17162408

Monteil, C., Simmons, P., & Hicks, A. (2020). Post-disaster recovery and socio-cultural change: Rethinking social capital development for the new social fabric. *International Journal of Disaster Risk Reduction*, *42*, 101356. Advance online publication. DOI: 10.1016/j.ijdrr.2019.101356

Moomaw, W. R., Chmura, G. L., Davies, G. T., Finlayson, C. M., Middleton, B. A., Natali, S. M., Perry, J. E., Roulet, N., & Sutton-Grier, A. E. (2018). Wetlands in a changing climate: Science, policy and management. *Wetlands*, *38*(2), 183–205. DOI: 10.1007/s13157-018-1023-8

Moore, I. D., & Burch, G. J. (1986). Physical Basis of the Length-Slope Factor in the Universal Soil Loss Equation. *Soil Science Society of America Journal*, *50*(5), 1294–1298. https://doi.org/https://doi.org/10.2136/sssaj1986.03615995005000050042x. DOI: 10.2136/sssaj1986.03615995005000050042x

Moreno, J., & Møller, A. P. (2023). Extreme climatic events in relation to global change and their impact on life histories. *Current Zoology*, *57*(3), 375–389. DOI: 10.1093/czoolo/57.3.375

Moreno, J., & Shaw, D. (2018). Women's empowerment following disaster: A longitudinal study of social change. *Natural Hazards*, *2018*(92), 205–224. DOI: 10.1007/s11069-018-3204-4

Morrissey, S. A., & Reser, J. P. (2007). National Research Council, Division on Earth, Life Studies, Committee on Disaster Research in the Social Sciences, FutureChallenges, &Opportunities. (2006). Facinghazards and disasters: Understandinghumandimensions. NationalAcademiesPress. Natural disasters, climate change and mental health considerations for rural Australia. *The Australian Journal of Rural Health*, *15*(2), 120–125. DOI: 10.1111/j.1440-1584.2007.00865.x PMID: 17441821

Mou, N., Yuan, R., Yang, T., Zhang, H., Tang, J., & Makkonen, T. (2020). Exploring spatio-temporal changes of city inbound tourism flow: The case of Shangai, China. *Tourism Management*, *76*, 1–14. DOI: 10.1016/j.tourman.2019.103955

MoWA. (2018). *Master plan on gender and climate change (2018-2030)*. Phnom Penh: Ministry of Women's Affairs-Royal Government of Cambodia. https://ncsd.moe.gov.kh/sites/default/files/2019-06/MASTER%20PLAN%20ON%20GENDER%20AND%20CLIMATE%20CHANGE.pdf

Mukhametzyanov, I., & Pamucar, D. (2018). A sensitivity analysis in MCDM problems: A statistical approach. *Decision Making: Applications in Management and Engineering*, *1*(2), 51–80. DOI: 10.31181/dmame1802050m

Mulesa, O., & Bohdan, Y. (2024). DEVELOPMENT OF A FUZZY PRODUCTION MODEL FOR ASSESSING THE DEGREE OF INFORMATION SECURITY IN INTERNATIONAL COOPERATION. *Technology Audit and Production Reserves*, *6*(2), 6–10. DOI: 10.15587/2706-5448.2024.318446

Mundhe, N. N., & Jaybhaye, R. G. (2023). Land Suitability Analysis for In Situ Slum Redevelopment of Pune City Using AHP, Remote Sensing and GIS Techniques. *Photonirvachak (Dehra Dun)*, *51*(8), 1777–1795. DOI: 10.1007/s12524-023-01731-y

Murayama, Y. (2021). Use of IT for situation awareness for disaster risk reduction. In *Disaster risk reduction* (pp. 145–160). Springer., DOI: 10.1007/978-981-16-0360-0_9

Musa, S. S., Bouri, T., & Manirambona, E. (2022). How climate change and insecurity pushed 5 million people to hunger in Chad, Africa. *Public Health Challenges*, *1*(2), 1–8. DOI: 10.1002/puh2.47 PMID: 40496685

Mwinjuma, M., Wang, R., Mtupili, M., & Twaha, M. (2026). Comparisons of SPI and SPEI in capturing drought dynamics: A Global assessment across arid and humid regions. *Atmospheric Research*, *329*, 108475. Advance online publication. DOI: 10.1016/j.atmosres.2025.108475

Myers, N. (2002). Environmental refugees: A growing phenomenon of the 21st century. *Philosophical Transactions of the Royal Society of London. Series B, Biological Sciences*, *357*(1420), 609–613. DOI: 10.1098/rstb.2001.0953 PMID: 12028796

Nagarajan, C., Vivekananda, J., Pham-Duc, B., Sylvestre, F., Pohl, B., & Munoz, H. (2024). Peace in an extreme climate: How climate-related security risks affect prospects for stability in Lake Chad. *PLOS Climate*, *3*(10), e0000314. Advance online publication. DOI: 10.1371/journal.pclm.0000314

NASA/METI. (2009). ASTER Global Digital Elevation Model V2 [30m resolution DEM]. https://doi.org/DOI: 10.5067/ASTER/ASTGTM.002

NASA's Earth observing system. (n.d.). https://eospso.nasa.gov/content/nasas-earth-observing-system-project-science-office

Nastev, M., & Todorov, N. (2013). Hazus: A standardized methodology for flood risk assessment in Canada. Canadian Water Resources Journal / Revue Canadienne Des Ressources Hydriques, 38(3), 223–231. https://doi.org/DOI: 10.1080/07011784.2013.801599

Nawrotzki, R. J., DeWaard, J., Bakhtsiyarava, M., & Ha, J. T. (2017). Climate shocks and rural-urban migration in Mexico: Exploring nonlinearities and thresholds. *Climatic Change*, *140*(2), 243–258. DOI: 10.1007/s10584-016-1849-0 PMID: 28435176

Ndimele, P. E., Ojewole, A. E., Mekuleyi, G. O., Badmos, L. A., Agosu, C. M., Olatunbosun, E. S., Lawal, O. O., Shittu, J. A., Joseph, O. O., Ositimehin, K. M., Ndimele, F. C., Ojewole, C. O., Abdulganiy, I. O., & Ayodele, O. T. (2024). Vulnerability, resilience and adaptation of Lagos coastal communities to flooding. *Earth Science Systems and Society*, *4*(1), 10087. DOI: 10.3389/esss.2024.10087

Nel, J. L., Le Maitre, D. C., Roux, D. J., Colvin, C., Smith, J. S., Smith-Adao, L. B., Maherry, A., & Sitas, N. (2017). Strategic water source areas for urban water security: Making the connection between protecting ecosystems and benefiting from their services. *Ecosystem Services*, *28*, 251–259. DOI: 10.1016/j.ecoser.2017.07.013

Nelson-Smith, A. (1978). Effects of dispersant use on shore life. In *Chemical dispersants for the control of oil spills*. ASTM International. DOI: 10.1520/STP35725S

Nene, G., & Taivan, A. (2017). Causality between tourism and economic growth: Evidence from Sub Saharan African (SSA). *Journal of Developing Areas*, *51*(2), 155–169. DOI: 10.1353/jda.2017.0037

Nesanır, M. O., Zenezini, G., Senvar, O., & Ottaviani, F. M. (2026). A Literature Review on the Integration of System Dynamics and Machine Learning/Artificial Intelligence in Addressing Environmental Matters. *IFIP Advances in Information and Communication Technology, 764 IFIPAI*, 370–384. DOI: 10.1007/978-3-032-03515-8_26

Niedenführ, H., & Steinbrink, M. (2019). *Africa on the move: Migration, translocal livelihoods and rural development in Sub-Saharan Africa*. https://www.amazon.com/Africa-Move-Livelihoods-Development-Sub-Saharan/dp/3030228436

Nieto, M. J. (2019). Banks, climate risk and financial stability. *Journal of Financial Regulation and Compliance*, *27*(2), 243–262. DOI: 10.1108/JFRC-03-2018-0043

Ningsih, R. D., Ariyaningsih, A., Sukhwani, V., & Shaw, R. (2022). Sendai Framework and its implementation. In *Handbook on Climate Change and Disasters* (pp. 30–41). Edward Elgar Publishing.

Nkwunonwo, . (2015). Relevance of social vulnerability assessment to flood risk reduction in the Lagos Metropolis of Nigeria. *British Journal of Applied Science and Technology*, *8*(4), 366–382. DOI: 10.9734/BJAST/2015/17518

Noureldeen, N., Mao, K., Mohmmed, A., Yuan, Z., & Yang, Y.NourEldeen. (2020). Spatiotemporal drought assessment over Sahelian countries from 1985 to 2015. *Journal of Meteorological Research*, *34*(4), 760–774. DOI: 10.1007/s13351-020-9178-7

Nuorteva, P., Keskinen, M., & Varis, O. (2010). Water, livelihoods and climate change adaptation in the Tonle Sap Lake area, Cambodia: Learning from the past to understand the future. *Journal of Water and Climate Change*, *1*(1), 87–101. DOI: 10.2166/wcc.2010.010

Nyaupane, G. P., & Chhetri, N. (2009). Vulnerability to climate change of nature-based tourism in the Nepalese Himalaya. *The Journal of Climate Change and Health*, *10*, 100208.

O'Brien, K., Eriksen, S., Nygaard, L. P., & Schjolden, A. (2011). Why different interpretations of vulnerability matter in climate change discourses. *Climate Policy*, *7*(1), 73–88. DOI: 10.1080/14693062.2007.9685639

Obokata, R., Veronis, L., & McLeman, R. (2014). Empirical research on international environmental migration: A systematic review. *Population and Environment*, *36*(1), 111–135. DOI: 10.1007/s11111-014-0210-7 PMID: 25132701

OECD. (2020) Tourism trends and policies 2020. https://www.oecd.org/cfe/tourism/OECD-Tourism-Trends-Policies%202020-Highlights-ENG.pdf

OECD. (2024) Employment by activity. https://data.oecd.org/emp/employment-by-activity.htm#indicator-chart

Ojiego-Okoro, C., Umeh, L., & Nwodom, D. (2024). Boko Haram-induced forced migration: Implications for social and economic development in the Lake Chad Basin. *International Journal of Social Science and Human Research*, *7*(5). Advance online publication. DOI: 10.47191/ijsshr/v7-i05-94

Ökten, Ş. (2017). Environmental justice, dams and displacement in Southeastern Anatolia region, Turkey. *Journal of International Social Research*, *10*(50), 414–419. DOI: 10.17719/jisr.2017.1674

Okutan, P., & Otay, E. N. (2025). Climate change risk perception, adaptive capacity and psychological distance in urban vulnerability: A district-level case study in Istanbul, Türkiye. *Sustainability (Basel)*, *17*(12), 5358. DOI: 10.3390/su17125358

Olowoyeye, O., & Kanwar, R. (2023). Water and food sustainability in the riparian countries of Lake Chad in Africa. *Sustainability (Basel), 15*(13), 10009. Advance online publication. DOI: 10.3390/su151310009

Ornes, S. (2018). How does climate change influence extreme weather? Impact attribution research seeks answers. *Proceedings of the National Academy of Sciences of the United States of America, 115*(33), 8232–8235. DOI: 10.1073/pnas.1811393115 PMID: 30108186

Ortakaya, Ü. İ. (2023). Examining climate migration risks through the lens of social equality: An evaluation of Germany, France and Türkiye. *Hacettepe Üniversitesi Sosyal Bilimler Dergisi, 5*(2), 186–210.

Ostrom, E. (2012). Polycentric systems for coping with collective action and global environmental change. In Barry, C. (Ed.), *Global justice* (pp. 423–430). Routledge., DOI: 10.4324/9781315254210

Owonikoko, S. B., & Momodu, J. A. (2020). Environmental degradation, livelihood, and the stability of Chad Basin Region. *Small Wars & Insurgencies, 31*(6), 1295–1322. DOI: 10.1080/09592318.2020.1776092

Ozcelik, M. (2022). Comparison of the environmental impact and production cost rates of aggregates produced from stream deposits and crushed rock quarries (Boğaçay Basin/Antalya/Turkey). *Geoheritage, 14*(1), 18. DOI: 10.1007/s12371-022-00659-y

Ozcelik, M., & Tuzlu, F. (2022). Environmental impacts of abandoned marina construction: An example Boğaçay River and Marina Project (Antalya, Turkey). *Arabian Journal of Geosciences, 15*(23), 1721. DOI: 10.1007/s12517-022-11026-0

Özkan, G. G. (2024). Deprem döneminde belirsizliğe tahammülsüzlük düzeyinin çalışanların iş stresi üzerindeki etkisinde psikolojik iyi oluşun aracılık rolü (Master's thesis, Hasan Kalyoncu Üniversitesi Lisansüstü Eğitim Enstitüsü İşletme Ana Bilim Dalı).

Özkazanç, S., & Yüksel, Ü. (2015). Evaluation of disaster awareness and sensitivity level of higher education students. *Procedia: Social and Behavioral Sciences, 197*, 745–753. DOI: 10.1016/j.sbspro.2015.07.168

Palmer, G., Platts, P. J., Brereton, T., Chapman, J. W., Dytham, C., Fox, R., Pearce-Higgins, J. W., Roy, D. B., Hill, J. K., & Thomas, C. D. (2017). Climate change, climatic variation and extreme biological responses. *Philosophical Transactions of the Royal Society B: Biological Sciences, 372*(1723), 20160144. DOI: 10.1098/rstb.2016.0144

Panagos, P., Borrelli, P., Meusburger, K., Alewell, C., Lugato, E., & Montanarella, L. (2015). Estimating the Soil Erosion Cover-Management Factor at the European Scale. *Land Use Policy*, *48*, 38–50. DOI: 10.1016/j.landusepol.2015.05.021

Panagos, P., Borrelli, P., Poesen, J., Ballabio, C., Lugato, E., Meusburger, K., Montanarella, L., & Alewell, C. (2015). The New Assessment of Soil Loss by Water Erosion in Europe. *Environmental Science & Policy*, *54*, 438–447. DOI: 10.1016/j.envsci.2015.08.012

Panagos, P., Meusburger, K., Ballabio, C., Borrelli, P., & Alewell, C. (2014). Science of the Total Environment Soil Erodibility in Europe : A high-Resolution Dataset Based on LUCAS. *The Science of the Total Environment*, *479*(480), 189–200. DOI: 10.1016/j.scitotenv.2014.02.010 PMID: 24561925

Pang, T., Wang, X., Nawaz, R. A., Keefe, G., & Adekanmbi, T. (2023). Coastal erosion and climate change: A review on coastal-change process and modeling. *Ambio*, *52*(12), 2034–2052. DOI: 10.1007/s13280-023-01901-9 PMID: 37405570

Park, A.-S., & Yang, J.-H. (2017). The effects of disaster awareness, attitude, and preparedness on the basic competencies of disaster nursing among nursing students. *International Journal of Advanced Nursing Education and Research*, *2*(1), 59–64. DOI: 10.21742/IJANER.2017.2.1.11

Parkash, S. (2023). Lessons Learned from Landslides of Socio-economic and Environmental Significance in India. In I. Alcántara-Ayala, Ž. Arbanas, D. Huntley, K. Konagai, M. Mikoš, K. Sassa, S. Sassa, H. Tang, & B. Tiwari (Eds.), *Progress in Landslide Research and Technology, Volume 1 Issue 2, 2022* (pp. 309–315). Springer International Publishing. DOI: 10.1007/978-3-031-18471-0_23

Parry, L., Davies, G., Almeida, O., Frausin, G., de Moraés, A., Rivero, S., Filizola, N., & Torres, P. (2018). Social vulnerability to climatic shocks is shaped by urban accessibility. *Annals of the American Association of Geographers*, *108*(1), 125–143. DOI: 10.1080/24694452.2017.1325726

Patel, R. K., & Kermanshachi, S. (2020). *A socioeconomic-based analysis of disaster preparedness, awareness and education. ResearchGate PDF*. Retrieved from. [ResearchGate]

Pecoraro, G., Calvello, M., & Piciullo, L. (2019). Monitoring strategies for local landslide early warning systems. *Landslides*, *16*(2), 213–231. DOI: 10.1007/s10346-018-1068-z

Peduzzi, P. (2019). The disaster risk, global change, and sustainability nexus. *Sustainability (Basel)*, *11*(4), 957. Advance online publication. DOI: 10.3390/su11040957

Pelek, D., Yıldızcan, C., & Turhan, E. (2025). Permanently temporary: Unveiling the im/mobility and intersecting vulnerabilities of migrant seasonal agricultural workers in disaster-affected areas of Türkiye. *Agriculture and Human Values*, *42*(4), 2541–2557. DOI: 10.1007/s10460-025-10761-w

Pelling, M. (2010). *Adaptation to climate change: From resilience to transformation.* Routledge., DOI: 10.4324/9780203889046

Peña-Mora, F., Aziz, Z., Chen, A. Y., Plans, A., & Foltz, S. (2008). Building assessment during disaster response and recovery. *Proceedings of the Institution of Civil Engineers. Urban Design and Planning*, *161*(4), 183–195. DOI: 10.1680/udap.2008.161.4.183

Penuel, K. B., & Statler, M. (2010). "Encyclopedia of disaster relief." SAGE, No. 2, California, USA.

Pepin, M., Bouloy, M., Bird, B. H., Kemp, A., & Paweska, J. (2010). Rift Valley fever virus (Bunyaviridae: Phlebovirus): an update on pathogenesis, molecular epidemiology, vectors, diagnostics and prevention. *Veterinary Research*, *41*(6), 61. DOI: 10.1051/vetres/2010033

Perkins-Kirkpatrick, S. E., Stone, D. A., Mitchell, D. M., Rosier, S., King, A. D., Lo, Y. T. E., Pastor-Paz, J., Frame, D., & Wehner, M. (2022). On the attribution of the impacts of extreme weather events to anthropogenic climate change. *Environmental Research Letters*, *17*(2), 024009. DOI: 10.1088/1748-9326/ac44c8

Perrings, C. (2005). Mitigation and adaptation strategies for the control of biological invasions. *Ecological Economics*, *52*(3), 315–325. DOI: 10.1016/j.ecolecon.2004.08.001

Perry, R. W., & Lindell, M. K. (2003). Preparedness for Emergency Response: Guidelines for the Emergency Planning Process. *Disasters*, *27*(4), 336–350. DOI: 10.1111/j.0361-3666.2003.00237.x PMID: 14725091

Peychev, V., & Stancheva, M. (2009). Changes of sediment balance at the Bulgarian Black Sea coastal zone influenced by anthropogenic impacts. *Compt. Rend. Acad. Bulg. Sci*, *62*(2), 277–285.

Phan, L. T., Jou, S. C., Chiou, C. R., & Yang, C. J. (2024). Is it because of climate change? Social-ecological system analysis of wetland protected area in Vietnamese Mekong Delta. *Wetlands Ecology and Management*, *32*(1), 33–57. DOI: 10.1007/s11273-023-09960-1

Piguet, E. (2022). Linking climate change, environmental degradation, and migration: An update after 10 years. *Wiley Interdisciplinary Reviews: Climate Change, 13*(1), e746. DOI: 10.1002/wcc.746

Pineda, V. G., Valencia-Arias, A., Giraldo, F. E. L., & Zapata-Ochoa, E. A. (2026). Integrating artificial intelligence and quantum computing: A systematic literature review of features and applications. *International Journal of Cognitive Computing in Engineering, 7*, 26–39. DOI: 10.1016/j.ijcce.2025.08.002

Poggio, L., Sousa, L. M. D., Batjes, N. H., Heuvelink, G. B. M., Kempen, B., & Ribeiro, E. (2021). SoilGrids 2. 0 : Producing Soil Information for the Globe with Quantified SpatialUncertainty. SOIL, 7(1995), 217–240. https://doi.org/https://doi .org/10.5194/soil-7-217-2021

Powell, N., Kløcker Larsen, R., de Bruin, A., Powell, S., & Elrick-Barr, C. (2017). Water Security in Times of Climate Change and Intractability: Reconciling Conflict by Transforming Security Concerns into Equity Concerns. *Water (Basel), 9*(12), 934. DOI: 10.3390/w9120934

Pradhan, S. P., & Siddique, T. (2019). Mass Wasting: An Overview. In Pradhan, S. P., Vishal, V., & Singh, T. N. (Eds.), *Landslides: Theory, Practice and Modelling* (Vol. 50, pp. 3–20). Springer International Publishing., DOI: 10.1007/978-3-319-77377-3_1

Pratiwi, A. P., Rohmat, F. I. W., Nugroho, E. O., Farid, M., & Kusuma, M. S. B. (2024, March). Flood hazard comparison based on geomorphic flood index and hydraulic HEC-RAS (Case study in Ciliwung Watershed, Jakarta). [). IOP Publishing.]. *IOP Conference Series. Earth and Environmental Science, 1314*(1), 012084. DOI: 10.1088/1755-1315/1314/1/012084

Pullin, A. S., & Knight, T. M. (2009). Doing more good than harm–Building an evidence-base for conservation and environmental management. *Biological Conservation, 142*(5), 931–934. DOI: 10.1016/j.biocon.2009.01.010

Qin, Y., Luo, Y., Zhao, Y., & Zhang, J. (2018). Research on relationship between tourism income and economic growth based on meta-analysis. *Applied Mathematics and Nonlinear Sciences, 3*(1), 105–114. DOI: 10.21042/AMNS.2018.1.00008

Quandt, A., Neufeldt, H., & Gorman, K. (2023). Climate change adaptation through agroforestry: Opportunities and gaps. *Current Opinion in Environmental Sustainability, 60*, 101244. DOI: 10.1016/j.cosust.2022.101244

Queiroz, H. M., Nóbrega, G. N., Ferreira, T. O., Almeida, L. S., Romero, T. B., Santaella, S. T., Bernardino, A. F., & Otero, X. L. (2018). The Samarco mine tailing disaster: A possible time-bomb for heavy metals contamination? *The Science of the Total Environment*, *637*, 498–506. DOI: 10.1016/j.scitotenv.2018.04.370

Quinn, P., Williamson, B., & Gibbs, L. (2022). Indigenous-informed disaster recovery: Addressing collective trauma using a healing framework. *Progress in Disaster Science*, *16*, 100257. Advance online publication. DOI: 10.1016/j.pdisas.2022.100257

Rahman, M., Goel, N., & Arya, D. (2012). Study of early flood warning dissemination system in Bangladesh. *Journal of Flood Risk Management*, *6*(4), 290–301. DOI: 10.1111/jfr3.12012

Reed, M. S. (2008). Stakeholder participation for environmental management: A literature review. *Biological Conservation*, *141*(10), 2417–2431. DOI: 10.1016/j.biocon.2008.07.014

Renard, K. G., Foster, G. R., Weesies, G. A., McCool, D. K., & Yoder, D. C. (1997). *Predicting Soil Erosion by Water: a Guide to Conservation Planning With the Revised Universal Soil Loss Equation (RUSLE)*. US Department of Agriculture, Agricultural Research Service.

Renard, K. G., Foster, R., Weesies, G., & Porter, G. (1991). RUSLE: Revised Universal Soil Loss Equation. *Journal of Soil and Water Conservation*, *46*(1), 30–33. DOI: 10.1080/00224561.1991.12456571

Republic of Chad. (2021). *First national climate change adaptation plan of Chad*. United Nations Framework Convention on Climate Change (UNFCCC). https://unfccc.int/sites/default/files/resource/CHAD-NAP_EN-web.pdf

Reyes, D. D., & Lu, J. L. (2016). Gender dimension in disaster situations: A case study of flood prone women in Malabon City, Metro Manila. *International Journal of Disaster Risk Reduction*, *15*, 162–168. DOI: 10.1016/j.ijdrr.2015.11.001

RGC. (2019). *National srategic development plan*. Phnom Penh. http://cdc-crdb.gov.kh/en/strategy/documents/nsdp-2019-2023_en.pdf

Rigaud, K. K., de Sherbinin, A., Jones, B., Bergmann, J., Clement, V., Ober, K., Schewe, J., Adamo, S., Kuriakose, A., Weber, U., & McClure, B. (2018). *Groundswell: Preparing for internal climate migration*. World Bank. DOI: 10.1596/29461

Rijsberman, F. R. (2006). Water scarcity: Fact or fiction? *Agricultural Water Management*, *80*(1-3), 5–22. DOI: 10.1016/j.agwat.2005.07.001

Rising, J., Tedesco, M., Piontek, F., & Stainforth, D. A. (2022). The missing risks of climate change. *Nature, 610*(7933), 643–651. DOI: 10.1038/s41586-022-05243-6 PMID: 36289386

Ritchie, B. W., & Jiang, Y. (2019). A review of research on tourism risk, crisis and disaster management: Launching the annals of tourism research curated collection on tourism risk, crisis and disaster management. *Annals of Tourism Research, 79*, 102812. DOI: 10.1016/j.annals.2019.102812

Ritchie, J. R. B., & Goeldner, C. R. (1994). *Travel, tourism, and hospitality research: A handbook for managers and researchers* (2nd ed.). Wiley.

Rittger, K., Bair, E., Racoviteanu, A., Brodzik, M. J., Armstrong, R. L., Jodha Khalsa, S., Raup, B., Painter, T. H., & Dozier, J. (2017). "Quantifying the individual contributions of melt from snow and glaciers in High Mountain Asia river basins: Syr Darya, Amu Darya, Indus, Ganges, and Brahmaputra." *Proceedings of European Geosciences Union General Assembly Conference Abstracts*, Vienna, Austria, April 23–28.

Rjoub, D., Alsharoa, A., & Masadeh, A. (2023). Unmanned-Aircraft-System-Assisted Early Wildfire Detection with Air Quality Sensors. *Electronics (Basel), 12*(5), 1239. DOI: 10.3390/electronics12051239

Röbke, B. R., & Vött, A. (2017). The tsunami phenomenon. *Progress in Oceanography, 159*, 296–322. DOI: 10.1016/j.pocean.2017.09.003

Rodríguez-Rojo, C. N., Pérez-Cayeiro, M. L., & Chica-Ruiz, J. A. (2025). Adaptation to Climate Change in Coastal Countries of the European Union—An Evaluation of Plans and Strategies. *Applied Sciences (Basel, Switzerland), 15*(11), 6281. DOI: 10.3390/app15116281

Roelvink, D., Reniers, A., van Dongeren, A., van Thiel de Vries, J., McCall, R., & Lescinski, J. (2009). Modeling storm impacts on beaches, dunes and barrier islands. *Coastal Engineering, 56*(11–12), 1133–1152. DOI: 10.1016/j.coastaleng.2009.08.006

Rohanadeera, S. L., & Hewagamage, K. P. (2026). Generative-AI-Powered TA BOT for Enhanced Personalized Support in Software Engineering Education. *Communications in Computer and Information Science, 2380 CCIS*, 197–204. DOI: 10.1007/978-981-96-6291-3_16

Romero-Lankao, P., & Gnatz, D. M. (2016). Conceptualizing urban water security in an urbanizing world. *Current Opinion in Environmental Sustainability, 21*, 45–51. DOI: 10.1016/j.cosust.2016.11.002

Romm, J. (2018). Extreme Weather and Climate Change. In *Climate Change*. Oxford University Press., DOI: 10.1093/wentk/9780190866112.003.0002

Ronan, K. R., & Johnston, D. M. (2005). Community resilience to disasters: Introduction to theory and review of research. In Ronan, K. R., & Johnston, D. M. (Eds.), *Promoting community resilience in disasters: The role for schools, youth, and families* (pp. 9–48). Springer., DOI: 10.1007/0-387-23821-2_2

Roy, P., & Pal, S. C. (2024). Modeling soil erosion susceptibility considering morphometric analysis and SWAT application: Policy recommendation to achieve SDGs. *Modeling Earth Systems and Environment*, *10*(4), 5735–5752. DOI: 10.1007/s40808-024-02089-x

Ruan, L., Yan, M., Zhang, L., Fan, X., & Yang, H. (2022). Spatial-temporal NDVI pattern of global mangroves: A growing trend during 2000–2018. *The Science of the Total Environment*, *844*, 157075. DOI: 10.1016/j.scitotenv.2022.157075

Ruppel, O., & Funteh, M. (2019). Chapter 5: Climate change, human security and the humanitarian crisis in the Lake Chad Basin region: selected legal and developmental aspects with a special focus on water governance. *Law | Environment | Africa*. https://doi.org/DOI: 10.5771/9783845294605-105

Sağlam, R., Uyguner, Z., & Küp, F. (2021). Şanlıurfa Bölgesinde Farklı Yıllarda Çırçırlanmış Pamuğun Lif Kalitesine Etki Eden Faktörlerin Belirlenmesi. *Journal of the Institute of Science and Technology*, *11*(1), 709–721.

Şahin, Ş., & Kurum, E. (2002). Erosion risk analysis by GIS in environmental impact assessments: A case study—Seyhan Köprü Dam construction. *Journal of Environmental Management*, *66*(3), 239–247. DOI: 10.1016/S0301-4797(02)90574-8 PMID: 12448403

Sakarya, A., Akın, O., & Kılıç, A. (2024). Investigating the damage from the February 6, 2023 earthquakes in terms of building characteristics in the Antakya and Defne districts of Hatay province. *International Journal of Disaster Risk Reduction*, *112*, 104783. DOI: 10.1016/j.ijdrr.2024.104783

Salari, M., Abbasiyazdi, A., & Shabanikiya, H. (2025). Geospatial analysis for identifying socially vulnerable areas to natural disasters. *Disaster Medicine and Public Health Preparedness*, *19*(e208), 1–11. DOI: 10.1017/dmp.2025.123 PMID: 40709522

Saleh, K., Mabrouk, W. M., & Metwally, A. M. (2025). Convolutional neural network approach for automated well zonation in the Lower Bahariya member north Western Desert Egypt. *Scientific Reports*, *16*(1), 625. DOI: 10.1038/s41598-025-32119-2 PMID: 41453993

Sánchez, A., Favara, M., Sheridan, M., & Behrman, J. (2024). Does early nutrition predict cognitive skills during later childhood? Evidence from two developing countries. *World Development, 176*, 106480. Advance online publication. DOI: 10.1016/j.worlddev.2023.106480 PMID: 38249341

Sánchez-Almodóvar, E., Gómez-Trigueros, I. M., & Olcina-Cantos, J. (2022). Climate Change and Extreme Weather Events in the Education of the Citizens of the Twenty-First Century: The Perception of Secondary Education Students. *Social Sciences (Basel, Switzerland), 12*(1), 27. DOI: 10.3390/socsci12010027

Sánchez-Cañizares, S. M., Cabeza-Ramírez, L. J., Muñoz-Fernández, G., & Fuentes-García, F. J. (2021). Impact of the perceived risk from Covid-19 on intention to travel. *Current Issues in Tourism, 24*(7), 970–984. DOI: 10.1080/13683500.2020.1829571

Sardar, A., Kıanı, A. K., Kuslu, Y., & Bılgıc, A. (2020). Examining the role of livelihood diversification as a part of climate-smart agriculture (CSA) strategy. *Atatürk Üniversitesi Ziraat Fakültesi Dergisi, 51*(1), 79–87. DOI: 10.17097/ataunizfd.604937

Sarı Karademir, B., & Şahin, S. B. (2025). Gender and disaster management in Turkey: A strategic-relational analysis of the state response to the February 2023 earthquakes. *Southeast European and Black Sea Studies*, 1–19. Advance online publication. DOI: 10.1080/14683857.2025.2473169

Sayan, R. C. (2019). Exploring place-based approaches and energy justice: Ecology, social movements, and hydropower in Turkey. *Energy Research & Social Science, 57*, 101234. DOI: 10.1016/j.erss.2019.101234

Scarpa, G. M., Davison, S., Manfè, G., Lorenzetti, G., Zaggia, L., & Braga, F. (2025). Suspended sediment dynamics at the inlets of Venice Lagoon: Unraveling the effects of storm surges and mobile barrier operations. *Journal of Hydrology (Amsterdam), 651*, 132588. DOI: 10.1016/j.jhydrol.2024.132588

SCDF. (2020). *Annual report FY2019-2020*. Singapore Civil Defence Force. https://www.scdf.gov.sg/home/about-scdf/media-room/publications/annual-report/FY2019-2020

Scheffers, A. M., May, S. M., & Kelletat, D. H. (2015). Mass Movements: Landforms Shaped Under the Force of Gravity. In A. M. Scheffers, S. M. May, & D. H. Kelletat, *Landforms of the World with Google Earth* (pp. 163–182). Springer Netherlands. DOI: 10.1007/978-94-017-9713-9_8

Schmid, A. P., & Jongman, A. J. (2017). "Political terrorism: A new guide to actors, authors, concepts, data bases, theories, and literature." Routledge, No. 1, New York, USA.

Schuster, R. L., & Highland, L. M. (2007). Overview of the Effects of Mass Wasting on the Natural Environment. *Environmental & Engineering Geoscience, 13*(1), 25–44. DOI: 10.2113/gseegeosci.13.1.25

Scott, D., & Lemieux, C. (2010). Weather and climate information for tourism. *Procedia Environmental Sciences, 1*, 146–183. DOI: 10.1016/j.proenv.2010.09.011

Scott, M., Gkartzios, M., & Halfacree, K. (2024). Introducing climate-related counterurbanisation: Individual adaptation or societal maladaptation? *Habitat International, 143*, 102970. DOI: 10.1016/j.habitatint.2023.102970

Sebesvari, Z., Sudmeier-Rieux, K., & Walz, Y. (2023). Towards a typology of nature-based solutions for disaster risk reduction. *Nature-Based Solutions, 3*, 100057. DOI: 10.1016/j.nbsj.2023.100057

Sebesvari, Z., Woelki, J., Walz, Y., Sudmeier-Rieux, K., Sandholz, S., Tol, S., Ruíz García, V., Blackwood, K., & Renaud, F. G. (2019). Opportunities for considering green infrastructure and ecosystems in the Sendai Framework Monitor. *Progress in Disaster Science, 2*, 100021. DOI: 10.1016/j.pdisas.2019.100021

Secretariat, A. S. E. A. N. (2021). *AADMER work programme 2021-2025*. ASEAN Secretariat., https://asean.org/wp-content/uploads/2021/08/AADMER-Work-Programme-2021-2025.pdf

Seddon, N., Chausson, A., Berry, P., Girardin, C. A. J., Smith, A., & Turner, B. (2020). Understanding the value and limits of nature-based solutions to climate change and other global challenges. *Philosophical Transactions of the Royal Society of London. Series B, Biological Sciences, 375*(1794), 20190120. DOI: 10.1098/rstb.2019.0120

Şekercioğlu, Ç. H., Primack, R. B., & Wormworth, J. (2012). The effects of climate change on tropical birds. *Biological Conservation, 148*(1), 1–18. DOI: 10.1016/j.biocon.2011.10.019

Sekti, B. A. (2025). Machine learning and artificial intelligence (AI) in manufacturing. In *Integrating Digital Innovation and Integrated Frameworks in Manufacturing* (pp. 255–292). DOI: 10.4018/979-8-3373-1082-4.ch011

Selby, J., Dahi, O. S., Fröhlich, C., & Hulme, M. (2017). Climate change and the Syrian civil war revisited. *Political Geography, 60*, 232–244. DOI: 10.1016/j.polgeo.2017.05.007

Seltzer, N. (2019). The evolution of tribalism: A social-ecological model of cooperation and inter-group conflict under pastoralism. *Journal of Artificial Societies and Social Simulation, 22*(2), 6. Advance online publication. DOI: 10.18564/jasss.3963

Semenza, J., Rocklöv, J., & Ebi, K. (2022). Climate change and cascading risks from infectious disease. *Infectious Diseases and Therapy*, *11*(4), 1371–1390. DOI: 10.1007/s40121-022-00647-3 PMID: 35585385

Sert, et. al. (2024). The disaster of the century, Kahramanmaras earthquake: A qualitative study on the experiences of the earthquake victims and search and rescue teams, *International Journal of Disaster Risk Reduction*, 107(2024) 104462, https://doi.org/DOI: 10.1016/j.ijdrr.2024.104462

Servetti, M., Caramia, M., Parodi, G., Loiacono, F., Nano, E., Biddau, G., Ferrando, L., Morinelli, L., Valente, P., Martinoia, S., Escelsior, A., Serafini, G., Tamburro, S., Baldassari, S., Fassio, A., Benfenati, F., Corradi, A., & Sterlini, B. (2025). Optimization of Transcription Factor-Driven Neuronal Differentiation from Human Induced Pluripotent Stem Cells for Disease Modelling and Drug Screening. *Stem Cell Reviews and Reports*, *21*(3), 816–833. DOI: 10.1007/s12015-025-10845-4 PMID: 39888571

Sever, B. C. (2023). An evaluation of the possibilities of human mobility and immobility in the context of the climate crisis in Türkiye. *İnsan Hareketliliği Uluslararası Dergisi (International Journal of Human Mobility),* Special Issue, 34–55.

Sevgi, , M. AYulu, , A. (2025). The old rains are gone, so are we: Kurdish Nomads of Mount Agri (Ararat). *British Journal of Middle Eastern Studies, 52*(2), 503-529.

Sevgi, M. A., & Yulu, A. (2024). "The old rains are gone, so are we": Kurdish nomads of Mount Ağrı (Ararat). *British Journal of Middle Eastern Studies*, *52*(2), 503–529. DOI: 10.1080/13530194.2024.2313752

Shaeri, S., Etemad-Shahidi, A., & Tomlinson, R. (2020). Revisiting longshore sediment transport formulas. *Journal of Waterway, Port, Coastal, and Ocean Engineering*, *146*(4), 04020009. DOI: 10.1061/(ASCE)WW.1943-5460.0000557

Shah, M. M. (2008). "Sustainable Development." Academic Press, No. 1, Oxford, England.

Shahbazbegian, M. R., Turton, A., & Mousavi Shafaee, S. M. (2016). Hydropolitical Self-Organization theory; system dynamics to analyse hydropolitics of Helmand transboundary river. *Water Policy*, *18*(5), 1088–1119. DOI: 10.2166/wp.2016.204

Shaheemath Suhara, K. K., Varughese, A., Sunny, A. C., & Krishna, A. P. R. (2023). Erosivity Factor of the Revised Universal Soil Loss Equation (RUSLE) - A Systematized Review. *Current World Environment*, *18*(2), 433–445. https://www.cwejournal.org/pdf/Vol18No2/CWE_Vol18_No2_p_433-445.pdf. DOI: 10.12944/CWE.18.2.02

Sharaunga, S., & Mudhara, M. (2016). The impact of improved 'water-use security'on women's reliance on agricultural incomes in KwaZulu-Natal Province, South Africa. *Food Security*, *8*(6), 1039–1052. DOI: 10.1007/s12571-016-0622-6

Sharma, P., Kumari, A., Shivali, S., & Sharma, V. (2024). Impact of food and health in 21st century era: A global economic crises. *International Journal of Technology*. https://doi.org/DOI: 10.52711/2231-3915.2024.00016

Sharma, P., Wang, J., Zhang, M., Woods, C., Kar, B., Bausch, D., Chen, Z., Tiampo, K., Glasscoe, M., Schumann, G., Pierce, M., & Eguchi, R. (2020). DisasterAWARE – a GLOBAL ALERTING PLATFORM FOR FLOOD EVENTS. ISPRS Annals of the Photogrammetry, Remote Sensing and Spatial Information Sciences, VI-3/W1-2020, 107–113. https://doi.org/DOI: 10.5194/isprs-annals-VI-3-W1-2020-107-2020

Shaw, R., Takeuchi, Y., & Rouhban, B. (2009). Education and public awareness for disaster reduction. In *Landslides – Disaster Risk Reduction*. Springer., DOI: 10.1007/978-3-540-69970-5_26

Shih, S. S., Hsu, W. C., & Hsu, Y. W. (2023). Waterline digital elevation model development to quantify inundation duration and coastal protection of tidal wetlands. *The Science of the Total Environment*, *874*, 162519. DOI: 10.1016/j.scitotenv.2023.162519

Shi, X., Tan, J., Guo, Z., & Liu, Q. (2013). A review of risk assessment of storm surge disaster. *Diqiu Kexue Jinzhan*, *28*(8), 866.

Shrestha, S., Aihara, Y., Bhattarai, A. P., Bista, N., Kondo, N., Futaba, K., Nishida, K., & Shindo, J. (2018). Development of an objective water security index and assessment of its association with quality of life in urban areas of developing countries. *SSM - Population Health*, *6*, 276–285. DOI: 10.1016/j.ssmph.2018.10.007

Siddiqi, Y. (2016). "Asian Water Development Outlook 2016: Strengthening Water Security in Asia and the Pacific." Asian Development Bank, Report No. 9292575449, Mandaluyong, Philippine.

Singh, A. (2022). Environment–conflict nexus in the Sahel region: Lake Chad as a case in point. *Journal of African Conflicts and Peace Studies*, *5*(1). Advance online publication. DOI: 10.54103/2612-6672/19469

Sisco, M. R., Bosetti, V., & Weber, E. U. (2017). When do extreme weather events generate attention to climate change? *Climatic Change*, *143*(1–2), 227–241. DOI: 10.1007/s10584-017-1984-2

Si-Tayeb, K., Noto, F. K., Sepac, A., Sedlic, F., Bosnjak, Z. J., Lough, J. W., & Duncan, S. A. (2010). Generation of human induced pluripotent stem cells by simple transient transfection of plasmid DNA encoding reprogramming factors. *BMC Developmental Biology*, *10*(1), 81. DOI: 10.1186/1471-213X-10-81 PMID: 20682060

Siwila, J., Mwaba, F., Chidumayo, N., & Mubanga, C. (2020). Food and waterborne protozoan parasites: The African perspective. *Food and Waterborne Parasitology*, *20*, e00088. Advance online publication. DOI: 10.1016/j.fawpar.2020.e00088 PMID: 32995582

Slater, D. (2010). *Ordering power: Contentious politics and authoritarian leviathans in Southeast Asia*. Cambridge University Press. DOI: 10.1017/CBO9780511760891

Slatton, K. C., Carter, W. E., Shrestha, R. L., & Dietrich, W. (2007). Airborne Laser Swath Mapping: Achieving the resolution and accuracy required for geosurficial research. *Geophysical Research Letters*, *34*(23), 2007GL031939. Advance online publication. DOI: 10.1029/2007GL031939

Slick, J., & Hertz, G. (2024). Gender and gender-based violence in disaster contexts in Canada: A systematic review of the literature. *International Journal of Disaster Risk Reduction*, *108*, 104529. Advance online publication. DOI: 10.1016/j.ijdrr.2024.104529

Smith, P., Pappenberger, F., Wetterhall, F., Del Pozo, J. T., Krzeminski, B., Salamon, P., Muraro, D., Kalas, M., & Baugh, C. (2016). On the Operational Implementation of the European Flood Awareness System (EFAS). In Elsevier eBooks (pp. 313–348). https://doi.org/DOI: 10.1016/B978-0-12-801884-2.00011-6

Smith, K., & Petley, D. N. (2009). *Environmental Hazards: Assessing Risk and Reducing Disaster*. Routledge. DOI: 10.4324/9780203884805

Smith, M. D. (2023). An ecological perspective on extreme climatic events: A synthetic definition and framework to guide future research. *Journal of Ecology*, *99*(3), 656–663. DOI: 10.1111/j.1365-2745.2011.01798.x

Smith, P. G. (2014). *Healthy, Resilient, and SustainableCommunities After Disasters: Strategies*. Opportunities, and Planning for Recovery.

Smith, P. J. (2007). Climate Change, Mass Migration and the Military Response. *Orbis*, *51*(4), 617–633. DOI: 10.1016/j.orbis.2007.08.006

Soest, C. V. (2020). *A heated debate: Climate change and conflict in Africa* (Working Paper). Hamburg: GIGA German Institute of Global and Area Studies – Leibniz-Institut für Globale und Regionale Studien, Institut für Afrika Studien.

Solow, A. R. (2017). On detecting ecological impacts of extreme climate events and why it matters. *Philosophical Transactions of the Royal Society B: Biological Sciences, 372*(1723), 20160136. DOI: 10.1098/rstb.2016.0136

Song, K., Choi, Y. E., Han, H. J., & Chon, J. (2021). Adaptation and transformation planning for resilient social-ecological system in coastal wetland using spatial-temporal simulation. *The Science of the Total Environment, 789*, 148007. DOI: 10.1016/j.scitotenv.2021.148007

Song, X., Zhang, L., Lu, Z., Liang, H., & Ren, W. (2025). Optimizing DEM error mitigation in multi-temporal InSAR: A detection-and-estimation strategy based on phase gradient direction consistency. *Remote Sensing of Environment, 331*, 115028. Advance online publication. DOI: 10.1016/j.rse.2025.115028

Son, J., Aziz, Z., & Peña-Mora, F. (2008). Supporting disaster response and recovery through improved situation awareness. *Structural Survey, 26*(5), 411–425. DOI: 10.1108/02630800810922757

Sonopant, J. (2015). Nurses' awareness and perceptions towards disaster situations. *Nursing Journal of India, 106*(4), 147–150. DOI: 10.48029/NJI.2015.CVI404 PMID: 30650932

Sözen, B., & Kılıç, S. E. (2025). Tourism-led rural gentrification in multi-conservation rural settlements: Yazıköy/Datça case. *Sustainability (Basel), 17*(18), 8439. DOI: 10.3390/su17188439

Spalding, M. D., McIvor, A. L., Beck, M. W., Koch, E. W., Möller, I., Reed, D. J., Rubinoff, P., Spencer, T., Tolhurst, T. J., Wamsley, T. V., van Wesenbeeck, B. K., Wolanski, E., & Woodroffe, C. D. (2014). Coastal ecosystems: A critical element of risk reduction. *Conservation Letters, 7*(3), 293–301. DOI: 10.1111/conl.12074

Srinivas, H., & Nakagawa, Y. (2008). Environmental implications for disaster preparedness: Lessons learnt from the Indian Ocean Tsunami. *Journal of Environmental Management, 89*(1), 4–13. DOI: 10.1016/j.jenvman.2007.01.054

Srivastava, P., & Yadav, U. (2023). Mode of the mechanism of biogenic nanomaterials involved in the adsorption of pollutants. In Adetunji, C. O., Singh, R. P., Singh, J., & Singh, K. R. B. (Eds.), *Nanobiotechnology for bioremediation* (pp. 273–296). Elsevier., DOI: 10.1016/B978-0-323-91767-4.00009-4

Ssentongo, P., Ssentongo, A., Ba, D., Ericson, J., Na, M., Gao, X., Fronterrè, C., Chinchilli, V., & Schiff, S. (2021). Global, regional and national epidemiology and prevalence of child stunting, wasting and underweight in low- and middle-income countries, 2006–2018. *Scientific Reports*, *11*(1), 5204. Advance online publication. DOI: 10.1038/s41598-021-84302-w PMID: 33664313

Staff, J. (2019). "DoD dictionary of military and associated terms." Joint Publication, No. 1, USA.

Stancheva, M., Ramieri, E., Stanchev, H., Bocci, M., Markovic, M., & Roberts, T. (2025). Exploring integration of land-sea interactions in maritime spatial planning in the Bulgarian context, Black Sea. *Marine Policy*, *171*, 106416. DOI: 10.1016/j.marpol.2024.106416

States, S. (2011). "Security and Emergency Planning for Water and Wastewater Utilities." American Water Works Association, No. 1, Mumbai, India.

Staudinger, T. J., England, E. C., & Bleckmann, C. (2006). Comparative analysis of water vulnerability assessment methodologies. *Journal of Infrastructure Systems*, *12*(2), 96–106. DOI: 10.1061/(ASCE)1076-0342(2006)12:2(96)

Stefanescu, C., Herrando, S., & Páramo, F. (2004). Butterfly species richness in the north-west Mediterranean Basin: The role of natural and human-induced factors. *Journal of Biogeography*, *31*(6), 905–915. DOI: 10.1111/j.1365-2699.2004.01088.x

Stevens, R. (2023, 27 July). Wildfires raging across southern Europe and North Africa kill more than 40 people. World Socialist Web Site. https://www.wsws.org/en/articles/2023/07/27/rkzb-j27.html

Stive, M. J. F., De Schipper, M. A., Luijendijk, A. P., Ranasinghe, R. W. M. R. J. B., Van Thiel de Vries, J., Aarninkhof, S., . . . Marx, S. (2013, June). The sand engine: a solution for vulnerable deltas in the 21st century. In Proceedings, Coastal Dynamics 2013: 7th International conference on coastal dynamics.

Stoffel, M., & Huggel, C. (2012). Effects of climate change on mass movements in mountain environments. *Progress in Physical Geography*, *36*(3), 421–439. DOI: 10.1177/0309133312441010

Stoffel, M., Tiranti, D., & Huggel, C. (2014). Climate change impacts on mass movements—Case studies from the European Alps. *The Science of the Total Environment*, *493*, 1255–1266. DOI: 10.1016/j.scitotenv.2014.02.102 PMID: 24630951

Stott, P. A., Christidis, N., Otto, F. E. L., Sun, Y., Vanderlinden, J., van Oldenborgh, G. J., Vautard, R., von Storch, H., Walton, P., Yiou, P., & Zwiers, F. W. (2023). Attribution of extreme weather and climate-related events. *Wiley Interdisciplinary Reviews: Climate Change*, *7*(1), 23–41. DOI: 10.1002/wcc.380 PMID: 26877771

Strauss, D., Murray, T., Harry, M., & Todd, D. (2017). Coastal data collection and profile surveys on the Gold Coast: 50 years on. In *Australasian Coasts & Ports 2017: Working with Nature: Working with Nature* (pp. 1030–1036). Engineers Australia, PIANC Australia and Institute of Professional Engineers New Zealand.

Stroebel, J., & Wurgler, J. (2021). *Nber Working Paper Series What Do You Think About Climate Finance?* https://www.nber.org/papers/w29136

Strupler, M., Hilbe, M., Kremer, K., Danciu, L., Anselmetti, F. S., Strasser, M., & Wiemer, S. (2018). Subaqueous landslide-triggered tsunami hazard for Lake Zurich, Switzerland. *Swiss Journal of Geosciences*, *111*(1–2), 353–371. DOI: 10.1007/s00015-018-0308-5

Suding, K. N. (2011). Toward an era of restoration in ecology: Successes, failures, and opportunities ahead. *Annual Review of Ecology, Evolution, and Systematics*, *42*(1), 465–487. DOI: 10.1146/annurev-ecolsys-102710-145115

Sullivan, C. A., Meigh, J. R., Simon, S., Lawrence, P., Calow, R., McKenzie, A., Acreman, M. C., & Moore, R. V. (2000). "The Development of a Water Poverty Index." Department for International Development, Report No. 1, Wallingford, UK.

Sun, D., Yang, T., Li, S., Goldberg, M., Kalluri, S., Helfrich, S., & Miralles, F. R. (2024). Hazard or non-hazard flood: Post analysis for paddy rice, wetland, and other potential non-hazard flood extraction from the VIIRS flood products. *ISPRS Journal of Photogrammetry and Remote Sensing*, *209*, 415–431. DOI: 10.1016/j.isprsjprs.2024.02.013

Sutton-Grier, A. E., & Sandifer, P. A. (2019). Conservation of wetlands and other coastal ecosystems: A commentary on their value to protect biodiversity, reduce disaster impacts, and promote human health and well-being. *Wetlands*, *39*(6), 1295–1302. DOI: 10.1007/s13157-018-1039-0

SYGM. (2018). Ceyhan Havzasi Taşkın Yönetim Planı. https://www.tarimorman.gov.tr/SYGM/Belgeler/Taşkın Yönetim Planları 26.12.2022/CEYHAN HAVZASI TAŞKIN YONETIM PLANI.pdf

Tan, K. W., Huang, Y. F., & Loh, P. N. (2023). Toward climate-resilient development and management: Developing indicators of climate hazards in Malaysia. (163-180). Editor: Ronni Alexander & Siriporn Wajjwalku, *Making Disaster: Safer A Gender and Vulnerability Approach*, Singapore: Springer Pte Ltd.

Tao, Y., Tian, B., Adhikari, B. R., Zuo, Q., Luo, X., & Di, B. (2024). A review of Cutting-Edge Sensor Technologies for improved flood monitoring and damage assessment. *Sensors (Basel)*, *24*(21), 7090. DOI: 10.3390/s24217090 PMID: 39517986

Taşeli, B. K. (2007). Nature conservation efforts set for specially protected areas of Turkey. *The Environmentalist*, *27*(3), 315–327. DOI: 10.1007/s10669-007-9030-y

Tavakoli, H.R., Sarafpour, R., and Samadi, M. (2005). "Water, Food and Bioterrorism." Journal of military medicine, 7(1), 75-82.

Taylor, A. L., Dessai, S., & Bruine de Bruin, W. (2024). Public perception of climate risk and adaptation in the UK: A review of the literature. *Climate Risk Management*, *4–5*, 1–16. DOI: 10.1016/j.crm.2014.09.001

Taylor, J., & Salvucci, G. (2025). Estimating contrasting soil moisture-precipitation feedbacks across global landmass using data from the Soil Moisture Active Passive satellite mission. *Science of Remote Sensing*, *12*, 100247. Advance online publication. DOI: 10.1016/j.srs.2025.100247

TEEB. (2010). *The Economics of Ecosystems and Biodiversity: Mainstreaming the Economics of Nature: A Synthesis of the Approach, Conclusions and Recommendations of TEEB*. Progress Press.

Tekin, M. K., & Deniz, O. (2025). Climate change-induced migration potential: A multi-dimensional field research on Türkiye. [Journal of Geographical Sciences]. *Coğrafi Bilimler Dergisi*, *23*(1), 236–263. DOI: 10.33688/aucbd.1517859

Teo, M., Goonetilleke, A., Ahankoob, A., Deilami, K., & Lawie, M. (2018). Disaster awareness and information seeking behaviour among residents from low socio-economic backgrounds. *International Journal of Disaster Risk Reduction*, *31*, 1121–1131. DOI: 10.1016/j.ijdrr.2018.09.008

Thapa, V. & Pathranarakul, P. (2019). Gender inclusiveness in disaster risk governance for sustainable recovery of 2015 Gorkha Earthquake, Nepal, International Journal of Disaster Risk Reduction, 34 (2019), 209–219. https://doi.org/DOI: 10.1016/j.ijdrr.2018.11.019

The Fund for Peace. (2024). *Country dashboard | Fragile States Index*. Retrieved October 20, 2025, from https://fragilestatesindex.org/country-data/

The Global Economy. (2025). *Chad: Percent agricultural land.* Retrieved September 17, 2025, from https://www.theglobaleconomy.com/Chad/Percent_agricultural_land

Thissen, P. (2019). Legacy institutions and political order in weak states: Evidence from Chad. SSRN *Electronic Journal.* https://doi.org/DOI: 10.2139/ssrn.3630086

Tichenor, V. J. (1963). Earning more and getting less. New Brunswick, New Jersey, and London: Rutgers University Press

Tierney, J., & Worth, R. F. (2003). "Attacks in iraq may be signals of new tactics." <www.nytimes.com> (Sep. 27, 2019).

Titz, A. (2021). Geographies of doing nothing–internal displacement and practices of post-disaster recovery in urban areas of the Kathmandu Valley, Nepal. *Social Sciences (Basel, Switzerland)*, *10*(3), 110. DOI: 10.3390/socsci10030110

TOKB. (1987). Tarım, Orman ve Köyişleri Bakanlığı (1987) Türkiye'nin genel yönetim planlaması (toprak koruma ana planı).

Tøraasen, M. (2020). Women's status in Haiti ten years after the earthquake, CMI BRIEF 2020:07.

Toros, H., Mokari, M., & Abbasnia, M. (2019). Regional variability of temperature extremes in the maritime climate of Turkey: A case study to develop agricultural adaptation strategies under climate change. *Modeling Earth Systems and Environment*, *5*(3), 857–865. DOI: 10.1007/s40808-019-00572-4

Torri, D., Poesen, J., & Borselli, L. (1997). Predictability and Uncertainty of the Soil Erodibility Factor Using a Global Dataset. *Catena*, *31*(1–2), 1–22. https://doi.org/https://doi.org/10.1016/S0341-8162(97)00036-2. DOI: 10.1016/S0341-8162(97)00036-2

Tourism, U. N. (2020) UNWTO launches a call for action for tourism's COVID-19 mitigation and recovery. https://www.unwto.org/news/unwto-launches-a-call-for-action-for-tourisms-covid-19-mitigation-and-recovery

Transparency International. (2025). *Chad.* https://www.transparency.org/en/countries/chad

Trogrlić, R. Š., Van Den Homberg, M., Budimir, M., McQuistan, C., Sneddon, A., & Golding, B. (2022). Early warning systems and their role in disaster risk reduction. In Springer eBooks (pp. 11–46). https://doi.org/DOI: 10.1007/978-3-030-98989-7_2

True, J. (2013). Gendered violence in natural disasters: Learning from New Orleans, Haiti and Christchurch. *Aotearoa New Zealand Social Work*, *25*(2), 78–89. DOI: 10.11157/anzswj-vol25iss2id83

TSKB (2024). The fault line deep down: Gender inequality, March-2024.

Tufail, M., Nasir, M. J., Tariq, A., Ali, Z., & Alarifi, S. S. (2025). Assessing access to safe drinking water in flood-affected areas of District Nowshera, Pakistan: A case study towards achieving sustainable development goal 6.1. *Ecohydrology & Hydrobiology*, *25*(2), 422–437. DOI: 10.1016/j.ecohyd.2024.07.001

Turan, M., Bulut, Y., Öztürk, G., & Göktekin, Z. (2017). The knowledge and behavior levels of the students taking disaster awareness training: The example of Tekirda province, Turkey. *Nepal Journal of Environmental Science*, *5*, 57–60. DOI: 10.3126/njes.v5i0.22716

Turetsky, M. R., Benscoter, B., Page, S., Rein, G., van der Werf, G. R., & Watts, A. (2015). Global vulnerability of peatlands to fire and carbon loss. *Nature Geoscience*, *8*(1), 11–14. DOI: 10.1038/ngeo2325

Turhan, E., Cerit Mazlum, S., Şahin, Ü., Şorman, A. H., & Cem Gündoğan, A. (2016). Beyond special circumstances: Climate change policy in Turkey 1992–2015. *Wiley Interdisciplinary Reviews: Climate Change*, *7*(3), 448–460. DOI: 10.1002/wcc.390

Türksever, Ö. (2021). Analysis of disaster awareness perception levels of students in social studies teaching undergraduate program. *OSF Preprints*. https://doi.org/ DOI: 10.31219/osf.io/4zq8v

Ummenhofer, C. C., & Meehl, G. A. (2017). Extreme weather and climate events with ecological relevance: a review. *Philosophical Transactions of the Royal Society B: Biological Sciences, 372*(1723), 20160135. DOI: 10.1098/rstb.2016.0135

UNDP. (2014). *Cambodia post-flood early recovery needs assessment report 2013*. Royal Government of Cambodia-United Nations Development Programme.

UNDP. (2023). *Disaster financial preparedness analysis report for Cambodia*. Phnom Penh: United Nations Development Programme. https://www.undp.org/sites/g/files/zskgke326/files/2023-05/01_Cambodia_Financial%20Preparedness%20Plan_Final_May2023.pdf

UNDP. (2025) Social protection, risk finance and insurance. https://irff.undp.org/sites/default/files/2025/Sep/Social-Protection-Risk-Finance-and-Insurance.pdf

UNDP. (2025). Kahramanmaraş'ta kadın mikro işletmelerinin iyileşme süreci ve afet dayanıklılığı: afet sonrası ihtiyaç analizi. Ocak 2025, İstanbul.

UNDRR. (2015). *Sendai framework for disaster risk reduction 2015–2030*. United Nations Office for Disaster Risk Reduction. https://www.undrr.org/publication/sendai-framework-disaster-risk-reduction-2015-2030

UNDRR. (2020). *Words into action: Multi-hazard early warning systems.* United Nations Office for Disaster Risk Reduction. https://www.undrr.org/words-into-action/guide-multi-hazard-early-warning

UNICEF. (2023). *Chad: Humanitarian situation report.* https://www.unicef.org/chad

United Nations (2015). Transforming our world: the 2030 Agenda for Sustainable Development.

United Nations Environment Programme. (n.d.). *Chad | Interactive Country Fiches.* https://dicf-unepgrid-ch.translate.goog/chad?_x_tr_sl=en&_x_tr_tl=tr&_x_tr_hl=tr&_x_tr_pto=tc

United Nations Office for Disaster Risk Reduction (UNDRR). (2019). *Global assessment report on disaster risk reduction 2019.* Geneva: UNDRR. https://www.undrr.org/publication/global-assessment-report-disaster-risk-reduction-2019

United Nations Office for Disaster Risk Reduction (UNDRR). (2020). *Global assessment report on disaster risk reduction.* UNDRR.

UN-Water (2019, 2023). World Water Development Report.

Ural, , Yalçın, E., İlkin Aydın, E., & Höbek Akarsu, R. (2024). Women's experiences in the aftermath of the Kahramanmaraş earthquakes in Türkiye: A phenomenological study. *International Journal of Disaster Risk Reduction, 110*, 104617. DOI: 10.1016/j.ijdrr.2024.104617

Uzelgun, M. A., & Castro, P. (2014). The voice of science on climate change in the mainstream Turkish press. *Environmental Communication, 8*(3), 326–344. DOI: 10.1080/17524032.2014.898674

van de Pol, M., Jenouvrier, S., Cornelissen, J. H. C., & Visser, M. E. (2017). Behavioural, ecological and evolutionary responses to extreme climatic events: challenges and directions. *Philosophical Transactions of the Royal Society B: Biological Sciences, 372*(1723), 20160134. DOI: 10.1098/rstb.2016.0134

Van Eck, N. J., & Waltman, L. (2010). Software survey: VOSviewer, a computer program for bibliometric mapping. *Scientometrics, 84*(2), 523–538. DOI: 10.1007/s11192-009-0146-3

Vanhove, N. (1981). Tourism and employment. *International Journal of Tourism Management, 2*(3), 162–175. DOI: 10.1016/0143-2516(81)90003-7

Varol, N., & Gültekin, T. (2016). Afet antropolojisi. *Elektronik Sosyal Bilimler Dergisi, 15*(59).

Veettil, A. V., & Mishra, A. K. (2016). Water security assessment using blue and green water footprint concepts. *Journal of Hydrology (Amsterdam)*, *542*, 589–602. DOI: 10.1016/j.jhydrol.2016.09.032

Veil, S. R., Buehner, T., & Palenchar, M. J. (2011). A work-in-process literature review: Incorporating social media in risk and crisis communication. *Journal of Contingencies and Crisis Management*, *19*(2), 110–122. DOI: 10.1111/j.1468-5973.2011.00639.x

Verloo, M. (2007). *Multiple meanings of gender equality: A critical frame analysis of gender policies in Europe*. Central European University Press. DOI: 10.1515/9786155211393

Vishvakarma, M. D. P. (2019). THE ROLE OF GEOSPATIAL TECHNOLOGIES IN DISASTER MANAGEMENT: a REVIEW. https://psychologyandeducation.net/pae/index.php/pae/article/view/8666

Vlasenko, T., Vlasovets, V., Kovalyshyn, O., & Bilovod, O. (2024). ANALYSIS OF THE PROJECT OF A MARKETING CAMPAIGN TO PROMOTE ROBOTIC SOLUTIONS USING RANDOM FOREST CLASSIFICATION. *Technology Audit and Production Reserves*, *6*(4), 43–50. DOI: 10.15587/2706-5448.2024.320367

Vörösmarty, C. J., McIntyre, P. B., Gessner, M. O., Dudgeon, D., Prusevich, A., Green, P., Glidden, S., Bunn, S. E., Sullivan, C. A., Liermann, C. R., & Davies, P. M. (2010). Global threats to human water security and river biodiversity. *Nature*, *467*(7315), 555–561. DOI: 10.1038/nature09440

Vousdoukas, M. I., Mentaschi, L., Voukouvalas, E., Bianchi, A., Dottori, F., & Feyen, L. (2018). Climatic and socioeconomic controls of future coastal flood risk in Europe. *Nature Climate Change*, *8*(9), 776–780. DOI: 10.1038/s41558-018-0260-4

Vrba, E. S. (2007). 4 Role of Environmental Stimuli in Hominid Origins. In W. Henke & I. Tattersall, *Handbook of Paleoanthropology* (pp. 1441–1481). Springer Berlin Heidelberg. DOI: 10.1007/978-3-540-33761-4_47

Walters, S. (2025). Recommendations for Managing Loans and Grants: A Policymaker Perspective. In Aid Dynamics in Jamaican Development: Empowering Communities through Loans and Grants (pp. 135-162). Palgrave Macmillan, Cham: Springer Nature Switzerland. DOI: 10.1007/978-3-031-94885-5_5

Walz, Y., Janzen, S., Narvaez, L., Ortiz-Vargas, A., Woelki, J., Doswald, N., & Sebesvari, Z. (2021). Disaster-related losses of ecosystems and their services. Why and how do losses matter for disaster risk reduction? *International Journal of Disaster Risk Reduction*, *63*, 102425. DOI: 10.1016/j.ijdrr.2021.102425

Wang, L., Cheng, J., Jiang, Y., Liu, N., & Wang, K. (2024). Prediction of Adaptability of Typical Vegetation Species in Flood Storage Areas under Future Climate Change: A Case in Hongze Lake FDZ, China. *Sustainability (Basel)*, *16*(15), 6331. DOI: 10.3390/su16156331

Wang, L., Zhang, F., Shi, X., Zeng, C., Ahmad, I., Wang, G., Thapa, S., & Xu, X. (2023). Water resources system vulnerability in high mountain areas under climate change. *Journal of Cleaner Production*, *403*, 136789. DOI: 10.1016/j.jclepro.2023.136789

Ward, C., & Ruckstuhl, S. (2017). *Water scarcity, climate change and conflict in the Middle East: Securing livelihoods, building peace*. Bloomsbury Publishing. DOI: 10.5040/9781350989719

Warlina, L., & Afredi, A. (2026). Geographic Information System (GIS) Application to Identify the Distribution of Tourist Attractions and Supporting Facilities Related to the Sustainable Development Goals (SDGs). *Indonesian Journal of Science and Technology*, *11*(2), 341–356. DOI: 10.17509/ijost.v11i2.89803

Warren, P. D. (2016). Forced migration after Paris COP21: Evaluating the "Climate Change Displacement Coordination Facility.". *Columbia Law Review*, *116*, 2103–2139.

Watts, N., Amann, M., Arnell, N., Ayeb-Karlsson, S., Belesova, K., Berry, H., Bouley, T., Boykoff, M., Byass, P., Cai, W., Campbell-Lendrum, D., Chambers, J., Daly, M., Dasandi, N., Davies, M., Depoux, A., Dominguez-Salas, P., Drummond, P., Ebi, K. L., & Costello, A. (2018). The 2018 report of the Lancet Countdown on health and climate change: Shaping the health of nations for centuries to come. *Lancet*, *392*(10163), 2479–2514. DOI: 10.1016/S0140-6736(18)32594-7 PMID: 30503045

WEF. (2020). *Cambodia Annual Country Report 2020*. Phnom Penh: World Food Programme. https://wfp.tind.io/record/130856/files/ACR%20Cambodia%202020.pdf

Wegerich, K., Van Rooijen, D., Soliev, I., & Mukhamedova, N. (2015). Water security in the Syr Darya basin. *Water (Basel)*, *7*(9), 4657–4684. DOI: 10.3390/w7094657

Wehner, M. (2023). Connecting extreme weather events to climate change. *Physics Today*, *76*(9), 40–46. DOI: 10.1063/PT.3.5309

Wheater, H. S., & Gober, P. (2015). Water security and the science agenda. *Water Resources Research*, *51*(7), 5406–5424. DOI: 10.1002/2015WR016892

Whiteford, S., & Witter, S. G. (1999). Water security: The issues and policy challenges. *International Review of Comparative Public Policy*, *11*, 1–25.

Wischmeier, W. H., & Smith, D. D. (1978). *Predicting Rainfall Erosion Losses: a Guide to Conservation Planning. Science and Education Administration*. US Department of Agriculture.

Wisner, B., Blaikie, P., Cannon, T., & Davis, I. (2014). *At risk: Natural hazards, people's vulnerability and disasters* (2nd ed.). Routledge.

Wissink, M., Düvell, F., & Van Eerdewijk, A. (2013). Dynamic migration intentions and the impact of socio-institutional environments: A transit migration hub in Turkey. *Journal of Ethnic and Migration Studies, 39*(7), 1087–1105. DOI: 10.1080/1369183X.2013.778026

Witmer, F. D., Linke, A. M., O'Loughlin, J., Gettelman, A., & Laing, A. (2017). Subnational violent conflict forecasts for sub-Saharan Africa, 2015–65, using climate-sensitive models. *Journal of Peace Research, 54*(2), 175–192. DOI: 10.1177/0022343316682064

Women, U. N. (2023a). Un Women brief on earthquake in Türkiye: impacts and priorities for women and girls, April 2023, Türkiye

Women, U. N. (2023b). Depremin ardından kadınlar ve kız çocukları: Türkiye'deki depremlerin etkisi: kadın ve kız çocuklarının hakları için çalışan sivil toplum kuruluşlarının deneyim ve bakış açıları, Mayıs 2023, Türkiye

World Bank & GFDRR. (2014) Financial protection against natural disasters. https://www.gfdrr.org/sites/default/files/documents/Financial%20Protection.pdf

World Bank Group. (2022). *G5 Sahel Region: Country Climate and Development Report.* Washington, D.C.: World Bank Group. https://documents.worldbank.org/curated/en/099150006302219454

World Bank. (2016). *Building regulation for resilience: Toward high performance building regulatory systems for disaster risk reduction* Washington DC: World Bank. https://documents1.worldbank.org/curated/en/326581468337788007/pdf/ACS15966-WP-PUBLIC-BRR-report-002.pdf

World Bank. (2021). *Chad poverty assessment: Investing in rural income growth, human capital, and resilience to support sustainable poverty reduction.* https://documents1.worldbank.org/curated/en/642041633397593711/pdf/Chad-Poverty-Assessment-Investing-in-Rural-Income-Growth-Human-Capital-and-Resilience-to-Support-Sustainable-Poverty-Reduction.pdf

World Bank. (2023a). *Africa Country Policy and Institutional Assessment (CPIA): Assessing Africa's policies and institutions – Country tables (English).* Washington, D.C.: World Bank Group. https://documents.worldbank.org/curated/en/099634410132333843

World Bank. (2023b). *The World Bank in Chad: Overview.* https://www.worldbank.org/en/country/chad/overview

World Bank. (2024) World Bank national accounts data, and OECD national accounts data files. https://databank.worldbank.org/metadataglossary/world-development-indicators/series/NY.GDP.MKTP.KD.ZG

World Health Organization. (2023). *Health emergency and disaster risk management framework.*https://apps.who.int/iris/handle/10665/326106

World Health Organization. (2024). *Chad health profile.* https://www.who.int/countries/tcd

World Health Organization. Health Cluster. (2025). *Chad [Data dashboard].* WHO Health Cluster. https://healthcluster.who.int/countries-and-regions/chad

World Meteorogiçal Organization. (2024*). Africa faces disproportionate burden from climate change and adaptation costs.* https://wmo.int/news/media-centre/africa-faces-disproportionate-burden-from-climate-change-and-adaptation-costs https://www.indyturk.com/node/438166/t%C3%BCrki%CC%87yeden-sesler/i%CC%87klim-de%C4%9Fi%C5%9Fikli%C4%9Fi-ve-afrika

World Travel and Tourism Council. (2019) Benchmarking research trends 2019: How does travel & tourism compare to other sectors? https://wttc.org/Portals/0/Documents/Reports/2019/Benchmarking%20Trends%20Report%202019.pdf?ver=2021-02-25-182807-540

Worlddata.info. (2023). *Climate: Chad.* https://www.worlddata.info/africa/chad/climate.php

Wu, W., Zhi, C., Gao, Y., Chen, C., Chen, Z., Su, H., Lu, W., & Tian, B. (2022). Increasing fragmentation and squeezing of coastal wetlands: Status, drivers, and sustainable protection from the perspective of remote sensing. *The Science of the Total Environment, 811*, 152339. DOI: 10.1016/j.scitotenv.2021.152339

Xia, J., & Bantin, A. B. (2017). The impact of hydro climatic variability on water resources of Lake Chad hydrographic basin. *Hydrology: Current Research, 8*(3), 1–10. DOI: 10.4172/2157-7587.1000282

Xu, T., Xiao, S., Reuschen, S., Wildt, N., Hendricks Franssen, H.-J., & Nowak, W. (2024). Towards a community-wide effort for benchmarking in subsurface hydrological inversion: Benchmarking cases, high-fidelity reference solutions, procedure, and first comparison. *Hydrology and Earth System Sciences*, *28*(24), 5375–5400. DOI: 10.5194/hess-28-5375-2024

Yalçıner, A. C., Kuran, U., Mınoura, K., Imamura, F., Takahashı, T., Papadopoulos, G., & Ersoy, Ş. (2005). *Türkiye Mühendislik Haberleri Sayı 438 - 2005/4.Türkiye Kıyılarında Depreşim Dalgası (Tsunami).* İzleri.

Yalçin, G. (2020). *Doğal afetlerin etkilerine ve afet risk yönetimine toplumsal cinsiyet perspektifinden bakış: Türk ve Japon kadınların duruş noktasından doğal afetler.* Master'sthesis, Sosyal Bilimler Enstitüsü.

Yamin, M. (2020). Counting the cost of Covid-19. *International Journal of Information Technology : an Official Journal of Bharati Vidyapeeth's Institute of Computer Applications and Management*, *12*(2), 311–317. DOI: 10.1007/s41870-020-00466-0 PMID: 32412538

Yang, Z., & Wang, J. (2017). A new air quality monitoring and early warning system: Air quality assessment and air pollutant concentration prediction. *Environmental Research*, *158*, 105–117. DOI: 10.1016/j.envres.2017.06.002 PMID: 28623745

Yavuz, K., & Çelik, A. (2023). Exploring the intersection of climate change and social policy in Türkiye: Challenges and opportunities for a sustainable future. *Journal of Management and Economics Research*, *21*(1), 244–265.

Yazdani, M., Zarate, P., Kazimieras Zavadskas, E., & Turskis, Z. (2019). A combined compromise solution (CoCoSo) method for multi-criteria decision-making problems. *Management Decision*, *57*(9), 2501–2519. DOI: 10.1108/MD-05-2017-0458

Yenal, Z. (2023). Buckling under ageing and climate crisis: Migration flows in Turkish agriculture. *International Migration (Geneva, Switzerland)*, *61*(6), 349–353. DOI: 10.1111/imig.13205

Yıldız, D., & Atmiş, E. (2023). Conflict management in protected areas in Turkey: Küre Mountains National Park example. *Forestist, 73*(2).

Yilmaz, A. G. (2015). The effects of climate change on historical and future extreme rainfall in Antalya, Turkey. *Hydrological Sciences Journal*, *60*(12), 2148–2162. DOI: 10.1080/02626667.2014.945455

Yucesoy, E., Balcik, B., & Coban, E. (2024). The role of drones in disaster response: A literature review of operations research applications. *International Transactions in Operational Research*. Advance online publication. DOI: 10.1111/itor.13484

Yulu, A., Albayrak, F., Işık, Z., Aras, S., & Taşçı, H. E. (2023). Türkiye'nin Çatısında Çevresel Kirlilik: Balıkgöl Örneği (Ağrı). *Doğu Coğrafya Dergisi*, *28*(49), 94–104. DOI: 10.5152/EGJ.2023.22046

Yumarni, T., Amaratunga, D., & Haigh, R. (2014). Assessing gender vulnerability within post-earthquake reconstruction: Case study from Indonesia. *Procedia Economics and Finance*, *18*, 763–771. DOI: 10.1016/S2212-5671(14)01000-4

Yusuf, A. A., & Francisco, H. (2009). *Climate change vulnerability mapping for Southeast Asia*. Economy and Environment Program for Southeast Asia (EEPSEA). https://www.preventionweb.net/files/7865_12324196651MappingReport1.pdf

Zare Bidaki, m. and Balali Mood, m. (2015). "Bioterrorism and biological warfare, from past to the present: A classic review." Journal Of Birjand University Of Medical Sciences, 22(3), 182-198. https://sid.ir/paper/95436/en

Zedler, J. B., & Kercher, S. (2005). Wetland resources: Status, trends, ecosystem services, and restorability. *Annual Review of Environment and Resources*, *30*(1), 39–74. DOI: 10.1146/annurev.energy.30.050504.144248

Zeitoun, M., Elaydi, H., Dross, J. P., Talhami, M., de Pinho-Oliveira, E., & Cordoba, J. (2017). Urban Warfare Ecology: A Study of Water Supply in Basrah. *International Journal of Urban and Regional Research*, *41*(6), 904–925. DOI: 10.1111/1468-2427.12546

Zeitoun, M., Lankford, B., Krueger, T., Forsyth, T., Carter, R., Hoekstra, A. Y., Taylor, R., Varis, O., Cleaver, F., Boelens, R., Swatuk, L., Tickner, D., Scott, C. A., Mirumachi, N., & Matthews, N. (2016). Reductionist and integrative research approaches to complex water security policy challenges. *Global Environmental Change*, *39*, 143–154. DOI: 10.1016/j.gloenvcha.2016.04.010

Zeleňáková, M., & Zvijáková, L. (2017). Risk analysis within environmental impact assessment of proposed construction activity. *Environmental Impact Assessment Review*, *62*, 76–89. DOI: 10.1016/j.eiar.2016.10.003

Zhang, Y., Tachibana, S., & Nagata, S. (2006). Impact of socio-economic factors on the changes in forest areas in China. *Forest Policy and Economics*, *9*(1), 63–76. DOI: 10.1016/j.forpol.2005.02.005

Zhang, Z. (2023). Weather and Climate Extreme Events in a Changing Climate. In *Climate Change 2021 – The Physical Science Basis* (pp. 1513–1766). Cambridge University Press., DOI: 10.1017/9781009157896.013

Zhou, M., Li, C., Li, Z., Yu, Z., & Zhou, X. (2023). Stratification of vertical canopy structure to improve estimation of forest inventory attributes using airborne LiDAR data in a large subtropical region of China. *Annals of Forest Research*, *66*(2), 101–120. DOI: 10.15287/afr.2023.3361

Zhou, Y., Liu, Y., Wu, W., & Li, N. (2015). Integrated risk assessment of multi-hazards in China. *Natural Hazards*, *78*(1), 257–280. DOI: 10.1007/s11069-015-1713-y

Zickgraf, C. (2019). Keeping people in place: Political factors of (im)mobility and climate change. *Social Sciences (Basel, Switzerland)*, *8*(8), 228. DOI: 10.3390/socsci8080228

Zscheischler, J., Westra, S., van den Hurk, B. J. J. M., Seneviratne, S. I., Ward, P. J., Pitman, A., AghaKouchak, A., Bresch, D. N., Leonard, M., Wahl, T., & Zhang, X. (2018). Future climate risk from compound events. *Nature Climate Change*, *8*(6), 469–477. DOI: 10.1038/s41558-018-0156-3

About the Contributors

İsmet Güney received his PhD in Human Geography from Ankara University. He is currently a faculty member at Recep Tayyip Erdoğan University. His academic work focuses on human-environment interactions, spatial representations, environmental discourse, and cultural geography. Over the years, Dr. Güney has contributed to the field through numerous peer-reviewed journal articles, book chapters, and interdisciplinary research projects. His research often combines qualitative and quantitative approaches, with particular emphasis on bibliometric analysis, phenomenological methods, and discourse theory. Dr. Güney continues to participate in both national and international collaborations addressing contemporary geographic challenges.

Mehmet Ali Çelik earned his Ph.D. in Philosophy and Religion from the Graduate School of Social Sciences at Çukurova University in 2018, and holds a second doctoral degree in Geography from the Graduate School of Social Sciences at Kahramanmaraş Sütçü İmam University (2016). He also completed his Master's in Geographic Information Systems at Ege University and a second Master's in Geography at Kahramanmaraş Sütçü İmam University, after receiving his initial B.A. in Geography from Süleyman Demirel University . He is currently serving as an Assistant Professor in the Physical Geography Division at the Department of Geography, Faculty of Arts & Sciences, Iğdır University . His research spans remote sensing, drought and climate variability assessment, land-use/land-cover change, ecological crisis studies, and environmental ethics. Dr. Çelik has authored and co-authored over 100 scholarly works, including articles in both national and international peer-reviewed journals, book chapters, and project reports. Notable publications include investigations of vegetation response to drought using satellite-derived indices (NDVI, EVI, VCI), multi-criteria risk assessments of natural hazards, and bibliometric analyses of drought research. He has also contributed to edited volumes on ecological crisis and participated actively in international symposia on geography, ecology, and environmental studies. His interdisciplinary methodology combines GIS-based spatial analysis with environmental ethics and socio-ecological inquiry, particularly focused on regions of Eastern Anatolia. At Iğdır University, Dr. Çelik teaches advanced courses in remote sensing, GIS, and environmental geography, while also supervising graduate theses. He continues to develop collaborative projects that address contemporary challenges at the intersection of climate change, land use, and ethical-environmental considerations.

***

Melahat Batu Agirkaya is currently working as Assoc. Prof. Dr. at Igdir University Banking and Finance Program. Bachelor's degree from Anadolu University, Faculty of Economics and Administrative Sciences, Department of Finance in 1990, In 2013, she received her master's degree in Economics from Kafkas University and in 2020 she received her doctorate degree from Atatürk University. She is currently working as the Head of the Department of Banking, Finance and Insurance at Igdir University. She is the author/co-author of books and book chapters, many journal and conference articles. She is an active researcher. Her research interests include economic growth, development, systemic risks, banks, migration, energy studies.

Muhammed Akiner graduated from the Civil Engineering Department of Eastern Mediterranean University in 2002. The author completed his MS study on Structural Mechanics at the same university in 2004. He held another MS degree in Environmental Engineering from SUNY at Buffalo in 2007. The author obtained his Ph.D. from the Civil Engineering Department of Bogazici University. He published several scientific publications on engineering, mainly in the field of environmental engineering. His research interests include environmental engineering, water resources, ecology, civil engineering, construction & building technology.

Asha Bala is an accomplished academician and researcher in the field of Mass Communication and Media Studies. She holds a doctoral degree and has extensive experience in teaching, research, and academic administration. Her core scholarly interests include Electronic Media, Multimedia Communication, New Media, and Environmental Communication. She has rich research experience, with active involvement in interdisciplinary and media-related research projects focusing on media impact, environmental awareness, and digital communication practices. Dr. Asha Bala has made significant academic contributions through the publication of research papers in reputed national and international journals, as well as chapters in edited books published by recognized academic publishers. She has also contributed to books in the areas of media studies and communication, reflecting her strong research orientation and subject expertise. Dr. Asha Bala has presented her research at various national and international conferences, seminars, and workshops, and her work emphasizes research-based learning and critical media engagement. She is deeply involved in mentoring undergraduate, postgraduate, and research scholars, encouraging analytical thinking and ethical media practices

Yaver Batu is a student at department of geography. He is interested in physical geography.

Emrah Çelik is an Assistant Professor at Iğdır University, Technical Sciences Vocational School, Forestry Department, Hunting and Wildlife Program, Türkiye. His academic and applied research focuses on ornithology, wetland ecology, biodiversity, avifauna, bird migration, bird population ecology, and conservation, with additional interests in GIS-based ecological studies and bibliometric analyses. Dr. Çelik has contributed to numerous national and international research and monitoring projects, particularly on wetland biodiversity and ornithological assessment reports for infrastructure and energy projects. He has published many peer-reviewed articles, conference papers, and book chapters in the fields of wildlife, ecology, and environmental sciences.

Sovannaroth Chheang earned her bachelor's degree in English from Build Bright University, Phnom Penh, in 2016, where she built a solid foundation in language studies, communication, and pedagogy. She began her teaching career in 2013 as a Tutor at the Kindergarten Association, conducting both part-time and full-time English programs. Since 2018, she has served as a Teacher of English and General Assistant at Sovanna Elton Foreign Language School. Through these roles, she has developed broad expertise in classroom management, language instruction, curriculum design, and student engagement, supporting the school's mission of advancing foreign language learning. Ms. Chheang's academic interests center on pursuing a master's degree in English education or related fields within the humanities and social sciences. Passionate about integrating global research methodologies into her practice, she aims to enhance pedagogy, foster classroom innovation, and promote effective, research-informed teaching. Her commitment lies in bridging academic research with real-world practice to advance English education in Cambodia and beyond.

Filiz Ekici is currently a Full Lecturer in Finance, Banking and Insurance Program in Iğdır University and a Ph.D. candidate in Civil Aviation Program. She holds an MSc in Finance (Kütahya Dumlupınar University) program which studied on finance-civil aviation-sustainability-externality relations, also holds a BSc in

Department of Finance. She has authored/co-authored books and book chapters, many journal and conference papers. Her research interests include public finance, economics, taxation, aviation environmental impacts and externalities, sustainable airports.

Manasi Vyankatesh Ghamande has done PhD from Singhania University,Rajasthan in 2021. She is working in Vishwakarma Institute of Technology, Pune since 1991 as an Assistant Professor in First Year Engineering Department . She has more than eight papers published in Scopus Indexed Journal and forty patents granted on her name and six book chapters written on her name. She has worked as session chair and chief geust in many conferences. Recently she has received an international award of Outstanding Women in Academia in the conference held at Vellore in November 2025

Jaishri Gothania is an academician, author, researcher, and data scientist with over 9 years of experience. She holds a Ph.D. in Computer Science Engineering, an M.Tech in Computer Science, a B.Tech in Computer Science and Engineering, and a PGD in Education. She has published 1 book, 15 book chapters, and 12 research articles; presented 6 conference papers; and holds multiple patents. Her work bridges data science and innovation, contributing significantly to academia, research, and technological advancement.

Oner Gumus graduated from Anadolu University Department of Economics in English in 2010, from Anadolu University Department of Public Finance in 2015, from Anadolu University Justice Associate Degree Program in 2018, and from Atatürk University Civil Air Transportation Management Associate Degree Program in 2024. He received a Master's degree in Public Finance from Kütahya Dumlupınar University in 2014, a Master's degree in Economics from Kütahya Dumlupınar University in 2018, and a doctorate degree from the same university in 2020. The author continues his studies as an Assistant Professor at Kütahya Dumlupınar University Tavşanlı Vocational School, and also continues his doctoral studies in the field of Civil Aviation at Eskişehir Technical University. He has authored/co-authored books and book chapters, many journal and conference papers. His research areas include public finance, economics, taxation and aviation.

Sovanna Huot, currently enrolled as a PhD candidate in Political Science at the University of Delhi, New Delhi, India, is also a vice principal at a Government Public High School of Cambodia, where he holds the responsibilities of the school's managerial and academic institutional leadership. An accomplished scholar in a vast array of fields, he has undergone a two-year teacher training course from 2010-2012 and has earned bachelor's degrees in English and Educational Administration.

Moreover, he holds two master's degrees, which he earned from a highly selective scholarship program, in TEFL and a Master of Education in School Leadership, Upper Secondary School Principalship, from the National Institute of Education (NIE), Phnom Penh, Cambodia, 2021. His areas of focus are on educational reform and the practice-policy nexus of digit pedagogy, alongside glonacal (global–national–local) education systems. His scholarly work encompasses both political science and education; he has formal training in both fields and has published in both, including Scopus-indexed articles (Author ID: 59316244400).

Caglar Karamasa is an Associate Professor at Anadolu University, the Faculty of Business Administration, the Accounting and Finance Department, Turkey. Caglar Karamasa obtained his MSc and PhD degrees in Quantitative Methods from Anadolu University in Eskişehir, Turkey. His research interest includes the fields of decision analysis, multi-criteria decision making, fuzzy set theory, and data mining. He has also been serving on the review board, and the editorial board for several international journals. He has published many research papers in indexed journals or books.

Muhittin Kulak is a faculty member at a state university in Türkiye, affiliated with departments related to agriculture, field crops, and plant sciences. His research focuses on plant stress physiology, particularly responses to salinity, drought, waterlogging, and heavy metal stress, as well as medicinal and aromatic plants such as basil, mint, and Hypericum species. He investigates secondary metabolites including phenolics, essential oils, and antioxidants, and explores nanoparticle–plant interactions involving Fe_2O_3, TiO_2, zinc, and other nanomaterials. His work also encompasses seed priming strategies for climate-resilient agriculture and the application of artificial intelligence to predict and optimize plant stress responses. Actively engaged in academic scholarship, he publishes in SCI/SCI-Expanded journals, supervises MSc and PhD students, authors review papers and original research articles, prepares TÜBİTAK and BAP research projects, teaches courses in plant biology, biochemistry, and occupational health and safety, and participates in academic administration and research project coordination.

Sanjeev Kumar Is a Professor in Lovely Professional University Punjab India. He received Undergraduate in Hotel Management from Osmania University, Postgraduate in Tourism Management from Madurai Kamaraj University. He completed Doctor of Philosophy in Hospitality Management in 2019 from Amity University Jaipur India. His research areas lies in food Service Industry, his research focuses on Alcoholic beverages, Event management and Sustainable Management Practices, Metaverse, AI Machine Learning and Artificial Intelligence. He published more than 115 research papers, articles and chapters in Scopus Indexed, UGC Approved and peer reviewed Journals and books. Dr. Sanjeev Kumar participated and acted as resource person in various National and International conferences, seminars, session chair, research workshops and industry talks and his work has been widely cited. Dr. Sanjeev also edited more than 20 books with various International Publishers such as IGI Global, Cambridge Scholars, Emerald, CABI, Bentham Science, Taylor & Francis. Dr Sanjeev is an active researcher and has published numerous research articles in prestigious National and International journals and books.

Osman Pala received the B.S. degree in econometrics from Dokuz Eylul University, İzmir, Turkey, in 2008, the M.S. degree in operations research from Dokuz Eylul University, İzmir, Turkey, in 2013, and the Ph.D. degree in operations research from Dokuz Eylul University, İzmir, Turkey, in 2019. He is currently researching at the Karamanoglu Mehmetbey University. He is a member of many journals and associations. He has authored/co-authored, many journal and conference papers. He is an active reviewer and author for numerous top quality international journals. He is an active researcher and supervisor. His research areas include data mining, decision making, optimization, social development and environment. Also has teaching experience on the undergraduate, graduate and postgraduate levels on artificial intelligence and computer programming.

Sindy Rahmahsari is an academic researcher focusing on artificial intelligence, data analytics, and information systems development. Her work explores the integration of emerging technologies into education, digital management, and smart infrastructure. She has participated in various research projects related to AI implementation and machine learning applications for institutional innovation. Her academic interests include intelligent systems, ethical technology design, and digital transformation in public and educational environments.

Sunil K. Sansaniwal brings extensive experience in research and academia within the energy and environment sector. His work is centered on fostering a sustainable built environment, with expertise spanning sustainable cooling, energy efficiency, clean energy adoption, indoor environmental quality and policy advocacy. Dr. Sansaniwal earned his Ph.D. with a focus on enhancing indoor environmental quality through occupants' adaptive actions in buildings. He is also a qualified Mechanical Engineer and a certified Energy Professional, further solidifying his technical expertise in the field. Throughout his career, Dr. Sansaniwal has received several awards and recognitions from prestigious organizations worldwide, underscoring his contributions to the field. His exceptional science communication skills have garnered affiliations with various international journals, where he has served in multiple capacities. He has authored numerous research articles, which have accumulated significant citations, reflecting the impact and reach of his work. Dr. Sansaniwal has played a pivotal role in several high-profile research projects aimed at improving sectorial energy efficiency, reducing building energy demand and advancing sustainable cooling technologies. Currently, Dr. Sansaniwal is associated as a Consultant with NITI Aayog, Government of India. In this role, he continues to make significant contributions to the field of Energy, Green Transition and Climate Change, driving forward innovative solutions and promoting sustainable practices within the energy and environment sectors.

Binastya Anggara Sekti has been a lecturer at the Faculty of Computer Science, Esa Unggul University, Jakarta, since 2016. He is actively involved in the Artificial Intelligence and Information Security Study Group. He holds a Bachelor's degree in Informatics Engineering from the Bandung Institute of Technology and a Master's degree in Information Systems Management from Trisakti University, Jakarta. Prior to entering academia, he accumulated over 25 years of professional experience as an IT practitioner in various companies across Jakarta. His work focused on developing data centers for internet service providers and delivering IT solutions for the manufacturing and oil & gas sectors. His research interests include artificial intelligence, cybersecurity, and the development of IT infrastructure.

Pradosh Kumar Sharma is working as an associate professor in the Department of Physics of Chinmaya Degree College, BHEL, Haridwar, affiliated to HNB Garhwal, University (A Central University), Srinagar, Garhwal. He is working as the Head of the Department . He has 37 years of teaching experience at UG and PG level. He is a life member of IAPT. He has prestigious research work. He combined a deep knowledge of the subject with an ability to present the concepts in an interesting and lucid style .Dr Sharma has co-authored ten books, which are nationally aclaimed.He has published 16 patents in the field of science. He received the Academic Excellence Award 2022 by SIPH TN and Bharat Bhushan Award 2023 by NAHFand Best Inspiring Faculty Award in 2025 by Edutech Research Academy Madurai.

Sitesh Kumar Singh is an accomplished academic and Chartered Engineer with a Ph.D. in Civil Engineering and over nine years of diverse experience in education, research, consultancy, and educational administration. Currently serving as an Assistant Professor in the Department of Civil and Environmental Engineering at the National University of Science and Technology, Muscat, Sultanate of Oman. Dr. Singh holds multiple advanced degrees including M.Tech, B.Tech, PGDIPR,

PGDUPDL and PGDE. His career encompasses curriculum development, research supervision, quality assurance, and strategic academic planning. He has supervised over 40 postgraduate and 2 doctoral students, contributing to academic leadership through workshops, conferences, and editorial roles, and has published over 5 books, 18 book chapters, 58 research articles & conference presentations, and 7 patents.

S. V. Subrahmanyam is working as an Associate Professor of Mathematics in Koneru Lakshmaiah Education Foundation (Deemed to be University), Guntur, Andhra Pradesh. He did M.Sc. (Applied Mathematics) from Acharya Nagarjuna University, Guntur, Andhra Pradesh in 1998 and Ph.D. from Dravidian University, Kuppam, Andhra Pradesh in 2016. He has 25 years of teaching experience. Three scholars were awarded Ph.D. under his guidance. He authored around 50 research publications and presented a good number of research papers in reputed international conferences.

Varadharajan Bhoopathy is a Professor at Sree Rama Engineering College. He holds a Ph.D. in Information and Communication Engineering from Anna University and has over 16 years of academic experience. Specializing in wireless sensor networks and IoT, he has published numerous research papers and holds several international patents. Dr. Bhoopathy is a recipient of prestigious awards for his contributions to education and research. His commitment to academia continues to drive innovation and excellence in the field of computer science.

Index

A

B

C

D

E

F

G

H

I

L

M

N

P

R

S

T

V

W

Y